HYDRODYNAMIC AND MAGNETOHYDRODYNAMIC TURBULENT FLOWS

FLUID MECHANICS AND ITS APPLICATIONS
Volume 48

Aims and Scope of the Series

The purpose of this series is to focus on subjects in which fluid mechanics plays a fundamental role.

As well as the more traditional applications of aeronautics, hydraulics, heat and mass transfer etc., books will be published dealing with topics which are currently in a state of rapid development, such as turbulence, suspensions and multiphase fluids, super and hypersonic flows and numerical modelling techniques.

It is a widely held view that it is the interdisciplinary subjects that will receive intense scientific attention, bringing them to the forefront of technological advancement. Fluids have the ability to transport matter and its properties as well as transmit force, therefore fluid mechanics is a subject that is particulary open to cross fertilisation with other sciences and disciplines of engineering. The subject of fluid mechanics will be highly relevant in domains such as chemical, metallurgical, biological and ecological engineering. This series is particularly open to such new multidisciplinary domains.

The median level of presentation is the first year graduate student. Some texts are monographs defining the current state of a field; others are accessible to final year undergraduates; but essentially the emphasis is on readability and clarity.

For a list of related mechanics titles, see final pages.

Hydrodynamic and Magnetohydrodynamic Turbulent Flows

Modelling and Statistical Theory

by

AKIRA YOSHIZAWA

Institute of Industrial Science,
University of Tokyo, Japan

KLUWER ACADEMIC PUBLISHERS

DORDRECHT / BOSTON / LONDON

A C.I.P. Catalogue record for this book is available from the Library of Congress.

ISBN 0-7923-5225-4

Published by Kluwer Academic Publishers,
P.O. Box 17, 3300 AA Dordrecht, The Netherlands.

Sold and distributed in the North, Central and South America
by Kluwer Academic Publishers,
101 Philip Drive, Norwell, MA 02061, U.S.A.

In all other countries, sold and distributed
by Kluwer Academic Publishers,
P.O. Box 322, 3300 AH Dordrecht, The Netherlands.

Printed on acid-free paper

Printed in the Netherlands.

PREFACE

Turbulence modeling encounters mixed evaluation concerning its importance. In engineering flow, the Reynolds number is often very high, and the direct numerical simulation (DNS) based on the resolution of all spatial scales in a flow is beyond the capability of a computer available at present and in the foreseeable near future. The spatial scale of energetic parts of a turbulent flow is much larger than the energy dissipative counterpart, and they have large influence on the transport processes of momentum, heat, matters, etc. The primary subject of turbulence modeling is the proper estimate of these transport processes on the basis of a bold approximation to the energy-dissipation one. In the engineering community, the turbulence modeling is highly evaluated as a mathematical tool indispensable for the analysis of real-world turbulent flow.

In the physics community, attention is paid to the study of small-scale components of turbulent flow linked with the energy-dissipation process, and much less interest is shown in the foregoing transport processes in real-world flow. This research tendency is closely related to the general belief that universal properties of turbulence can be found in small-scale phenomena. Such a study has really contributed much to the construction of statistical theoretical approaches to turbulence. The estrangement between the physics community and the turbulence modeling is further enhanced by the fact that the latter is founded on a weak theoretical basis, compared with the study of small-scale turbulence.

In astro/geophysical phenomena, fluid flow often plays a central role through its dominant contribution to the transport processes. The geometrical scale of such a flow is huge, and the Reynolds number is overwhelmingly larger than that of engineering laboratory flow. As a result, the flow is intrinsically turbulent, and some kind of turbulence modeling is indispensable for the estimate of transport processes there. Nevertheless, the turbulence modeling in astro/geophysical fields remains at a premature stage, except the meteorological one. This is partly due to the circumstance that a small portion of research efforts is allocated to the study of turbulence effects since a lot of important physical processes are included in those phenomena. This situation makes sharp contrast with the mechanical engineering field where a simple flow of constant-density fluid can still play an important role and the turbulence modeling has been developed highly. Therefore it is academically important to transfer the turbulence-modeling accomplishments in the engineering fields to astro/geophysical fields, in addition to making more efforts towards the consolidation of the theoretical basis of the turbulence modeling.

This book is written from my belief that turbulence modeling is not merely a practical make-shift approach to the analysis of real-world flow. It is the attempt to abstract the universal characteristics of energetic components of a turbulent flow subject to various external effects that generate inhomogeneity and anisotropy of flow properties. The complexity is a great stumbling block for performing the turbulence modeling on the theoretical basis as firm as the study of small-scale turbulence. The difficulty does not lower the academic importance of turbulence modeling, but it signifies that turbulence modeling is a challenging research subject in hydrodynamics.

The primary goal of this book lies in the following two points. The first is to elucidate the mathematical structures of the current turbulence modeling and give a firmer statistical theoretical basis to it. This effort is helpful to seeking further theoretical developments in turbulence modeling. The second is to clarify generation mechanisms of magnetic fields in turbulent motion of electrically conducting fluids in astro/geophysical phenomena. The theoretical study of the mechanisms called turbulent dynamo is shown to share many common mathematical aspects with the turbulence modeling of hydrodynamic flow. In this course, turbulence modeling is confirmed to be an academic approach to abstracting the primary characteristics of energetic flow components that are common to hydrodynamic and magnetohydrodynamic turbulence.

Akira Yoshizawa
Tokyo, March 1998

ACKNOWLEDGMENTS

This book covers a wide range of topics in turbulence modeling such as the mathematical framework of the conventional turbulence modeling, its statistical theoretical formulation, and the turbulence modeling of magnetohydrodynamic flow or the turbulent dynamo with the application to astrophysical and controlled-fusion phenomena. Some of the primary parts of this book are dependent on my works. In their course, I owe much to the invaluable discussions and the collaborative works with a number of researchers.

One of the theoretical cornerstones of this book is a two-scale direct-interaction approximation (TSDIA). It is a combination of a multiple-scale perturbation method with a statistical theory of isotropic turbulence, the DIA. I became familiar with the perturbation method in the study of laminar flow for about ten years since I entered the Graduate School of Physics, University of Tokyo. During this period, the discussions with Drs Isao Imai and Fumio Naruse were encouraging. The refinement of the TSDIA and the application to turbulence modeling were made through the direct or indirect collaboration with my former graduate students, Drs Yutaka Shimomura, Fujihiro Hamba, Nobumitsu Yokoi, Masayoshi Okamoto, and my colleague, Mr Shoiti Nisizima. Parts of the works on the derivation of turbulence models using the TSDIA were made during my several stays at the ICOMP, NASA Lewis Research Center, and the discussions with Drs Meng-Sing Liou, Tsan-Hsing Shih, Robert Rubinstein, and William W. Liou were beneficial. The significance of the TSDIA findings about the Reynolds stress was pointed out, earlier than me, by Dr Charles G. Speziale, who continued to give me invaluable incentives to turbulence-modeling research.

My interest in turbulence modeling was enhanced through the discussions with the members of the IIS (Institute of Industrial Science) Research Group "NST (Numerical Simulation of Turbulence)" organized by Drs. Shuzo Murakami and Toshio Kobayashi. My understanding of turbulence modeling was deepened in the discussions with the nation-wide research group on turbulence modeling, whose active members are Drs Nobuhide Kasagi, Hiroshi Kawamura, Yutaka Miyake, Yasutaka Nagano, Nobuyuki Shima, and Kiyosi Horiuti.

To the understanding of basic ingredients of turbulence theories, the discussions with Drs Robert H. Kraichnan, Jackson R. Herring, Tohru Nakano, and Yoshifumi Kimura were helpful. In the context of dynamo research, my incentives to the study of magnetohydrodynamic turbulence originated in the study of reversed-field pinches of plasma. The discussions with Drs Leaf Turner, Kenro Miyamoto, and Nobuyuki Inoue were stimulating. The discussions with Drs Kimitaka Itoh and Sanae Itoh on H-modes of tokamak's fusion plasma were helpful to the serious consideration on the limit of applicability of turbulent dynamo. Concerning the understanding of solar magnetic fields and accretion disks, I owe much to the discussions with Dr Shoji Kato. To the understanding of fundamental aspects of geomagnetic fields, the discussion with Dr Hirofumi Kato was beneficial.

Late Dr Nobumasa Takemitsu had large influence on my turbulence research. This book is dedicated to him, who died with his wife, Sayuri, in the tragic traffic accident of June 10, 1990 in Canada.

Finally, I am grateful to Dr Karel Nederveen of the Kluwer Academic Publishers, who offered me the opportunity of writing this book, and to Dr Nobumitsu Yokoi, without whose kind assistance this book would not have been written in the present LaTeX form. I am also grateful to Dr Taco Hoekwater of the Kluwer Academic Publishers for his kind assistance concerning the LaTeX style file.

TABLE OF CONTENTS

CHAPTER 1

INTRODUCTORY REMARKS

1.1. Necessity of Turbulence Modeling

Dynamical behaviors of an electrically non-conducting fluid flow are characterized by the Reynolds number R, which is given by

$$R = U_R L_R / \nu, \tag{1.1}$$

where U_R and L_R are the representative velocity and length of a flow, respectively, and ν is the kinematic viscosity. The Reynolds number R is usually defined as the ratio of the nonlinear term in the hydrodynamic momentum equation, $(\mathbf{u} \cdot \nabla)\mathbf{u}$, to the viscous one $\nu \Delta \mathbf{u}$, where $\mathbf{u}$ is the fluid velocity. Another interesting definition is the ratio of the viscous diffusion time scale L_R^2/ν to the advection counterpart L_R/U_R (Tennekes and Lumley 1972). Both of the definitions indicate that the nonlinear effects due to fluid flow become dominant for large R.

In laboratory experiments of engineering flows, the most familiar fluids are air and water. The magnitude of ν is about $0.1\mathrm{cm}^2\mathrm{s}^{-1}$ for air and $0.01\mathrm{cm}^2\mathrm{s}^{-1}$ for water. The larger value of ν for air means that viscous effects have greater influence on air flow since air is much lighter than water (note that ν is defined as the viscosity divided by fluid density). The typical size of experimental instruments is from a few cm to a few m, and the velocity counterpart is from a few $10\mathrm{cms}^{-1}$ to a few $10\mathrm{ms}^{-1}$. Then R ranges from $O(10^2)$ to $O(10^6)$ for air and from $O(10^3)$ to $O(10^7)$ for water.

The foregoing magnitude of R implies that nonlinear effects are much more important in most of engineering flow phenomena than viscous ones. This statement produces no misunderstanding under the definition of R based on the time-scale ratio. On the other hand, the definition based on the term ratio needs some caution. Under it, we should understand that viscous effects are not important at the spatial scale L_R, but not that they are negligible at all scales. As will be shown in Chapter 3, the energy-dissipation scale (the scale at which viscous effects become dominant), ℓ_D, is related to L_R as

$$\ell_D / L_R = O\left(R^{-3/4}\right), \tag{1.2}$$

at high R. Equation (1.2) indicates that increasing R leads to the occurrence of smaller-scale components of motion in the case of fixed L_R.

Equation (1.2) gives $\ell_D/L_R = O(10^{-3})$ for R of $O(10^4)$. This fact imposes an intolerably heavy burden on the direct numerical simulation (DNS) of a turbulent flow that is based on the straightforward use of the hydrodynamic equations. In the DNS resolving all the spatial scales in a flow, the finest computational grid size should be of $O(\ell_D)$. The total number of grid points necessary for the DNS is of $O(10^9)$ because of the three-dimensionality of flow. In addition, the temporal variation of fine-scale components of motion is very rapid, compared with the large-scale counterpart, and a very short time step is required for the proper treatment of the temporal evolution of the hydrodynamic equations. From the viewpoint of the capability of a computer available at present and in the foreseeable near future, the possibility of the DNS of a flow with $O(10^4)$ is very little, except some special flows whose geometry is simple (for example, a flow between two parallel walls).

As the theoretical methods of avoiding the foregoing difficulty, we have two approaches. One is to focus attention on small-scale components of a turbulent flow and replace the energetic parts with a hypothetical energy reservoir. Through this simplification, we may seek universal properties of small-scale turbulence. The other approach is to proceed in a directly opposite direction concerning spatial scales. We pay attention to the energetic parts of a flow that seem to possess less universal properties, and replace the small-scale components with a proper model. This approach is usually called turbulence modeling, and a resulting mathematical expression is called a turbulence model.

The difficulty associated with R is enhanced in aerodynamical flows, mainly due to high flow speed. We consider a flow around a cruising aircraft, where L_R is from a few 10cm to a few 10m, depending on a wing or a fuselage, and U_R is a few 100ms^{-1}. The resulting Reynolds number R ranges from $O(10^6)$ to $O(10^8)$. Such a flow is beyond the reach of the DNS in the near future. In this case, we also encounter the phenomena intrinsic to high speed flows. Their typical example is the occurrence of a shock wave. The velocity component normal to the shock wave decreases steeply from the supersonic to subsonic state across it, and fluid is highly compressed, generating large fluctuations of density. The interaction of turbulent flow with shock waves is one of the most challenging subjects in turbulence modeling. The other example of flows in which prominent compressibility effects appear is the mixing of two parallel streams with different velocity, density, and temperature. With the increase in the Mach number based on the difference of the two stream speeds (more exactly, the convective Mach number), the growth rate of the mixing layer is suppressed drastically. The explanation of this phenomenon based on the turbulence modeling is very interesting from both the practical and academic viewpoints.

As an example of huge R, we can mention flows in natural sciences such as astrophysics and geophysics. One of the primary factors making R huge is their geometrical scales. This situation is easily understandable by considering a typhoon or a hurricane in which the geometrical scale is a few 10km and the moving speed is a few $10\mathrm{ms}^{-1}$. In astrophysical flows associated with stars and galaxies, their Reynolds number may become literally astronomically large. In these phenomena, turbulence modeling is indispensable for the proper treatment of their small-scale components ("small" means relatively small, compared with huge geometrical scales). In reality, various turbulence models are incorporated into their studies through the concepts of turbulent viscosity, diffusivity, and resistivity. Those turbulence models are generally crude, compared with the counterparts in the engineering fields such as mechanical engineering. The reason is clear. Astrophysical and geophysical phenomena include a number of important physical processes. Turbulence effects on them are important, but they are related only to a small portion of those processes, unlike the mechanical-engineering field.

One of the interesting astrophysical and geophysical phenomena in which turbulence effects play a critical role is the generation processes of magnetic fields in the Earth, the Sun, and galaxies. It is widely accepted that these magnetic fields are generated by the turbulent or highly asymmetric motion of electrically conducting fluids. The study of the generation mechanism with attention focused on energetic components of turbulent motion is turbulent dynamo, which may be called the turbulence modeling of magnetohydrodynamic flow. This study is behind the hydrodynamic counterpart in the engineering field. In this sense, it is significant to transfer the knowledge about engineering turbulence modeling to the natural-science fields.

1.2. Basic Stance on Turbulence Modeling

Turbulent flow shows two prominent features, compared with the laminar state of flow. One is the enhancement of the energy-dissipation effect giving rise to the conversion of kinetic energy to thermodynamic one. The other is the enhancement of transport or mixing effects on the momentum and scalars such as temperature and matter concentration. At high Reynolds numbers, the energy dissipation arises from the fine-scale components of turbulent motion whose spatial length is given by ℓ_D [Eq. (1.2)], whereas the momentum and scalar transports are due to the energetic components. These two processes are not independent of each other, but one does not occur without the other. When our primary interest lies in the transport or mixing process due to turbulent motion, the energy-dissipation effect originating in the fine-scale components may be regarded as an energy sink for the energetic components. If the energy transfer from the energetic to fine-scale components of motion be estimated properly on the basis of the knowledge about the former components, the momentum and scalar transports may be examined without delving into the fine-scale energy-dissipation process. This is the very starting point of turbulence modeling.

Energetic components of turbulent motion tend to behave much less universally than the fine-scale counterparts, for they are greatly influenced by the difference of energy-input mechanisms. Specifically, a solid boundary becomes a big stumbling block to the theoretical study of turbulent flows. Almost all of well-developed turbulence theories are based on the use of the Fourier representation of fluctuating quantities. Such theories are originally applicable only to homogeneous turbulence whose properties are statistically uniform in space, and turbulent shear flow with spatially varying mean velocity is beyond the scope of their straightforward application.

The study of turbulence modeling was started about three decades ago, and turbulence models have been constructed mainly on the basis of simple tensor analysis and deep insight into observational results. The models at the early stage contained many factors that cannot be deduced from the fundamental hydrodynamic equations. This tendency still remains in the present highly sophisticated turbulence models. This is one of the primary reasons why theoretical researchers in the hydrodynamics and physics communities have raised the doubt about the academic importance of turbulence modeling. In reality, the turbulence modeling is sometimes regarded as a makeshift tool for engineering flow analysis.

In the Reynolds- or ensemble-mean turbulence modeling, the velocity $\mathbf{u}$ is divided into the mean $\langle \mathbf{u} \rangle$ and the fluctuation around it, $\mathbf{u}'$. The representative turbulence models are the turbulent-viscosity and second-order models. In the former, the Reynolds stress in the mean-velocity equation, $\langle u'_i u'_j \rangle$, is expressed in an algebraic form in terms of $\langle \mathbf{u} \rangle$ and the characteristic turbulence quantities such as the turbulent energy $\langle \mathbf{u}'^2/2 \rangle$. In the latter, the equation for $\langle u'_i u'_j \rangle$ is directly dealt with, and the third-order correlation functions in it are modeled. These two types of modeling have been developed rather independently since the early stage of their studies, but their close connection has recently began to be recognized.

In this book, we aim at gaining a systematic understanding of turbulence modeling from a theoretical viewpoint. Turbulence modeling is an effort towards the understanding of turbulent shear flow greatly influenced by external conditions, whereas homogeneous-turbulence study focuses attention on small-scale properties including the energy-dissipation process. As a natural consequence of such a difference, turbulence modeling cannot be free from much more approximations and assumptions, compared with the study of homogeneous turbulence. We do not consider that this situation impairs the academic value of turbulence modeling. Our basic stance on turbulence modeling is that it is a challenging task of abstracting as many universal properties as possible from real-world complicated flows. In Sec. 1.1, we referred to the turbulent dynamo or magnetohydrodynamic turbulence modeling related to natural sciences. The other aim of this book is to establish a systematic connection between hydrodynamic and magnetohydrodynamic turbulence modeling. Such an effort will clearly show an academic aspect of turbulence modeling.

1.3. Outline of Book and Guide for Readers

Readers are expected to have the basic knowledge about fluid mechanics in the undergraduate course. This book, however, is not premised on the detailed knowledge about turbulence. The important properties of turbulence, which are necessary for turbulence modeling, will be explained in a self-contained manner.

The following parts of this book are organized as follows. In Chapter 2, we shall give a compact derivation of the hydrodynamic equations and explain their fundamental properties associated with the conservation laws.

In Chapter 3, we shall discuss the important properties of small-scale turbulent motion. Special attention will be paid to the inertial-range properties of fluctuations and the direct-interaction approximation (DIA) method as the prototype of homogeneous-turbulence theories.

In Chapter 4, we shall discuss two kinds of Reynolds- or ensemble-mean turbulence modeling from the conventional viewpoint based on tensor analysis and invariance principles. One is the algebraic modeling, and the other is the second-order modeling making full use of turbulence transport equations. The connection between these two kinds of modeling subject to rather different development courses will be viewed from a systematic standpoint.

In Chapter 5, we shall discuss the subgrid-scale modeling based on the spatial filtering procedure, which is the key ingredient in large eddy simulation of turbulent flow. Reference will be also made to the relationship with the Reynolds-mean modeling.

In Chapter 6, we shall present a statistical theory of turbulent shear flow that is called a two-scale direct-interaction approximation (TSDIA). It is based on the DIA and a multiple-scale perturbational method. In the combination of this method with the inertial-range properties, we shall discuss the conventional turbulence modeling from a statistical theoretical viewpoint. Emphasis will be placed on the close relationship between the higher-order algebraic and second-order modeling.

The TSDIA method is not straightforwardly applicable to complicated turbulent shear flow such as aerodynamical compressible flow and astrophysical magnetohydrodynamic flows. In Chapter 7, we shall give a much simpler method called a Markovianized two-scale (MTS) one for the turbulence modeling of such flows.

In Chapter 8, we shall discuss compressible turbulence modeling that is one of the most challenging subjects in turbulence research. With the help of the TSDIA and MTS methods, a compressible turbulence model with effects of density fluctuation as a key ingredient will be presented with special emphasis on the application to a free-shear flow.

Chapters 9 and 10 are devoted to the discussions on magnetohydrodynamic turbulent flow at high magnetic Reynolds number. In Chapter 9, we shall clarify the effects of fluctuations on the equations for the mean velocity and magnetic field with the aid of some different theoretical methods. Specifically, a new dynamo concept called the cross-helicity effect will be introduced. In Chapter 10, we shall discuss the generation mechanisms of global magnetic fields in some typical flow phenomena associated with astrophysics and controlled fusion.

Except Chapters 2 and 3 (Secs. 3.1-3.4), the contents of this book are classified, according to the interest of each reader, as follows.

(a) Interested in conventional hydrodynamic modeling: Chapters 4, 5, 7, and 8.
(b) Interested in statistical hydrodynamic modeling: Chapters 3, 4, and 6.
(c) Interested in turbulent dynamo: Chapters 6 (Secs. 6.1-6.3), 7, 9, and 10.

The citation in each chapter is not exhaustive, but the books, review articles, and papers including many references of related topics are preferentially selected to avoid long lists of references.

References

Tennekes, H. and Lumley, J. L. (1972), *A First Course in Turbulence* (The MIT, Cambridge).

CHAPTER 2

HYDRODYNAMIC EQUATIONS

2.1. General Conservation Form

In the Cartesian coordinate system, we write a position vector $\mathbf{x}$ as (x, y, z) or $(x_i; i = 1 - 3)$. Similarly, the velocity vector $\mathbf{u}$ is denoted as (u, v, w) or $(u_i; i = 1 - 3)$. We consider the conservation rule concerning a physical quantity per unit volume, f, in a fluid blob of a rectangular prism with three infinitesimal sides Δx, Δy, and Δz, (see Fig. 2.1). The center of the prism is located at a point (x, y, z). Later, we shall adopt the mass, momentum, and internal energy as f.

The conservation rule for f may be written as

$$\frac{\partial}{\partial t} f \Delta V = \sum_{i=1}^{3} M_{f_i} + S_f, \tag{2.1}$$

with $\Delta V = \Delta x \Delta y \Delta z$. Here, M_{f_i} expresses the rate of temporal increase in f that is due to the flow across the two faces normal to the i axis. For instance, M_{f_x} is given by

$$\begin{aligned} M_{f_x} = {} & f\left(x - \frac{\Delta x}{2}, y, z\right) u\left(x - \frac{\Delta x}{2}, y, z\right) \Delta y \Delta z \\ & - f\left(x + \frac{\Delta x}{2}, y, z\right) u\left(x + \frac{\Delta x}{2}, y, z\right) \Delta y \Delta z. \end{aligned} \tag{2.2}$$

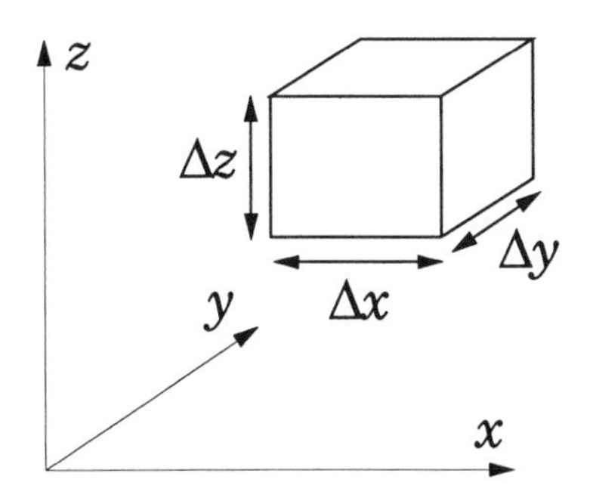

Figure 2.1. Infinitesimal rectangular prism.

In Eq. (2.2), $u\Delta y\Delta z$ means the flow rate across the face normal to the x axis. The first term expresses the influx of f per unit time that is induced by the flow in the x direction, whereas the second does the outflux counterpart. The second term of Eq. (2.1), S_f, is a source generating f in the rectangular-prism region.

Equation (2.2) is reduced to

$$M_{f_x} = -\frac{\partial}{\partial x} fu\Delta V, \tag{2.3}$$

from the Taylor expansion concerning infinitesimal Δx. From Eqs. (2.1) and (2.3), we have

$$\frac{\partial f}{\partial t} + \frac{\partial}{\partial x_i} fu_i = \lim_{\Delta V \to 0} \frac{S_f}{\Delta V}, \tag{2.4}$$

where repeated subscripts obey the summation rule

$$\frac{\partial}{\partial x_i} fu_i \equiv \sum_{i=1}^{3} \frac{\partial}{\partial x_i} fu_i \equiv \nabla \cdot (f\mathbf{u}). \tag{2.5}$$

Note: We refer to the Gauss' integral theorem

$$\int_S \mathbf{A} \cdot \mathbf{n} dS \equiv \int_S A_i n_i dS \equiv \int_S A_n dS = \int_V \nabla \cdot \mathbf{A} dV, \tag{2.6}$$

where $\mathbf{A}$ is an arbitrary vector, S is an arbitrary closed surface, V is the region surrounded by S, and $\mathbf{n}$ is the outward unit vector normal to S. We calculate the left-hand side of Eq. (2.6) at the surface of the infinitesimal rectangular prism in Fig. 2.1. Then we have

$$\int_S \mathbf{A} \cdot \mathbf{n} dS = -\left(M_{A_x} + M_{A_y} + M_{A_z}\right), \tag{2.7}$$

where M_{A_x} denotes the expression obtained by replacing fu in Eq.(2.2) with A_x (the x component of $\mathbf{A}$). Using the M_{A_x} counterpart of Eq.(2.3), the right-hand side of Eq.(2.7) is reduced to

$$\frac{\partial A_i}{\partial x_i}\Delta V = \int_V \nabla \cdot \mathbf{A} dV; \tag{2.8}$$

namely, the Gauss' integral theorem has been proved in the infinitesimal region. This result may be easily extended to an arbitrary region V. The region may be divided into an infinite number of infinitesimal rectangular prisms. At the interface of two prisms, one surface integral cancels with the other owing to the opposite direction of $\mathbf{n}$. By repeating this procedure, we can reach Eq. (2.6) for arbitrary S and V.

2.2. Mass Equation

We take the density of fluid, ρ, as f and put

$$f = \rho. \tag{2.9}$$

Within the framework of classical physics, we have no generation and annihilation mechanisms of mass, which lead to $S_\rho = 0$ in Eq. (2.4). Therefore we have

$$\frac{\partial \rho}{\partial t} + \nabla \cdot (\rho \mathbf{u}) = 0, \tag{2.10}$$

which is also called the equation of continuity. Equation (2.10) is rewritten as

$$\frac{D\rho}{Dt} + \rho \nabla \cdot \mathbf{u} = 0 \tag{2.11a}$$

or

$$\frac{1}{\rho}\frac{D\rho}{Dt} = -\nabla \cdot \mathbf{u}, \tag{2.11b}$$

where D/Dt is the Lagrange derivative defined as

$$\frac{D}{Dt} \equiv \frac{\partial}{\partial t} + \mathbf{u} \cdot \nabla. \tag{2.12}$$

One of the prominent features of the hydrodynamic equations is the occurrence of the Lagrange derivative D/Dt. Its importance comes from the fact that fluid can move freely without experiencing binding constraints, unlike an elastic continuum. The physical meaning of D/Dt may be understood by writing

$$\frac{Df}{Dt} = \lim_{\Delta t \to 0} \frac{f(\mathbf{x} + \mathbf{u}\Delta t, t + \Delta t) - f(\mathbf{x}, t)}{\Delta t}, \tag{2.13}$$

for arbitrary f. Namely, D/Dt signifies the temporal change of f that is measured by an observer in the reference frame moving with the local fluid velocity $\mathbf{u}$. This is called the Lagrange frame. In the usual or Euler frame, an observer is fixed to it.

In the Lagrange frame, Df/Dt is caused by the source or sink effect, except the advection one. In the case of ρ, the sole effect leading to its change is the volume expansion or compression represented by $\nabla \cdot \mathbf{u}$. Positive and negative $\nabla \cdot \mathbf{u}$ signify the volume expansion and compression rates per unit time, respectively. In the absence of $\nabla \cdot \mathbf{u}$, ρ is constant along a flow path; namely, we have

$$\frac{D\rho}{Dt} = 0. \tag{2.14}$$

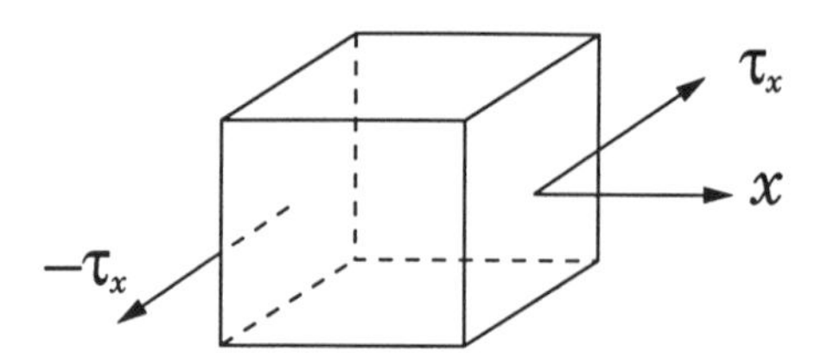

Figure 2.2. Stress acting at the face normal to the x axis.

In the case of constant ρ, which is an important subject of turbulence research, Eq. (2.10) is greatly simplified as

$$\nabla \cdot \mathbf{u} = 0. \tag{2.15}$$

2.3. Momentum Equation

2.3.1. CONCEPT OF STRESS

As f in Eq. (2.4), we adopt the i component of the momentum per unit volume

$$f = \rho u_i. \tag{2.16}$$

This case is not so simple as the ρ counterpart, for we have a source term for the momentum, $S_{\rho u_i}$, that originates in the molecular motions smoothed out under the continuum approximation. The molecules outside a closed region may exert two kinds of forces to the inner fluid across the surface. They are the pressure and frictional forces normal to and along the surface, respectively, which may be related to $S_{\rho u_i}$ with the help of the concept of stress.

The stress is defined as the surface force per unit area that is exerted to the fluid in a closed region by the surrounding fluid. We denote the stress acting at the face normal to the x axis by $\boldsymbol{\tau}_x$, as in Fig. 2.2. The subscript x of $\boldsymbol{\tau}_x$ indicates that the face is normal to the x axis. The components of $\boldsymbol{\tau}_x$ are given by

$$\boldsymbol{\tau}_x = \left(\tau_{xx}, \tau_{xy}, \tau_{xz}\right), \tag{2.17}$$

which implies that the stress is represented by a tensor of the second rank.

More generally, τ_{ij} expresses the j component of the stress that acts at the surface normal to the i axis. This second-rank tensor is symmetric; namely, we have

$$\tau_{ij} = \tau_{ji}. \tag{2.18}$$

This symmetry is a consequence of the conservation of angular momentum and may be proved as follows. From the law of action and counteraction, the stresses, which act at the two faces located at $(x + (\Delta x/2), y, z)$ and $(x - (\Delta x/2), y, z)$ in Fig. 2.2, are nearly equal in the magnitude since Δx is infinitesimal, but they are opposite in the direction. They give the torque around the y axis

$$(\tau_{xz}\Delta y\Delta z)\,\Delta x. \tag{2.19}$$

Similarly, the torque produced by the stress acting at the two faces normal to the z axis, τ_{zx}, is written as

$$(\tau_{zx}\Delta x\Delta y)\,\Delta z. \tag{2.20}$$

The conservation of angular momentum around the y axis requires the balance between these two torques, resulting in $\tau_{xz} = \tau_{zx}$.

Using the stress τ_{ij}, the source for the i component of the momentum, $S_{\rho u_i}$, may be given in the form

$$S_{\rho u_i} = \sum_{j=1}^{3} C_{ji}. \tag{2.21}$$

For instance, C_{xi} is the contribution of the i component of the stress acting at the two faces normal to the x axis and is given by

$$C_{xi} = \tau_{xi}\left(x + \frac{\Delta x}{2}, y, z\right)\Delta y\Delta z - \tau_{xi}\left(x - \frac{\Delta x}{2}, y, z\right)\Delta y\Delta z, \tag{2.22}$$

which is reduced to

$$C_{xi} = \frac{\partial \tau_{xi}}{\partial x}\Delta V, \tag{2.23}$$

for infinitesimal Δx.

From Eqs. (2.4), (2.21), and (2.23), we have

$$\frac{\partial}{\partial t}\rho u_i + \frac{\partial}{\partial x_j}\rho u_j u_i = \frac{\partial \tau_{ji}}{\partial x_j}, \tag{2.24a}$$

which is also written as

$$\rho\frac{Du_i}{Dt} = \frac{\partial \tau_{ji}}{\partial x_j}, \tag{2.24b}$$

using the mass equation (2.10). In the presence of an external force such as the gravitational force per unit mass, $\mathbf{K}$, ρK_i is added to the right-hand side of Eq. (2.24).

2.3.2. MODELING OF STRESS

We are in a position to write the stress τ_{ij} explicitly. Within the framework of the continuum mechanics, τ_{ij} is the quantity to be modeled on the basis of tensor analysis and some plausible physical assumptions, but not to be deduced mathematically. In this sense, the construction of τ_{ij} is rather similar to the turbulence modeling by the conventional approach that will be detailed in Chapter 4.

We have already mentioned that τ_{ij} comes from the smoothed-out molecular motion. It is most simply modeled using two effects. One is the static-pressure effect that survives in the absence of a macroscopic fluid motion $\mathbf{u}$. The other is the resistive-force effect in the presence of a spatially nonuniform fluid motion or spatially varying $\mathbf{u}$. In this context, constant $\mathbf{u}$ is equivalent to vanishing $\mathbf{u}$ from the Galilean invariance of Eq. (2.24) that will be referred to later.

The second-rank tensor not dependent on $\mathbf{u}$ is the Kronecker delta symbol δ_{ij} (it is one for $i = j$ and zero otherwise). The nonuniformity of $\mathbf{u}$ may be expressed using its spatial derivatives. Specifically, the first derivative of $\mathbf{u}$, $\partial u_j/\partial x_i$, is a tensor of the second rank and is a promising candidate for the modeling elements. We consider the symmetry of τ_{ij} and write

$$\tau_{ij} = -p\delta_{ij} + \mu\left(\frac{\partial u_j}{\partial x_i} + \frac{\partial u_i}{\partial x_j}\right) + \mu'\nabla\cdot\mathbf{u}\delta_{ij}. \tag{2.25}$$

In Eq. (2.25), the first term gives $-p$ to all of τ_{xx}, τ_{yy}, and τ_{zz}, and p corresponds to the so-called static pressure. The primary feature of the second term can be seen by considering a unidirectional flow in the x direction, $u = u(y)$. The component τ_{yx} is given by $\mu\partial u/\partial y$ and is the drag force that the upper fluid exerts to the lower one. The coefficient μ is the viscosity. The last term expresses the stress normal to the surface that occurs from the expansion or compression of fluid volume. The coefficient μ' is called the second viscosity.

In the case of constant ρ or vanishing $\nabla\cdot\mathbf{u}$, p in Eq. (2.25) obeys the relation

$$p = -(1/3)\tau_{ii}. \tag{2.26}$$

In the compressible case of nonvanishing $\nabla\cdot\mathbf{u}$, however, the first and third terms are combined to generate an isotropic effect. One of these terms is not distinguishable from the other from the observational viewpoint, and μ' cannot be defined uniquely. A method of removing this ambiguity is to use Eq. (2.26) to define p (Imai 1973). Then we have the relationship between μ and μ' as

$$\mu' = -(2/3)\mu. \tag{2.27}$$

Finally, we have

$$\tau_{ij} = -p\delta_{ij} + \mu\left(\frac{\partial u_j}{\partial x_i} + \frac{\partial u_i}{\partial x_j} - \frac{2}{3}\nabla\cdot\mathbf{u}\delta_{ij}\right). \tag{2.28}$$

2.3.3. RESULTING EQUATIONS

We substitute Eq. (2.28) into Eq. (2.24) and have

$$\frac{\partial}{\partial t}\rho u_i + \frac{\partial}{\partial x_j}\rho u_j u_i = -\frac{\partial p}{\partial x_i} + \frac{\partial}{\partial x_j}\left(\mu\left(\frac{\partial u_j}{\partial x_i} + \frac{\partial u_i}{\partial x_j} - \frac{2}{3}\nabla\cdot\mathbf{u}\delta_{ij}\right)\right) \tag{2.29a}$$

or

$$\rho\frac{Du_i}{Dt} = -\frac{\partial p}{\partial x_i} + \frac{\partial}{\partial x_j}\left(\mu\left(\frac{\partial u_j}{\partial x_i} + \frac{\partial u_i}{\partial x_j} - \frac{2}{3}\nabla\cdot\mathbf{u}\delta_{ij}\right)\right). \tag{2.29b}$$

This equation is called the Navier-Stokes equation.

In the simplest case of constant ρ, the change of the viscosity μ is neglected. Equation (2.29) is reduced to

$$\frac{\partial u_i}{\partial t} + \frac{\partial}{\partial x_j}u_j u_i = -\frac{1}{\rho}\frac{\partial p}{\partial x_i} + \nu\Delta u_i \tag{2.30a}$$

or

$$\frac{D\mathbf{u}}{Dt} = -\frac{1}{\rho}\nabla p + \nu\Delta\mathbf{u}, \tag{2.30b}$$

where ν is the kinematic viscosity defined by

$$\nu = \mu/\rho. \tag{2.31}$$

Equation (2.30) is combined with Eq. (2.15) to constitute a closed system of equations for the motion of a constant-density fluid.

Note: For air and water, μ and ν are given as follows:

$$\mu = 1.8\times 10^{-4}\ \mathrm{g\ cm^{-1}s^{-1}(air)},\ \ 0.01\ \mathrm{g\ cm^{-1}s^{-1}(water)}; \tag{2.32a}$$

$$\rho = 1.2\times 10^{-3}\ \mathrm{g\ cm^{-3}(air)},\ \ 1.0\ \mathrm{g\ cm^{-3}(water)}. \tag{2.32b}$$

Therefore, we have

$$\nu = 0.1\ \mathrm{cm^2s^{-1}(air)},\ \ 0.01\ \mathrm{cm^2s^{-1}(water)}. \tag{2.32c}$$

The kinematic viscosity for air is larger by one order than for water. This fact means that air flow is more sensitive to viscous effects than water one since air is much lighter.

2.4. Energy Equation

2.4.1. LOSS OF KINETIC ENERGY

In order to see the role of viscous effects, we discuss the equation for the kinetic energy $\mathbf{u}^2/2$. For simplicity of discussion, we consider the constant-density case. Using Eq. (2.30), we have

$$\frac{\partial}{\partial t}\frac{\mathbf{u}^2}{2} = \frac{\partial}{\partial x_j}\left(-\frac{\mathbf{u}^2 u_j}{2} + \nu\frac{\partial}{\partial x_j}\frac{\mathbf{u}^2}{2} - \frac{1}{\rho}pu_j\right) - \nu\left(\frac{\partial u_i}{\partial x_j}\right)^2. \tag{2.33}$$

We integrate Eq. (2.33) in a fluid region V and use the Gauss' integral theorem (2.6). Then we have

$$\frac{\partial}{\partial t}\int_V \frac{\mathbf{u}^2}{2}dV = \int_S \left(-\frac{\mathbf{u}^2\mathbf{u}}{2} + \nu\nabla\frac{\mathbf{u}^2}{2} - \frac{1}{\rho}p\mathbf{u}\right)\cdot\mathbf{n}dS - \int_V \nu\left(\frac{\partial u_i}{\partial x_j}\right)^2 dV, \tag{2.34}$$

where S is the surface surrounding the region V, and $\mathbf{n}$ is the outward unit vector normal to S. In the surface-integral part of Eq. (2.34), the first term is the energy transport due to fluid motion, the second is the molecular counterpart, and the third is the work done by the surrounding fluid through the pressure (the work by pressure will be explained in more detail in Sec. 2.6.2.A). The volume-integral part including the minus sign is always non-positive and expresses the loss of kinetic energy due to viscous effects. This lost energy is converted to heat and appears in the conservation law of internal energy, as will be seen below.

In the absence of molecular effects, the total amount of kinetic energy is conserved so long as there is no energy injection across S. It will be confirmed later that this property will play a very important role in turbulence modeling.

2.4.2. INTERNAL ENERGY

As a quantity expressing thermal energy, we introduce the internal energy per unit mass, e, which is related to the temperature θ as

$$e = C_V\theta, \tag{2.35}$$

where C_V is the specific heat at constant volume. Using e, the total energy e_T is given by

$$e_T = e + \mathbf{u}^2/2, \tag{2.36}$$

and ρe_T will be taken as f in Eq. (2.4) (Lagerstrom 1964).

As the source giving rise to the change of e_T, S_{e_T}, in the infinitesimal region ΔV (see Fig. 2.1), we may mention the following two factors. One is the inflow of heat across the surrounding surface ΔS, which is written as

$$-\int_{\Delta S} q_n dS = -\int_{\Delta V} \nabla \cdot \mathbf{q} dV \cong -\nabla \cdot \mathbf{q} \Delta V, \tag{2.37}$$

using the Gauss' integral theorem (2.6). Here $\mathbf{q}$ is the heat flux per unit time and surface, which is most simply modeled as

$$\mathbf{q} = -\kappa \nabla \theta, \tag{2.38}$$

where κ is the heat conductivity.

The other source is the work done on the fluid of the volume ΔV by the surrounding fluid through the stress. This work may be written as

$$\int_{\Delta S} \boldsymbol{\tau}_n \cdot \mathbf{u} dS = \int_{\Delta S} \tau_{ni} u_i dS = \int_{\Delta V} \frac{\partial}{\partial x_j} \tau_{ji} u_i dV \cong \Delta V \frac{\partial}{\partial x_j} \tau_{ji} u_i \tag{2.39}$$

since $\tau_{ni} = \tau_{in} = \tau_{ij} n_j$ [$\tau_{ij} u_j$ has been adopted as A_i in Eq. (2.6)].

From Eqs. (2.37)-(2.39), we have

$$\frac{S_{e_T}}{\Delta V} = \nabla \cdot (\kappa \nabla \theta) + \tau_{ji} \frac{\partial u_i}{\partial x_j} + u_i \frac{\partial \tau_{ji}}{\partial x_j}. \tag{2.40}$$

Using Eq. (2.28) for τ_{ij}, the second term may be rewritten as

$$\begin{aligned} \tau_{ji} \frac{\partial u_i}{\partial x_j} &= \frac{1}{2} \tau_{ji} \left(\frac{\partial u_j}{\partial x_i} + \frac{\partial u_i}{\partial x_j} \right) \\ &= -p \nabla \cdot \mathbf{u} + \frac{1}{2} \mu \left(\left(\frac{\partial u_j}{\partial x_i} + \frac{\partial u_i}{\partial x_j} \right)^2 - \frac{4}{3} (\nabla \cdot \mathbf{u})^2 \right). \end{aligned} \tag{2.41}$$

Equation (2.4) with ρe_T as f is combined with Eqs. (2.40) and (2.41) to lead to

$$\begin{aligned} \frac{\partial}{\partial t} \rho \left(e + \frac{\mathbf{u}^2}{2} \right) + \nabla \cdot \left(\rho \left(e + \frac{\mathbf{u}^2}{2} \right) \mathbf{u} \right) &= \nabla \cdot (\kappa \nabla \theta) - p \nabla \cdot \mathbf{u} \\ + \frac{1}{2} \mu \left(\left(\frac{\partial u_j}{\partial x_i} + \frac{\partial u_i}{\partial x_j} \right)^2 - \frac{4}{3} (\nabla \cdot \mathbf{u})^2 \right) &+ u_i \frac{\partial \tau_{ji}}{\partial x_j}. \end{aligned} \tag{2.42}$$

From Eq. (2.42), we subtract

$$\frac{\partial}{\partial t} \rho \frac{\mathbf{u}^2}{2} + \frac{\partial}{\partial x_j} \rho \frac{\mathbf{u}^2 u_j}{2} = u_i \frac{\partial \tau_{ji}}{\partial x_j} \tag{2.43}$$

that is obtained by multiplying Eq. (2.24) by u_i. Then we have

$$\frac{\partial}{\partial t}\rho e + \nabla \cdot (\rho e \mathbf{u}) = \nabla \cdot (\kappa \nabla \theta) - p \nabla \cdot \mathbf{u} + \phi \tag{2.44a}$$

or

$$\rho \frac{De}{Dt} = \nabla \cdot (\kappa \nabla \theta) - p \nabla \cdot \mathbf{u} + \phi, \tag{2.44b}$$

where ϕ is called the dissipation function and is given by

$$\phi = \frac{1}{2}\mu \left(\left(\frac{\partial u_j}{\partial x_i} + \frac{\partial u_i}{\partial x_j} \right)^2 - \frac{4}{3} (\nabla \cdot \mathbf{u})^2 \right). \tag{2.45}$$

The role of each term on the right-hand side of Eq. (2.44) is clear. The first term expresses the contribution from the heat flow due to the temperature difference. The second term is the work due to the expansion or compression of fluid. In the compression case, $\nabla \cdot \mathbf{u}$ is negative, and the term expresses the work done by the surrounding one, resulting in the gain of the internal energy. The third term, which corresponds to the last term on the right-hand side of Eq. (2.33) in the constant-density case, originates in the loss of kinetic energy due to viscous effects.

Summarizing the foregoing discussions, we have reached three equations for ρ, $\mathbf{u}$, and e, (2.10), (2.29), and (2.44). In the present system consisting of ρ, $\mathbf{u}$, e, and p, the last relation to be designated is the p-related one. In order to close this system, we usually add the thermodynamic relation for a perfect gas, which is given by

$$p = R_G \rho \theta, \ R_G = C_P - C_V, \tag{2.46}$$

where R_G is the gas constant, and C_P is the specific heat at constant pressure. In the case of a perfect gas, both C_P and C_V are functions of θ. From Eqs. (2.35) and (2.46), p is related to e as

$$p = (\gamma - 1)\, \rho e, \tag{2.47}$$

where γ is the ratio of specific heats, which is defined by

$$\gamma = C_P / C_V. \tag{2.48}$$

In the case of constant or nearly constant ρ, we can neglect the second and third terms in (2.44) and regard C_V and κ as constant. Then Eq. (2.44b) is reduced to the simple heat conduction equation

$$\frac{D\theta}{Dt} = \lambda \Delta \theta, \tag{2.49}$$

where the temperature diffusivity λ is defined by

$$\lambda = \kappa / (\rho C_V) . \tag{2.50}$$

2.5. Fundamental Hydrodynamic Concepts

2.5.1. VELOCITY-STRAIN AND VORTICITY TENSORS

The spatial nonuniformity of fluid motion is characterized by the first derivative of velocity, $\partial u_j / \partial x_i$. It is a tensor of the second rank and is divided into the symmetric and anti-symmetric parts as

$$\frac{\partial u_j}{\partial x_i} = \frac{1}{2} (s_{ij} + \omega_{ij}) , \tag{2.51}$$

where

$$s_{ij} = \frac{\partial u_j}{\partial x_i} + \frac{\partial u_i}{\partial x_j}, \tag{2.52}$$

$$\omega_{ij} = \frac{\partial u_j}{\partial x_i} - \frac{\partial u_i}{\partial x_j}. \tag{2.53}$$

These two quantities, s_{ij} and ω_{ij}, are called the velocity-strain and vorticity tensors, respectively.

In what follows, we shall consider a two-dimensional fluid motion

$$\mathbf{u} = (u(x, y), v(x, y), 0) , \tag{2.54}$$

and clarify the physical meanings of s_{ij} and ω_{ij}.

A. Vorticity

As a simple example of fluid rotation, we take the solid rotation around the z axis, whose angular velocity is Ω_0 (see Fig. 2.3). The motion is described by

$$x = r \cos\theta, \ y = r \sin\theta \ \ (\theta = \Omega_0 t + \theta_0) , \tag{2.55}$$

where θ_0 is an initial value of θ. The corresponding velocity is given by

$$u = \frac{dx}{dt} = -\Omega_0 y, \ v = \frac{dy}{dt} = \Omega_0 x. \tag{2.56}$$

From Eq. (2.56), we have

$$\boldsymbol{\omega} \equiv \nabla \times \mathbf{u} = (0, \ 0, \ 2\Omega_0) . \tag{2.57}$$

Equation (2.57) signifies that in the presence of $\boldsymbol{\omega}$ at location $\mathbf{x}_0$, the fluid near $\mathbf{x}_0$ is rotating with the angular velocity $|\boldsymbol{\omega}|/2$ as if it were a spinning

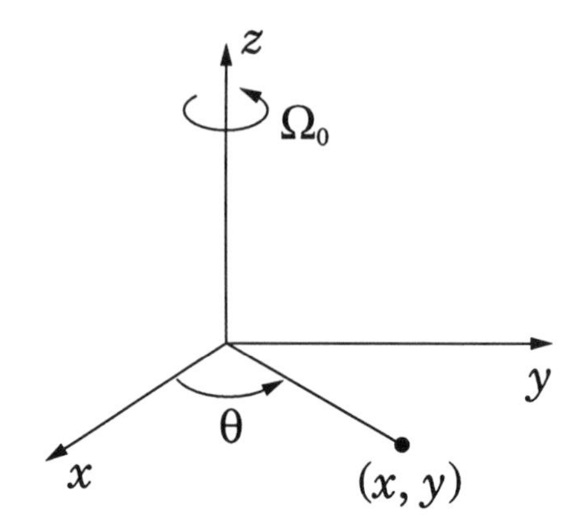

Figure 2.3. Solid rotation around the z axis.

top, and its rotation axis is in the direction of $\boldsymbol{\omega}$. The vorticity $\boldsymbol{\omega}$ is one of the central concepts in hydrodynamics.

The component of $\boldsymbol{\omega}$ is written as

$$\omega_i = \epsilon_{ij\ell} \frac{\partial u_\ell}{\partial x_j}. \tag{2.58}$$

Here $\epsilon_{ij\ell}$ is the alternating tensor, which is defined as

$$\epsilon_{ij\ell} = 1 \text{ for the even permutation of } (1,2,3); \tag{2.59a}$$

$$\epsilon_{ij\ell} = -1 \text{ for the odd permutation of } (1,2,3); \tag{2.59b}$$

$$\epsilon_{ij\ell} = 0 \text{ otherwise.} \tag{2.59c}$$

For instance, we have $\epsilon_{231} = 1$, $\epsilon_{132} = -1$, and $\epsilon_{112} = 0$. The alternating tensor $\epsilon_{ij\ell}$ has the following important properties:

$$\epsilon_{ij\ell}\epsilon_{ikm} = \delta_{jk}\delta_{\ell m} - \delta_{jm}\delta_{lk}, \ \epsilon_{ij\ell}\epsilon_{ijk} = 2\delta_{\ell k}. \tag{2.60}$$

Using Eq. (2.60), we may show that

$$\omega_{ij} = \epsilon_{ij\ell}\omega_\ell \tag{2.61a}$$

or

$$\omega_i = (1/2)\epsilon_{ij\ell}\omega_{j\ell}. \tag{2.61b}$$

The vorticity tensor ω_{ij} is physically the same quantity as the vorticity vector $\boldsymbol{\omega}$.

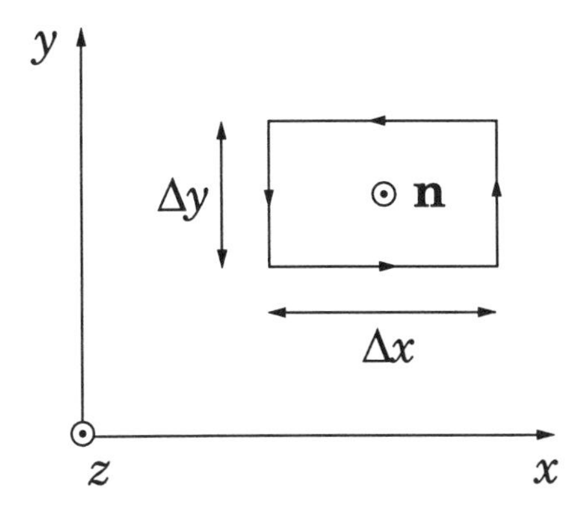

Figure 2.4. Infinitesimal rectangle.

Note: In relation to the vorticity, we refer to the Stokes' integral theorem

$$\oint_C \mathbf{u} \cdot \mathbf{ds} = \int_S (\nabla \times \mathbf{u}) \cdot \mathbf{n} dS, \tag{2.62}$$

where C is a closed line, S is any surface spanning on C, and $\mathbf{n}$ is the unit vector normal to S. First, we shall prove Eq. (2.62) in the infinitesimal rectangular plane region, whose center is located at (x, y, z) and whose two sides are Δx and Δy (see Fig. 2.4). The left-hand side of Eq. (2.62) is

$$\begin{aligned}\oint_C u \cdot ds &= u\left(x, y - \frac{\Delta y}{2}, z\right) \Delta x - u\left(x, y + \frac{\Delta y}{2}, z\right) \Delta x \\ &\quad + v\left(x + \frac{\Delta x}{2}, y, z\right) \Delta y - v\left(x - \frac{\Delta x}{2}, y, z\right) \Delta y \\ &\cong \left(-\frac{\partial u}{\partial y} + \frac{\partial v}{\partial x}\right) \Delta x \Delta y. \end{aligned} \tag{2.63}$$

The last term may be rewritten as

$$(\nabla \times \mathbf{u})_z \, \Delta x \Delta y = \int_S (\nabla \times \mathbf{u}) \cdot \mathbf{n} dS \tag{2.64}$$

since $\mathbf{n}$ is in the z direction. Then we have proved the Stokes' integral theorem in an infinitesimal surface S. Any surface is divided into an infinite number of such surfaces. Summing up their contributions, we can obtain the Stokes' integral theorem at an arbitrary surface in the manner entirely similar to the proof of the Gauss' integral theorem (2.6).

B. Strain

Under the solid rotation given by Eq. (2.56), the rectangular fluid region rotates without any deformation, as in Fig. 2.5, and s_{ij} vanishes. Then s_{ij}

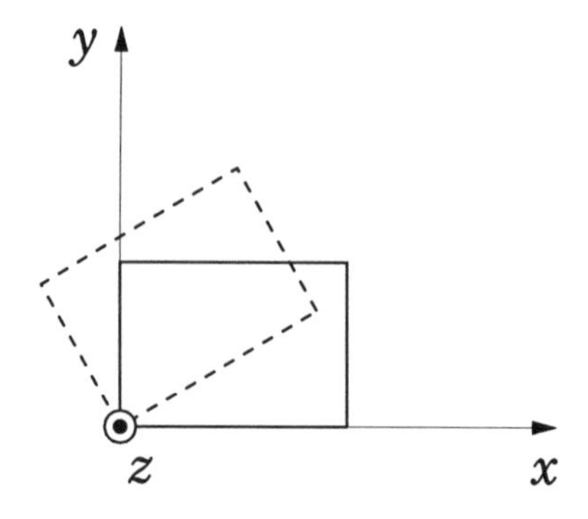

Figure 2.5. Solid rotation of a rectangle.

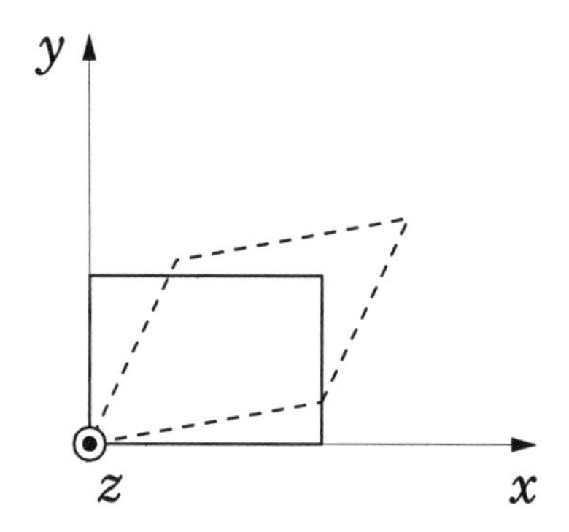

Figure 2.6. Pure straining motion of a rectangle.

is an indicator of the degree of deformation of a fluid region. In order to see this point, we consider a two-dimensional motion of a constant-density fluid

$$u = Sy, \ v = Sx \ (S \geq 0), \tag{2.65}$$

in correspondence to Eq. (2.56). This motion deforms a rectangular region into a parallelogram, as in Fig. 2.6. In this case, we have

$$s_{12} = s_{21} = S, \ s_{ij} = 0 \text{ (otherwise)}; \ \omega_{ij} = 0. \tag{2.66}$$

Vanishing of ω_{ij} signifies that the torque due to u is balanced with that due to v, and the rotational motion does not occur, resulting in the pure straining motion.

The motion represented by

$$u = -Dx, \ v = Dy \ (D \geq 0) \tag{2.67}$$

gives

$$s_{11} = -D, \; s_{22} = D, \; s_{ij} = 0 \text{ (otherwise)}; \; \omega_{ij} = 0. \tag{2.68}$$

This fluid motion compresses the rectangular region in the x direction, while expanding it in the y direction.

Both the motions given by Eqs. (2.65) and (2.67) deform a fluid region, unlike the solid rotational motion. In the modeling of τ_{ij} expressing effects of molecular motion, s_{ij} plays a central role, as in Eq. (2.28). This point is easily understandable from the fact that the deformation of a fluid region exerts influence on the molecular motion inside it, resulting in the occurrence of τ_{ij}. The stress caused by Eq. (2.66) is specifically called the shear stress since two blades of a shear exert such a force.

2.5.2. NONINERTIAL FRAMES

We consider Eq. (2.11) in the other reference frame moving with the relative constant velocity $\mathbf{U}_\infty$. These two frames are connected with each other by the relation

$$\mathbf{x}' = \mathbf{x} - \mathbf{U}_\infty t, \; t' = t, \; \mathbf{u}' = \mathbf{u} - \mathbf{U}_\infty. \tag{2.69}$$

Equation (2.69) gives

$$\frac{\partial}{\partial t} = \frac{\partial}{\partial t'} - \mathbf{U}_\infty \cdot \nabla', \; \nabla = \nabla'. \tag{2.70}$$

As a result, we have

$$\frac{D\rho}{Dt} = \frac{D\rho}{Dt'} \equiv \left(\frac{\partial}{\partial t'} + \mathbf{u}' \cdot \nabla' \right) \rho; \tag{2.71}$$

namely, the mathematical form for the mass equation is independent of the frame. This property is called the Galilean invariance and holds also for the momentum and energy equations (2.29) and (2.44). These two frames are called inertial ones. In turbulence modeling, any turbulence models proposed should not violate this invariance.

Effects of rotation become important in engineering machinery represented by a turbine. The similar situation holds in natural sciences associated with the Earth and stars, and some of their essential properties are generated and sustained by effects of frame rotation. We consider two frames, F and F', that share their origin. The frame F is rotating with the constant angular velocity relative to F', $\boldsymbol{\Omega}_F$. The velocity in F', $\mathbf{u}'$, is related to the counterpart in F, $\mathbf{u}$, as

$$\mathbf{u}' = \mathbf{u} + \boldsymbol{\Omega}_F \times \mathbf{x}. \tag{2.72}$$

The relationship between the accelerations in the two frames is given by

$$\frac{D\mathbf{u}'}{Dt'} = \frac{D\mathbf{u}}{Dt} + 2\mathbf{\Omega}_F \times \mathbf{u} + \mathbf{\Omega}_F \times (\mathbf{\Omega}_F \times \mathbf{x}). \tag{2.73}$$

Here the last term may be rewritten as

$$\begin{aligned}(\mathbf{\Omega}_F \times (\mathbf{\Omega}_F \times \mathbf{x})) &= (\mathbf{\Omega}_F \cdot \mathbf{x})\,\mathbf{\Omega}_F - \mathbf{\Omega}_F^2\mathbf{x} \\ = \nabla\left(\frac{1}{2}(\mathbf{\Omega}_F \cdot \mathbf{x})^2 - \frac{1}{2}\mathbf{\Omega}_F^2\mathbf{x}^2\right) &= -\nabla\left(\frac{1}{2}(\mathbf{\Omega}_F \times \mathbf{x})^2\right).\end{aligned} \tag{2.74}$$

In the cylindrical coordinates (r, θ, z) with the z axis coincident with the $\mathbf{\Omega}_F$ direction, we have

$$\nabla\left(\frac{1}{2}(\mathbf{\Omega}_F \times \mathbf{x})^2\right) = \nabla\left(\frac{1}{2}\Omega_F^2 r^2\right), \tag{2.75}$$

which expresses the centrifugal force.

We apply the foregoing transformation to the momentum equation with constant density, (2.30). Then we have

$$\frac{D\mathbf{u}}{Dt} + 2\mathbf{\Omega}_F \times \mathbf{u} = -\frac{1}{\rho}\nabla p + \nu\Delta\mathbf{u}, \tag{2.76}$$

where the centrifugal-force term, expression (2.75), has been absorbed into the pressure p. By this transformation, a new term called the Coriolis effect occurs on the left-hand side. These two frames are called noninertial. We use the relation

$$(\mathbf{u} \cdot \nabla)\,\mathbf{u} = \nabla\left(\frac{\mathbf{u}^2}{2}\right) + \boldsymbol{\omega} \times \mathbf{u}, \tag{2.77}$$

and take the curl of Eq. (2.76). The resulting equation is

$$\frac{\partial \boldsymbol{\omega}}{\partial t} = \nabla \times (\mathbf{u} \times (\boldsymbol{\omega} + 2\mathbf{\Omega}_F)) + \nu\Delta\boldsymbol{\omega}. \tag{2.78}$$

Equation (2.78) signifies that the vorticity in the two frames F and F', $\boldsymbol{\omega}$ and $\boldsymbol{\omega}'$, are related to each other as

$$\boldsymbol{\omega}' = \boldsymbol{\omega} + 2\mathbf{\Omega}_F, \tag{2.79a}$$

in correspondence to Eq. (2.72) for the velocity. The vorticity tensor ω_{ij}, which is defined by Eq. (2.61a), obeys the transformation rule

$$\omega'_{ij} = \omega_{ij} + 2\epsilon_{ijk}\Omega_{Fk}. \tag{2.79b}$$

Equation (2.79) will become important in the context of the turbulence modeling of Chapter 4. Once that the dependence of a model on $\boldsymbol{\omega}$ has been found by some method, the $\boldsymbol{\Omega}_F$ dependence in a rotating frame can be obtained through the replacement of $\boldsymbol{\omega}$ with Eq. (2.79a). On the contrary, the $\boldsymbol{\omega}$ dependence can be derived from the $\boldsymbol{\Omega}_F$ one.

2.5.3. HELICITY

In the vorticity equation

$$\frac{\partial \boldsymbol{\omega}}{\partial t} = \nabla \times (\mathbf{u} \times \boldsymbol{\omega}) + \nu \Delta \boldsymbol{\omega} \tag{2.80a}$$

or

$$\frac{D\boldsymbol{\omega}}{Dt} = (\boldsymbol{\omega} \cdot \nabla)\,\mathbf{u} + \nu \Delta \boldsymbol{\omega}, \tag{2.80b}$$

the nonlinearity intrinsic to the momentum equation appears in the first term. If $\mathbf{u} \times \boldsymbol{\omega}$ should vanish, that is, $\mathbf{u}$ should be aligned with $\boldsymbol{\omega}$, Eq. (2.80) would become

$$\frac{\partial \boldsymbol{\omega}}{\partial t} = \nu \Delta \boldsymbol{\omega}. \tag{2.81}$$

Under Eq. (2.81), the vortical structure of fluid motion established once hardly changes at high Reynolds number (R). Real-world flows, however, are turbulent at high R, and their vortical structures change rapidly in time. This fact signifies that $\mathbf{u}$ is not aligned with $\boldsymbol{\omega}$ there, and that its degree of deviation from the alignment is closely related to the duration of vortical structures.

The magnitude of $\mathbf{u} \times \boldsymbol{\omega}$ is related to that of $\mathbf{u} \cdot \boldsymbol{\omega}$ as

$$\frac{|\mathbf{u} \times \boldsymbol{\omega}|^2}{\mathbf{u}^2 \boldsymbol{\omega}^2} + \frac{|\mathbf{u} \cdot \boldsymbol{\omega}|^2}{\mathbf{u}^2 \boldsymbol{\omega}^2} = 1. \tag{2.82}$$

From Eq. (2.82), the increase in $|\mathbf{u} \cdot \boldsymbol{\omega}|$ leads to the decrease in $|\mathbf{u} \times \boldsymbol{\omega}|$, and the nonlinearity of Eq. (2.80) is controlled by $\mathbf{u} \cdot \boldsymbol{\omega}$ that expresses the degree of alignment between $\mathbf{u}$ and $\boldsymbol{\omega}$. The quantity $\mathbf{u} \cdot \boldsymbol{\omega}$ is called the helicity. Nonvanishing $\mathbf{u} \cdot \boldsymbol{\omega}$ signifies that fluid makes a helical motion; namely, a fluid blob flows while rotating. The physical meaning of helicity from a topological viewpoint is discussed by Moffatt (1969).

We may mention two prominent features of $\mathbf{u} \cdot \boldsymbol{\omega}$. One is that it is a pseudo-scalar quantity. Familiar scalars such as the energy $\mathbf{u}^2/2$ are invariant under the reflection of the coordinate system, that is, the transformation from the right- to left-handed system: $\mathbf{x} \rightarrow -\mathbf{x}$. On the other hand, $\mathbf{u} \cdot \boldsymbol{\omega}$ changes its sign. The clockwise helical motion represented by negative $\mathbf{u} \cdot \boldsymbol{\omega}$ becomes the counter-clockwise motion under the transformation. In this sense, the helicity is a useful indicator expressing the relationship between the directions of flow and rotation. Another feature is that the total amount of helicity is conserved in the absence of the viscosity. From Eqs. (2.30) and (2.80), we have

$$\frac{\partial}{\partial t}\mathbf{u}\cdot\boldsymbol{\omega} = \frac{\partial}{\partial x_i}\left(-\mathbf{u}\cdot\boldsymbol{\omega}u_i + \left(\frac{\mathbf{u}^2}{2} - \frac{p}{\rho}\right)\omega_i + \nu\frac{\partial}{\partial x_i}\mathbf{u}\cdot\boldsymbol{\omega}\right) - 2\nu\frac{\partial u_j}{\partial x_i}\frac{\partial \omega_j}{\partial x_i}. \tag{2.83}$$

This equation has essentially the same mathematical structure as Eq. (2.33) for the kinetic energy, and the entirely similar discussion can be made. We integrate Eq. (2.83) in the whole fluid region and derive the equation for the total amount of helicity. The first term on the right-hand side is reduced to the integral at the surface surrounding the volume. In the absence of viscous effects, the total amount is conserved so long as no helicity is injected across the surface by an external force. In hydrodynamic and magnetohydrodynamic turbulence modeling, it will be shown that the helicity is one of the key concepts in the sustainment mechanisms of global vortical and magnetic-field structures in turbulent motion.

2.5.4. BUOYANCY EFFECTS

The temperature in the case of constant or nearly constant ρ is governed by Eq. (2.49). The concentration of matter in fluid motion, which is denoted by C, obeys the same type of equation

$$\frac{DC}{Dt} = \lambda_C \Delta C, \tag{2.84}$$

where λ_C is the matter diffusivity. Scalars such as temperature and matter concentration are called passive scalar so long as they have no influence on fluid motion.

The representative effect of a scalar quantity on fluid motion is the buoyancy force due to density change. We add the buoyancy force $\rho\mathbf{g}$ to Eq. (2.29b), resulting in

$$\frac{D\mathbf{u}}{Dt} = -\frac{1}{\rho}\nabla p + \mathbf{g} + \mathbf{F}_V, \tag{2.85}$$

where $\mathbf{g}$ is the gravitational acceleration vector, and $\mathbf{F}_V$ is the viscosity-related part. We assume the small change of density, and write

$$\frac{1}{\rho} = \frac{1}{\rho_R + \Delta\rho} \cong \frac{1}{\rho_R} - \frac{\Delta\rho}{{\rho_R}^2}, \tag{2.86}$$

where

$$\Delta\rho = \rho - \rho_R, \tag{2.87}$$

and ρ_R is the reference density. We substitute Eq. (2.86) into Eq. (2.85) and have

$$\frac{D\mathbf{u}}{Dt} = -\frac{1}{\rho_R}\nabla p + \mathbf{g} + \frac{\Delta\rho}{{\rho_R}^2}\nabla p + \mathbf{F}_V. \tag{2.88}$$

In the Cartesian coordinates (x, y, z), we write

$$\mathbf{g} = (0, 0, -g). \tag{2.89}$$

We combine p with the gravity term to introduce

$$p_g = p + \rho_R\phi, \tag{2.90}$$

where ϕ is the gravitational potential

$$\phi = gz. \tag{2.91}$$

Using Eqs. (2.88) and (2.90), we have

$$\frac{D\mathbf{u}}{Dt} = -\frac{1}{\rho_R}\nabla p_g + \frac{\Delta\rho}{\rho_R}\mathbf{g} + \frac{\Delta\rho}{{\rho_R}^2}\nabla p_g + \mathbf{F}_V. \tag{2.92}$$

The density change $\Delta\rho$ is mainly caused by the temperature counterpart $\Delta\theta$, which is defined as

$$\Delta\theta = \theta - \theta_R, \tag{2.93}$$

where θ_R is the reference temperature corresponding to ρ_R.

We expand $\Delta\rho$ with respect to $\Delta\theta$ and retain the leading contribution only. Then we have

$$\frac{\Delta\rho}{\rho_R} = -\alpha\Delta\theta, \tag{2.94}$$

where α is the thermal expansion coefficient. Then Eq. (2.92) is reduced to

$$\frac{D\mathbf{u}}{Dt} = -\frac{1}{\rho_R}\nabla p_g - \alpha\left(\theta - \theta_R\right)\mathbf{g} - \frac{\alpha}{\rho_R}\left(\theta - \theta_R\right)\nabla p_g + \mathbf{F}_V. \tag{2.95}$$

In the familiar Boussinesq approximation, effects of density change are taken into account through only the second term on the right-hand side of Eq. (2.95) (Phillips 1977). The third term as well as the effect of density change on $\mathbf{F}_V$ is neglected. The resulting equation is

$$\frac{D\mathbf{u}}{Dt} = -\frac{1}{\rho_R}\nabla p_g - \alpha\left(\theta - \theta_R\right)\mathbf{g} + \nu\Delta\mathbf{u}, \tag{2.96}$$

where $\nu = \mu/\rho_R$. We combine Eq. (2.96) with the solenoidal condition on $\mathbf{u}$, (2.15), to reach a closed system of equation for fluid motion subject to buoyancy effects. Here we should note that the neglected third term on the right-hand side of Eq. (2.95) is of the first order in $\theta-\theta_R$, just as the retained second term. Therefore some caution is necessary for this treatment in the presence of strong pressure gradient.

2.6. Equations for Turbulent Motion

2.6.1. MEAN EQUATIONS

At high Reynolds number, flow quantities such as the velocity fluctuate highly in time and space. Their long-lasting properties are one of the primary interests in turbulence phenomena. As a method for abstracting these properties, we usually resort to the averaging procedures such as the ensemble, time, and volume ones. Of these three, the ensemble averaging procedure is clearest at least from the theoretical viewpoint. In a temporally decaying or growing flow, the time interval appropriate for the averaging cannot be chosen uniquely. The similar situation holds for the volume averaging in the presence of the spatial change of averaged values. In what follows, we shall adopt the ensemble averaging procedure, except some special cases such as the subgrid-scale modeling in large eddy simulation (Chapter 5) and the mass-weighted averaging in compressible turbulence modeling (Chapter 8). In this section, we shall focus attention on fluid motion with constant density.

We use the ensemble averaging $\langle\cdot\rangle$ to divide a quantity f into the mean F and the fluctuation around it, f':

$$f = F + f', \ F = \langle f \rangle , \tag{2.97}$$

where

$$f = (\mathbf{u}, p, \theta), \ F = (\mathbf{U}, P, \Theta), \ f' = (\mathbf{u}', p', \theta'). \tag{2.98}$$

Here p denotes the original pressure divided by constant density.

We apply the ensemble averaging to Eqs. (2.15) and (2.30) and have

$$\nabla \cdot \mathbf{U} = 0, \tag{2.99}$$

$$\frac{DU_i}{Dt} \equiv \left(\frac{\partial}{\partial t} + \mathbf{U}\cdot\nabla \right) U_i = -\frac{\partial P}{\partial x_i} + \frac{\partial}{\partial x_j}\left(-R_{ij} + \nu \frac{\partial U_i}{\partial x_j} \right), \tag{2.100}$$

where R_{ij} is the so-called Reynolds stress, which is defined as

$$R_{ij} = \left\langle u_i' u_j' \right\rangle . \tag{2.101}$$

Strictly speaking, $-R_{ij}$ should be called the Reynolds stress. Throughout this book, however, R_{ij} is simply called the Reynolds stress since this terminology brings no confusion. In what follows, the Lagrange derivative D/Dt is based on the mean velocity, unlike Eq. (2.12). The importance of R_{ij} is clear from the fact that it is the sole quantity distinguishing the original momentum equation (2.30) from the ensemble-mean counterpart (2.100). Equation (2.101) expresses the momentum transport by fluctuations; namely, R_{ij} is the transport rate of the i-component momentum in the j direction and vice versa. The ν-related term on the right-hand side is the molecular transport effect. In turbulent flow at high Reynolds number, the momentum transport due to turbulence is much larger than the molecular one.

Similarly, the mean temperature Θ obeys

$$\frac{D\Theta}{Dt} = \nabla \cdot \left(-\mathbf{H}_\theta + \lambda \nabla \Theta\right) \tag{2.102}$$

from Eq. (2.49), where $\mathbf{H}_\theta$ is the turbulent heat flux due to velocity fluctuations

$$\mathbf{H}_\theta = \left\langle \mathbf{u}'\theta' \right\rangle . \tag{2.103}$$

In order to close this equation as well as the velocity counterpart (2.100), we have to introduce some additional relations concerning R_{ij} and $\mathbf{H}_\theta$. This is the very goal of turbulence modeling.

2.6.2. TURBULENCE EQUATIONS

A. Reynolds-Stress Equation

We subtract Eq. (2.100) from the full counterpart (2.30), and have

$$\frac{Du'_i}{Dt} + \frac{\partial}{\partial x_j}\left(u'_i u'_j - R_{ij}\right) + u'_j \frac{\partial U_i}{\partial x_j} = -\frac{\partial p'}{\partial x_i} + \nu \Delta u'_i. \tag{2.104}$$

From Eq. (2.104), the equation for R_{ij} is given by

$$\frac{DR_{ij}}{Dt} = P_{ij} + \Pi_{ij} - \varepsilon_{ij} + \frac{\partial T_{ij\ell}}{\partial x_\ell}, \tag{2.105}$$

where

$$P_{ij} = -R_{i\ell}\frac{\partial U_j}{\partial x_\ell} - R_{j\ell}\frac{\partial U_i}{\partial x_\ell}, \tag{2.106}$$

$$\Pi_{ij} = \left\langle p' \left(\frac{\partial u'_j}{\partial x_i} + \frac{\partial u'_i}{\partial x_j} \right) \right\rangle, \tag{2.107}$$

$$\varepsilon_{ij} = 2\nu \left\langle \frac{\partial u'_i}{\partial x_\ell}\frac{\partial u'_j}{\partial x_\ell} \right\rangle, \tag{2.108}$$

$$T_{ij\ell} = -\left(\left\langle u_i' u_j' u_\ell' \right\rangle + \left\langle p' u_j' \right\rangle \delta_{i\ell} + \left\langle p' u_i' \right\rangle \delta_{j\ell}\right) + \nu \frac{\partial R_{ij}}{\partial x_\ell}. \tag{2.109}$$

Before discussing the physical meanings of Eq. (2.105), we consider the equation for the turbulent energy K defined by

$$K = \left\langle \mathbf{u}'^2/2 \right\rangle. \tag{2.110}$$

We put $i = j$ in Eq. (2.105) to have

$$\frac{DK}{Dt} = P_K - \varepsilon + \nabla \cdot \mathbf{T}_K, \tag{2.111}$$

where

$$P_K = -R_{ij} \frac{\partial U_j}{\partial x_i}, \tag{2.112}$$

$$\varepsilon = \nu \left\langle \left(\frac{\partial u_j'}{\partial x_i} \right)^2 \right\rangle, \tag{2.113}$$

$$\mathbf{T}_K = -\left\langle \left(\frac{\mathbf{u}'^2}{2} + p' \right) \mathbf{u}' \right\rangle + \nu \nabla K. \tag{2.114}$$

On the right-hand side of Eq. (2.111), only the first term is explicitly dependent on the mean velocity gradient $\nabla \mathbf{U}$ with the definition

$$(\nabla \mathbf{U})_{ij} = \frac{\partial U_j}{\partial x_i}. \tag{2.115}$$

The second term ε is the energy dissipation rate and is always non-negative. The third term is written in the divergence form. In Eq. (2.114) for $\mathbf{T}_K$, $\langle (\mathbf{u}'^2/2)\mathbf{u}' \rangle$ is the K transport due to fluctuations, $\langle p'\mathbf{u}' \rangle$ expresses the work done by the fluctuating pressure, and $\nu \nabla K$ is the K transport rate due to molecular effects that is similar to the heat flow occurring in the counter direction of temperature gradient.

On integrating Eq. (2.111) over a fluid region V, the last term is reduced to the integral of $\mathbf{T}_K$ at the surface surrounding V from the Gauss' integral theorem (2.6). At a solid boundary, $\mathbf{T}_K$ vanishes in general. So long as no inflow of the turbulent energy occurs across a boundary, nonvanishing K is sustained by the mean velocity gradient in P_K. This point can be easily understood from many observational findings that K is largest near solid boundaries where large mean velocity gradients are generated owing to the noslip velocity condition.

In order to see the origin of the turbulent energy, we consider the equation for the mean-velocity energy. Using Eq. (2.100), we have

$$\frac{D}{Dt}\frac{\mathbf{U}^2}{2} = -P_K - \nu\left\langle\left(\frac{\partial U_j}{\partial x_i}\right)^2\right\rangle + \frac{\partial}{\partial x_j}\left(-PU_j - R_{ij}U_i + \nu\frac{\partial}{\partial x_j}\frac{\mathbf{U}^2}{2}\right). \tag{2.116}$$

Here we should note that P_K in the K equation (2.111) occurs with the minus sign attached. In the case of positive P_K generating K, $-P_K$ results in the drain of the mean-velocity energy. In other words, the energy is supplied to the fluctuating field in the form of the loss of the mean-velocity energy. This process is a kind of energy cascade from large to small-scale components of motion. The concept of energy cascade at small scales will be referred to in Sec. 3.2.

What is the source of the mean-velocity energy? The representative source is the work done by pressure effects. We integrate Eq. (2.116) over a region V and pay special attention to the mean pressure P. Then we have

$$\int_V \nabla\cdot(-P\mathbf{U})\,dV = \int_S \mathbf{U}\cdot(-P\mathbf{n})\,dS \tag{2.117}$$

from the Gauss' integral theorem (2.6). Here $-P\mathbf{n}$ is the pressure force that the fluid outside the region V exerts to the inner fluid ($\mathbf{n}$ is the outward unit vector). Considering that fluid travels a distance $\mathbf{U}$ in unit time, $\mathbf{U}\cdot(-P\mathbf{n})\,dS$ is the work done on the inner fluid across the surface element dS.

The foregoing clear physical picture of energy supply to fluctuations comes from the fact that the total amount of kinetic energy is conserved in the absence of molecular effects, as is stated in the context of Eq. (2.34). Here we should stress that the entirely similar physical picture holds for any quantities whose total amounts are conserved in the absence of molecular effects. Their turbulence parts obey the same type of equations. Such representative quantities are the helicity $\mathbf{u}\cdot\boldsymbol{\omega}$ and the scalar intensity θ^2. On the contrary, it is difficult to abstract a clear physical picture from a governing equation for the turbulence part of a quantity not subject to the conservation rule. This point becomes a great stumbling block for turbulence modeling.

We return to Eq. (2.105) for R_{ij}. On the right-hand side, the first, third, and fourth terms survive as P, ε, and $\mathbf{T}_K$, respectively, after taking the contraction $i = j$. Therefore their properties are essentially the same as the latter counterparts. The sole exception is Π_{ij} [Eq. (2.107)]. This term does not contribute to the temporal change of K, but it exerts influence to the evolution of each turbulent intensity such as $\langle {u'_1}^2\rangle$. In case that the main flow is in the x_1 direction, energy is first supplied to $\langle {u'_1}^2\rangle$ from the main flow, resulting in largest $\langle {u'_1}^2 \rangle$ of three intensities. The energy of $\langle {u'_1}^2 \rangle$ goes to the remaining two through Π_{ij}. For this role, Π_{ij} is often called the redistribution term.

In the noninertial frame rotating with the angular velocity $\mathbf{\Omega}_F$, the counterpart of Eq. (2.105) is given by

$$\frac{DR_{ij}}{Dt} + 2\left(\epsilon_{i\ell k}\Omega_{F\ell}R_{kj} + \epsilon_{j\ell k}\Omega_{F\ell}R_{ki}\right) = P_{ij} + \Pi_{ij} - \varepsilon_{ij} + \frac{\partial T_{ij\ell}}{\partial x_\ell}. \quad (2.118)$$

Here the $\mathbf{\Omega}_F$-related terms vanish for $i = j$ and do not enter the K equation explicitly. This fact indicates that the frame rotation is closely associated with the degree of anisotropy of three turbulent intensities such as $R_{11} (= \langle {u'_1}^2 \rangle)$. The situation is understandable since the axis of rotation brings a preferred direction and tends to destroy the isotropy of turbulence state.

B. Heat-Flux and Heat-Variance Equations

The temperature fluctuation θ' obeys

$$\frac{D\theta'}{Dt} + \nabla \cdot \left(\mathbf{u}'\theta' - \mathbf{H}_\theta\right) + \mathbf{u}' \cdot \nabla\Theta = \lambda\Delta\theta'. \quad (2.119)$$

From Eqs. (2.104) and (2.119), the turbulent heat flux $\mathbf{H}_\theta$ obeys

$$\frac{DH_{\theta i}}{Dt} = P_{\theta i} + \Pi_{\theta i} - \varepsilon_{\theta i} + \frac{\partial T_{\theta ij}}{\partial x_j}, \quad (2.120)$$

where

$$P_{\theta i} = -H_{\theta_j}\frac{\partial U_i}{\partial x_j} - R_{ij}\frac{\partial \Theta}{\partial x_j}, \quad (2.121)$$

$$\Pi_{\theta i} = \left\langle p'\frac{\partial \theta'}{\partial x_i}\right\rangle, \quad (2.122)$$

$$\varepsilon_{\theta i} = (\nu + \lambda)\left\langle \frac{\partial u'_i}{\partial x_j}\frac{\partial \theta'}{\partial x_j}\right\rangle, \quad (2.123)$$

$$T_{\theta ij} = -\left\langle u'_i u'_j \theta'\right\rangle - \left\langle p'\theta'\right\rangle\delta_{ij} + \nu\left\langle \theta'\frac{\partial u'_i}{\partial x_j}\right\rangle + \lambda\left\langle u'_i\frac{\partial \theta'}{\partial x_j}\right\rangle. \quad (2.124)$$

On the right-hand side of Eq. (2.120), each term corresponds to the counterpart in Eq. (2.105) for R_{ij}.

From Eq. (2.49), we can easily see that the total amount of the scalar intensity θ^2 is conserved in the absence of the molecular temperature diffusivity λ. Therefore the discussion similar to that on K can be made on the heat or scalar diffusion. The scalar variance K_θ denoted by

$$K_\theta = \left\langle \theta'^2 \right\rangle \quad (2.125)$$

obeys

$$\frac{DK_\theta}{Dt} = P_\theta - \varepsilon_\theta + \nabla \cdot \mathbf{T}_\theta, \tag{2.126}$$

where

$$P_\theta = -2\mathbf{H}_\theta \cdot \nabla\Theta, \tag{2.127}$$

$$\varepsilon_\theta = 2\lambda \left\langle \left(\frac{\partial \theta'}{\partial x_i} \right)^2 \right\rangle, \tag{2.128}$$

$$\mathbf{T}_\theta = -\left\langle \theta'^2 \mathbf{u}' \right\rangle + \lambda \nabla K_\theta. \tag{2.129}$$

C. Turbulent-Helicity Equation

In the case of constant density, the total amount of helicity is conserved in the absence of molecular effects, as is seen from Eq. (2.83). In hydrodynamic flows, the role of the helicity is not so clear as those of the energy and the temperature intensity, but this conservation property suggests to us its importance in turbulence analysis. From the vorticity equation (2.80b), the fluctuation part $\boldsymbol{\omega}'$ obeys

$$\frac{D\omega_i'}{Dt} + \frac{\partial}{\partial x_j}\left(u_j'\omega_i' - u_i'\omega_j' - \left\langle u_j'\omega_i' - u_i'\omega_j' \right\rangle \right) + u_j'\frac{\partial \Omega_i}{\partial x_j} - \omega_j'\frac{\partial U_i}{\partial x_j} - \Omega_j\frac{\partial u_i'}{\partial x_j} = \nu\Delta\omega_i'. \tag{2.130}$$

The turbulent part of helicity, H, which is defined by

$$H = \left\langle \mathbf{u}' \cdot \boldsymbol{\omega}' \right\rangle, \tag{2.131}$$

is governed by

$$\frac{DH}{Dt} = P_H - \varepsilon_H + \nabla \cdot \mathbf{T}_H, \tag{2.132}$$

where

$$P_H = -\left(\mathbf{V}_M - \nabla K\right) \cdot \boldsymbol{\Omega} - R_{ij}\frac{\partial \Omega_i}{\partial x_j} = \frac{\partial R_{ij}}{\partial x_j}\Omega_i - R_{ij}\frac{\partial \Omega_i}{\partial x_j}, \tag{2.133}$$

$$\varepsilon_H = 2\nu \left\langle \frac{\partial u_j'}{\partial x_i}\frac{\partial \omega_j'}{\partial x_i} \right\rangle, \tag{2.134}$$

$$\mathbf{T}_H = -\left\langle \left(\mathbf{u}' \cdot \boldsymbol{\omega}'\right)\mathbf{u}' \right\rangle + \left\langle \left(\frac{\mathbf{u}'^2}{2} - p' \right) \boldsymbol{\omega}' \right\rangle + \nu\nabla H, \tag{2.135}$$

with

$$\mathbf{V}_M = \left\langle \mathbf{u}' \times \boldsymbol{\omega}' \right\rangle. \tag{2.136}$$

In the derivation of the second relation of Eq. (2.133), use has been made of Eq. (2.77), that is,

$$\left(\mathbf{u}' \times \boldsymbol{\omega}'\right)_i = -\frac{\partial u'_i u'_j}{\partial x_j} + \frac{\partial}{\partial x_i}\left(\frac{\mathbf{u}'^2}{2}\right). \tag{2.137}$$

Equation (2.132) is very similar to Eq. (2.111) for K. The prominent feature of the former, however, is that $\boldsymbol{\Omega}$ and $\nabla\boldsymbol{\Omega}$ occur in Eq. (2.133) in place of $\nabla\mathbf{U}$ in Eq. (2.112) for P_K. This fact indicates that the helicity may become important in phenomena associated with global vortical fluid motions.

References

Imai, I. (1973), *Fluid Mechanics* (Shokabo, Tokyo).
Lagerstrom, P. A. (1964), in *Theory of Laminar Flows*, edited by F. K. Moore (Princeton U. P., Princeton), p. 20.
Moffatt, H. K. (1969), J. Fluid Mech. **35**, 117.
Phillips, O. M. (1977), *The Dynamics of the Upper Ocean* (Cambridge U. P., Cambridge).

CHAPTER 3

SMALL-SCALE TURBULENCE

3.1. Fundamental Concepts

The most prominent feature of turbulence at high Reynolds number lies in a number of scales included in fluid motion. The injection of energy is usually made through the components of motion related to the reference spatial length L_R, whereas the energy is dissipated at the fine energy-dissipation scale ℓ_D. These two scales are related to each other as Eq. (1.2) that will be derived in this chapter. The components of motion whose spatial scales are much smaller than L_R tend to be independent of the details of the energy-injection process. In a long history of the theoretical study of turbulence, such small-scale components have attracted much attention. In their study, the energy-injection mechanism connected with the mean velocity gradient is not a primary interest and is replaced with the concept of energy-containing components of velocity fluctuations. These components play the role of a reservoir of externally injected energy, and the transport process of energy from this reservoir to the small-scale components is the main subject.

In the study of turbulence by the statistical theoretical methods represented by a direct-interaction approximation (DIA) method of Kraichnan (1959), attention has been focused on turbulence with constant density. In this chapter, we shall discuss its small-scale properties as a starting point towards the study of real-world turbulent flows.

3.1.1. HOMOGENEITY

As the simplest but most important turbulence statistics, we consider the velocity correlation functions of the second and third orders

$$Q_{ij}\left(\mathbf{x},\mathbf{x}';t,t'\right) = \left\langle u'_i(\mathbf{x};t)u'_j(\mathbf{x}';t')\right\rangle, \tag{3.1}$$

$$Q_{ij\ell}\left(\mathbf{x},\mathbf{x}',\mathbf{x}'';t,t',t''\right) = \left\langle u'_i(\mathbf{x};t)u'_j(\mathbf{x}';t')u'_\ell(\mathbf{x}'';t'')\right\rangle. \tag{3.2}$$

We introduce the relative position vector of two locations as

$$\mathbf{r} = \mathbf{x}' - \mathbf{x},\ \ \mathbf{r}' = \mathbf{x}'' - \mathbf{x}, \tag{3.3}$$

and rewrite Eqs. (3.1) and (3.2) as

$$Q_{ij}\left(\mathbf{x},\mathbf{x}+\mathbf{r};t,t'\right) \equiv Q_{ij}\left(\mathbf{x}|\mathbf{r};t,t'\right) = \left\langle u_i'(\mathbf{x};t)u_j'(\mathbf{x}+\mathbf{r};t')\right\rangle, \tag{3.4}$$

$$\begin{aligned} Q_{ij\ell}\left(\mathbf{x},\mathbf{x}+\mathbf{r},\mathbf{x}+\mathbf{r}';t,t',t''\right) &\equiv Q_{ij\ell}\left(\mathbf{x}|\mathbf{r}|\mathbf{r}';t,t',t''\right) \\ &= \left\langle u_i'(\mathbf{x};t)u_j'(\mathbf{x}+\mathbf{r};t')u_\ell'(\mathbf{x}+\mathbf{r}';t'')\right\rangle. \end{aligned} \tag{3.5}$$

In case that statistical quantities such as Eqs. (3.4) and (3.5) are independent of the reference position $\mathbf{x}$, this turbulence state is called statistically homogeneous. Entirely similarly, the state is called statistically stationary when Eqs. (3.4) and (3.5) depend only on the time differences $|t'-t|$ and $|t''-t|$. In order to see the relationship of the homogeneity with mean velocity gradients, we consider the turbulent energy defined by Eq. (2.110), K, which is also written as

$$K\left(\mathbf{x};t\right) = (1/2)\,Q_{ii}\left(\mathbf{x},\mathbf{x};t,t\right) \equiv (1/2)\,Q_{ii}\left(\mathbf{x}|\mathbf{0};t,t\right). \tag{3.6}$$

Under the condition of homogeneity, Eq. (2.111) is reduced to

$$\frac{\partial K}{\partial t} = -Q_{ij}\left(\mathbf{0}|\mathbf{0};t,t\right)\frac{\partial U_i}{\partial x_j} - \varepsilon, \tag{3.7}$$

where we should note that $(\mathbf{U}\cdot\nabla)K$ and $\nabla\cdot\mathbf{T}_K$ vanish.

From the $\mathbf{x}$ independence of K, $Q_{ij}(\mathbf{0}|\mathbf{0};t,t)$, and ε, we have

$$\frac{\partial U_i}{\partial x_j} = C_{ij}(t), \tag{3.8}$$

which leads to

$$U_i = C_{ij}x_j. \tag{3.9}$$

We can add $C_i(t)$ to Eq. (3.9), but it may be eliminated using the Galilean transformation more general than Eq. (2.69),

$$\mathbf{x}' = \mathbf{x} - \int^t \mathbf{C}(s)ds,\ t' = t,\ \mathbf{u}' = \mathbf{u} - \mathbf{C}. \tag{3.10}$$

The constancy of fluid density, $\nabla\cdot\mathbf{U} = 0$, requires

$$C_{ii} = 0. \tag{3.11}$$

From Eqs. (2.100) for $\mathbf{U}$ and (3.9), the mean pressure divided by constant density, P, obeys

$$\Delta P = -C_{ij}C_{ji}. \tag{3.12}$$

A typical instance of homogeneous turbulence subject to constant mean velocity gradient is the so-called homogeneous-shear turbulence. Its mean velocity is given by

$$\mathbf{U} = (Sy, 0, 0) \tag{3.13}$$

in the Cartesian coordinates (x, y, z), resulting in

$$C_{12} = S, \; C_{ij} = 0 \text{ (otherwise)} \tag{3.14}$$

(S is the shear rate). Equation (3.13) with constant P obeys Eq. (2.100) identically. The importance of this flow lies in the fact that its geometrical structure is very simple, but that it still retains the energy-generation mechanism due to the mean velocity gradient. For Eq. (3.13), Eq. (2.104) for the velocity fluctuation $\mathbf{u}'$ becomes

$$\frac{\partial u_i'}{\partial t} + \frac{\partial}{\partial x_j} u_i' u_j' + S\left(x_2 \frac{\partial u_i'}{\partial x_1} + \delta_{i1} u_2' \right) = -\frac{\partial p'}{\partial x_i} + \nu \Delta u_i', \tag{3.15}$$

with the solenoidal condition

$$\nabla \cdot \mathbf{u}' = 0, \tag{3.16}$$

where δ_{ij} is the Kronecker delta symbol.

3.1.2. FOURIER REPRESENTATION

Homogeneity is premised on the infinity of fluid region since the reference position $\mathbf{x}$ in $Q_{ij}(\mathbf{x}|\mathbf{r}; t, t')$ and $Q_{ij}(\mathbf{x}|\mathbf{r}|\mathbf{r}'; t, t', t'')$ can be chosen freely. In this case, the Fourier representation is a useful mathematical tool for describing the properties of a fluctuating quantity $f'(\mathbf{x})$. Its Fourier representation is defined as

$$f'(\mathbf{x}) = \int f'(\mathbf{k}) \exp(-i\mathbf{k} \cdot \mathbf{x})\, d\mathbf{k}, \tag{3.17}$$

and $f'(\mathbf{k})$ corresponds to the component of $f'(\mathbf{x})$ with the spatial scale $2\pi/|\mathbf{k}|$. The inverse representation of Eq. (3.17) is

$$f'(\mathbf{k}) = \frac{1}{(2\pi)^3} \int f'(\mathbf{x}) \exp(i\mathbf{k} \cdot \mathbf{x})\, d\mathbf{x}. \tag{3.18}$$

In Eqs. (3.17) and (3.18), the ranges of integration with respect to $\mathbf{k}$ and $\mathbf{x}$ cover each whole domain, and we should note the formula for the Dirac's delta function

$$\delta(\mathbf{x}) = \frac{1}{(2\pi)^3} \int \exp(\pm i\mathbf{k} \cdot \mathbf{x})\, d\mathbf{k}. \tag{3.19}$$

We apply Eq. (3.18) to Eqs. (3.15) and (3.16), and have

$$\frac{\partial}{\partial t}u_i'(\mathbf{k};t) - ik_j \iint u_i'(\mathbf{p};t)u_j'(\mathbf{q};t)\delta(\mathbf{k}-\mathbf{p}-\mathbf{q})\,d\mathbf{p}d\mathbf{q}$$
$$-S\left(k_1\frac{\partial}{\partial k_2}u_i'(\mathbf{k};t) - \delta_{i1}u_2'(\mathbf{k};t)\right) = ik_i p'(\mathbf{k};t) - \nu k^2 u_i'(\mathbf{k};t), \quad (3.20)$$

$$k_i u_i'(\mathbf{k};t) = 0. \quad (3.21)$$

The first part of the S-related terms on the left-hand side of Eq. (3.20) has been derived using

$$\int x_2(-ik_1)u_i'(\mathbf{k};t)\exp(-i\mathbf{k}\cdot\mathbf{x})d\mathbf{k}$$
$$= \int k_1 u_i'(\mathbf{k};t)\frac{\partial}{\partial k_2}\exp(-i\mathbf{k}\cdot\mathbf{x})d\mathbf{k}$$
$$= \int \frac{\partial}{\partial k_2}\left(k_1 u_i'(\mathbf{k};t)\exp(-i\mathbf{k}\cdot\mathbf{x})\right)d\mathbf{k}$$
$$- \int \left(k_1\frac{\partial}{\partial k_2}u_i'(\mathbf{k};t)\right)\exp(-i\mathbf{k}\cdot\mathbf{x})d\mathbf{k}. \quad (3.22)$$

The first part in the second relation of Eq. (3.22) vanishes after the integration with respect to k_2, for vanishing of $|\mathbf{k}|^n u_i'(\mathbf{k};t)$ in the limit of $|\mathbf{k}| \to \infty$ for any integer n is guaranteed under the concept of generalized functions (Lighthill 1970). Such a mathematical property will be used fully in the following Fourier analysis. The Fourier transformation of the second part leads to the $\partial/\partial k_2$-related term in Eq. (3.20).

We apply Eq. (3.21) to Eq. (3.20), and have

$$p'(\mathbf{k};t) = -\frac{k_j k_\ell}{k^2}\iint u_j'(\mathbf{p};t)u_\ell'(\mathbf{q};t)\delta(\mathbf{k}-\mathbf{p}-\mathbf{q})\,d\mathbf{p}d\mathbf{q}$$
$$+iS\left(\frac{k_1 k_j}{k^2}\frac{\partial}{\partial k_2}u_j'(\mathbf{k};t) - \frac{k_1}{k^2}u_2'(\mathbf{k};t)\right). \quad (3.23)$$

We use Eq. (3.23) to eliminate p' from Eq. (3.20). As a result, we have

$$\frac{\partial}{\partial t}u_i'(\mathbf{k};t) + \nu k^2 u_i'(\mathbf{k};t)$$
$$-iM_{ij\ell}(\mathbf{k})\iint u_j'(\mathbf{p};t)u_\ell'(\mathbf{q};t)\delta(\mathbf{k}-\mathbf{p}-\mathbf{q})\,d\mathbf{p}d\mathbf{q}$$
$$-S\left(k_1 D_{ij}(\mathbf{k})\frac{\partial}{\partial k_2}u_j'(\mathbf{k};t) - D_{i1}(\mathbf{k})u_2'(\mathbf{k};t)\right) = 0, \quad (3.24)$$

where

$$D_{ij}(\mathbf{k}) = \delta_{ij} - \left(k_i k_j/k^2\right), \quad (3.25)$$

$$M_{ij\ell}(\mathbf{k}) = (1/2)\left(k_j D_{i\ell}(\mathbf{k}) + k_\ell D_{ij}(\mathbf{k})\right). \tag{3.26}$$

Here $D_{ij}(\mathbf{k})$ is the so-called solenoidal projection operator coming from Eq. (3.21), and we have the relations

$$\begin{gathered} k_i D_{ij}(\mathbf{k}) = k_i M_{ij\ell}(\mathbf{k}) = 0, \\ D_{i\ell}(\mathbf{k}) D_{\ell j}(\mathbf{k}) = D_{ij}(\mathbf{k}), \;\; D_{im}(\mathbf{k}) M_{mj\ell}(\mathbf{k}) = M_{ij\ell}(\mathbf{k}). \end{gathered} \tag{3.27}$$

A great merit of the use of the Fourier representation is that the spatially-differential operation in physical space is replaced with the algebraic one in wavenumber space. However, owing to the presence of the $\partial/\partial k_2$-related term in Eq. (3.24), the differential equation in physical space has merely been converted to the counterpart in wavenumber space, although its rank is reduced from two to one. The retention of the mean velocity leads to the preferential directivity due to the direction in which the mean velocity changes. This directivity makes velocity fluctuations deviate from the statistical isotropy, and their mathematical description becomes very complicated (we shall later refer to the statistical isotropy in the context of the energy spectrum).

From the foregoing two reasons, turbulence with no mean velocity has been a primary subject in the theoretical study of homogeneous turbulence. The simplicity stemming from no mean velocity is very helpful to the exploitation of a statistical theoretical formalism. In this case, the governing equation is

$$\begin{aligned} \frac{\partial}{\partial t} u_i(\mathbf{k};t) &= -\nu k^2 u_i(\mathbf{k};t) \\ &\quad + i M_{ij\ell}(\mathbf{k}) \iint u_j(\mathbf{p};t) u_\ell(\mathbf{q};t) \delta(\mathbf{k}-\mathbf{p}-\mathbf{q}) d\mathbf{p} d\mathbf{q}. \end{aligned} \tag{3.28}$$

Here and in the following part of this chapter, we replace $u'_i(\mathbf{k};t)$ with $u_i(\mathbf{k};t)$ for simplicity of writing. On the right-hand side of Eq. (3.28), the first molecular-viscosity term annihilates the eddy or wave with the wavenumber $\mathbf{k}$ or the wavelength $2\pi/|\mathbf{k}|$. The second term, which comes from the nonlinearity of the Navier-Stokes equation, expresses that the eddy with the wavenumber $\mathbf{p}$ is combined with the $\mathbf{q}$ counterpart to generate a new eddy with the wavenumber $\mathbf{k}$. The fact that $M_{ij\ell}(\mathbf{k})$ is of $O(|\mathbf{k}|)$ comes from the advection operator $\mathbf{u} \cdot \nabla$.

3.1.3. ISOTROPY AND ENERGY SPECTRUM

In order to see some of the basic properties of homogeneous turbulence, we consider the two-time, two-point correlation function, $Q(\mathbf{x}|\mathbf{r};t,t')$, given by

Eq. (3.4). First, we use Eq. (3.18) to write

$$\langle u_i(\mathbf{k};t)u_j(\mathbf{k}';t')\rangle = \frac{1}{(2\pi)^6}\iint \langle u_i(\mathbf{x};t)u_j(\mathbf{x}';t')\rangle \times \exp\left(i\left(\mathbf{k}\cdot\mathbf{x}+\mathbf{k}'\cdot\mathbf{x}'\right)\right)d\mathbf{x}d\mathbf{x}', \tag{3.29a}$$

which is rewritten as

$$\langle u_i(\mathbf{k};t)u_j(\mathbf{k}';t')\rangle = \frac{1}{(2\pi)^6}\iint Q_{ij}\left(\mathbf{x}|\mathbf{r};t,t'\right)\exp\left(i\mathbf{k}'\cdot\mathbf{r}\right) \times \exp\left(i\left(\mathbf{k}+\mathbf{k}'\right)\cdot\mathbf{x}\right)d\mathbf{x}d\mathbf{r}. \tag{3.29b}$$

Under the condition of homogeneity, $Q(\mathbf{x}|\mathbf{r};t,t')$ is independent of $\mathbf{x}$. Therefore Eq. (3.29b) leads to

$$\langle u_i(\mathbf{k};t)u_j(\mathbf{k}';t')\rangle = \frac{1}{(2\pi)^3}\delta\left(\mathbf{k}+\mathbf{k}'\right)\int Q_{ij}\left(\mathbf{0}|\mathbf{r};t,t'\right)\exp\left(i\mathbf{k}'\cdot\mathbf{r}\right)d\mathbf{r}, \tag{3.30}$$

where use has been made of Eq. (3.19). The inverse of Eq. (3.30) gives

$$Q_{ij}(\mathbf{0}|\mathbf{r};t,t') = \int \frac{\langle u_i(\mathbf{k};t)u_j(\mathbf{k}';t')\rangle}{\delta\left(\mathbf{k}+\mathbf{k}'\right)}\exp\left(-i\mathbf{k}'\cdot\mathbf{r}\right)d\mathbf{k}'. \tag{3.31}$$

Here we should remark that the description of $1/\delta(\mathbf{k}+\mathbf{k}')$ is not correct in the strict mathematical sense of generalized functions. Throughout this book, we shall use it as if it were a usual function since it brings no errors to final results. The turbulent energy K defined by Eq. (3.6) is written as

$$K = \frac{1}{2}\int \frac{\langle u_i(\mathbf{k};t)u_i(\mathbf{k}';t)\rangle}{\delta\left(\mathbf{k}+\mathbf{k}'\right)}d\mathbf{k}. \tag{3.32}$$

Equation (3.28) contains no factors generating the preferential directivity, unlike Eq. (3.24) for homogeneous-shear turbulence. Therefore the statistics of $\mathbf{u}(\mathbf{k};t)$ continue to keep a state free from any preferential directivity. Such a state is called statistical isotropy. As an instance of statistical isotropy, we write down the isotropic form for $\langle u_i(\mathbf{k};t)u_j(\mathbf{k}';t')\rangle$. The tensor quantities that have no preferential directivity are the Kronecker delta symbol δ_{ij}, the alternating tensor $\epsilon_{ij\ell}$, and the tensors made from the wavevector $\mathbf{k}$. With the aid of these quantities, $\langle u_i(\mathbf{k};t)u_j(\mathbf{k}';t')\rangle$ may be written as

$$\frac{\langle u_i(\mathbf{k};t)u_j(\mathbf{k}';t')\rangle}{\delta\left(\mathbf{k}+\mathbf{k}'\right)} = \delta_{ij}Q(k;t,t') + Q'(k;t,t')k_ik_j + \frac{i}{2}\frac{k_\ell}{k^2}\epsilon_{ij\ell}H(k;t,t'), \tag{3.33}$$

with $k = |\mathbf{k}|$. The velocity fluctuation $\mathbf{u}(\mathbf{k};t)$ obeys the solenoidal condition (3.21). We multiply Eq. (3.33) by k_i and have

$$Q'(k;t,t') = -\frac{1}{k^2}Q(k;t,t'), \tag{3.34}$$

where we should note that $k_i k_\ell \epsilon_{ij\ell} = 0$ from Eq. (2.59). Then Eq. (3.33) is reduced to

$$\frac{\langle u_i(\mathbf{k};t)u_j(\mathbf{k}';t')\rangle}{\delta(\mathbf{k}+\mathbf{k}')} = D_{ij}(\mathbf{k})Q(k;t,t') + \frac{i}{2}\frac{k_\ell}{k^2}\epsilon_{ij\ell}H(k;t,t'). \tag{3.35}$$

From Eqs. (3.32) and (3.35), we have the isotropic expression

$$K = \int_0^\infty E(k)dk, \tag{3.36}$$

where the energy spectrum $E(k)$ comes from the first part of Eq. (3.35) and is given by

$$E(k) = 4\pi k^2 Q(k;t,t). \tag{3.37}$$

Under the reflection $\mathbf{k} \to -\mathbf{k}$, $Q(k;t,t)$ is invariant and is called a pure scalar. On the other hand, $H(k;t,t')$, which does not contribute to K, changes its sign under the reflection owing to the dependence on k_ℓ. Such a quantity is called a pseudo-scalar. In Sec. 2.5.3, we have already referred to the helicity as a representative pseudo-scalar quantity in physical space. Its turbulence part $\langle \mathbf{u}\cdot\boldsymbol{\omega}\rangle$ is written as

$$\langle \mathbf{u}\cdot\boldsymbol{\omega}\rangle = \int \frac{\langle u_i(\mathbf{k};t)\omega_i(\mathbf{k}';t)\rangle}{\delta(\mathbf{k}+\mathbf{k}')}d\mathbf{k} = \int (ik_j)\,\epsilon_{ij\ell}\frac{\langle u_i(\mathbf{k};t)u_\ell(\mathbf{k}';t)\rangle}{\delta(\mathbf{k}+\mathbf{k}')}d\mathbf{k}. \tag{3.38}$$

We substitute Eq. (3.35) into Eq. (3.38) to have

$$\langle \mathbf{u}\cdot\boldsymbol{\omega}\rangle = \int H(k)d\mathbf{k}; \tag{3.39}$$

namely, the second term of Eq. (3.35) is related to the helical property of flow. We simply call $H(k)$ the helicity spectrum, although $4\pi k^2 H(k)$ should be called so in the context of $E(k)$.

Note: As a simple example possessing some preferential directivity, we consider the velocity fluctuation $\mathbf{u}$ that is statistically axisymmetric around the x_3 axis. In this case, the first part of Eq. (3.35) is replaced with $D_{i3}(\mathbf{k})D_{3j}(\mathbf{k})Q(k;t,t')$. This quantity fulfills the requirements such as the symmetry with respect to i and j and the solenoidal condition.

3.2. Kolmogorov's Power Law

From Eq. (3.28), we have

$$\frac{\partial}{\partial t} E(k) = -2\nu k^2 E(k) + S(k), \tag{3.40}$$

where $S(k)$ is given by

$$S(k) = 2\pi i k^2 M_{ij\ell}(\mathbf{k}) \left[\iint \langle u_i(\mathbf{k}'; t) u_j(\mathbf{p}; t) u_\ell(\mathbf{q}; t) \rangle \, \delta(\mathbf{k} - \mathbf{p} - \mathbf{q}) d\mathbf{p} d\mathbf{q} \right.$$
$$\left. - \iint \langle u_i(\mathbf{k}; t) u_j(\mathbf{p}; t) u_\ell(\mathbf{q}; t) \rangle \, \delta(\mathbf{k}' - \mathbf{p} - \mathbf{q}) d\mathbf{p} d\mathbf{q} \right] / \delta \left(\mathbf{k} + \mathbf{k}' \right), \tag{3.41}$$

and expresses the transfer rate of energy to $E(k)$ from the components of motion with other wavenumbers. The transfer rate $S(k)$ obeys

$$\int_0^\infty S(k) dk = 0. \tag{3.42}$$

This relation is a direct consequence of the homogeneity, that is, $\nabla \cdot \mathbf{T}_K = 0$ [see Eq. (2.114)]. From Eqs. (3.40) and (3.42), we have

$$\frac{\partial K}{\partial t} = -\varepsilon \equiv - \int_0^\infty D(k) dk, \tag{3.43}$$

where $D(k)$ is called the energy dissipation spectrum and is related to $E(k)$ as

$$D(k) = 2\nu k^2 E(k). \tag{3.44}$$

3.2.1. INERTIAL RANGE

Let us consider the energy spectrum of $E(k)$, whose profile is depicted schematically in Fig. 3.1. The wavenumber k_C is called the energy-containing wavenumber and is closely associated with the spatial scale characterizing the turbulent-energy production mechanism that is represented by P_K [Eq. (2.112)]. In homogeneous isotropic turbulence, however, such a mechanism is not dealt with directly, and the spectrum range near k_C or the energy-containing one plays the role of an energy reservoir for the range with larger k. At high R (Reynolds number) corresponding to very small ν, $D(k)$ is negligible near k_C and becomes important for k much larger than k_C. In Fig. 3.1, k_D characterizes the dissipation range where the kinetic energy is converted to heat, and is named the dissipation wavenumber. Increasing R leads to the larger separation between k_C and k_D. In the situation, the spectrum range between them is associated with neither the

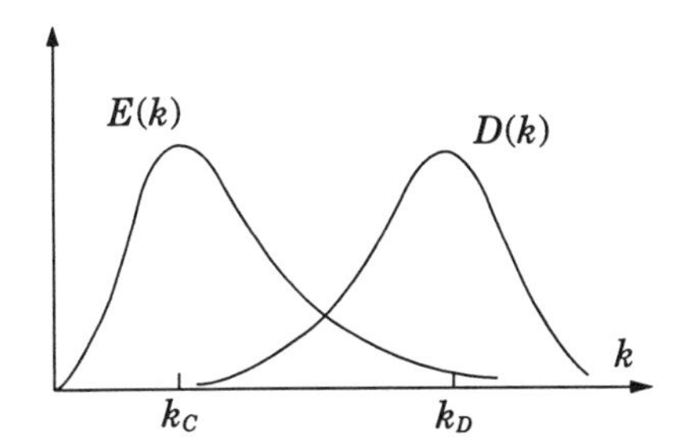

Figure 3.1. Energy and dissipation spectra.

energy-injection process nor the energy-dissipation one. This intermediate range, which is called the inertial one, receives some amount of energy from the energy-containing range, and compensates for the energy lost in the dissipation range. The energy transfer from the low- to high-wavenumber ranges is the so-called energy cascade, which originates in the conservation property of the kinetic energy in the absence of ν.

In the foregoing picture of the inertial range, the most important parameter is the rate of energy passing there. How is the rate determined? Is it determined mainly by the amount of energy injected by the energy-containing range or by the amount lost by the dissipation range? In the equilibrium state in which these two rates balance with each other, it is not necessary to distinguish between them. In real-world flows, the turbulent energy is generated by P_K [Eq. 2.112)], and its representative source is the mean pressure, as is stated in Sec. 2.6.2.A. For instance, the amount of energy supplied to a pipe flow is determined by the magnitude of the mean pressure gradient along the axis. The profile of the mean velocity is determined through the balance between the energy supply and the energy drain due to P_K that is dependent on the mean velocity gradient [recall Eq. (2.116)]. Therefore we may consider that the energy-transfer rate in the inertial range is determined mainly by the energy-production mechanism linked with the energy-containing range, except the isotropic case without such a mechanism. This statement does not lower the important effects of the energy dissipation range on the inertial one. Such effects become dominant at the dissipation-range side of the inertial range. This point will be discussed later.

Assuming that $E(k)$ in the inertial range is determined by the wavenumber magnitude k and the energy-injection rate ε (which is equal to the dissipation rate in the equilibrium state), we may write

$$E(k) = K_O \varepsilon^{2/3} k^{-5/3}, \tag{3.45}$$

using simple dimensional analysis, where it should be noted that the dimension of ε is L^2T^{-3} (L and T express the dimension of length and time, respectively). Equation (3.45) is the well-known Kolmogorov's $-5/3$ power law (Kolmogorov 1941a, b), and the Kolmogorov constant K_O scatters around 1.5 in observations (Paquin and Pond 1971; Williams and Paulson 1977; see also the works cited therein).

The energy spectrum $E(k)$ gives an important information on the intensity of velocity fluctuations with the spatial dimension $2\pi/k$. The enhancement of mixing or diffusion effects by velocity fluctuations is also a prominent feature of turbulence. This property may be detected from two-time observations. A quantity characterizing the two-time properties of turbulence is the correlation time between $\mathbf{u}(\mathbf{k};t)$ and $\mathbf{u}(\mathbf{k}';t')$, $\tau(k)$. The time scale $\tau(k)$ may be written as

$$\tau(k) = C_\tau \varepsilon^{-1/3} k^{-2/3}, \tag{3.46}$$

using dimensional analysis, where C_τ is a numerical factor. Equation (3.46) may be regarded as the life time of the eddies with the spatial dimension $2\pi/k$, and smaller eddies corresponding to larger k have shorter life time. The wave breaking due to blocks along a shore is partly based on this property.

3.2.2. INDICATIONS OF THE INERTIAL RANGE

The role of Eq. (3.45) seems rather small from the viewpoint that the inertial range merely passes energy over to the dissipation range. In the context of turbulence modeling dealing with real-world flows, it is not the case. The intermediate property of the inertial range brings much more merits than what it seems. In fluid mechanics, it is rare that the intermediate properties such as the inertial range can be compactly captured in an analytical form. For example, the analytical properties of laminar flows at low and high R's can be found by solving the Navier-Stoke equation by singular perturbation methods, but the counterparts at intermediate R may be obtained numerically only through the computer simulation.

In turbulent flow with mean velocity, the energy-containing eddies with k close to k_C are much affected by external factors such as boundaries. Therefore it is difficult to capture their intrinsic properties. Some of them, however, can be abstracted with the aid of the knowledge about the Kolmogorov spectrum, Eq. (3.45), at the lower limit of $k \to 0$. We substitute Eq. (3.45) into Eq. (3.36). The divergence of the integral at the lower limit comes from the existence of a characteristic spatial scale related to energy-containing eddies. We use k_C as the lower limit and calculate K as

$$K = (3/2) K_O \varepsilon^{2/3} k_C^{-2/3}. \tag{3.47}$$

Using Eq. (3.47), the spatial scale of energy-containing eddies, ℓ_C, is given by

$$\ell_C \equiv \frac{2\pi}{k_C} = 2\pi \left(\frac{2}{3}\right)^{3/2} K_O^{-3/2} \frac{K^{3/2}}{\varepsilon}. \tag{3.48}$$

Corresponding to ℓ_C, we define the intensity of energy-containing eddies by

$$u_C = \sqrt{2K}. \tag{3.49}$$

From Eqs. (3.48) and (3.49), the time scale of energy-containing eddies is estimated as

$$\tau_C \equiv \frac{\ell_C}{u_C} = \frac{4\pi}{3^{3/2}} K_O^{-3/2} \frac{K}{\varepsilon}. \tag{3.50}$$

A turbulence time scale may be found originally from Eq. (3.46). In reality, Eq. (3.50) is obtained by substituting $k = k_C$ into Eq. (3.46). The derivation of Eq. (3.50) indicates that the characteristic time scale of energy-containing eddies may be estimated using the energy spectrum only. This fact will be used fully in turbulence modeling of complex flows such as compressible turbulence (Chapter 8).

Next, we consider Eq. (3.43). We substitute Eq. (3.45) into Eq. (3.43), and replace the upper limit with the dissipation wavelength k_D. Then we have

$$\varepsilon = \nu \int_0^{k_D} D(k) dk, \tag{3.51}$$

which is reduced to

$$k_D = \left(\frac{2}{3K_O}\right)^{3/4} \left(\frac{\varepsilon}{\nu^3}\right)^{1/4} \tag{3.52a}$$

or

$$\ell_D \equiv \frac{2\pi}{k_D} = 2\pi \left(\frac{3K_O}{2}\right)^{3/4} \left(\frac{\nu^3}{\varepsilon}\right)^{1/4}. \tag{3.52b}$$

We see the relationship of ℓ_D with the reference velocity U_R and the reference length L_R. For this purpose, we need to estimate the magnitude of ε. As has already been stated, the amount of energy to be dissipated is determined primarily by the rate of energy supplied from the mean flow. Then ε may be estimated using the quantities characterizing the mean flow, that is, U_R and L_R, resulting in

$$\varepsilon = O\left(U_R^3 / L_R\right). \tag{3.53}$$

From Eqs. (3.52b) and (3.53), we have

$$\ell_D / L_R = O\left(R^{-3/4}\right), \tag{3.54}$$

which has already been given as Eq. (1.2), and R is defined by Eq. (1.1). The difficulty with the numerical simulation of turbulent flow at high R and the necessity of turbulence modeling are discussed using Eq. (1.2) in Sec. 1.1.

From Eqs. (3.48) and (3.52b), we have

$$\ell_D / \ell_C = O\left(R_T^{-3/4}\right), \tag{3.55}$$

where

$$R_T = K^2 / \nu\varepsilon . \tag{3.56}$$

Equation (3.56) may be also expressed as

$$R_T \propto u_C \ell_C / \nu , \tag{3.57}$$

using Eqs. (3.48) and (3.49). Namely, R_T is the Reynolds number based on the turbulence characteristics of energy-containing eddies, and may be called the turbulent Reynolds number. This Reynolds number is smaller than the global counterpart R [Eq. (1.1)], but it is larger than unity, as will be noted below. Equation (3.55) shows that R_T is an indicator of the degree of separation between energy-containing and dissipation ranges.

We write

$$\nu_T = K^2 / \varepsilon \propto u_C \ell_C, \tag{3.58}$$

which has the dimension of viscosity and is called the turbulent viscosity. In Chapter 4, it will be shown that ν_T is a quantity closely associated with the enhancement of dissipation and diffusion effects due to turbulent motion. Using Eq. (3.58), we may rewrite Eq. (3.56) as

$$R_T = \nu_T / \nu . \tag{3.59}$$

In the case of strong turbulence state, R_T is much larger than unity.

3.3. Passive-Scalar Spectrum

We consider the diffusion of a scalar quantity such as temperature and matter concentration in turbulent motion. In case that temperature variation is not so large, effects of temperature on fluid motion through the buoyancy force can be neglected. Similarly, effects of matter on fluid motion are negligible in a dilute-concentration case. In these cases, such scalars are called passive and obey Eq. (2.49).

As the simplest passive scalar diffusion, we examine the case of no mean temperature and velocity. Specifically, we assume that velocity fluctuations are statistically homogeneous and isotropic. In this situation, Eq. (2.49) may be transformed into the Fourier representation

$$\frac{\partial}{\partial t}\theta(\mathbf{k};t) = -\lambda k^2\theta(\mathbf{k};t) + ik_i \iint u_i(\mathbf{p};t)\theta(\mathbf{q};t)\delta\left(\mathbf{k}-\mathbf{p}-\mathbf{q}\right)d\mathbf{p}d\mathbf{q}. \quad (3.60)$$

The intensity of scalar fluctuations is characterized by the scalar-variance spectrum $E_\theta(k)$, which is defined as

$$\left\langle \theta^2 \right\rangle = \int_0^\infty E_\theta(k)dk, \quad (3.61)$$

with

$$E_\theta(k) = 4\pi k^2 \left\langle \theta(\mathbf{k};t)\theta(\mathbf{k}';t)\right\rangle / \delta\left(\mathbf{k}+\mathbf{k}'\right). \quad (3.62)$$

The scalar variance $\langle \theta^2 \rangle$ obey

$$\frac{\partial}{\partial t}\left\langle \theta^2 \right\rangle = -\varepsilon_\theta \equiv -\int_0^\infty D_\theta(k)dk, \quad (3.63)$$

where the scalar destruction rate ε_θ is given by Eq. (2.128), and the destruction spectrum $D_\theta(k)$ is defined as

$$D_\theta(k) = 2\lambda k^2 E_\theta(k), \quad (3.64)$$

in a manner entirely similar to the velocity counterpart (3.44).

In fluid motion, the relative importance of the inertia to viscous effects is characterized by the Reynolds number, Eq. (1.1). The counterpart in scalar diffusion is given by

$$R_\lambda = U_R L_R / \lambda\,, \quad (3.65)$$

which is called the Peclet and Schmidt numbers in the temperature and matter diffusion, respectively. In the case of large R_λ, the dominant range of $E_\theta(k)$ is separated from that of $D_\theta(k)$, as is similar to velocity fluctuations. As a result, $E_\theta(k)$ may be divided into three parts: the variance-dominant range including the peak of $E_\theta(k)$, the scalar destruction range, and the intermediate inertial range. In this situation, we can draw a picture similar to the Kolmogorov law for $E(k)$, but the picture differs depending on the relative magnitude of two diffusion effects represented by ν and λ. The magnitude is characterized by

$$P_r = \nu/\lambda, \quad (3.66)$$

and is called the Prandtl number.

In case that P_r is near unity, each range of $E_\theta(k)$ overlaps that of $E(k)$. As a result, the inertial range of $E_\theta(k)$ may be described using the inertial-range properties of velocity field, in addition to ε_θ that characterizes the scalar cascade. Considering that the inertial range of $E(k)$ is characterized by ε, $E_\theta(k)$ may be written as

$$E_\theta(k) = B_A \varepsilon_\theta \varepsilon^{-1/3} k^{-5/3}, \tag{3.67}$$

from dimensional analysis (Batchelor 1959), where the numerical coefficient B_A corresponds to K_O in Eq. (3.45), and is often called the Batchelor constant. The observational value of B_A scatters around 0.7 (Williams and Paulson 1977; Hill 1978; see also the works cited therein).

In the case of P_r far from unity, $E_\theta(k)$ deviates much from Eq. (3.67). For instance, under $P_r \ll 1$, the molecular effects due to λ may become important in the wavenumber region where $E(k)$ is given by the inertial-range form (3.45) (Batchelor et al. 1959). As a result, $E_\theta(k)$ in the region is dependent on λ in addition to k, ε, and ε_θ. The similar discussion can be made for $P_r \gg 1$.

3.4. Breakage of Kolmogorov's Scaling

The essence of the Kolmogorov spectrum (3.45) lies in the dependence on ε only, except k. This scaling, which is called Kolmogorov's one, has been tested using observational results at very high Reynolds numbers. The advance of a computer is also making possible its numerical test at moderate Reynolds numbers. These texts are affirmative for Eq. (3.45), although the small deviation from the $k^{-5/3}$ form cannot be ruled out. On the other hand, the Kolmogorov scaling is invalidated clearly for the higher-order statistics of velocity fluctuations. In what follows, we shall pay special attention to $E(k)$ and refer to the breakage of the scaling. The deviation from the $k^{-5/3}$ form may occur twofold; one is the deviation at the high-wavenumber side of the inertial range, and the other is the low-wavenumber counterpart. We shall examine the degree of their importance from the viewpoint of turbulence modeling that is the subject of this book.

3.4.1. INTERMITTENCY EFFECTS

In order to see the breakage of the Kolmogorov scaling with the increasing order of statistics, we examine the auto-correlation of the velocity difference

$$V_n = \langle V'_n \rangle,\ V'_n = |\mathbf{u}(\mathbf{x}+\mathbf{r}) - \mathbf{u}(\mathbf{x})|^n, \tag{3.68}$$

where n is a positive integer. In case that $\mathbf{u}$ is statistically homogeneous and isotropic, V_n is independent of $\mathbf{x}$. Under the condition that the Reynolds number of turbulence is high and that r falls in the inertial range, that is,

$$\ell_D \ll r \ll \ell_C, \tag{3.69}$$

the Kolmogorov scaling gives

$$V_{Kn} = C_{Vn} \left(\varepsilon r\right)^{n/3}, \tag{3.70}$$

where subscript K denotes the Kolmogorov scaling, and C_{Vn} is a numerical coefficient. Equation (3.70) with $n = 2$ corresponds to Eq. (3.45).

As a cause of the discrepancy between observational results and the scaling of Eq. (3.70), we can mention the relationship between the motion of highly-fluctuating fine-scale eddies and the energy-dissipation process. We introduce the instantaneous dissipation rate ε' as

$$\varepsilon' = \nu \left(\frac{\partial u_j}{\partial x_i} \right)^2, \tag{3.71}$$

using which the usual dissipation rate ε is given by $\varepsilon = \langle \varepsilon' \rangle$.

We assume that r satisfies Eq. (3.69), and write

$$V'_n = C_{V2n} \left(\varepsilon' r\right)^{n/3}, \tag{3.72}$$

similar to Eq. (3.70). Taking the ensemble mean of Eq. (3.72), we have

$$V_n = C_{V2n} r^{2n/3} \left\langle \varepsilon'^{n/3} \right\rangle. \tag{3.73}$$

Under the approximation

$$\left\langle \varepsilon'^{n/3} \right\rangle \cong \langle \varepsilon' \rangle^{n/3} = \varepsilon^{n/3}, \tag{3.74}$$

the Kolmogorov law (3.70) is recovered.

Equation (3.74) is accurate only when the fluctuation of ε' is weak. In the presence of its strong fluctuation, it is not sufficient to take into account only the mean value ε, and the deviation from it becomes important. The statistically highly nonuniform state of ε' is called intermittent. In order to estimate Eq. (3.73), we need to assume the probability density distribution of ε', $P\{\varepsilon'\}$. Its typical example is the so-called log-normal distribution by Oboukhov (1962). In the distribution, the variable is $\log \varepsilon'$, which obeys the normal or Gaussian distribution

$$P\{\log \varepsilon'\} = \frac{1}{\sqrt{2\pi}\sigma_{\ell\varepsilon}} \exp\left(-\frac{\left(\log \varepsilon' - m_{\ell\varepsilon}\right)^2}{2\sigma_{\ell\varepsilon}^2} \right), \tag{3.75}$$

where $m_{\ell\varepsilon}(=\langle\log\varepsilon'\rangle)$ and $\sigma_{\ell\varepsilon}$ are the mean and the standard deviation of $\log\varepsilon'$, respectively. The physical meaning of the log-normal distribution will be briefly explained in *Note*.

We apply Eq. (3.75) to Eq. (3.73) to have

$$\left\langle\varepsilon'^{n/3}\right\rangle=\int_{-\infty}^{\infty}\varepsilon'^{n/3}P\left\{\log\varepsilon'\right\}d\left(\log\varepsilon'\right)=\exp\left(\frac{n}{3}m_{\ell\varepsilon}+\frac{n^2}{18}\sigma_{\ell\varepsilon}^2\right). \tag{3.76}$$

As this special case, the mean dissipation rate $\varepsilon(=\langle\varepsilon'\rangle)$ is given by

$$\varepsilon=\int_{-\infty}^{\infty}\varepsilon'P\left\{\log\varepsilon'\right\}d\left(\log\varepsilon'\right)=\exp\left(m_{\ell\varepsilon}+\frac{\sigma_{\ell\varepsilon}^2}{2}\right). \tag{3.77}$$

We use Eq. (3.77) to eliminate $m_{\ell\varepsilon}$ from Eq. (3.76). As a result, we have

$$\frac{V_n}{V_{Kn}}=\exp\left(\frac{n(n-3)}{18}\sigma_{\ell\varepsilon}^2\right). \tag{3.78}$$

Equation (3.78) shows that the deviation from the Kolmogorov scaling is related to $\sigma_{\ell\varepsilon}$ characterizing the strength of ε' fluctuation. It also increases with the order of statistics, n. In order to know the relationship of the deviation with spatial scales, we need the information on the dependence of $\sigma_{\ell\varepsilon}$ on them. In real-world turbulent flow, the magnitude of ε is greatly affected by that of the energy production rate P_K [Eq. (2.112)] that is closely associated with the low-wavenumber components of motion. Highly probably, the fluctuation of ε' around ε is free from such low-wavenumber contributions, and it is dominated by the high-wavenumber components intrinsic to the dissipation process. Considering that the effect of $\sigma_{\ell\varepsilon}$ becomes strong as r deviates from the scale of energy-containing eddies, ℓ_C [Eq. (3.48)], we assume

$${\sigma_{\ell\varepsilon}}^2=\mu\log\left(\frac{\ell_C}{r}\right)+C_\sigma\left(\ell_C\right) \tag{3.79}$$

(Oboukhov 1962; Kolmogorov 1962). Here μ is a positive universal constant, and C_σ is a quantity dependent on ℓ_C. In the inertial range where $r\ll\ell_C$, the first part plays a key role.

We substitute Eq. (3.79) into Eq. (3.78) and have

$$V_n/V_{Kn}\propto\left(\ell_C/r\right)^{\mu n(n-3)/18}. \tag{3.80}$$

Equation (3.80) with $n=2$ corresponds to the energy spectrum

$$E(k)\propto\varepsilon^{2/3}k^{-5/3}\left(k\ell_C\right)^{-\mu/9}. \tag{3.81}$$

Observations suggest $\mu \cong 0.2$, and the deviation of the exponent from $-5/3$ is very small and does not seem so important. What should be noted, however, is the occurrence of the large spatial scale ℓ_C. Its definition is not always unique, resulting in the nonuniqueness of the proportional coefficient in Eq. (3.81).

From Eq. (3.80), we can see that the deviation of V_n from V_{Kn} becomes prominent with increasing n, as is consistent with observations. For $n > 15$, however, the discrepancy of Eq. (3.80) from observations is large. For overcoming the difficulty, the more elaborate modeling based on the concept of multifractals is under the intensive study. We do not enter such a discussion from the following reason. In turbulence modeling, the energy-containing eddies whose spatial scale ℓ_C is close to the mean-field counterpart have dominant influence on the transports of momentum and scalar. The foregoing deviation disappears as $k^{-1} \to \ell_C$ or at the energy-containing side of the inertial range. In this sense, the breakage of the Kolmogorov scaling at the energy-containing side is more important, as will be discussed below. The recent developments in the study of intermittency effects are detailed by Meneveau and Screenivasan (1991) and Frisch (1995).

Note: A theoretical support to the log-normal hypothesis may be given as follows [Gurvich and Yaglom 1967; this point is compactly reviewed by Leslie (1973)]. In order to express a number of scales included in turbulent motions, we use the parameter s much smaller than unity and write

$$\ell_0 \left(\cong \ell_C\right), \; \ell_1(= s\ell_0), \; \ell_2(= s^2\ell_0), \; \cdots, \; \ell_M(= s^M\ell_0) \cong \ell_D, \tag{3.82}$$

where the dissipation scale ℓ_D is defined by Eq. (3.52b).

By ε_m, we denote the volume average of ε' in an eddy of the size ℓ_m. Then the mean dissipation rate ε corresponds to ε_0. From the condition that $\ell_M \cong \ell_D$, ε_M may be regarded as the instantaneous dissipation rate ε'. Using this relation and the identity

$$\varepsilon_M = \varepsilon \prod_{m=1}^{M} d_m, \; d_m = \varepsilon_m / \varepsilon_{m-1} \, , \tag{3.83}$$

we have

$$\log \varepsilon' = \varepsilon + \sum_{m=1}^{M} \log d_m. \tag{3.84}$$

For very small s, we assume that the direct energy cascade from an eddy of size ℓ_m to one of size ℓ_{m+1} is very rare. Under this condition, d_m may be regarded as a random variable, and the distribution of $\log \varepsilon'$ for large M becomes Gaussian from the central limit theorem.

3.4.2. NONEQUILIBRIUM EFFECTS

In the foregoing discussions on intermittency effects, we assumed the balance between the mean injection and dissipation rates of turbulent energy, that is, the equilibrium state. In turbulent flows with mean velocity, the assumption is not always satisfied. Such an example is the homogeneous-shear turbulence that was referred to in Sec. 3.1.1. In this case, turbulent energy continues to be injected from the mean flow, and the turbulence state is statistically nonstationary, as is seen from Eqs. (3.7) and (3.13). In Chapters 4 and 6, we shall show that the turbulent viscosity corresponding to the equilibrium energy spectrum (3.45) cannot describe the proper development of homogeneous-shear turbulence.

The quantity $\partial\varepsilon/\partial t$ is a typical nonstationary quantity and is proper for being included in an expression for the inertial-range energy spectrum; namely, we write

$$E(k) = F\left\{k, \varepsilon, \partial\varepsilon/\partial t\right\}, \tag{3.85}$$

where $F\{A\}$ denotes a function or functional of A. Using $\partial\varepsilon/\partial t$, we can introduce a new length scale ℓ_P defined by

$$\ell_P = \varepsilon^2 \left| \frac{\partial\varepsilon}{\partial t} \right|^{-3/2} . \tag{3.86}$$

Equation (3.86) may be called the energy-production scale since the occurrence of $\partial\varepsilon/\partial t$ is a consequence of the nonstationary energy-production process. The length ℓ_P plays a role of ℓ_C in the intermittency effect expressed by Eq. (3.81).

As will be derived in Chapter 6, nonvanishing $\partial\varepsilon/\partial t$ leads to the correction to the Kolmogorov spectrum as

$$E(k) = K_O \varepsilon^{2/3} k^{-5/3} \left(1 - C_N \mathrm{sgn}\left(\frac{\partial\varepsilon}{\partial t}\right) (k\ell_P)^{-2/3}\right), \tag{3.87}$$

in case that $|\partial\varepsilon/\partial t|$ is not so large (Yoshizawa 1994). Here $C_N (\cong 0.6)$ is a numerical factor, and $\mathrm{sgn}(A)$ is defined as 1 and -1 for $A > 0$ and $A < 0$, respectively. The prominent feature of Eq. (3.87) is that the correction part becomes important as $k^{-1} \to \ell_P$, whereas it vanishes as $k^{-1} \to \ell_D$. Namely, this correction is significant at the energy-containing side of the inertial range. From the standpoint of turbulence modeling, such a type of correction is linked more tightly with the transports of momentum and scalar, compared with the foregoing intermittency counterpart.

3.5. Homogeneous Turbulence Theory

3.5.1. HISTORICAL DEVELOPMENTS

In Sec. 3.2 we discussed the Kolmogorov's inertial-range spectrum in a little detail. In Chapter 6, we shall use this information as one of the essential theoretical ingredients and proceed to a statistical theoretical study of turbulence modeling. For this purpose, we need a method that can derive the Kolmogorov's spectrum on a mathematically well-founded basis. What is further important is the applicability to turbulence with arbitrary mean velocity that is generally called turbulent shear flow. At least at present, we have no theory that can meet these two requirements completely.

A pioneering statistical theoretical study of homogeneous isotropic turbulence was done by Kraichnan (1959), who presented the so-called direct-interaction approximation (DIA) method in the Eulerian framework. This theory, however, fails in reproducing the Kolmogorov spectrum owing to the overestimate of advection effects of energy-containing eddies on inertial-range ones. In order to rectify this critical shortfall, Kraichnan further proposed a DIA in the Lagrangian framework and succeeded in its derivation with the reasonable estimate of the Kolmogorov constant (1965, 1977). The reader can know the developments in the Lagrangian method from the works by Leslie (1973), Kaneda (1981), and McComb (1990). These formalisms are mathematically complicated, compared with the Eulerian counterparts, and the difficulty of the application to turbulent shear flow is enhanced inevitably.

One of the prominent features of the DIA's in both the Eulerian and Lagrangian frameworks is the introduction of Green's or response functions. In the Eulerian framework, the Green's function encounters the foregoing overestimate of effects of energy-containing eddies. This shortfall is closely associated with two-time statistical properties of turbulence, and occurs as the divergence of integral at the low-wavenumber limit ($k \to 0$). A method of avoiding this difficulty in the Eulerian framework is the use of one-time quantities such as the energy spectrum, without introducing the Green's function. The representative method is the eddy-damped quasi-normal Markovian (EDQNM) method (Orszag 1970; Leith 1971; André and Lesieur 1977; Lesieur 1997). The other method falling in this category is the renormalization group (RNG) method. This formalism was originally developed in the study of statistical physics such as phase transition (Wilson and Kogut 1974), and was applied later to turbulence by Yakhot and Orszag (1986). In this course, some alternations beyond the original scope of the RNG method were introduced. These points are discussed in the works by Dannevik et al. (1987), McComb (1990), Smith and Reynolds (1992), Nakano (1992), Eyink (1994), and Nagano and Itazu (1997).

In turbulent shear flows, the mean velocity gradient $\nabla \mathbf{U}$ enters the governing equation for the fluctuation $\mathbf{u}'$, as in Eq. (2.104). The temporal and spatial variations of $\mathbf{U}$ are much slower in general than those of $\mathbf{u}'$. Therefore the inclusion of the $\nabla \mathbf{U}$ effects on $\mathbf{u}'$ in a perturbational manner is a significant first step towards the construction of a statistical theory of turbulent shear flow. In the treatment, the concept of Green's or response functions plays a central role, as is well recognized through the study of transport effects in statistical physics. In this sense, the Eulerian DIA method possesses a great merit, apart from the shortfall of overestimating energy-containing eddies.

In this section, we shall give a detailed explanation of the Eulerian DIA. Its preferential use is due to the applicability to turbulent shear flow, as is detailed in Chapter 6. Provided that small-scale properties of turbulence represented by the inertial range are primary interest, it is needless to say that the other methods such as the Lagrangian DIA should be adopted.

3.5.2. FORMULATION OF DIA

A prominent feature of turbulence is that the nonlinear part represented by the second term on the right-hand side of Eq. (3.28) is dominant, compared with the first linear part. The ratio of the nonlinear to linear parts is characterized by the turbulent Reynolds number R_T that is defined by Eq. (3.57), where u_C and ℓ_C defined by Eqs. (3.48) and (3.49) are adopted as the representative length and velocity, respectively, since the mean flow is absent. The inertial-range characteristics can be clearly detected for large R_T from Eq. (3.55). Therefore a simple perturbation method with the linear viscous part as the leading term cannot be applied to their investigation at all. A method of alleviating this difficulty is to combine such a perturbation method with the so-called renormalization procedure. This is an essential ingredient of the DIA.

A. Perturbational Expansion

We first integrate Eq. (3.28) formally with the nonlinear part regarded as known. Here we assume that turbulence is in a Gaussian or random state with high turbulence intensity in the infinite past, and that it is statistically stationary or quasi-stationery at present. Then we have

$$u_i(\mathbf{k};t) = v_i(\mathbf{k};t) + iM_{ij\ell}(\mathbf{k}) \iint \delta\left(\mathbf{k}-\mathbf{p}-\mathbf{q}\right) d\mathbf{p}d\mathbf{q} \times \int_{-\infty}^{t} dt_1 g\left(k;t,t_1\right) u_j\left(\mathbf{p};t_1\right) u_\ell\left(\mathbf{q};t_1\right), \tag{3.88}$$

with

$$\mathbf{v}(\mathbf{k};t) = \mathbf{A}(\mathbf{k}) \exp\left(-\nu k^2 t\right), \tag{3.89}$$

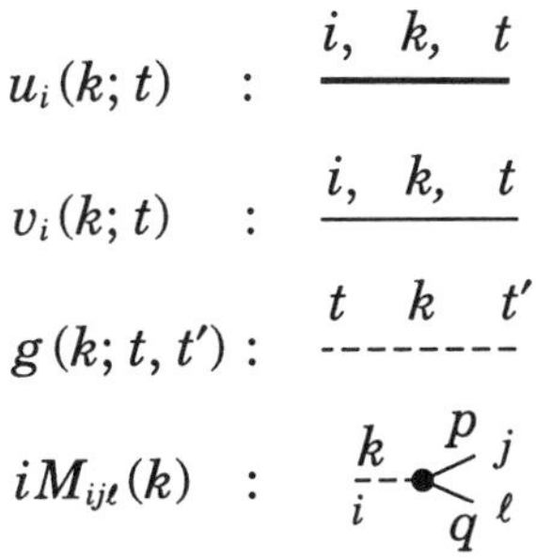

$u_i(k;t)$: $i,\ k,\ t$

$v_i(k;t)$: $i,\ k,\ t$

$g(k;t,t')$: $t\ \ k\ \ t'$

$iM_{ij\ell}(k)$: k, i, p, j, q, ℓ

Figure 3.2. Symbolical representation of the constituent elements in Eq. (3.88).

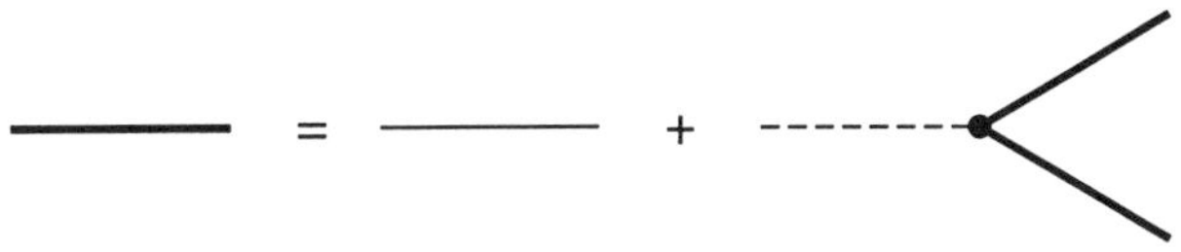

Figure 3.3. Diagrammatic representation of Eq. (3.88).

$$g(k;t,t') = H(t-t')\exp\left(-\nu k^2(t-t')\right). \tag{3.90}$$

Here $\mathbf{A}(\mathbf{k})$ is a random variable, and $H(t)$ is the unit step function that is 1 and 0 for $t > 0$ and $t < 0$, respectively. We can replace the foregoing assumption about the infinite past with the Gaussianity at an initial time and make the discussion entirely similar to the following one. What is important under these circumstances is to find a turbulent state that is independent of the initial or past condition, for the assumption about the Gaussianity is a big one.

Let us solve Eq. (3.88) in a perturbational manner with $\mathbf{v}(\mathbf{k};t)$ as the leading term. A method of performing this perturbation is to use its diagrammatic representation (Wyld 1961). We express $\mathbf{u}(\mathbf{k};t)$, $\mathbf{v}(\mathbf{k};t)$, $g(k;t,t')$, and $iM_{ij\ell}(\mathbf{k})$ symbolically, as in Fig. 3.2. Then Eq. (3.88) may be represented by Fig. 3.3. The part $iM_{ij\ell}(\mathbf{k})$ is usually called the vertex, specifically, the bare vertex (the meaning of "bare" will be clear later). The time integration such as $\int_{-\infty}^{t} dt_1$ is attached to the vertex part. The perturbational solution of Fig. 3.3 may be given by Fig. 3.4. Here the numerical factor "2" in the third term comes from the fact that the nonlinear part in Fig. 3.3 is symmetric with respect to $u_j(\mathbf{p};t_1)$ and $u_\ell(\mathbf{q};t_1)$.

B. Calculation of Velocity Variance

We use Fig. 3.4 and calculate the velocity variance $\langle u_i(\mathbf{k};t)u_j(\mathbf{k}';t')\rangle$. Here we make full use of the assumption that $\mathbf{A}(\mathbf{k})$ or, equivalently, $\mathbf{v}(\mathbf{k};t)$ is a

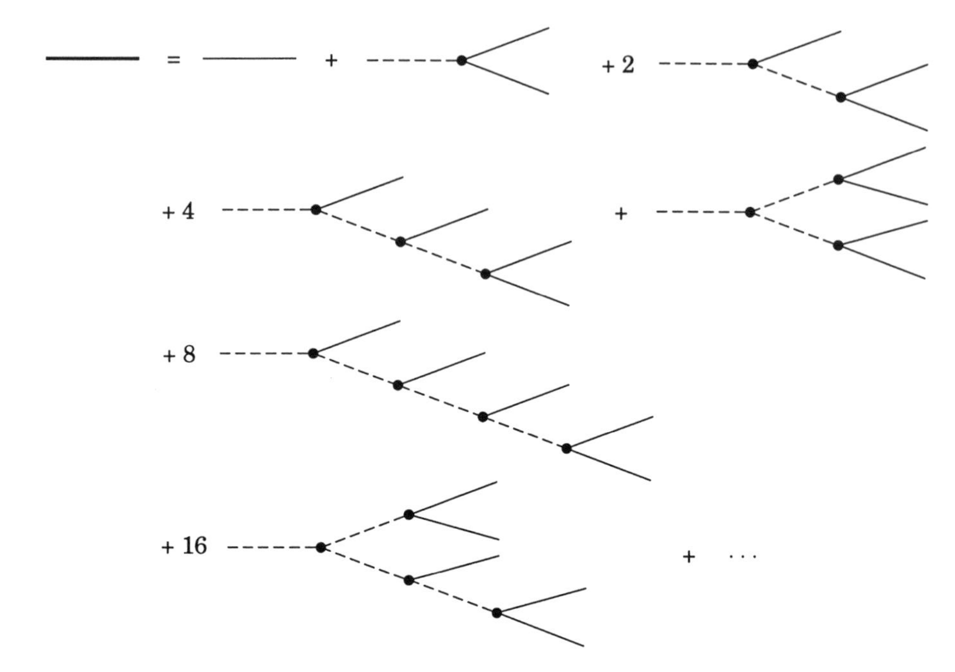

Figure 3.4. Perturbational expansion of Eq. (3.88).

random variable with zero mean. For instance, the fourth-order correlation concerning $\mathbf{v}(\mathbf{k};t)$ can be divided into the sum of the products concerning its variance; namely, we have

$$\begin{aligned}&\langle v_i(\mathbf{k}_1)\, v_j(\mathbf{k}_2)\, v_\ell(\mathbf{k}_3)\, v_m(\mathbf{k}_4)\rangle \\ &\quad = \langle v_i(\mathbf{k}_1)\, v_j(\mathbf{k}_2)\rangle \langle v_\ell(\mathbf{k}_3)\, v_m(\mathbf{k}_4)\rangle + \langle v_i(\mathbf{k}_1)\, v_\ell(\mathbf{k}_3)\rangle \langle v_j(\mathbf{k}_2)\, v_m(\mathbf{k}_4)\rangle \\ &\quad + \langle v_i(\mathbf{k}_1)\, v_m(\mathbf{k}_4)\rangle \langle v_j(\mathbf{k}_2)\, v_\ell(\mathbf{k}_3)\rangle . \end{aligned} \tag{3.91}$$

In general, the nth-order correlation may be written as

$$\begin{aligned}&\langle v_{n_1}(\mathbf{k}_1)\, v_{n_2}(\mathbf{k}_2)\, v_{n_3}(\mathbf{k}_3)\cdots v_{n_m}(\mathbf{k}_m)\rangle \\ &\quad = \sum \langle v_{n_i}(\mathbf{k}_i)\, v_{n_j}(\mathbf{k}_j)\rangle \cdots \langle v_{n_\ell}(\mathbf{k}_\ell)\, v_{n_m}(\mathbf{k}_m)\rangle \ \text{for even } m; \end{aligned} \tag{3.92a}$$

$$= 0 \text{ for odd } m. \tag{3.92b}$$

In Eq. (3.92a), the summation is taken over all the possible combinations of the products concerning the variance.

Corresponding to Fig. 3.2, we express the second-order correlations

$$Q_{ij}(\mathbf{k};t,t') = \langle u_i(\mathbf{k};t)u_j(\mathbf{k}';t')\rangle / \delta(\mathbf{k}+\mathbf{k}') , \tag{3.93}$$

$$Q_{Vij}(\mathbf{k};t,t') = \langle v_i(\mathbf{k};t)v_j(\mathbf{k}';t')\rangle / \delta(\mathbf{k}+\mathbf{k}') , \tag{3.94}$$

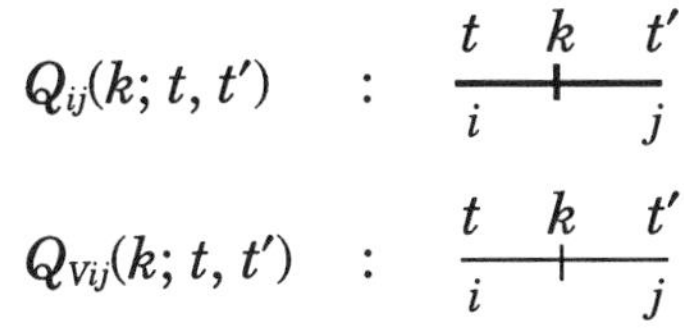

$Q_{ij}(k; t, t')$: $\underset{i \qquad j}{\overset{t \quad k \quad t'}{\longrightarrow}}$

$Q_{Vij}(k; t, t')$: $\underset{i \qquad j}{\overset{t \quad k \quad t'}{\longrightarrow}}$

Figure 3.5. Symbolical representation of the velocity variances.

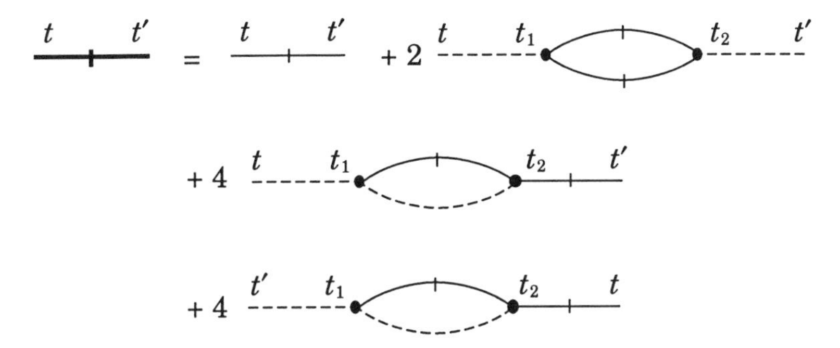

Figure 3.6. Perturbational expansion of $Q_{ij}(\mathbf{k}; t, t')$ up to $O(M^2)$.

using the symbolical representation of Fig. 3.5, where the subscript of $Q_{Vij}(\mathbf{k}; t, t')$, V, means the correlation based on the linear viscous solution for $\mathbf{u}(\mathbf{k}; t)$. Using Fig. 3.5 and Eq. (3.92), we calculate $Q_{ij}(\mathbf{k}; t, t')$ for $t > t'$. The resulting expression for $Q_{ij}(\mathbf{k}; t, t')$ up to the second order concerning $M_{ij\ell}(\mathbf{k})$ is given in Fig. 3.6, where the time sequence is written explicitly for the clear understanding of the features of each diagram. The higher-order counterparts may be given similarly, although the number of diagrams increases rapidly with the order of $M_{ij\ell}(\mathbf{k})$. In what follows, we shall simply denote the nth order of $M_{ij\ell}(\mathbf{k})$ by $O(M^n)$. Using Fig. 3.7, the second term on the right-hand side in Fig. 3.6 may be converted to the usual expression

$$-2 \iint \delta\left(\mathbf{k}-\mathbf{p}-\mathbf{q}\right) d\mathbf{p} d\mathbf{q} M_{iab}(\mathbf{k}) M_{jcd}(-\mathbf{k})$$
$$\times \int_{-\infty}^{t} dt_1 \int_{-\infty}^{t'} dt_2 g\left(k; t, t_1\right) g\left(k; t', t_2\right) Q_{Vac}\left(\mathbf{p}; t_1, t_2\right) Q_{Vbd}\left(\mathbf{q}; t_1, t_2\right). \quad (3.95)$$

C. Renormalization

In the straightforward perturbational expansion of Eq. (3.88) with the viscous term as the unperturbed one, the ratio of the nonlinear or perturbed

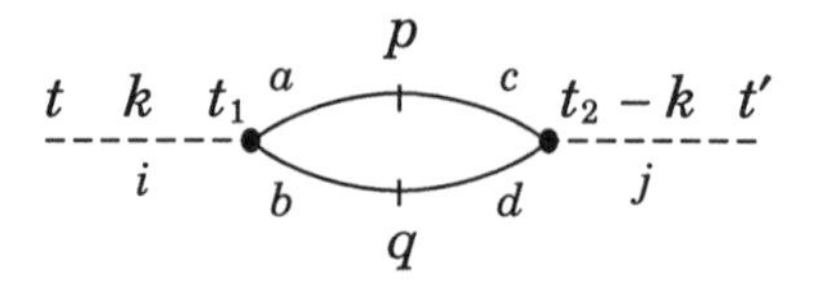

Figure 3.7. Dependence on wavenumbers $(\mathbf{k}, \mathbf{p}, \mathbf{q})$ and components (i, j, a, b, c, d).

term to the viscous one is of $O(R_T)$, where R_T is defined by Eq. (3.56) or (3.57). In the case of strong turbulence with $R_T \gg 1$, Fig. 3.6 leads to a very poor solution of Eq. (3.93). We take one simple instance and see this situation. We consider

$$f(x) = 1 + x + x^2 + x^3 + \cdots \equiv \sum_{n=0}^{\infty} x^n. \tag{3.96}$$

The radius of convergence of this series is $|x| = 1$, and the correct value for $|x| \geq 1$ cannot be obtained as long as the number of the terms to be retained is finite. Equation (3.96) is the expansion of

$$f(x) = 1/(1-x) \tag{3.97}$$

for $|x| < 1$. In order to abstract the correct behaviors of $f(x)$ for $|x| \geq 1$, we need to find Eq. (3.97) or its proper approximate form using the original series (3.96). We rewrite Eq. (3.96) as

$$f(x) = 1 + x\left(1 + x + x^2 + \cdots\right) = 1 + xf(x), \tag{3.98}$$

which leads to Eq. (3.97)

In the present diagrammatic perturbation, the mathematical structure of the solution becomes complicated with the increasing order of $M_{ij\ell}(\mathbf{k})$. Therefore it is necessary to infer some reasonable expression for large R_T from a small number of terms that are obtained explicitly. A method of performing this procedure is the so-called renormalization. The above example is too simple, but it still keeps some important features of the renormalization procedure.

In Fig. 3.6, we replace $Q_{V_{ij}}(\mathbf{k}; t, t')$ with the exact counterpart $Q_{ij}(\mathbf{k}; t, t')$, resulting in the integral equation in Fig. 3.8. In order to see what type of terms are included there, we perform the perturbational expansion with the first or viscous correlation term as the unperturbed one. By the first perturbation, we return to Fig. 3.6. In the second perturbation, we substitute this solution into Fig. 3.8, and have the $O(M^4)$ terms. The diagrams in Fig. 3.9 arise from the combination of the third term in Fig. 3.6

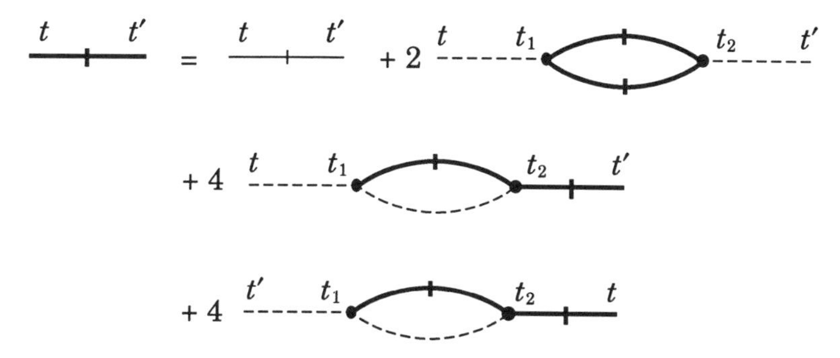

Figure 3.8. Renormalization of Fig. 3.6.

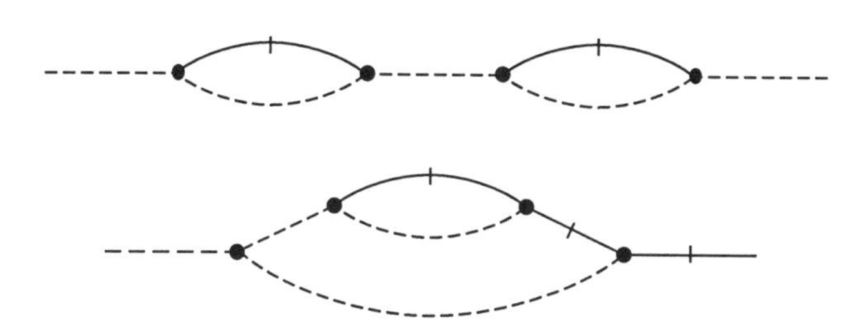

Figure 3.9. Terms of $O(M^4)$ arising from the third diagram in Fig. 3.8.

with the third in Fig. 3.8, where numerical factors have been dropped. In the context of the perturbational expansion of Fig. 3.4, the first diagram in Fig. 3.9 is constructed from the first and sixth ones, whereas the second is from the first and seventh ones. The reader may see these more details in the work by Wyld (1961).

Using the first and sixth diagrams in Fig. 3.4, we obtain Fig. 3.10. Its feature lies in the mathematical structure of the vertex that has two g's, as in Fig. 3.11. This diagram cannot be reproduced from the perturbational expansion of Fig. 3.8. In order to overcome this difficulty, we need the concept of Green's function.

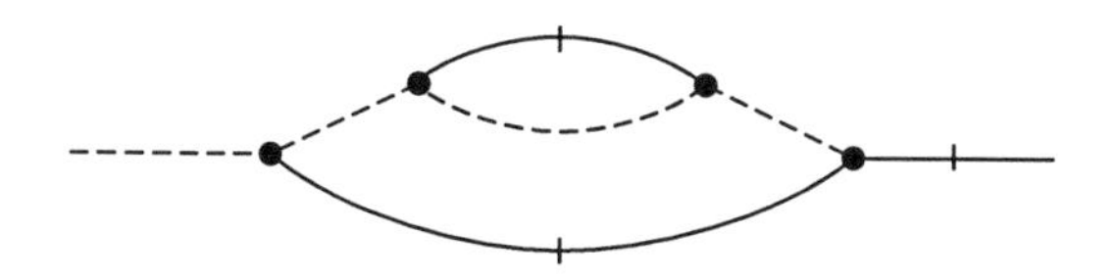

Figure 3.10. Terms of $O(M^4)$ arising from the first and sixth diagrams in Fig. 3.4.

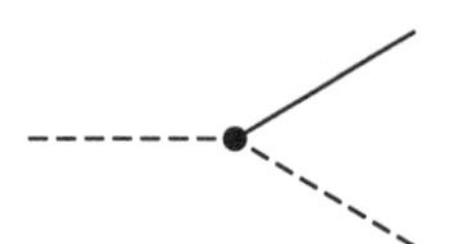

Figure 3.11. Vertex with two g's.

D. Introduction of Green's Function

From Eq. (3.90), $g(k;t,t')$ is the Green's or response function for Eq. (3.28) with the nonlinear part regarded as known; namely, it is the Green's function at low R_T. We consider the counterpart at high R_T, $G'_{ij}(\mathbf{k},\mathbf{k}';t,t')$. By $\delta\mathbf{u}(\mathbf{k};t)$, we denote the response generated by adding an infinitesimal disturbance $\delta\mathbf{f}(\mathbf{k};t)$ to Eq. (3.28). Then we may write

$$\delta u_i(\mathbf{k};t) = \int d\mathbf{k}' \int_{-\infty}^{t} G'_{ij}\left(\mathbf{k},\mathbf{k}';t,t_1\right)\delta f_j\left(\mathbf{k}';t_1\right)dt_1, \tag{3.99}$$

where $G'_{ij}(\mathbf{k},\mathbf{k}';t,t')$ obeys

$$\frac{\partial}{\partial t}G'_{ij}(\mathbf{k},\mathbf{k}';t,t') + \nu k^2 G'_{ij}(\mathbf{k},\mathbf{k}';t,t') - D_{ij}(\mathbf{k}')\delta\left(\mathbf{k}-\mathbf{k}'\right)\delta\left(t-t'\right)$$
$$= 2iM_{i\ell m}(\mathbf{k})\iint u_\ell(\mathbf{p};t)G'_{mj}(\mathbf{q},\mathbf{k}';t,t')\delta\left(\mathbf{k}-\mathbf{p}-\mathbf{q}\right)d\mathbf{p}d\mathbf{q}. \tag{3.100}$$

Here the solenoidal operator defined by Eq. (3.25), $D_{ij}(\mathbf{k})$, has been inserted to guarantee that any disturbances coming from the nonlinear effects of Eq. (3.88) are subject to the solenoidal condition. We put

$$G'_{ij}(\mathbf{k},\mathbf{k}';t,t') = G'_{ij}(\mathbf{k};t,t')\delta\left(\mathbf{k}-\mathbf{k}'\right), \tag{3.101}$$

and integrate Eq. (3.100) with respect to $\mathbf{k}'$. As a result, $G'_{ij}(\mathbf{k};t,t')$ is governed by

$$\frac{\partial}{\partial t}G'_{ij}(\mathbf{k};t,t') + \nu k^2 G'_{ij}(\mathbf{k};t,t') - D_{ij}(\mathbf{k})\delta\left(t-t'\right)$$
$$= 2iM_{i\ell m}(\mathbf{k})\iint u_\ell(\mathbf{p};t)G'_{mj}(\mathbf{q};t,t')\delta\left(\mathbf{k}-\mathbf{p}-\mathbf{q}\right)d\mathbf{p}d\mathbf{q}. \tag{3.102}$$

In the following, we shall use $G'_{ij}(\mathbf{k};t,t')$ as the Green's function for Eq. (3.28).

Similar to Eq. (3.88), we formally integrate Eq. (3.102) with the right-hand side regarded as known, and have

$$G'_{ij}(\mathbf{k};t,t') = G_{Vij}(\mathbf{k};t,t') + 2iM_{n\ell m}(\mathbf{k})\iint \delta\left(\mathbf{k}-\mathbf{p}-\mathbf{q}\right)d\mathbf{p}d\mathbf{q}$$
$$\times \int_{t'}^{t} dt_1 G_{Vin}\left(\mathbf{k};t,t_1\right)u_\ell\left(\mathbf{p};t_1\right)G'_{mj}\left(\mathbf{q};t_1,t'\right), \tag{3.103}$$

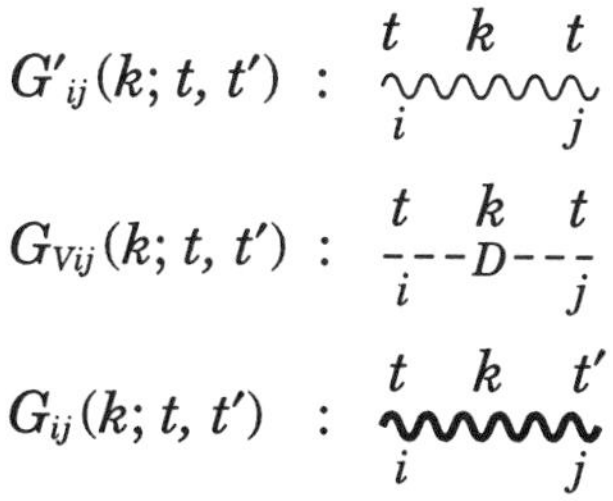

$G'_{ij}(k; t, t')$: $t\ k\ t$, $i\ j$

$G_{Vij}(k; t, t')$: $t\ k\ t$, ---D---, $i\ j$

$G_{ij}(k; t, t')$: $t\ k\ t'$, $i\ j$

Figure 3.12. Symbolical representation of the Green's functions.

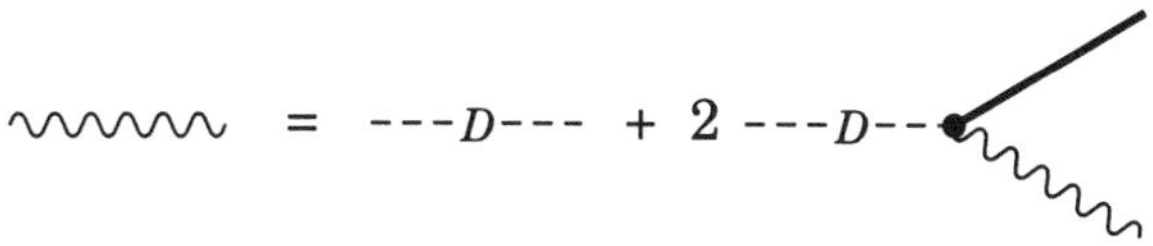

= ---D--- + 2 ---D---

Figure 3.13. Diagrammatic representation of Eq. (3.103).

where

$$G_{V_{ij}}(\mathbf{k}; t, t') = D_{ij}(\mathbf{k})g(k; t, t'). \tag{3.104}$$

In Eq. (3.103), we should note that the lower limit of the time integral is not $-\infty$ but t', owing to $\delta(t - t')$ in Eq. (3.102). Moreover, we rewrote the right-hand side of Eq. (3.102) as

$$2iD_{in}(\mathbf{k})M_{n\ell m}(\mathbf{k}) \iint u_\ell(\mathbf{p}; t)G'_{mj}(\mathbf{q}; t, t')\delta(\mathbf{k} - \mathbf{p} - \mathbf{q})d\mathbf{p}d\mathbf{q}, \tag{3.105}$$

using the last of Eq. (3.27).

For $G'_{ij}(\mathbf{k}; t, t')$, $G_{Vij}(\mathbf{k}; t, t')$, and the ensemble mean of the former defined by

$$G_{ij}(\mathbf{k}; t, t') = \left\langle G'_{ij}(\mathbf{k}; t, t')\right\rangle, \tag{3.106}$$

we also introduce the symbolical representations, as in Fig. 3.12. Corresponding to Fig. 3.3, Eq. (3.103) may be expressed as Fig. 3.13, and its perturbational expansion up to $O(M^2)$ is given by Fig. 3.14. Here we should note that the viscous Green's function in Fig. 3.4 for $\mathbf{u}(\mathbf{k}; t)$ is $g(k; t, t')$ but not $G_{Vij}(\mathbf{k}; t, t')$. The sole difference between them, however, is the solenoidal operator $D_{ij}(\mathbf{k})$, as is seen from Eqs. (3.90) and (3.104). In Fig. 3.4, $g(k; t, t')$ can be replaced by $G_{Vij}(\mathbf{k}; t, t')$ with the aid of the last of Eq. (3.27).

The calculation of $G_{ij}(\mathbf{k}; t, t')$ up to $O(M^2)$ is simple, which leads to Fig. 3.15. From the renormalization based on both the exact velocity variance and Green's function, $Q_{ij}(\mathbf{k}; t, t')$ and $G_{ij}(\mathbf{k}; t, t')$, we have Fig. 3.16.

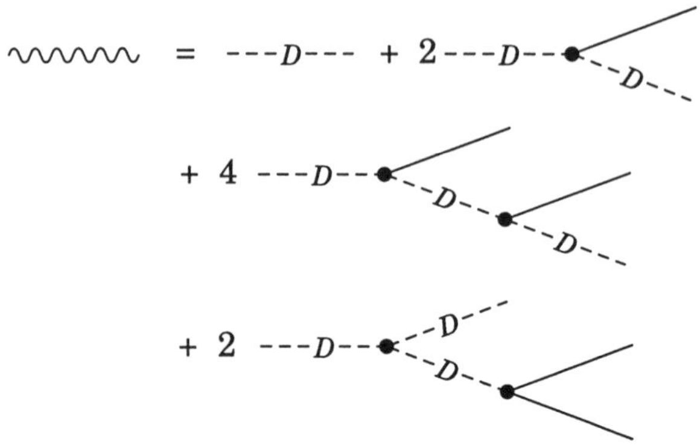

Figure 3.14. Perturbational expansion of Eq. (3.103) up to $O(M^2)$.

Corresponding to Fig. 3.16, we renormalize Fig. 3.8 concerning $G_{ij}(\mathbf{k};t,t')$, and reach Fig. 3.17. The resulting system of equations keeps the terms that cannot be included through the renormalization of only the velocity variance. Really, we have Fig. 3.10 by substituting the second term of Fig. 3.15 into the third one of Fig. 3.17.

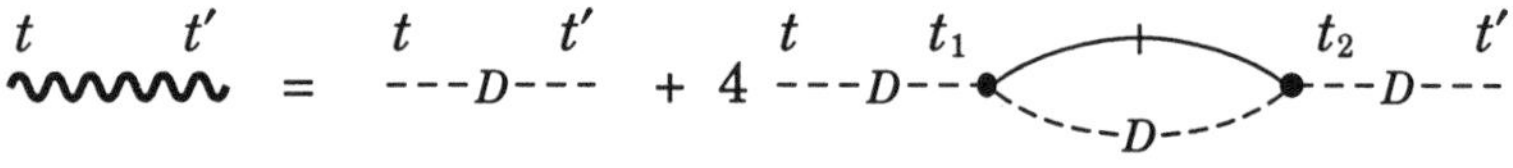

Figure 3.15. Mean Green's function up to $O(M^2)$.

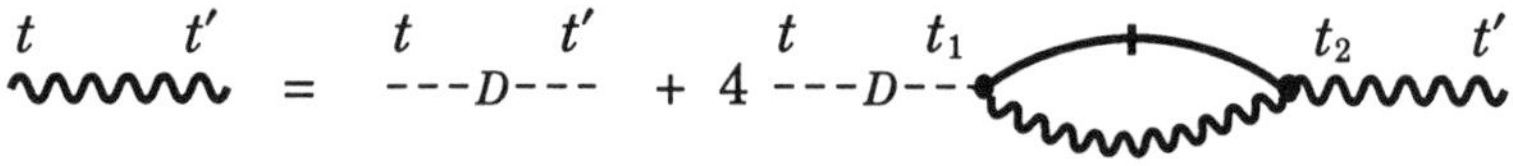

Figure 3.16. Renormalization of Fig. 3.15.

E. DIA System of Equations

The diagrammatic equations in Figs. 3.16 and 3.17 are written explicitly as

$$Q_{ij}(\mathbf{k};t,t') = Q_{Vij}(\mathbf{k};t,t') + 2\iint \delta(\mathbf{k}-\mathbf{p}-\mathbf{q})d\mathbf{p}d\mathbf{q}$$

$$\times M_{abc}(\mathbf{k})M_{fde}(\mathbf{k})\int_{-\infty}^{t}dt_1\int_{-\infty}^{t'}dt_2 G_{Via}(\mathbf{k};t,t_1)$$

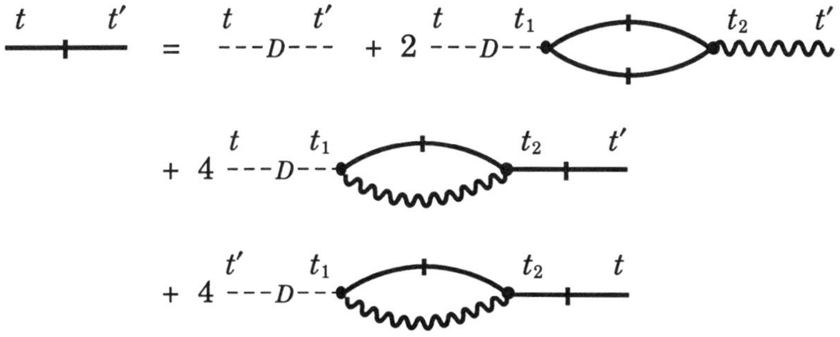

Figure 3.17. Renormalization of the Green's-function parts in Fig. 3.8.

$$
\begin{aligned}
&\times G_{jf}\left(-\mathbf{k};t',t_2\right)Q_{bd}\left(\mathbf{p};t_1,t_2\right)Q_{ce}\left(\mathbf{q};t_1,t_2\right)\\
&-4\iint\delta\left(\mathbf{k}-\mathbf{p}-\mathbf{q}\right)d\mathbf{p}d\mathbf{q}M_{abc}(\mathbf{k})M_{edf}(\mathbf{q})\int_{-\infty}^{t}dt_1\int_{-\infty}^{t_1}dt_2\\
&\quad\times G_{Via}\left(\mathbf{k};t,t_1\right)G_{ce}\left(\mathbf{q};t_1,t_2\right)Q_{bd}\left(\mathbf{p};t_1,t_2\right)Q_{fj}\left(-\mathbf{k};t_2,t'\right)\\
&-4\iint\delta\left(\mathbf{k}-\mathbf{p}-\mathbf{q}\right)d\mathbf{p}d\mathbf{q}M_{abc}(\mathbf{k})M_{edf}(\mathbf{q})\int_{-\infty}^{t'}dt_1\int_{-\infty}^{t_1}dt_2\\
&\quad\times G_{Via}\left(\mathbf{k};t',t_1\right)G_{ce}\left(\mathbf{q};t_1,t_2\right)Q_{bd}\left(\mathbf{p};t_1,t_2\right)Q_{fj}\left(-\mathbf{k};t_2,t\right),
\end{aligned}
\tag{3.107}
$$

$$
\begin{aligned}
G_{ij}(\mathbf{k};t,t') &= G_{Vij}(\mathbf{k};t,t')-4\iint\delta\left(\mathbf{k}-\mathbf{p}-\mathbf{q}\right)d\mathbf{p}d\mathbf{q}\\
&\times M_{abc}(\mathbf{k})M_{edf}(\mathbf{q})\int_{t'}^{t}dt_1\int_{t'}^{t_1}dt_2G_{Via}\left(\mathbf{k};t,t_1\right)G_{ce}\left(\mathbf{q};t_1,t_2\right)\\
&\quad\times G_{fj}\left(-\mathbf{k};t_2,t'\right)Q_{bd}\left(\mathbf{p};t_1,t_2\right).
\end{aligned}
\tag{3.108}
$$

In order to transform Eqs. (3.107) and (3.108) into a differential form in time, we apply the linear or viscous operator $L(=\partial/\partial t+\nu k^2)$ to them. From Eq. (3.107), we have

$$
\begin{aligned}
LQ_{ij}(\mathbf{k};t,t') &= 2\iint\delta\left(\mathbf{k}-\mathbf{p}-\mathbf{q}\right)d\mathbf{p}d\mathbf{q}M_{iab}(\mathbf{k})M_{ecd}(\mathbf{k})\\
&\quad\times\int_{-\infty}^{t'}dt_1G_{je}\left(-\mathbf{k};t',t_1\right)Q_{ac}\left(\mathbf{p};t,t_1\right)Q_{bd}\left(\mathbf{q};t,t_1\right)\\
&-4\iint\delta\left(\mathbf{k}-\mathbf{p}-\mathbf{q}\right)d\mathbf{p}d\mathbf{q}M_{iab}(\mathbf{k})M_{dce}(\mathbf{q})\\
&\quad\times\int_{-\infty}^{t}dt_1G_{bd}\left(\mathbf{q};t,t_1\right)Q_{ac}\left(\mathbf{p};t,t_1\right)Q_{ej}\left(-\mathbf{k};t_1,t'\right)\\
&-4\iint\delta\left(\mathbf{k}-\mathbf{p}-\mathbf{q}\right)d\mathbf{p}d\mathbf{q}M_{abc}(\mathbf{k})M_{edf}(\mathbf{q})\int_{-\infty}^{t'}dt_1\int_{-\infty}^{t_1}dt_2
\end{aligned}
$$

$$\times G_{Via}(\mathbf{k};t',t_1)\,G_{ce}(\mathbf{q};t_1,t_2)\,Q_{bd}(\mathbf{p};t_1,t_2)\,L\left[Q_{fj}(-\mathbf{k};t_2,t)\right], \tag{3.109}$$

where use has been made of

$$LQ_{Vij}(\mathbf{k};t,t') = 0, \tag{3.110a}$$

$$LG_{Vij}(\mathbf{k};t,t') = D_{ij}(\mathbf{k})\delta(t-t'), \tag{3.110b}$$

$$\int_{-\infty}^{t} \delta\,(t-s)ds = 0. \tag{3.110c}$$

The last term on the right-hand side of Eq. (3.109) contains the L-related part. We neglect this term and use the resulting equation to estimate the L-dependent term. Then it becomes of $O(M^4)$. Originally, Fig. 3.17 leading to Eq. (3.107) was obtained through the renormalization of the $O(M^2)$ terms. Under this renormalization procedure, we can neglect the L-dependent term. Finally, we have

$$\begin{aligned}\left(\frac{\partial}{\partial t}+\nu k^2\right)Q_{ij}(\mathbf{k};t,t') &= 2\iint \delta\,(\mathbf{k}-\mathbf{p}-\mathbf{q})d\mathbf{p}d\mathbf{q}M_{iab}(\mathbf{k})M_{ecd}(\mathbf{k})\\ &\quad\times\int_{-\infty}^{t'} dt_1 G_{je}\,(-\mathbf{k};t',t_1)\,Q_{ac}\,(\mathbf{p};t,t_1)\,Q_{bd}\,(\mathbf{q};t,t_1)\\ &\quad-4\iint \delta\,(\mathbf{k}-\mathbf{p}-\mathbf{q})d\mathbf{p}d\mathbf{q}M_{iab}(\mathbf{k})M_{dce}(\mathbf{q})\\ &\quad\times\int_{-\infty}^{t} dt_1 G_{bd}\,(\mathbf{q};t,t_1)\,Q_{ac}\,(\mathbf{p};t,t_1)\,Q_{ej}\,(-\mathbf{k};t_1,t')\,.\end{aligned} \tag{3.111}$$

Entirely similarly, Eq. (3.108) may be rewritten as

$$\begin{aligned}\left(\frac{\partial}{\partial t}+\nu k^2\right)G_{ij}(\mathbf{k};t,t') &= D_{ij}(\mathbf{k})\delta\,(t-t') - 4\iint\delta\,(\mathbf{k}-\mathbf{p}-\mathbf{q})d\mathbf{p}d\mathbf{q}\\ &\quad\times M_{iab}(\mathbf{k})M_{dce}(\mathbf{q})\int_{t'}^{t} dt_1 G_{bd}(\mathbf{q};t,t_1)\,G_{ej}(-\mathbf{k};t_1,t')\,Q_{ac}(\mathbf{p};t,t_1)\,.\end{aligned} \tag{3.112}$$

Equations (3.111) and (3.112) constitute a closed system of equations in the DIA method presented by Kraichnan (1959).

F. Features of DIA

The two-time statistics such as $Q_{ij}(\mathbf{k};t,t')$ and $G_{ij}(\mathbf{k};t,t')$ are called propagators in general. Then the procedure for rewriting Figs. 3.6 and 3.15 as Figs. 3.16 and 3.17 may be called the propagator renormalization. The resulting DIA system includes the contributions of the perturbational solution of Fig. 3.4 up to the infinite order of $M_{ij\ell}(\mathbf{k})$, but the system cannot cover all kinds of terms occurring from the solution. The typical example of

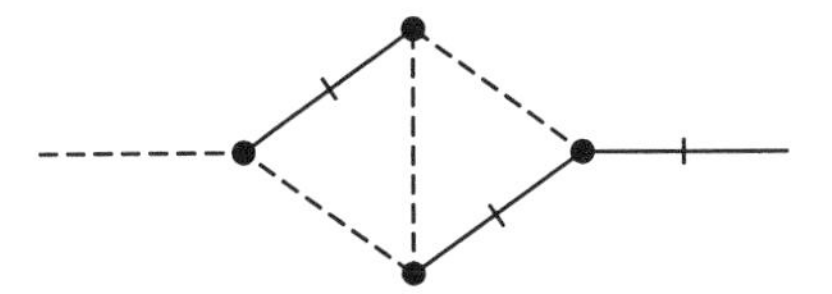

Figure 3.18. Terms of $O(M^4)$ not renormalized.

such terms is the diagram in Fig. 3.18, which is constructed from the first and sixth terms in Fig. 3.4. In order to include it, the propagator renormalization is not sufficient, and the vertex $M_{ij\ell}(\mathbf{k})$ needs to be renormalized [this is the reason why $M_{ij\ell}(\mathbf{k})$ is called the bare vertex]. The expressions resulting from the vertex renormalization are very complicated, and the procedure is not applicable to the study of turbulent shear flow that is a primary interest of this book. Therefore we do not enter the discussion on this point [the vertex renormalization is detailed in the work by Wyld (1961)].

Let us summarize the key steps of the renormalization procedure in the DIA since its understanding is helpful to the study of much more complicated flow such as turbulent shear one. The procedure consists of the following three steps.

(i) A quantity ϕ obeys

$$L\phi = N\{\phi\}, \tag{3.113}$$

where L is the linear operator, and $N\{\phi\}$ expresses the nonlinear functional of ϕ. We expand

$$\phi = \sum_{n=0}^{\infty} \phi_n, \tag{3.114}$$

where the lowest-order solution ϕ_0 satisfies $L\phi_0 = 0$.

(ii) Using this perturbational solution, we calculate the correlation functions of ϕ, for instance, $Q_{\phi\phi}(= \langle\phi^2\rangle)$, as

$$Q_{\phi\phi} = \langle\phi_0^2\rangle + 2\langle\phi_0\phi_1\rangle + \cdots. \tag{3.115}$$

In the presence of the nonvanishing contribution to $\langle\phi_0\phi_1\rangle$, we express it in terms of $\langle\phi_0^2\rangle$ and the mean of the Green's function related to the operator L, $G_{0\phi}$.

(iii) In the resulting expression for $\langle\phi_0\phi_1\rangle$, we replace $\langle\phi_0^2\rangle$ and $G_{0\phi}$ with their exact counterparts to have an integral equation for $Q_{\phi\phi}$.

The lowest-order solution in the renormalization procedure of the DIA is the linear viscous one. In the case of turbulence shear flow detailed in Chapter 6, the counterpart is the homogeneous, isotropic solution that is a solution of the DIA system of equations.

3.5.3. APPLICATION TO ISOTROPIC TURBULENCE

A. Isotropic DIA Equations

In order to see the relationship of the DIA system of equations, (3.111) and (3.112), with the Kolmogorov's $-5/3$ power law, we consider isotropic turbulence. In this case, we have

$$Q_{ij}(\mathbf{k};t,t') = D_{ij}(\mathbf{k})Q(k;t,t'), \tag{3.116}$$

$$G_{ij}(\mathbf{k};t,t') = D_{ij}(\mathbf{k})G(k;t,t'), \tag{3.117}$$

using Eq. (3.35). Here we have assumed the mirrorsymmetry of turbulence and dropped the helicity-related part since we have no factors causing and sustaining the breakage of such symmetry. We substitute Eqs. (3.116) and (3.117) into Eqs. (3.111) and (3.112), and put $i = j$. As a result, we have

$$\begin{aligned}\left(\frac{\partial}{\partial t} + \nu k^2\right) Q(k;t,t') &= k^2 \iint \delta\left(\mathbf{k}-\mathbf{p}-\mathbf{q}\right) d\mathbf{p}d\mathbf{q} \\ &\times \left(N_{Q1}(k,p,q) \int_{-\infty}^{t'} dt_1 G\left(k;t',t_1\right) Q\left(p;t,t_1\right) Q\left(q;t,t_1\right)\right. \\ &\left. - N_{Q2}(k,p,q) \int_{-\infty}^{t} dt_1 G\left(q;t,t_1\right) Q\left(p;t,t_1\right) Q\left(k;t_1,t'\right)\right), \end{aligned} \tag{3.118}$$

$$\begin{aligned}\left(\frac{\partial}{\partial t} + \nu k^2\right) G(k;t,t') &= \delta\left(t-t'\right) - k^2 \iint \delta\left(\mathbf{k}-\mathbf{p}-\mathbf{q}\right) d\mathbf{p}d\mathbf{q} \\ &\times N_G(k,p,q) \int_{t'}^{t} dt_1 G\left(q;t,t_1\right) G\left(k;t_1,t'\right) Q\left(p;t,t_1\right), \end{aligned} \tag{3.119}$$

where the geometrical factors, N_{Q1}, N_{Q2}, and N_G, are given by

$$N_{Q1}(k,p,q) = M_{eab}(\mathbf{k})M_{ecd}(\mathbf{k})D_{ac}(\mathbf{p})D_{bd}(\mathbf{q}), \tag{3.120a}$$

$$N_{Q2}(k,p,q) = 2M_{dab}(\mathbf{k})M_{bcd}(\mathbf{q})D_{ac}(\mathbf{p}), \tag{3.120b}$$

$$N_G(k,p,q) = N_{Q2}(k,p,q), \tag{3.120c}$$

from Eq. (3.27). These geometrical factors may be simplified further as

$$\begin{aligned}N_{Q1}(k,p,q) &= N_{Q2}(k,p,q) = N_G(k,p,q) \\ &\equiv N(k,p,q) = \frac{q}{k}\left(xz + y^3\right)\end{aligned} \tag{3.121}$$

(Leslie 1973), where x, y, and z are the cosines of angles defined by Fig. 3.19.

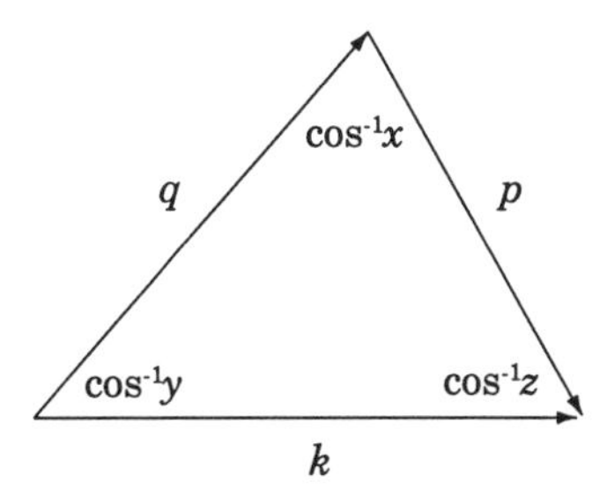

Figure 3.19. Triangle formed by k, p, and q.

From Eq. (3.118), the one-time covariance $Q(k;t,t)$ satisfies

$$\left(\frac{\partial}{\partial t}+2\nu k^2\right)Q(k;t,t)=2k^2\iint\delta(\mathbf{k}-\mathbf{p}-\mathbf{q})d\mathbf{p}d\mathbf{q}N(k,p,q)$$
$$\times\left(\int_{-\infty}^{t}dt_1G(k;t,t_1)Q(p;t,t_1)Q(q;t,t_1)\right.$$
$$\left.-\int_{-\infty}^{t}dt_1G(q;t,t_1)Q(p;t,t_1)Q(k;t,t_1)\right). \qquad (3.122)$$

Here we have used

$$\lim_{t'\to t}\frac{\partial}{\partial t}Q(k;t,t')=\frac{1}{2}\frac{\partial}{\partial t}Q(k;t,t), \qquad (3.123)$$

which comes from the symmetry of $Q(k;t,t')$ with respect to t and t'.

Note: In the numerical calculation of Eqs. (3.118) and (3.119), it is convenient to use the formula

$$\iint f(k,p,q)\delta(\mathbf{k}-\mathbf{p}-\mathbf{q})\,d\mathbf{p}d\mathbf{q}=2\pi\iint_{\Delta}\frac{pq}{k}f(k,p,q)dpdq \qquad (3.124)$$

(Leslie 1973). Here $\int\int_{\Delta}dpdq$ denotes the integration over the hatched region in Fig. 3.20, which obeys the relation

$$|p-q|<k<p+q. \qquad (3.125)$$

B. Quasi-Stationary State

We bear the inertial range in mind and seek a stationary solution of Eqs. (3.118) and (3.119), where $Q(k;t,t')$ and $G(k;t,t')$ are dependent on the time difference $t-t'$. In this case, we consider that the energy-containing

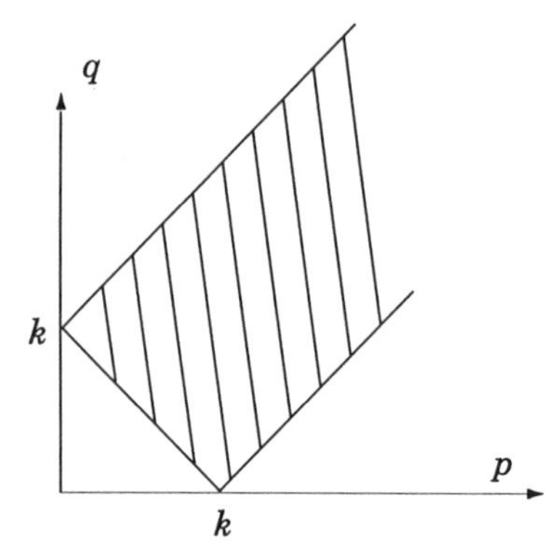

Figure 3.20. Region of integration.

range possessing sufficient energy is in a quasi-stationary state while supplying energy to the inertial range. For the feasibility of analytical discussion, we assume a simple mathematical form

$$Q(k;t,t') = \sigma(k)\exp\left(-\omega_Q(k)|t-t'|\right), \tag{3.126}$$

$$G(k;t,t') = H(t-t')\exp\left(-\omega_G(k)(t-t')\right), \tag{3.127}$$

with

$$\omega_Q(k) = \omega_G(k) = \omega(k). \tag{3.128}$$

Here $\omega_Q(k)^{-1}$ and $\omega_G(k)^{-1}$ express the correlation and response times, respectively, and they are originally different from each other in a quantitative sense. Equation (3.128), however, is plausible in the present approximate treatment based on the simplified form.

We substitute Eqs. (3.126) and (3.127) into Eq. (3.122), and have

$$2\nu k^2\sigma(k) = 2k^2\iint \delta\left(\mathbf{k}-\mathbf{p}-\mathbf{q}\right)d\mathbf{p}d\mathbf{q}N\left(k,p,q\right)\frac{\sigma(p)\left(\sigma(q)-\sigma(k)\right)}{\omega(k)+\omega(p)+\omega(q)}. \tag{3.129}$$

Equation (3.129) was also derived by Edwards (1964) and Herring (1965) using the Liouville equation for the probability distribution function for the velocity field. We integrate Eq. (3.129) as

$$2\nu\int_{r>k} r^2\sigma(r)d\mathbf{r} = 2\int_{r>k} r^2 d\mathbf{r}\iint \delta\left(\mathbf{r}-\mathbf{p}-\mathbf{q}\right)d\mathbf{p}d\mathbf{q} \times N(r,p,q)\frac{\sigma(p)\left(\sigma(q)-\sigma(r)\right)}{\omega(r)+\omega(p)+\omega(q)}. \tag{3.130}$$

Here we consider that k falls in the inertial range, that is,

$$k_C \ll k \ll k_D \tag{3.131}$$

[k_C and k_D are the wavenumbers characterizing the energy-containing and -dissipation ranges, which are given by Eqs. (3.48) and (3.52a), respectively]. Using Eq. (3.44) and

$$E(k) = 4\pi k^2 \sigma(k), \tag{3.132}$$

we rewrite

$$2\nu \int_{r>k} r^2 \sigma(r) d\mathbf{r} = 2\nu \int_{r>k} r^2 E(r) dr \cong \varepsilon \tag{3.133}$$

(Edwards 1964), where we should note that the energy-containing range makes no direct contribution to the energy dissipation process. Therefore we have

$$\varepsilon = 2 \int_{r>k} r^2 d\mathbf{r} \iint \delta(\mathbf{r} - \mathbf{p} - \mathbf{q}) d\mathbf{p} d\mathbf{q} N(r,p,q) \frac{\sigma(p)(\sigma(q) - \sigma(r))}{\omega(r) + \omega(p) + \omega(q)}. \tag{3.134}$$

Next, we consider the response equation (3.119) for $t > t'$. The simplified form for $Q(k;t,t')$ and $G(k;t,t')$, (3.126) and (3.127), cannot satisfy it exactly. These expressions are proper for examining some important long-time properties of $G(k;t,t')$. Taking this point into account, we seek a weak solution of Eq. (3.119) in the sense that

$$\int_{t'+0}^{\infty} dt \left(\left(\frac{\partial}{\partial t} + \nu k^2 \right) G(k;t,t') + k^2 \iint \delta(\mathbf{k} - \mathbf{p} - \mathbf{q}) d\mathbf{p} d\mathbf{q} \right.$$
$$\left. \times N(k,p,q) \int_{t'}^{t} dt_1 G(q;t,t_1) G(k;t_1,t') Q(p;t,t_1) \right) = 0. \tag{3.135}$$

For the short time interval, it is more adequate to adopt

$$Q(k;t,t') = \sigma(k) \exp\left(- (\omega_Q(k)(t-t'))^2 \right), \tag{3.136}$$

$$G(k;t,t') = H(t-t') \exp\left(- (\omega_G(k)(t-t'))^2 \right). \tag{3.137}$$

We substitute Eqs. (3.126) and (3.127) into Eq. (3.135), and have

$$\omega(k) = \nu k^2 + k^2 \iint \delta(\mathbf{k} - \mathbf{p} - \mathbf{q}) d\mathbf{p} d\mathbf{q} N(k,p,q) \frac{\sigma(p)}{\omega(p) + \omega(q)}. \tag{3.138}$$

C. Inertial-Range Form

From the discussions in Sec. 3.2.1, the inertial-range form for $\sigma(k)$ and $\omega(k)$ are written as

$$\sigma(k) = \sigma \varepsilon^{2/3} k^{-11/3}, \tag{3.139}$$

$$\omega(k) = \omega \varepsilon^{1/3} k^{2/3}, \tag{3.140}$$

where σ and ω are numerical factors. Equation (3.139) is combined with Eq. (3.132) to give the Kolmogorov spectrum (3.45) with

$$K_O = 4\pi\sigma. \tag{3.141}$$

On the other hand, Eq. (3.140) is proportional to the inverse of the life time of eddies with the spatial scale $2\pi/k$, $\tau(k)$, which is defined by Eq. (3.46).

We substitute Eqs. (3.139) and (3.140) into Eq. (3.134), and have

$$\frac{\omega}{\sigma^2} = 2\int_{r>1} r^2 d\mathbf{r} \iint \delta\left(\mathbf{r}-\mathbf{p}-\mathbf{q}\right) d\mathbf{p} d\mathbf{q} N(r,p,q) \frac{p^{-11/3}\left(q^{-11/3}-r^{-11/3}\right)}{p^{2/3}+q^{2/3}+r^{2/3}}. \tag{3.142}$$

In the limit of $p \to 0$, we have

$$p^{-11/3}\left(q^{-11/3}-r^{-11/3}\right) \to O\left(p^{-8/3}\right), \tag{3.143}$$

which guarantees the convergence of Eq. (3.142) in the lower limit of the $\mathbf{p}$ integration. Its numerical computation gives

$$\omega \Big/ \sigma^2 = 30.1 \tag{3.144}$$

(Kraichnan 1971).

On the other hand, the response equation (3.138) gives

$$\frac{\omega^2}{\sigma} = \iint \frac{p^{-11/3}}{p^{2/3}+q^{2/3}} N(1,p,q)\delta\left(\frac{\mathbf{k}}{k}-\mathbf{p}-\mathbf{q}\right) d\mathbf{p} d\mathbf{q}, \tag{3.145}$$

where the molecular viscosity ν has been neglected. Equation (3.145) encounters the divergence in the limit of $p \to 0$, which is called the infra-red divergence. This point is a critical difficulty of the DIA in the study of small-scale turbulence. The infra-red divergence means physically that small eddies in the inertial range are affected spuriously by the advection effect of much larger eddies.

In order to see the effects of small p on Eq. (3.138), we consider the situation

$$0 < p \leq k_C \ll k. \tag{3.146}$$

Under Eq. (3.146), Eq. (3.138) may be rewritten as

$$\omega(k) \cong \frac{2\pi k^2}{\omega(k)} \int_0^{k_C} p^2 \sigma(p) dp \int_{-1}^{1} \left(1-z^2\right) dz, \tag{3.147}$$

which gives

$$\omega(k)^2 \cong \frac{8\pi}{3} k^2 \int_0^{k_C} p^2 \sigma(p) dp. \tag{3.148}$$

In obtaining Eq. (3.147), we used

$$\lim_{p\to 0} N(k,p,q) = 1 - z^2, \tag{3.149}$$

from Eq. (3.121) (see Fig. 3.19).

Let us write

$$\frac{1}{2}U_C^2 = \int_{p\le k_C} \sigma(p)d\mathbf{p}. \tag{3.150}$$

The right-hand side of Eq. (3.150) denotes the energy of the energy-containing eddies whose wavenumber is characterized by k_C. Therefore U_C means the characteristic velocity of energy-containing eddies of the size $2\pi/k_C$. From Eqs. (3.148) and (3.150), we have

$$\omega(k) \propto U_C k, \tag{3.151}$$

in contrast with the Kolmogorov counterpart (3.140). We substitute Eq. (3.151) into Eq. (3.134), and pay attention to the dependence on U_C and k. As a result, we can easily see

$$\sigma(k) \propto (U_C\varepsilon)^{1/2} k^{-7/2}. \tag{3.152}$$

Equation (3.152) leads to the energy spectrum

$$E(k) \propto (U_C\varepsilon)^{1/2} k^{-3/2}, \tag{3.153}$$

using Eq. (3.132). The exponent $-3/2$ does not differ so much from the Kolmogorov counterpart, $-5/3$. The critical difference, however, is the dependence of Eq. (3.153) on U_C, that is, the effect of energy-containing eddies on smaller ones.

D. Advection Effects of Large Eddies

We seek a cause of the failure of the DIA in reproducing the Kolmogorov spectrum. For this purpose, we consider the contribution of small p or q in Eq. (3.28). Under the condition

$$0 < p \text{ or } q \le k_C \ll k, \tag{3.154}$$

we have

$$\frac{\partial}{\partial t}\mathbf{u}(\mathbf{k};t) + i\mathbf{k}\cdot\mathbf{u}_C\mathbf{u}(\mathbf{k};t) = 0, \tag{3.155}$$

with

$$\mathbf{u}_C = \int_{p\le k_C} \mathbf{u}(\mathbf{p};t)d\mathbf{p}. \tag{3.156}$$

Here we dropped the effect of ν, and used

$$M_{ij\ell}(\mathbf{k})u_j(\mathbf{k};t) = \frac{i}{2}k_\ell u_i(\mathbf{k};t) \tag{3.157}$$

from Eq. (3.26).

As is symbolized by Eq. (3.46), the temporal variation of large eddies is slow, compared with small ones. Under Eq. (3.154), we can neglect the time variation of $\mathbf{u}_C$ in Eq. (3.155), and have

$$\mathbf{u}(\mathbf{k};t) \propto \exp\left(-i\mathbf{k}\cdot\mathbf{u}_C t\right). \tag{3.158}$$

The contribution of Eq. (3.158) to Eq. (3.93), which is denoted by $Q_C(\mathbf{k};t,t')$, is written as

$$Q_C(\mathbf{k};t,t') \propto \left\langle \exp\left(-i\mathbf{k}\cdot\mathbf{u}_C(t-t')\right)\right\rangle. \tag{3.159}$$

Here we should note that $\mathbf{u}_C$ is a fluctuating quantity, although it is slowly varying. From observations, it is known well that the instantaneous velocity at one location in isotropic turbulence may be regarded as random. We write its distribution function density as

$$P\left\{\mathbf{u}_C\right\} = \frac{1}{\left(2\pi\sigma_u^2\right)^{3/2}} \exp\left(-\frac{\mathbf{u}_C^2}{2\sigma_u^2}\right), \tag{3.160}$$

where σ_u is the standard deviation of $\mathbf{u}_C$. From Eqs. (3.159) and (3.160), we have

$$Q_C(\mathbf{k};t,t') \propto \exp\left(-\omega_C^2(t-t')^2\right), \tag{3.161}$$

where

$$\omega_C(k) = \frac{1}{\sqrt{2}}\sigma_u k. \tag{3.162}$$

From Eq. (3.161), the correlation time of $Q_C(\mathbf{k};t,t')$, τ_C, is given by

$$\tau_C = \omega_C(k)^{-1} \propto \left(\sigma_u k\right)^{-1}. \tag{3.163}$$

The standard deviation of $\mathbf{u}_C$, σ_u, is the essentially the same as U_C in Eq. (3.150). Therefore the characteristic time scale derived by the DIA, which is given by the inverse of Eq. (3.151), corresponds to Eq. (3.163). These time scales are related to the disappearance of memory due to effects of advection, and their physical meaning is different from that of the Kolmogorov time scale, Eq. (3.46). This point may be explained as follows. We assume that at time t, a turbulent eddy of the spatial scale $2\pi/k$ is located at point $\mathbf{x}$. The eddy is swept away by the flow whose advection velocity is characterized by σ_u or U_C, and its effect disappears in the time interval of $(\sigma_u k)^{-1}$ or $(U_C k)^{-1}$.

On the other hand, the Kolmogorov spectrum comes from the deformation of eddies; namely, the eddies of similar sizes interact with one another and are deformed or destroyed. This process corresponds to the so-called energy-cascade process causing the transfer of energy to smaller scales. In order to catch the process, we need to pursue the true deformation of eddies themselves. In the Eulerian frame with an observer fixed, however, it is difficult to abstract such deformation since a sweeping flow generates an apparent big change of flow properties not related to the cascade process. This is the biggest reason why the DIA in the Eulerian frame fails in reproducing the Kolmogorov spectrum, as was first recognized by Kraichnan (1965) [this point is also detailed in the work by Nakano (1972)]. A method for overcoming this difficulty is the use of the Lagrange frame, that is, the frame moving with a local velocity. This point will be referred to below.

3.5.4. PASSIVE-SCALAR DIFFUSION

A. Scalar DIA System

The diffusion of a passive scalar (for instance, the temperature in the absence of a buoyancy force) is governed by Eq. (3.60). We have already referred to the scalar-variance spectrum similar to the Kolmogorov energy spectrum, (3.67). We examine the DIA system of equations for a passive-scalar diffusion. Its derivation can be made in the manner entirely similar to the procedure leading to Eqs. (3.118) and (3.119). First, we introduce the equation for the scalar response function $G'_\theta(\mathbf{k};t,t')$, which is governed by

$$\frac{\partial}{\partial t}G'_\theta(\mathbf{k};t,t') + \lambda k^2 G'_\theta(\mathbf{k};t,t') - \delta\left(t-t'\right) = ik_i \iint u_i(\mathbf{p};t)G'_\theta(\mathbf{q};t,t')\delta\left(\mathbf{k}-\mathbf{p}-\mathbf{q}\right)d\mathbf{p}d\mathbf{q}. \tag{3.164}$$

Here the fluctuating velocity $\mathbf{u}(\mathbf{k};t)$ is regarded as a given quantity.

We solve Eq. (3.164) in the perturbational manner with the right-hand side treated as a perturbation. In what follows, we shall assume the isotropy of $\mathbf{u}(\mathbf{k};t)$, as in Eq. (3.116), in addition to the Gaussianity. Then the mean scalar response function $G_\theta(k;t,t')$, which is defined by

$$G_\theta(k;t,t') = \left\langle G'_\theta(\mathbf{k};t,t')\right\rangle, \tag{3.165}$$

obeys

$$\left(\frac{\partial}{\partial t} + \lambda k^2\right) G_\theta(k;t,t') = \delta\left(t-t'\right) - k^2 \iint \delta\left(\mathbf{k}-\mathbf{p}-\mathbf{q}\right)d\mathbf{p}d\mathbf{q}$$

$$\times N_\theta(k,p,q)\int_{t'}^{t} G_\theta(q;t,t_1)G_\theta(k;t_1,t')Q(p;t,t_1)dt_1, \tag{3.166}$$

where

$$N_\theta(k,p,q) = 1 - z^2. \tag{3.167}$$

Next, we consider the scalar variance of $\theta(\mathbf{k};t)$ defined by

$$Q_\theta\,(k;t,t) = \langle \theta(\mathbf{k};t)\theta(\mathbf{k}';t')\rangle \,/\delta\,(\mathbf{k}+\mathbf{k}')\,. \tag{3.168}$$

We follow the same procedure as for the velocity-variance equation (3.111). Then we have

$$\begin{aligned}\left(\frac{\partial}{\partial t} + \lambda k^2\right) Q_\theta(\mathbf{k};t,t') &= k^2 \iint \delta\,(\mathbf{k}-\mathbf{p}-\mathbf{q})d\mathbf{p}d\mathbf{q}N_\theta(k,p,q) \\ &\times \left(\int_{-\infty}^{t'} G_\theta\,(k;t',t_1)\,Q\,(p;t,t_1)\,Q_\theta\,(q;t,t_1)dt_1\right. \\ &\left. - \int_{-\infty}^{t} G_\theta\,(q;t,t_1)\,Q\,(p;t,t_1)\,Q_\theta\,(k;t_1,t')dt_1\right).\end{aligned} \tag{3.169}$$

B. Quasi-Stationary State

As is in Sec. 3.5.3.B, we assume the quasi-stationary stare:

$$G_\theta(k;t,t') = H(t-t')\exp\left(-\omega_\theta(k)\,(t-t')\right), \tag{3.170}$$

$$Q_\theta(k;t,t') = \sigma_\theta(k)\exp\left(-\omega_{Q_\theta}(k)\,|t-t'|\right), \tag{3.171}$$

with Eq. (3.126) for the velocity variance $Q(k;t,t')$. From Eq. (3.166), we have

$$\omega_\theta(k) = \lambda k^2 + \iint \delta\,(\mathbf{k}-\mathbf{p}-\mathbf{q})d\mathbf{p}d\mathbf{q}N_\theta(k,p,q)\frac{\sigma(p)}{\omega(p)+\omega_\theta(q)}. \tag{3.172}$$

Equation (3.169) with $t=t'$ leads to the equation for the one-point scalar variance $Q_\theta(k;t,t)$:

$$\begin{aligned}2\lambda k^2\sigma_\theta(k) &= 2k^2 \iint \delta\,(\mathbf{k}-\mathbf{p}-\mathbf{q})d\mathbf{p}d\mathbf{q}N_\theta(k,p,q) \\ &\times \left(\frac{\sigma(p)\sigma_\theta(q)}{\omega_\theta(k)+\omega(p)+\omega_{Q_\theta}(q)} - \frac{\sigma(p)\sigma_\theta(k)}{\omega_\theta(q)+\omega(p)+\omega_{Q_\theta}(k)}\right).\end{aligned} \tag{3.173}$$

From Eq. (3.63), $\sigma_\theta(k)$ may be related to the scalar-variance destruction rate ε_θ as

$$\varepsilon_\theta \equiv 2\lambda \int r^2\sigma_\theta(r)d\mathbf{r}. \tag{3.174}$$

In case that k falls in the scalar inertial range similar to the velocity one, as in expression (3.131), the scalar destruction also occurs at the wavenumber much larger than k. Then we have

$$\varepsilon_\theta = 2\int_{r>k} r^2 d\mathbf{r} \iint \delta(\mathbf{r}-\mathbf{p}-\mathbf{q}) d\mathbf{p} d\mathbf{q} N_\theta(r,p,q)$$
$$\times \left(\frac{\sigma(p)\sigma_\theta(q)}{\omega_\theta(r)+\omega(p)+\omega_{Q_\theta}(q)} - \frac{\sigma(p)\sigma_\theta(r)}{\omega_\theta(q)+\omega(p)+\omega_{Q_\theta}(r)} \right), \tag{3.175}$$

which should be compared with Eq. (3.134) for the velocity field.

C. Inertial-Range Form

The inertial-range form for $\omega_\theta(k)$, $\sigma_\theta(k)$, and $\omega_{Q_\theta}(k)$ may be written as

$$\omega_\theta(k) = \omega_\theta \omega(k), \tag{3.176}$$

$$\sigma_\theta(k) = \sigma_\theta \varepsilon_\theta \varepsilon^{-1/3} k^{-11/3}, \tag{3.177}$$

$$\omega_{Q_\theta}(k) = \omega_{Q_\theta} \omega(k), \tag{3.178}$$

where $\omega(k)$ is defined by Eq. (3.140), and ω_θ, σ_θ, and ω_{Q_θ} are numerical factors. Equation (3.177) corresponds to the Kolmogorov counterpart (3.139) and comes from Eq. (3.67).

Using Eq. (3.176) as well as Eq. (3.139) for $\sigma(k)$, Eq. (3.172) is reduced to

$$\frac{\omega_\theta \omega^2}{\sigma} = \iint \frac{p^{-11/3}}{p^{2/3}+\omega_\theta q^{2/3}} N_\theta(1,p,q) \delta\left(\frac{\mathbf{k}}{k}-\mathbf{p}-\mathbf{q}\right) d\mathbf{p} d\mathbf{q}. \tag{3.179}$$

This integral diverges in the limit of $p \to 0$, as is entirely similar to Eq. (3.145) for fluid motion. Its cause was detailed in Sec. 3.5.3.D.

We substitute Eqs. (3.176)-(3.178) into Eq. (3.175), and have

$$\frac{\omega}{\sigma_\theta \sigma} = 2\int_{r>1} r^2 d\mathbf{r} \iint \delta(\mathbf{r}-\mathbf{p}-\mathbf{q}) d\mathbf{p} d\mathbf{q} N_\theta(r,p,q)$$
$$\times \left(\frac{p^{-11/3} q^{-11/3}}{\omega_\theta r^{2/3}+p^{2/3}+\omega_{Q_\theta} q^{2/3}} - \frac{p^{-11/3} r^{-11/3}}{\omega_\theta q^{2/3}+p^{2/3}+\omega_{Q_\theta} r^{2/3}} \right). \tag{3.180}$$

This equation encounters no difficulty in the limit of $p \to 0$ from the same reason as Eq. (3.142).

3.5.5. PATHS TOWARDS RECTIFYING INFRA-RED DIVERGENCE

In the foregoing discussions, the DIA, which is based on the straightforward use of the renormalization in the Eulerian frame, was shown to fail in reproducing the simple Kolmogorov scaling for the energy spectrum and the scalar counterpart. The cause is the incorrect estimate of the sweeping effect by large eddies on small ones. The remedy for rectifying this deficiency is the proper elimination of the effect. How this elimination is implemented is closely related to the target of turbulence theory. One primary target is the clarification of small-scale turbulence characteristics whose intermittent properties originate in the energy dissipation process. Its directly opposite position is the mathematical formulation of turbulent shear flow for providing a rational basis of turbulence modeling. In this case, attention is focused on spatial scales equal or larger than integral scales. Between these two extremities, there may be some alternatives following the goal of each research.

In what follows, we shall refer to some typical instances of such paths towards rectifying the infra-red divergence in the DIA system.

A. Lagrangian Formalism

A method of properly detecting the deformation of small eddies arising from the interaction among themselves is to observe these eddies in a frame moving with the flow sweeping them away. This orthodox method was presented by Kraichnan (1965, 1977) who is the founder of the DIA formalism. For this purpose, Kraichnan developed a Lagrangian description of turbulence. In the description, we pursue the time development of the path of a fluid particle. In fluid mechanics, its use is rather rare, and the representative example is the derivation of the Lagrange vortex theorem.

In Kraichnan's formalism, we seek the evolution of a fluid particle at time s, which was located at $\mathbf{x}$ at a previous time t. We denote its velocity by $\mathbf{u}(\mathbf{x};t|s)$, which is reduced to the usual Eulerian velocity $\mathbf{u}(\mathbf{x};t)$ for $s=t$. We consider that the particle is located at $\mathbf{x}'$ at an intermediate time t'; namely, we have

$$\mathbf{u}(\mathbf{x}';t'|s) = \mathbf{u}(\mathbf{x};t|s), \tag{3.181}$$

where $\mathbf{x}'$ and t' are related as

$$\mathbf{x}' = \mathbf{x} + \int_t^{t'} \mathbf{u}\left(\mathbf{x}, t\left|t_1\right.\right) dt_1. \tag{3.182}$$

Equation (3.181) is expanded in infinitesimal $|\mathbf{x}-\mathbf{x}'|$ and $|t-t'|$ and is reduced to the equation for the Lagrange velocity $\mathbf{u}(\mathbf{x};t|s)$

$$\left(\frac{\partial}{\partial t} + \mathbf{u}(\mathbf{x};t)\cdot\nabla\right) \mathbf{u}\left(\mathbf{x};t\left|s\right.\right) = 0. \tag{3.183}$$

The pivotal point of the Lagrangian formalism lies in the introduction of the Green's or response function and correlation functions concerning the Lagrange velocity $\mathbf{u}(\mathbf{x};t|s)$. This method may pave the way for the rational elimination of the spurious effects by large eddies. In reality, Kraichnan (1965) succeeded in reproducing the Kolmogorov's $-5/3$ power law with the proper estimate of its proportional constant. The formalism was also extended to the study of passive scalar diffusion in isotropic turbulence. The further developments in the turbulence theories in a Lagrange frame are given in the works by Kaneda (1981) and McComb (1990). The Lagrangian formalism is much more complicated than the Eulerian counterpart, the DIA, and has hardly been applied to the study of turbulent shear flow.

B. Markovianization of DIA

As was seen in Sec. 3.5.3.D, the spurious sweeping effects by large eddies occur in close relation to two-time statistics. In reality, we have encountered no difficulty in the equation for the turbulent energy that is a one-time quantity [recall the discussion below Eq. (3.142)]. This fact indicates the possibility that a theory consistent with the Kolmogorov scaling can be constructed from the DIA system as long as the response function is not dealt with directly.

We consider Eq. (3.122) for the turbulent-energy density $Q(k;t,t)$. The right-hand side is dependent on the two-time quantities such as $Q(k;t,t')$ and $G(k;t,t')$. First, we approximate $Q(k;t,t')$ as

$$\frac{Q(k;t,t')}{Q(k;t,t)} = G(k;t,t'). \tag{3.184}$$

We substitute Eq. (3.184) into Eq. (3.122), and have

$$\left(\frac{\partial}{\partial t} + 2\nu k^2\right) Q(k;t,t) = 2k^2 \iint \delta\left(\mathbf{k}-\mathbf{p}-\mathbf{q}\right) d\mathbf{p}d\mathbf{q} \\ \times N(k,p,q)\chi(k,p,q)Q(p;t,t)\left(Q(q;t,t) - Q(k;t,t)\right), \tag{3.185}$$

where $\chi(k,p,q)$ is defined as

$$\chi(k,p,q) = \int_{-\infty}^{t} G\left(k;t,t_1\right) G\left(p;t,t_1\right) G\left(q;t,t_1\right) dt_1. \tag{3.186}$$

The right-hand side of Eq. (3.185) depends on the latest time t only, apart from $\chi(k,p,q)$. This procedure is called the Markovianization and makes the original DIA much simpler, provided that $\chi(k,p,q)$ may be expressed in terms of one-time quantities.

As $G(k;t,t')$, we adopt the simple exponential form, Eq. (3.127), which means

$$\omega(k)^{-1} = \int_{-\infty}^{t} G\left(k;t,t_1\right) dt_1. \tag{3.187}$$

Then Eq. (3.186) is reduced to

$$\chi(k,p,q) = \frac{1}{\omega(k)+\omega(p)+\omega(q)}. \tag{3.188}$$

In Sec. 3.5.3.C, we noted that $\omega(k)$ is inversely proportional to the life time of eddies of the spatial scale $2\pi/k$. In their evolution, the eddies of the scale larger than $2\pi/k$ exert dominant influence on them (these effects are overestimated in the DIA). Then we write

$$\omega(k) = C'_\omega \left(\int_0^k r^4 Q(r;t,t)dr \right)^{1/2} + \nu k^2, \tag{3.189}$$

with the aid of dimensional analysis, where C'_ω is a numerical factor, and the molecular viscosity effect on $\omega(k)$ has been included.

Equation (3.185) combined with Eqs. (3.188) and (3.189) is called the eddy-damped quasi-normal Markovianized (EDQNM) approximation (Orszag 1970; Leith 1971; André and Lesieur 1977; Lesieur 1997). In this derivation, the physical meaning of "quasi-normal" is not clear. The original quasi-normal approximation corresponds to the retention of only the second viscous term in Eq. (3.189), whereas the first term expresses the shortening of the life time of eddies due to nonlinear effects [see Tatsumi (1980) for the Markovianization based on the former viscous term]. In the usual EDQNM formulation, $Q(k;t,t)$ etc. in Eqs. (3.185) and (3.189) are replaced with the energy spectrum $E(k)$ through Eq. (3.37). This method can reproduce the Kolmogorov's energy spectrum, but the magnitude of the Kolmogorov constant is dependent on the choice of C'_ω.

The EDQNM approximation based on an ad hoc choice of the response-function effects, $\chi(k,p,q)$, is a crude method, compared with the Lagrangian formalism. The simplicity coming from the Markovianization, however, brings a big merit of the applicability to anisotropic turbulence such as homogeneous-shear turbulence (Cambon et al. 1981). Moreover, effects of frame rotation on isotropic turbulence can be examined by incorporating these effects into $\chi(k,p,q)$ (Cambon and Jacquin 1989). The EDQNM approximation is not applicable to turbulent shear flow with arbitrary mean velocity since it is fully dependent on the use of the Fourier representation. Specifically, the response function is replaced with a special functional form such as Eq. (3.188) with (3.189), and loses its original capability in solving a differential equation. This point will become clear in the theoretical study of turbulent shear flow in Chapter 6.

The renormalization group (RNG) method developed by Yakhot and Orszag (1986) is rather similar to the EDQNM approximation in the sense that no direct use is made of a renormalized Green's function. In the method, we divide a quantity $f(\mathbf{x};t)$ into two parts as

$$f(\mathbf{x};t) = f(\mathbf{x};t)^{+} + f(\mathbf{x};t)^{-}, \tag{3.190}$$

where

$$f(\mathbf{x};t)^{+} = \int_{k<k_0} f(\mathbf{k};t) \exp\left(-i\mathbf{k}\cdot\mathbf{x}\right) d\mathbf{k}, \tag{3.191a}$$

$$f(\mathbf{x};t)^{-} = \int_{k>k_0} f(\mathbf{k};t) \exp\left(-i\mathbf{k}\cdot\mathbf{x}\right) d\mathbf{k}, \tag{3.191b}$$

where k_0 is a specific cut-off wavenumber. Roughly speaking, $f(\mathbf{x};t)^{+}$ consists of large eddies, whereas $f(\mathbf{x};t)^{-}$ does of small ones. We substitute Eq. (3.190) into Eq. (3.15) with vanishing S, and incorporate the nonlinear effects of $f(\mathbf{x};t)^{-}$ into $f(\mathbf{x};t)^{+}$ through a renormalized viscosity. The final form of the viscosity may be derived as a fixed point after repeating this procedure many times. The overestimate of energetic-eddy effects in the Eulerian DIA is avoided using Eq. (3.191b) whose lower limit is k_0. In the final stage of analysis, k_0 is replaced with arbitrary k. This result is combined with the EDQNM method to lead to the reasonable estimate of the Kolmogorov constant, although the replacement such as $k_0 = Ck$ cannot be ruled out (C is an arbitrary constant). As is noted in Sec. 3.5.1, many works have been done on the critique and refinement of this method. The relationship of the RNG method with turbulence modeling will be referred to in Chapter 6.

C. Artificial Cut-Off of Lower-Wavenumber Effects

The simplest way to eliminate the sweeping effects causing the infra-red divergence of the DIA response equation is to remove the low-wavenumber contribution in an artificial manner. Such an ad hoc method is of little value as a theory elucidating small-scale properties of turbulent motion and scalar diffusion. In the application to turbulent shear flows, however, this simple method retains a big merit in relation to response functions. For applying the statistical methods represented by the DIA to turbulent shear flow, some further approximations are indispensable. One of them is the use of a perturbation method, as is detailed in Chapter 6. In this case, response functions play a key role in solving the perturbational equations. In the study of turbulent shear flow, the anisotropy of fluctuations caused by mean velocity gradient is more important than the small-scale intermittency effects associated with the energy-dissipation mechanism. The anisotropy may be taken into account through the anisotropic effects on the inertial-range fluctuations, although it is not quite sufficient. In this context, the information about the isotropic part of fluctuations beyond the Kolmogorov law is not always necessary, and the simplicity of theory is a more important ingredient.

As the method of removing low-wavenumber contributions, we can mention the following two. One method is the simple truncation (Leslie 1973). First, we write the right-hand side of the response equation (3.138) as

$$I = \int N(k,p,q)\eta(p,q)\delta\left(\mathbf{k}-\mathbf{p}-\mathbf{q}\right)d\mathbf{p}d\mathbf{q}, \tag{3.192}$$

where

$$\eta(p,q) = \frac{\sigma(p)}{\omega(p)+\omega(q)}. \tag{3.193}$$

In this truncation method, Eq. (3.192) is rewritten as

$$I = \int_{p \geq C_K k} \int N\left(k,p,|\mathbf{k}-\mathbf{p}|\right)\eta\left(p,|\mathbf{k}-\mathbf{p}|\right)d\mathbf{p}. \tag{3.194}$$

Here C_K is a numerical factor and is chosen so that the resulting Kolmogorov constant should fall around 1.5. Another truncation is a smooth one, in which the continuous removal of low-wavenumber contributions is performed (Yoshizawa 1978, 1981). In the method, we introduce a new function $N_C(k,p,q)$ and rewrite Eq. (3.192) as

$$I = \int \left(N(k,p,q) - N_C(k,p,q)\right)\eta(p,q)\delta\left(\mathbf{k}-\mathbf{p}-\mathbf{q}\right)d\mathbf{p}d\mathbf{q}. \tag{3.195}$$

A method of determining $N_C(k,p,q)$ is referred to in the following *Note.*

Neither of these two methods brings anything new in the investigation of the energy spectrum. The numerical factors in the scalar response function and the scalar-variance spectrum, however, can be estimated on the same theoretical basis. They are indispensable for the estimate of model constants in scalar turbulence models.

Note: We simply refer to the determination of $N_C(k,p,q)$ in Eq. (3.195), for we shall use the estimated numerical constants in the theoretical study of turbulence modeling in Chapter 6. We impose the following two conditions on $N_C(k,p,q)$:

(C1) $N_C(k,p,q)$ coincides with $N(k,p,q)$ in the infra-red or $p \to 0$ limit, and the former magnitude does not exceed the latter in the limiting process;

(C2) $N_C(k,p,q)$ becomes much smaller than $N(k,p,q)$, as p increases.

In order to designate (C1) in mathematical terms, we define the magnitude of $f(k,p,q)$ by its integral at the spherical surface of radius p:

$$\overline{f(k,p,q)} = \frac{1}{4\pi p^2}\int_{S_p} f(k,p,q)dS \tag{3.196}$$

(S_p denotes the surface of radius p). Then (C1) may be expressed as

$$\lim_{p\to 0}\left(N(k,p,q) - N_C(k,p,q)\right) = 0, \tag{3.197a}$$

and

$$\overline{\left(N(k,p,q) - N_C(k,p,q)\right)^2} \cong 0,\ \ \overline{N(k,p,q)} \geq \overline{N_C(k,p,q)} \tag{3.197b}$$

for small p.

We introduce the test function

$$N_C(k,p,q) = \frac{q}{k}y^{\alpha}(xz+y), \tag{3.198}$$

where α is a numerical factor to be determined. We can easily see that Eq. (3.198) obeys Eq. (3.197a). The value of α larger than one leads to smaller $N_C(k,p,q)$ and is helpful to the fulfillment of (C2). For small p, we have

$$N(k,p,q) - N_C(k,p,q) \cong \frac{\alpha}{2}\left(1-z^2\right)^2\left(\frac{p}{k}\right)^2, \tag{3.199}$$

$$\overline{\left(N(k,p,q) - N_C(k,p,q)\right)^2} \cong \left(\frac{32}{315}\alpha^2 - \frac{16}{35}\alpha + \frac{8}{15}\right)\left(\frac{p}{k}\right)^4. \tag{3.200}$$

From Eq. (3.199), positive α guarantees the second of expression (3.197b). For fixed p/k, Eq. (3.200) becomes minimum at $\alpha = 2.25$, where the α-related part on the right-hand side is 0.019. From the realness and regularity of $N_C(k,p,q)$ for negative z, α should be a positive integer. Then we have $\alpha = 2$ or $\alpha = 3$, but the latter is more desirable in the context of the first of Eq. (3.197b). Finally, we have

$$N_C(k,p,q) = \frac{q}{k}y^3(xz+y). \tag{3.201}$$

We apply Eqs. (3.195) and (3.201) to Eq. (3.145), and have

$$\omega^2/\sigma = 1.49. \tag{3.202}$$

Equation (3.202) is combined with Eq. (3.144) to give

$$\sigma = 0.118,\ \omega = 0.42, \tag{3.203}$$

which leads to

$$K_O \cong 1.5. \tag{3.204}$$

The estimate of K_O is consistent with the observational results. This fact indicates that this method brings no critical error, although it does not justify the simple elimination of sweeping effects.

We apply this method to the passive scalar diffusion. In Eq. (3.179), we perform the elimination of sweeping effects by the replacement

$$N_\theta(k,p,q) \to N_\theta(k,p,q) - N_{\theta C}(k,p,q). \tag{3.205}$$

We make use of the procedure described above, and introduce

$$N_{\theta C}(k,p,q) = \frac{q}{k} y(xz + y). \tag{3.206}$$

As a result, Eq. (3.179) with Eq. (3.206) inserted gives

$$\omega_\theta = 1.6, \tag{3.207}$$

where Eq. (3.203) has been used for σ and ω.

Equation (3.180) contains two unknown factors, σ_θ and ω_{Q_θ}. We assume that the correlation time of scalar fluctuations, $\omega_{Q_\theta}(k)^{-1}$, is not far from the velocity counterpart $\omega(k)^{-1}$; namely, we put

$$\omega_{Q_\theta} \cong 1. \tag{3.208}$$

As a result, we have

$$\sigma_\theta = 0.066, \tag{3.209}$$

which gives

$$B_A \equiv 4\pi\sigma_\theta = 0.83, \tag{3.210}$$

in Eq. (3.67). The observational results scatter around 0.7 (Paquin and Pond 1971; Williams and Paulson 1977; Hill 1978), and the above estimate is reasonable.

Equation (3.207) indicates

$$\omega_\theta(k) > \omega(k). \tag{3.211}$$

In Chapter 6, this relation will be shown to lead to the conclusion that the turbulent Prandtl number, which is defined by the ratio of the turbulent viscosity to the turbulent diffusivity, is less than one. This finding is consistent with observations.

References

André, J. C. and Lesieur, M. (1977), J. Fluid Mech. **81**, 187.
Batchelor, G. K. (1959), J. Fluid Mech. **5**, 113.
Batchelor, G. K., Howells, I. D., and Townsend, A. A. (1959), J. Fluid Mech. **5**, 134.
Cambon, C., Jeandel, D., and Mathieu, J. (1981), J. Fluid Mech. **104**, 247.
Cambon, C. and Jacquin, L. (1989), J. Fluid Mech. **202**, 295.
Dannevik, W. P., Yakhot, V., and Orszag, S. A. (1987), Phys. Fluids **30**, 2021.
Edwards, S. F. (1964), J. Fluid Mech. **18**, 239.
Eyink, G. L. (1994), Phys. Fluids **6**, 3063.
Frisch, U. (1995), *Turbulence* (Cambridge U. P., Cambridge).
Gurvich, A. S. and Yaglom, A. M. (1967), Phys. Fluids **10** (Supplement), S59.
Herring, J. R. (1965), Phys. Fluids **8**, 2219.
Hill, R. (1978), J. Fluid Mech. **88**, 541.
Kaneda, Y. (1981), J. Fluid Mech. **107**, 131.
Kolmogorov, A. N. (1941a), C. R. Acad. Sci. USSR **30**, 301.
Kolmogorov, A. N. (1941b), C. R. Acad. Sci. USSR **32**, 16.
Kolmogorov, A. N. (1962), J. Fluid Mech. **13**, 82.
Kraichnan, R. H. (1959), J. Fluid Mech. **5**, 497.
Kraichnan, R. H. (1965), Phys. Fluids **8**, 575.
Kraichnan, R. H. (1971), J. Fluid Mech. **47**, 525.
Kraichnan, R. H. (1977), J. Fluid Mech. **83**, 349.
Leith, C. E. (1971), J. Atmos. Sci. **28**, 145.
Lesieur, M. (1997), *Turbulence in Fluids* (Kluwer, Dordrecht).
Leslie, D. C. (1973), *Developments in the Theory of Turbulence* (Clarendon, Oxford).
Lighthill, M. J. (1970), *Fourier Representation and Generalized Functions* (Cambridge U. P., Cambridge).
McComb, M. D. (1990), *The Physics of Fluid Turbulence* (Clarendon, Oxford).
Meneveau, C. and Screenivasan, K. R. (1991), J. Fluid Mech. **224**, 429.
Nagano, Y. and Itazu, Y. (1997), Phys. Fluids **9**, 143.
Nakano, T. (1972), Ann. Phys. (NY) **73**, 326.
Nakano, T. (1992), J. Phys. Soc. Jpn. **61**, 3994.
Oboukhov, A. M. (1962), J. Fluid Mech. **13**, 77.
Orszag, S. A. (1970), J. Fluid Mech. **41**, 363.
Paquin, J. E. and Pond, S. (1971), J. Fluid Mech. **50**, 257.
Smith, L. M. and Reynolds, W. C. (1992), Phys. Fluids A **4**, 364.

Tatsumi, T. (1980), Adv. in Appl. Mech. **20**, 39.
Wilson K. G. and Kogut, J. (1974), Phys. Rep. **12** C, 75.
Williams, R. M. and Paulson, C. A. (1977), J. Fluid Mech. **83**, 547.
Wyld, H. W. (1961), Ann. Phys. (NY) **14**, 143.
Yakhot, V. and Orszag, S. A. (1986), J. Sci. Comput. **1**, 3.
Yoshizawa, A. (1978), J. Phys. Soc. Jpn. **45**, 1734.
Yoshizawa, A. (1980), J. Phys. Soc. Jpn. **48**, 301.
Yoshizawa, A. (1994), Phys. Rev. E **49**, 4065.

CHAPTER 4

CONVENTIONAL TURBULENCE MODELING

4.1. Algebraic and Second-Order Modeling

The mean-velocity ($\mathbf{U}$) equation (2.100) and the mean-scalar (Θ) equation (2.102) do not constitute closed systems of equations owing to the presence of the Reynolds stress $R_{ij}(= \langle u_i' u_j' \rangle)$ and the turbulent scalar flux $\mathbf{H}_\theta(= \langle \mathbf{u}'\theta' \rangle)$. In turbulence modeling, we relate these quantities to $\mathbf{U}$ and Θ explicitly or implicitly. Here we have two primary approaches. One is the algebraic modeling represented by the turbulent-viscosity approximation to R_{ij}, and the other is the second-order modeling, in which the transport equations for R_{ij} and $\mathbf{H}_\theta$, (2.105) and (2.120), are treated directly. In the former, R_{ij} and $\mathbf{H}_\theta$ are expressed in terms of $\mathbf{U}$, Θ, and some one-point quantities characterizing turbulence, such as the turbulent energy K and its dissipation rate ε. In the latter, the third-order quantities such as the pressure-strain correlation function Π_{ij} [Eq. (2.107)] need to be modeled.

In this chapter, we shall confine ourselves to incompressible flow and discuss these two approaches in the conventional manner based on dimensional and tensor analyses. We shall start with a simple, linear turbulent-viscosity model and examine its fundamental model structures. In this course, the critical deficiencies of the model will be clarified. On the basis of these discussions, we shall proceed to the improvement of the turbulent-viscosity model by adding higher-order corrections. In the second-order modeling, we shall pay attention to its basic concept and the relationship with the higher-order modeling of the turbulent-viscosity type.

The goal of turbulence modeling is the construction of a one-point system of equations applicable to a variety of turbulent flows in engineering and natural sciences. Its study is in rapid progress in both the proposal of models and their test through applications. The aim of this chapter is not to follow those developments and discuss the performance of models. For that purpose, readers can consult the review articles by Speziale (1991), Shih (1996), Launder (1996), and the works cited therein. Turbulence models from a meteorological standpoint are detailed by Mellor and Yamada (1982). In what follows, we shall focus attention on the fundamental mathematical structures of models in the algebraic and second-order models. This stance is closely related to one of the primary subjects of this book, that is, the subject of providing a statistical theoretical foundation to various turbulent models that have been proposed using the conventional approach and confirmed to be useful in practical applications.

4.2. Linear Algebraic Modeling

4.2.1. TURBULENT VISCOSITY

The physical meaning of the Reynolds stress R_{ij} [Eq. (2.101)] may be understood clearly in the comparison with the Navier-Stokes equation (2.24). In the latter, the stress τ_{ij} is the remnant of the smoothed-out molecular motion and occurs in the form of $\partial \tau_{ij}/\partial x_j$. In the mean Navier-Stokes equation (2.100), R_{ij} appears in the same mathematical form as τ_{ij} (this is the reason why R_{ij} is called the Reynolds stress). In the special case that

$$\mathbf{u} = (u(y), 0, 0)\,, \ \tau_{yx} = \mu \frac{du}{dy} (> 0), \tag{4.1}$$

the larger-y region drags the smaller-y one in the x direction, and the x-component of the momentum is transferred from the former to the latter. On the other hand, R_{xy} is the transfer rate of the x component of the momentum due to the y component of the fluctuating velocity or vice versa. Then we may see the following correspondence:

$$\text{Fluctuating motion} \Longleftrightarrow \text{Molecular motion}; \tag{4.2a}$$

$$\text{Mean velocity} \Longleftrightarrow \text{Macroscopic flow}. \tag{4.2b}$$

This correspondence suggests that the momentum transfer due to turbulence, which is expressed by R_{ij}, occurs from the region with the higher mean velocity to the lower counterpart. The spatial nonuniformity of the mean velocity $\mathbf{U}$ is characterized by $\partial U_j/\partial x_i$. Considering that R_{ij} is a symmetric tensor, we write the turbulent-viscosity model

$$R_{ij} = \frac{2}{3} K \delta_{ij} - \nu_T S_{ij}, \tag{4.3}$$

where S_{ij} is the mean velocity-strain tensor defined by

$$S_{ij} = \frac{\partial U_j}{\partial x_i} + \frac{\partial U_i}{\partial x_j}. \tag{4.4}$$

The first term of Eq. (4.3) comes from the identity

$$R_{ii} = 2K. \tag{4.5}$$

The coefficient ν_T corresponds to the molecular kinematic viscosity in Eq. (2.30b), ν, and is called the turbulent viscosity. Positive ν_T generates the momentum transport from the high- to low-$\mathbf{U}$ region. Substituting Eq. (4.3) into Eq. (2.100), we have

$$\frac{DU_i}{Dt} = -\frac{\partial}{\partial x_i}\left(P + \frac{2}{3}K\right) + \frac{\partial}{\partial x_j}\left(\left(\nu_T + \nu\right) S_{ji}\right). \tag{4.6}$$

A big difference between ν and ν_T lies in the degree of scale separation. In the case of ν, the characteristic spatial scale of molecular motion, that is, the mean-free path of molecules is much shorter than the counterpart of the macroscopic fluid motion described by $\mathbf{u}$. This clear-cut separation of scales does not hold between $\mathbf{u}'$ and $\mathbf{U}$. The concept of viscosity becomes more accurate with the increasing scale separation. Therefore the turbulent-viscosity approximation is anticipated to suffer from various shortfalls.

For expressing ν_T, we need two independent physical quantities at least. Here it is important whether or not ν is chosen as one of them. In order to see the degree of the importance of ν effects, we use the turbulent Reynolds number R_T, which is defined by Eq. (3.59) or

$$R_T = \nu_T / \nu. \tag{4.7}$$

Large ν_T signifies the dominance of the momentum transfer due to turbulence effects over the molecular one, and a highly turbulent state is characterized by large ν_T. Effects of ν occur twofold. One is the effect of low Reynolds number coming from decaying or weakening of turbulence, and it is related to the relaminarization of turbulence, that is, the return to a laminar-flow state. The other is the ν effect coming from the low velocity near a solid wall. These points will be referred to in Sec. 4.7.

As two quantities except ν, the turbulent energy K [Eq. (2.110)] and its dissipation rate ε [Eq. (2.113)] are often chosen. The choice of K is natural since it is the quantity characterizing the intensity of velocity fluctuation and is tightly linked with the conservation of kinetic energy. The choice of ε is also natural since the K equation (2.111) contains ε. Moreover, the proper estimate of ε without delving into the fine scale given by Eq. (3.54) is indispensable for the one-point turbulence modeling, as was stated in Chapter 2. The method based on K and ε is usually called the K-ε modeling and is most familiar in turbulence modeling.

Using K and ε, the turbulent viscosity ν_T can be expressed as

$$\nu_T = C_\nu K^2 / \varepsilon\ , \tag{4.8}$$

where C_ν is a numerical factor. In place of K or ε, we can use the spatial scale ℓ_C that characterizes energy-containing parts of velocity fluctuations.

It is related to K and ε as Eq. (3.48). In terms of K and ℓ_C, we may write

$$\nu_T = C'_\nu K^{1/2} \ell_C, \tag{4.9}$$

where C'_ν is a numerical factor, and $K^{1/2}$ is a characteristic turbulent velocity. The physical meaning of Eq. (4.9) is clear from the fact that ν_T has the dimension of the product of velocity and length.

From Eq. (4.8), Eq. (4.7) is reduced to

$$R_T \propto K^2 / (\varepsilon \nu) \, . \tag{4.10}$$

For large R_T, effects of ν on ν_T may be neglected, resulting in Eq. (4.8) as the simplest model. Such a model is called a high-Reynolds-number version of ν_T,whereas an expression for ν_T with effects of ν incorporated is a low-Reynolds-number version.

4.2.2. MODELING OF TURBULENT TRANSPORT EQUATIONS

In order to close Eq. (4.6) with Eq. (4.8) as ν_T, we need to determine K and ε in a self-consistent manner. In what follows, we shall consider the modeling of their transport equations.

A. Modeling of Equation for K

The equation for K is given exactly by Eq. (2.111). On the right-hand side, no modeling is necessary for the first two terms, P_K and ε. The third term, which is called the diffusion term, consists of the triple velocity correlation and the pressure-velocity correlation. Of these two, the property of the latter is much more complicated.

In order to see the pressure effect, we consider the equation for p'. We take the divergence of Eq. (2.104), and have

$$\Delta p' = -2 \frac{\partial u'_j}{\partial x_i} \frac{\partial U_i}{\partial x_j} - \frac{\partial^2}{\partial x_i \partial x_j} \left(u'_i u'_j - R_{ij} \right) . \tag{4.11}$$

We introduce the Green's function for Eq. (4.11) as

$$\Delta G(\mathbf{x}, \mathbf{x}') = \delta(\mathbf{x} - \mathbf{x}'). \tag{4.12}$$

Then $G(\mathbf{x}, \mathbf{x}')$ is given by

$$G(\mathbf{x}, \mathbf{x}') = -1 / (4\pi r) \, , \tag{4.13}$$

with $r = |\mathbf{x} - \mathbf{x}'|$. We may use $G(\mathbf{x}, \mathbf{x}')$ and solve Eq. (4.11) formally (this point will be referred to in *Note*). As a result, we have

$$\begin{aligned} p'(\mathbf{x}) = & \frac{1}{2\pi} \int_V \frac{1}{r} \frac{\partial U_i(\mathbf{x}')}{\partial x_j} \frac{\partial u'_j(\mathbf{x}')}{\partial x_i} dV + \frac{1}{4\pi} \int_V \frac{1}{r} \frac{\partial^2 u'_i(\mathbf{x}') u'_j(\mathbf{x}')}{\partial x_i \partial x_j} dV \\ & + \frac{1}{4\pi} \int_S \left(\frac{1}{r} \frac{\partial p'(\mathbf{x}')}{\partial n} - p'(\mathbf{x}') \frac{\partial}{\partial n} \left(\frac{1}{r} \right) \right) dS, \end{aligned} \tag{4.14}$$

where V is the region occupied by fluid, S is its surface, and $\mathbf{n}$ is the outward unit vector normal to the surface. The contribution of R_{ij} has been dropped since it is not necessary for the following discussion. Using Eq. (4.14), the pressure-velocity correlation is written as

$$\begin{aligned}\langle p'(\mathbf{x})\mathbf{u}'(\mathbf{x})\rangle &= \frac{1}{2\pi}\int_V \frac{1}{r}\frac{\partial U_i(\mathbf{x}')}{\partial x_j}\left\langle \mathbf{u}'(\mathbf{x})\frac{\partial u_j'(\mathbf{x}')}{\partial x_i}\right\rangle dV \\ &+\frac{1}{4\pi}\int_V \frac{1}{r}\left\langle \mathbf{u}'(\mathbf{x})\frac{\partial^2 u_i'(\mathbf{x}')u_j'(\mathbf{x}')}{\partial x_i \partial x_j}\right\rangle dV \\ &+\frac{1}{4\pi}\int_S \left(\frac{1}{r}\left\langle \mathbf{u}'(\mathbf{x})\frac{\partial p'(\mathbf{x}')}{\partial n}\right\rangle - \langle \mathbf{u}'(\mathbf{x})p'(\mathbf{x}')\rangle \frac{\partial}{\partial n}\left(\frac{1}{r}\right)\right) dS. \quad (4.15)\end{aligned}$$

The complicated behaviors of the pressure-related correlation occur in the following two respects. One is that the pressure-velocity correlation contains the spatial derivatives of $\mathbf{u}'$, unlike the triple velocity correlation. This fact implies that the smaller-scale components of $\mathbf{u}'$ become important. Another is that the pressure-velocity correlation is dependent on two locations $\mathbf{x}$ and $\mathbf{x}'$, and that it is the sum of the contributions from all $\mathbf{x}'$. Then we can easily understood the difficulty in modeling the pressure-velocity correlation using only one-point or local quantities like $\mathbf{U}$, K, etc. In Eq. (4.15), the third part expresses the contribution from a boundary. It may be considered small so long as the location $\mathbf{x}$ is not near a boundary. Since our main interest lies in the high-Reynolds-number version of turbulence models valid not close to a boundary, we shall neglect such boundary effects and pay special attention to the first two parts.

The relative importance of Eq. (4.15) to the triple velocity correlation $\langle(\mathbf{u}'^2/2)\mathbf{u}'\rangle$ is not clear because of its complicated structure. From the viewpoint of observations, the estimate of $\langle p'\mathbf{u}'\rangle$ is difficult since it cannot be measured directly. Its magnitude is usually inferred from Eq. (2.111) by measuring all the remaining terms. This estimate is indirect and is subject to the accumulating errors coming from the measurement of those terms. With this reservation in mind, it is usually considered that $\langle p'\mathbf{u}'\rangle$ is smaller than $\langle(\mathbf{u}'^2/2)\mathbf{u}'\rangle$. In Chapter 6, a theoretical support will be given to this estimate.

The triple velocity correlation $\langle(\mathbf{u}'^2/2)\mathbf{u}'\rangle$ may be regarded as the transfer rate of the instantaneous turbulent energy $\mathbf{u}'^2/2$ due to $\mathbf{u}'$. From the analogy with the foregoing momentum transfer, the turbulent energy is supposed to be transferred from the larger-K region to the smaller-K one. Then we write

$$\left\langle \frac{\mathbf{u}'^2}{2}\mathbf{u}'\right\rangle = -\frac{\nu_T}{\sigma_K}\nabla K. \quad (4.16)$$

Here ν_T/σ_K is the diffusivity of K, which corresponds to the momentum diffusivity or turbulent viscosity, ν_T. In general, σ_K is not a numerical factor, but it is often chosen as a positive constant.

From the viewpoint of energy transport, Eq. (4.16) seems to be reasonable. In addition to K, however, ε has been adopted as another quantity characterizing the state of turbulence. It is more natural to consider that the spatial nonuniformity of ε, $\nabla\varepsilon$, also has influence on the transport rate of K. In this sense, we should write

$$\left\langle \frac{\mathbf{u}'^2}{2}\mathbf{u}' \right\rangle = -\frac{\nu_T}{\sigma_K}\nabla K + \frac{K}{\varepsilon}\frac{\nu_T}{\sigma_{K\varepsilon}}\nabla\varepsilon, \tag{4.17}$$

where $\sigma_{K\varepsilon}$ is a positive coefficient similar to σ_K. The second part may be called the cross-diffusion effect.

In Eq. (4.17), the sign of the second term is inferred to be opposite to the first one. This point may be explained as follows. Using this expression, we have

$$\nabla\cdot\left\langle \frac{\mathbf{u}'^2}{2}\mathbf{u}' \right\rangle = -\frac{\nu_T}{\sigma_K}\Delta K + \frac{K}{\varepsilon}\frac{\nu_T}{\sigma_{K\varepsilon}}\Delta\varepsilon, \tag{4.18}$$

where only the highest-order derivatives of K and ε have been retained for simplicity of discussion. The physical meaning of Eq. (4.18) can be understood more easily in its finite-difference form. We discuss the two-dimensional case, but its extension to the three-dimensional case is straightforward. We take four points near the reference location 0, as in Fig. 4.1. Then the finite-difference form of Eq. (4.18) is given by

$$\begin{aligned}\left(\nabla\cdot\left\langle \frac{\mathbf{u}'^2}{2}\mathbf{u}' \right\rangle\right)_0 &= -\left(\frac{\nu_T}{\sigma_K}\right)_0 \frac{1}{h^2}\left(\sum_{i=1}^{4} K_i - 4K_0\right) \\ &\quad + \left(\frac{K}{\varepsilon}\frac{\nu_T}{\sigma_{K\varepsilon}}\right)_0 \frac{1}{h^2}\left(\sum_{i=1}^{4} \varepsilon_i - 4\varepsilon_0\right),\end{aligned} \tag{4.19}$$

where h is the distance between the point 0 and one of the surrounding four points. In the combination with the equation for K, (2.111), the first part of Eq. (4.19) indicates that the location 0 is supplied with the energy by the surrounding four points when K_0 is smaller than the average of the surrounding four K's. On the other hand, the second part decreases K_0 in case that ε_0 is smaller than the average of the surrounding four ε's. Considering that increasing ε tends to decrease K, this opposite role seems reasonable. In reality, the occurrence of the second part will be shown by a statistical theory in Chapter 6.

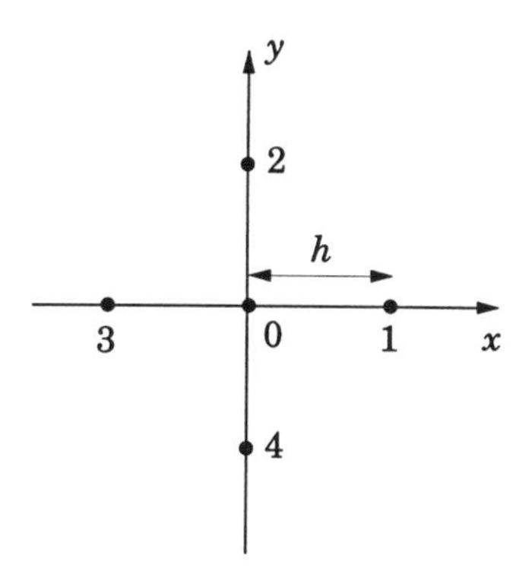

Figure 4.1. Two-dimensional grid points with equal distance.

Note: We give a formal solution of Eq. (4.11). We write it as

$$\Delta p' = F_p. \tag{4.20}$$

From Eqs. (4.12) and (4.20), we have

$$\begin{aligned} &G(\mathbf{x},\mathbf{x}')\Delta_{x'}p'(\mathbf{x}') - p'(\mathbf{x}')\Delta_{x'}G(\mathbf{x},\mathbf{x}') \\ &\quad = G(\mathbf{x},\mathbf{x}')F_p(\mathbf{x}') - p(\mathbf{x}')\delta(\mathbf{x}-\mathbf{x}'), \end{aligned} \tag{4.21}$$

with $\Delta_{x'} = \partial^2/\partial x_i'^2$. The left-hand side of Eq. (4.21) is rewritten as

$$\begin{aligned} &G(\mathbf{x},\mathbf{x}')\Delta_{x'}p'(\mathbf{x}') - p'(\mathbf{x}')\Delta_{x'}G(\mathbf{x},\mathbf{x}') \\ &= \nabla_{x'} \cdot \left(G(\mathbf{x},\mathbf{x}')\nabla_{x'}p'(\mathbf{x}') - p'(\mathbf{x}')\nabla_{x'}G(\mathbf{x},\mathbf{x}')\right). \end{aligned} \tag{4.22}$$

We integrate Eq. (4.21) in the region V, and apply the Green's integral theorem (2.6). Then we have

$$\begin{aligned} p'(\mathbf{x}) = &\int_V G(\mathbf{x},\mathbf{x}')F_p(\mathbf{x}')dV \\ &- \int_S \left(G(\mathbf{x},\mathbf{x}')\frac{\partial p'(\mathbf{x}')}{\partial n} - p'(\mathbf{x}')\frac{\partial G(\mathbf{x},\mathbf{x}')}{\partial n} \right) dS. \end{aligned} \tag{4.23}$$

Next, we consider a solution of Eq. (4.12). We take the origin of the coordinates at $\mathbf{x}'$, and write $\mathbf{s} = \mathbf{x} - \mathbf{x}'$ ($s = |\mathbf{s}|$). From the spherical symmetry of $G(s)$, Eq. (4.12) is reduced to

$$\frac{1}{s^2}\frac{d}{ds}s^2\frac{dG(s)}{ds} = \delta(\mathbf{s}). \tag{4.24}$$

We integrate Eq. (4.24) in a sphere of radius r, and have

$$4\pi r^2 \frac{dG(r)}{dr} = 1, \tag{4.25}$$

which leads to Eq. (4.13).

B. Modeling of Equation for ε

In the turbulence modeling of the K-ε type, the last equation to be given is the one for the dissipation rate ε [Eq. (2.113)]. From Eq. (2.104), we can derive the exact ε equation as

$$\frac{D\varepsilon}{Dt} = -2\nu\left\langle\frac{\partial u_i'}{\partial x_j}\frac{\partial u_i'}{\partial x_\ell}\frac{\partial u_j'}{\partial x_\ell}\right\rangle - 2\left\langle\left(\nu\frac{\partial^2 u_i'}{\partial x_j \partial x_\ell}\right)^2\right\rangle$$
$$-2\nu\left\langle\frac{\partial u_i'}{\partial x_\ell}\frac{\partial u_j'}{\partial x_\ell} + \frac{\partial u_\ell'}{\partial x_i}\frac{\partial u_\ell'}{\partial x_j}\right\rangle\frac{\partial U_i}{\partial x_j} - 2\nu\left\langle u_j'\frac{\partial u_i'}{\partial x_\ell}\right\rangle\frac{\partial^2 U_i}{\partial x_j \partial x_\ell}$$
$$+\frac{\partial}{\partial x_i}\left(-\nu\left\langle u_i'\left(\frac{\partial u_j'}{\partial x_\ell}\right)^2\right\rangle - 2\nu\left\langle\frac{\partial p'}{\partial x_j}\frac{\partial u_i'}{\partial x_j}\right\rangle\right) + \nu\Delta\varepsilon. \tag{4.26}$$

The mathematical structure of Eq. (4.26) is much more complicated, compared with the K equation (2.111). In the latter, the role of each part is clear, as was explained in Sec. 2.6.2.A. Entirely the same situation holds for the equation for the scalar variance $\langle\theta'^2\rangle$, (2.126), and the equation for the turbulent helicity $\langle\mathbf{u}'\cdot\boldsymbol{\omega}'\rangle$, (2.132). Their clear mathematical structures come from the fact that the total amounts of $\mathbf{u}^2$, θ^2, and $\mathbf{u}\cdot\boldsymbol{\omega}$ in a fluid region are conserved in the absence of molecular viscosity and diffusivity. On the other hand, ε does not possess such a counterpart. This is the reason for the complicatedness of Eq. (4.26).

We see the structure of Eq. (4.26) in more detail. We have already stated that the energy dissipation occurs at the spatial scale ℓ_D that is defined by Eq. (3.52b). Under the spatial scaling

$$|\mathbf{x}| = O(\nu^{3/4}\varepsilon^{-1/4}), \tag{4.27}$$

$\mathbf{u}'$ is scaled as

$$|\mathbf{u}'| = O(\nu^{1/4}\varepsilon^{1/4}). \tag{4.28}$$

Then each term of Eq. (4.26) is estimated as

$$\nu\left\langle\frac{\partial u_i'}{\partial x_j}\frac{\partial u_i'}{\partial x_\ell}\frac{\partial u_j'}{\partial x_\ell}\right\rangle = O(\varepsilon^{3/2}\nu^{-1/2}), \tag{4.29}$$

$$2\left\langle\left(\nu\frac{\partial^2 u_i'}{\partial x_j \partial x_\ell}\right)^2\right\rangle = O(\varepsilon^{3/2}\nu^{-1/2}), \tag{4.30}$$

$$\nu\left\langle\frac{\partial u_i'}{\partial x_\ell}\frac{\partial u_j'}{\partial x_\ell} + \frac{\partial u_\ell'}{\partial x_i}\frac{\partial u_\ell'}{\partial x_j}\right\rangle = O(\varepsilon), \tag{4.31}$$

$$\nu \left\langle u'_j \frac{\partial u'_i}{\partial x_\ell} \right\rangle = O(\nu^{3/4}\varepsilon^{3/4}). \tag{4.32}$$

The last two parts of Eq. (4.26) will be referred to below.

In the limit of high-Reynolds number, which corresponds to $\nu \to 0$, Eqs. (4.29) and (4.30) of $O(\nu^{-1/2})$ become infinite, whereas Eqs. (4.31) and (4.32) remain finite. Except the region close to a solid boundary, the spatial change of $\mathbf{U}$ is not large. Therefore the third and fourth parts of Eq. (4.26) are much smaller than the first two parts. This fact indicates that these two parts should cancel with each other at $O(\nu^{-1/2})$, and that the resulting contribution is of $O(\nu^0)$. This kind of discussion was first made by Tennekes and Lumley (1972) using the transport equation for the vorticity variance $\langle \boldsymbol{\omega}'^2 \rangle$. From Eq. (2.60), we can show

$$\left\langle \boldsymbol{\omega}'^2 \right\rangle = \left\langle \left(\epsilon_{ij\ell} \frac{\partial u'_\ell}{\partial x_j} \right)^2 \right\rangle = \left\langle \left(\frac{\partial u'_j}{\partial x_i} \right)^2 \right\rangle - \frac{\partial^2}{\partial x_i \partial x_j} \left\langle u'_i u'_j \right\rangle . \tag{4.33}$$

In homogeneous turbulence, the last term vanishes, and we have

$$\varepsilon = \nu \left\langle \boldsymbol{\omega}'^2 \right\rangle . \tag{4.34}$$

Therefore the equations for $\langle \boldsymbol{\omega}'^2 \rangle$ is identical to that for ε.

In the case of weakly anisotropic velocity fluctuations, the third term of Eq. (4.26) may be shown to be much smaller than the estimate based on Eq. (4.31). Using Eqs. (3.17) and (3.35), we have

$$\left\langle \frac{\partial u'_i}{\partial x_\ell} \frac{\partial u'_j}{\partial x_\ell} + \frac{\partial u'_\ell}{\partial x_i} \frac{\partial u'_\ell}{\partial x_j} \right\rangle = \frac{4}{3} \delta_{ij} \int k^2 Q(k;t,t) d\mathbf{k}, \tag{4.35}$$

where we should note

$$\int k_i k_j d\mathbf{k} = \frac{1}{3} \delta_{ij} \int k^2 d\mathbf{k}. \tag{4.36}$$

Then the third term vanishes because of the solenoidal condition $\nabla \cdot \mathbf{U} = 0$. This term, however, may become important near a solid boundary, where large mean velocity gradient and highly anisotropic fluctuations of velocity are encountered. In reality, the term becomes comparable to the first two terms near the wall in channel flow, as is shown by the direct numerical simulation (DNS) of Mansour et al. (1989). Of the last two terms in Eq. (4.26), the former is expressed in a divergence form and vanishes in homogeneous turbulence. In general, it does not contribute to the net generation and destruction of ε (the similar situation was discussed in Sec. 2.6.2.A in the context of the K equation). Therefore the role of the term may be considered small, compared with the first four parts. The last term of Eq. (4.26) comes from the viscous part of the Navier-Stokes equation, and it becomes important only near a solid wall.

From the foregoing discussions, the modeling of the ε equation based on the straightforward use of Eq. (4.26) needs the detailed knowledge about the fine-scale properties of velocity fluctuations that guarantee the exact cancellation of its first two terms at $O(\nu^{-1/2})$. This requirement is far from the original purpose of turbulence modeling based on one-point quantities. In it, we aim at estimating energetic parts of turbulent shear flow without delving into fine-scale components of motion.

In order to construct a model equation for ε without using Eq. (4.26), we need a different viewpoint of ε. In the equilibrium state of turbulence where the Kolmogorov scaling holds, ε is also regarded as the rate of energy injected by energy-containing eddies into the inertial-range counterparts. Such an equilibrium state ceases to hold in case that the energy injection rate changes rapidly, as in homogeneous-shear turbulence (this point was noted simply in Sec. 3.4.2). In a nonequilibrium case, the magnitude of ε is determined mainly by the amount of energy that is injected through the so-called production term P_K [Eq. (2.112)]. This fact implies that the primary part of ε may be estimated using the quantities related to the energetic parts of turbulent motion, such as $\mathbf{U}$, K, etc.

From the foregoing viewpoint, a model equation for ε, which is widely used in turbulence modeling, is written as

$$\frac{D\varepsilon}{Dt} = C_{\varepsilon 1}\frac{\varepsilon}{K}P_K - C_{\varepsilon 2}\frac{\varepsilon^2}{K} + \nabla \cdot \left(\frac{\nu_T}{\sigma_\varepsilon}\nabla\varepsilon\right), \tag{4.37}$$

where $C_{\varepsilon 1}$ and $C_{\varepsilon 2}$ are usually chosen to be numerical constants, and σ_ε is the counterpart of σ_K in Eq. (4.16). Equation (4.37) has essentially the same mathematical structure as the K equation (2.111). The first term is closely connected with the viewpoint that the magnitude of the energy dissipation rate or the energy transfer rate to fine-scale components is mainly dependent on that of the production rate of K. The second term, which is rewritten as $C_{\varepsilon 2}\varepsilon/(K/\varepsilon)$, corresponds to the second term of Eq. (2.111). It represents the destruction rate of ε per unit time since K/ε is the time scale characterizing energetic eddies, as is defined by Eq. (3.50). The last term expresses the diffusion effect. It is not guaranteed, however, that the effects of inhomogeneity of turbulence, such as $\nabla\varepsilon$, always occur in a divergence form similar to the last part of the K equation (2.111). Therefore we cannot rule out the possibility that the other terms related to ∇K and $\nabla\varepsilon$ are attached to the right-hand side in a nondivergent form. This point will be referred to later from the theoretical standpoint of Chapter 6.

4.2.3. MODEL CONSTANTS AND LOGARITHMIC-VELOCITY LAW

In the K-ε model presented in Secs. 4.2.1 and 4.2.2, we have five model constants; namely, they are

$$C_\nu, \ \sigma_K, \ C_{\varepsilon 1}, \ C_{\varepsilon 2}, \ \sigma_\varepsilon, \tag{4.38}$$

where only the first term of Eq. (4.17) has been adopted for the later discussion. We are in a position to examine the analytical properties of the model in some typical flow situations and discuss the relationship among these constants and their magnitude. In this course, we shall refer to the logarithmic-velocity law, which is one of the primary features in shear flow.

A. Grid Turbulence

As the simplest turbulent flow, we consider the so-called grid turbulence. It is generated by a uniform flow passing the screen normal to it that is composed of a number of thin rods perpendicularly intersecting one another. We pay attention to the flow region in which the features of the grid fade away, but the flow is turbulent enough for the K-ε model to be applicable. In this case, the mean velocity in the x direction, U, may be regarded as nearly constant. Then we write

$$\frac{dK}{d\tau} = -\varepsilon, \tag{4.39}$$

$$\frac{d\varepsilon}{d\tau} = -C_{\varepsilon 2}\frac{\varepsilon^2}{K}. \tag{4.40}$$

Here τ is defined as

$$\tau = x/U, \tag{4.41}$$

which is the time measured in the frame moving with the velocity U. Moreover, we have dropped the diffusion terms from the reason given below.

We put

$$\frac{K}{K_0} = \left(\frac{\tau}{\tau_0} - 1\right)^{-m}, \tag{4.42a}$$

$$\frac{\varepsilon}{\varepsilon_0} = \left(\frac{\tau}{\tau_0} - 1\right)^{-n}, \tag{4.42b}$$

where $\tau_0 = x_0/U$ (x_0 is the effective origin of x), and K_0 and ε_0 are the reference values of K and ε, respectively. We substitute Eq. (4.42) into Eqs. (4.39) and (4.40), and have

$$\frac{K_0}{\varepsilon_0 \tau_0} = \frac{1}{m}\left(\frac{\tau}{\tau_0} - 1\right)^{-n+m+1} = \frac{C_{\varepsilon 2}}{n}\left(\frac{\tau}{\tau_0} - 1\right)^{-n+m+1}. \tag{4.43}$$

Equation (4.43) gives

$$n = m + 1, \tag{4.44a}$$

$$C_{\varepsilon 2} = \frac{n}{m} = 1 + \frac{1}{m}. \tag{4.44b}$$

The turbulent viscosity ν_T, which is defined by Eq. (4.8), is written as

$$\nu_T \propto \frac{{K_0}^2}{\varepsilon_0} \left(\frac{\tau}{\tau_0} - 1 \right)^{-m+1}, \tag{4.45}$$

from Eqs. (4.42) and (4.44). Using Eq. (4.45), we may estimate the diffusion term in the K equation as

$$\frac{d}{dx} \left(\frac{\nu_T}{\sigma_K} \frac{dK}{dx} \right) \propto \left(\frac{\tau}{\tau_0} - 1 \right)^{-2m-1}. \tag{4.46}$$

As will be stated below, m is close to one. Then this term becomes much smaller for large τ/τ_0, compared with the terms in Eq. (4.39), and is negligible. The same discussion can be made on the diffusion term in the ε equation.

From Eq. (4.45), the turbulent Reynolds number R_T, which is given by Eq. (4.10), is written as

$$R_T \propto \frac{{K_0}^2}{\varepsilon_0 \nu} \left(\frac{\tau}{\tau_0} - 1 \right)^{-m+1}. \tag{4.47}$$

In the case of $m > 1$, R_T decreases with increasing x, and effects of low Reynolds number occur. Near the screen, the features of rods are alive. Therefore it is necessary to apply the model in the intermediate region. From observations, we have

$$m = 1.1 \sim 1.3 \tag{4.48}$$

(Comte-Bellot and Corrsin 1966), which gives

$$C_{\varepsilon 2} = 1.8 \sim 1.9. \tag{4.49}$$

The value 1.9 is often adopted in the K-ε model.

B. Channel Turbulence

As a typical bounded shear flow, we can mention a flow between two parallel plates, which is usually called channel turbulence. The mean velocity is written as

$$\mathbf{U} = (U(y), 0, 0), \tag{4.50}$$

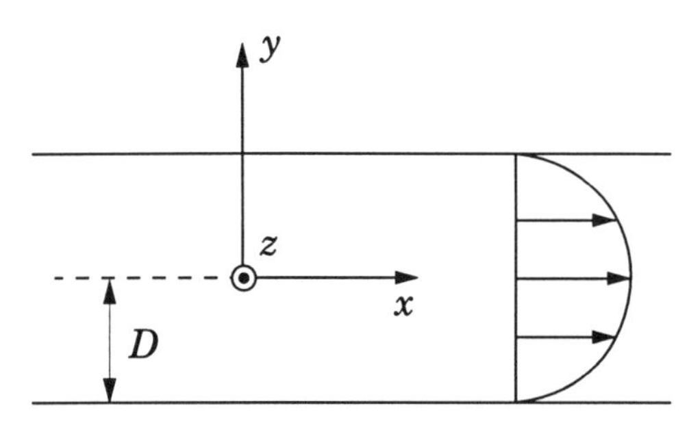

Figure 4.2. Channel turbulence.

and the half width of the channel is D (see Fig. 4.2). The mean velocity U obeys

$$\frac{d}{dy}\left(-R_{xy}+\nu\frac{dU}{dy}\right)+G=0, \tag{4.51}$$

where G is related to the pressure gradient as

$$G=-\frac{dP}{dx}, \tag{4.52}$$

which is taken as positive.

We integrate Eq. (4.51) from the lower to upper walls, and consider the symmetry of U. Then we have

$$\nu\left(\frac{dU}{dy}\right)_{y=-D}=GD. \tag{4.53}$$

This equation means the balance between the resistive force at the wall and the pressure force acting the cross section. Now we define the wall friction velocity

$$u_W\equiv\sqrt{\nu\frac{dU}{dy}_{y=-D}}=\sqrt{GD}. \tag{4.54}$$

Corresponding to Eq. (4.54), we introduce the coordinate in wall units, y_W, which is written as

$$y_W=u_W\left(y+D\right)/\nu\,. \tag{4.55}$$

Equation (4.55) may be interpreted as the Reynolds number based on u_W and the distance from the lower wall, $y+D$. These two wall-related quantities, u_W and y_W, play an important role in near-wall turbulence. Specifically, U in the close vicinity of the wall is easily seen to obey

$$U/u_W=y_W. \tag{4.56}$$

We integrate Eq. (4.51) from the lower wall to arbitrary y, and have

$$- R_{xy} + \nu \frac{dU}{dy} = -Gy. \tag{4.57}$$

The K-ε model discussed in Sec. 4.2.2 cannot be applied to the region where effects of ν become dominant. In the lower half of the channel, we consider the region specified as

$$(y + D)/D \ll 1, \ y_W \gg 1. \tag{4.58}$$

This condition singles out the region that is near the wall, but to which the high-Reynolds number version of a turbulence model is applicable. Under the condition (4.58), we can neglect the ν-related term in Eq. (4.57), and have

$$R_{xy} = -u_W^2. \tag{4.59}$$

In channel turbulence, the K equation may be written as

$$- R_{xy} \frac{dU}{dy} - \varepsilon + \frac{d}{dy} \left(\frac{\nu_T}{\sigma_K} \frac{dK}{dy} \right) = 0, \tag{4.60}$$

from Eqs. (2.111) and (4.16), where

$$R_{xy} = -C_\nu \frac{K^2}{\varepsilon} \frac{dU}{dy}. \tag{4.61}$$

In the following discussion, we drop the third diffusion term in Eq. (4.60) (this point will be referred to below). Equation (4.60) is combined with Eq. (4.59) to give

$$\frac{dU}{dy} = \frac{\varepsilon}{u_W^2}, \tag{4.62}$$

whereas the turbulent-viscosity model, Eq. (4.61), leads to

$$\frac{dU}{dy} = \frac{u_W^2 \varepsilon}{C_\nu K^2}. \tag{4.63}$$

From Eqs. (4.62) and (4.63), we have

$$\frac{K}{u_W^2} = \frac{1}{\sqrt{C_\nu}}. \tag{4.64}$$

Namely, K is nearly constant in the region specified by the condition (4.58). From observations, we know that

$$|R_{xy}|/K \cong 0.3, \tag{4.65}$$

in the region with which we are concerned. Then we have

$$C_\nu = 0.09, \tag{4.66}$$

from Eq. (4.64). This is the value that is widely used as C_ν in the K-ε model.

From Eq. (3.48), $K^{3/2}/\varepsilon$ is the length characterizing energy-containing parts of velocity fluctuations. Under the condition (4.58), the half width of the channel, D, is not a proper length for the scaling of $K^{3/2}/\varepsilon$, and the distance from the wall, $y + D$, is more relevant. This fact combined with Eq. (4.64) gives

$$\varepsilon = \frac{1}{\kappa} \frac{u_W^3}{y + D}. \tag{4.67}$$

Here κ is called the Karman constant, whose meaning will become clear below.

We substitute Eq. (4.67) into Eq. (4.62), and have

$$\frac{U}{u_W} = \frac{1}{\kappa} \log y_W + A. \tag{4.68}$$

This is the well-known logarithmic velocity profile, and is ubiquitous in bounded turbulent flow. From observations (Hussain and Reynolds 1975), we have

$$\kappa \cong 0.4. \tag{4.69}$$

Another factor A, which corresponds to an integral constant, depends on the details of the wall surface, such as smooth or rough. In the case of a smooth wall, we have $A \cong 5.0$. Two typical velocity profiles, Eqs. (4.56) and (4.68), are given in Fig. 4.3.

The remaining equation (4.37) for ε is reduced to

$$- C_{\varepsilon 1} \frac{\varepsilon}{K} R_{xy} \frac{dU}{dy} - C_{\varepsilon 2} \frac{\varepsilon^2}{K} + \frac{d}{dy} \left(\frac{\nu_T}{\sigma_\varepsilon} \frac{d\varepsilon}{dy} \right) = 0. \tag{4.70}$$

In Eq. (4.60), we neglected the diffusion term, but we cannot drop the counterpart of Eq. (4.70). In its absence, the consistency between these two transport equations requires $C_{\varepsilon 1} = C_{\varepsilon 2}$, resulting in the loss of the independence between the two equations. Therefore we retain the diffusion term in Eq. (4.70), and substitute Eqs. (4.64), (4.67), and (4.68). As a result, we find the relationship among the model constants

$$\frac{1}{\sigma_\varepsilon} + (C_{\varepsilon 1} - C_{\varepsilon 2}) \frac{\sqrt{C_\nu}}{\kappa^2} = 0. \tag{4.71}$$

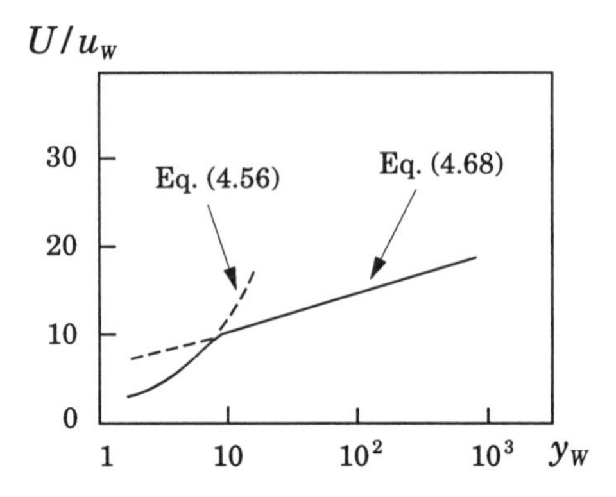

Figure 4.3. Mean velocity along a smooth solid wall.

At the centerline of the channel, dU/dy vanishes in the K equation (4.60), and the dissipation term needs to balance with the diffusion term. The similar situation holds in the ε equation (4.70), and the importance of diffusion effects become prominent there. No definite mathematical relation has not been found between σ_K and σ_ε. Both of them are usually chosen to be about one. Finally, a widely-used set of model constants is written as

$$C_\nu = 0.09, \ \sigma_K = 1, C_{\varepsilon 1} = 1.4, \ C_{\varepsilon 2} = 1.9, \ \sigma_\varepsilon = 1.3. \tag{4.72}$$

4.2.4. ASYMPTOTIC SOLUTION OF K-ε MODEL

A. Log-Law-Type Solution

In Sec. 4.2.3.B, we derived the logarithmic-velocity profile as a local solution of the K-ε model, and discussed the relationship among the model constants. In addition to its importance as a physical law of shear flow, it often plays a role of an artificial boundary condition in the numerical computation of a turbulent flow using turbulence models. In a wall-bounded flow, a number of grid points need to be taken near a solid boundary requiring the noslip velocity condition. The application of the condition to complicated real-world flows is very difficult because of the increasing computational burden. In order to overcome this difficulty, we often replace the noslip boundary condition with the analytic near-wall properties represented by the logarithmic-velocity law.

At the stage of Sec. 4.2.3.B, we cannot say that the logarithmic-velocity law is derived on a mathematically firm basis. For, the diffusion effect is dropped in the K equation (4.60), whereas it is retained in the ε equation (4.70) for the consistency of the logarithmic-velocity law with the K-ε model. In order to remove this ambiguity, we need to derive the law in a more well-defined manner; namely, it is important to relax the constraint (4.58) and show in what asymptotic series expansion the logarithmic-velocity law is given as the leading term. This problem was discussed by Takemitsu (1987, 1990) as follows.

In a channel flow, we seek an asymptotic series solution of the K-ε model near a wall except the close vicinity where effects of ν become dominant. For this purpose, we write

$$U\,/(u_W/\kappa) = \log y_W + \kappa A + \sum_{n=1}^{\infty} a_{Un} y_D^n, \tag{4.73}$$

$$K \Big/ \left(u_W^2/\sqrt{C_\nu}\right) = 1 + \sum_{n=1}^{\infty} a_{Kn} y_D^n, \tag{4.74}$$

$$\varepsilon \Big/ \left(u_W^3/(\kappa D)\right) = y_D^{-1}\left(1 + \sum_{n=1}^{\infty} a_{\varepsilon n} y_D^n\right), \tag{4.75}$$

where $y_D (= (y+D)/D)$ is the nondimensionalized distance from the lower wall (see Fig. 4.2). Moreover, we introduce the auxiliary expansion for

$$\nu_T\,/(\kappa u_W D) = y_D\left(1 + \sum_{n=1}^{\infty} a_{\nu n} y_D^n\right). \tag{4.76}$$

We substitute Eqs. (4.73)-(4.76) into the foregoing K-ε model, that is, Eqs. (4.51), (4.60), and (4.70) [the ν-related part in Eq. (4.51) is dropped). For the modeling of the transport term, however, we adopt Eq. (4.17) in place of the ordinary one, Eq. (4.16). The reason will become clear soon. As a result, Eq. (4.60) is replaced with

$$-R_{xy}\frac{dU}{dy} - \varepsilon + \frac{d}{dy}\left(\frac{\nu_T}{\sigma_K}\frac{dK}{dy}\right) - \frac{d}{dy}\left(\frac{\nu_T}{\sigma_{K\varepsilon}}\frac{K}{\varepsilon}\frac{d\varepsilon}{dy}\right) = 0. \tag{4.77}$$

This calculation is very tedious, but the final expressions can be written compactly using

$$\alpha = \kappa^2 \Big/ \sqrt{C_\nu}\,,\ \beta = C_{\varepsilon 2}/C_{\varepsilon 1}\,,\ \gamma = 1/(C_{\varepsilon 1}\sigma_\varepsilon)\,. \tag{4.78}$$

Among these three quantities, we have the important relation

$$\alpha\gamma + 1 - \beta = 0, \tag{4.79}$$

which corresponds to Eq. (4.71).

The first-order solution is given by

$$a_{U1} = -a_{\nu 1} - 2, \tag{4.80a}$$

$$a_{K1} = \frac{2\left(2-(1-\beta)\,a_{\varepsilon 1}\right)}{\beta - 3}, \tag{4.80b}$$

$$a_{\varepsilon 1} = \frac{2\left((\beta-1)\sigma_K - \alpha\left(1+3\sigma_K\delta\right)\right)}{(\beta-1)\left(\alpha - 2\sigma_K\right) + \alpha\beta\sigma_K\delta}, \tag{4.80c}$$

$$a_{\nu 1} = 2a_{K1} - a_{\varepsilon 1}, \tag{4.80d}$$

where δ is defined as

$$\delta = 1/\sigma_{K\varepsilon}\,, \tag{4.81}$$

which is related to the cross-diffusion term in Eq. (4.17). In the limit of $\delta \to 0$, we recover the corresponding solution of the usual K-ε model.

Similarly, the second-order solution may be written as

$$a_{U2} = \left(a_{\nu 1}\left(a_{\nu 1}+2\right) - a_{\nu 2}\right)/2\,, \tag{4.82a}$$

$$a_{K2} = \frac{(\alpha\gamma+1-\beta)\sigma_K\left((1-2\alpha\delta)\chi_1+\chi_2\right) + 3\alpha\sigma_K\delta\left((\alpha\gamma+1)\chi_1+\chi_3\right)}{2(\alpha\gamma+1-\beta)\left(2\alpha-\sigma_K\right) - 3\alpha(\beta+1)\sigma_K\delta}, \tag{4.82b}$$

$$a_{\varepsilon 2} = -\frac{(1-2\alpha\delta)\sigma_K\chi_1 + \sigma_K\chi_2 + 2\sigma_K - 2\alpha\left(2+3\sigma_K\delta\right)}{6\alpha\sigma_K\delta}, \tag{4.82c}$$

$$a_{\nu 2} = 2a_{K2} - a_{\varepsilon 2} + \chi_1, \tag{4.82d}$$

where

$$\chi_1 = \left(a_{K1} - a_{\varepsilon 1}\right)^2, \tag{4.83a}$$

$$\chi_2 = -\left(a_{\nu 1}+2\right)^2 - 2\alpha\left(\frac{a_{K1}a_{\nu 1}}{\sigma_K} + \delta\left(a_{\nu 1}-a_{\varepsilon 1}\right)\left(a_{\nu 1}-a_{\varepsilon 1}\right)\right), \tag{4.83b}$$

$$\chi_3 = -\left(a_{\nu 1}+2\right)^2 - \left(a_{K1}-a_{\varepsilon 1}\right)\left(a_{K1}+a_{\nu 1}+4-\beta\left(a_{K1}-a_{\varepsilon 1}\right)\right). \tag{4.83c}$$

In the solution (4.80), there is nothing unusual. On the other hand, the denominator of Eq. (4.82c) is linearly dependent on δ defined by Eq. (4.81). The familiar transport model in the K-ε model, Eq. (4.16), corresponds to the case of vanishing δ, and Eq. (4.82c) diverges. Namely, the standard K-ε model does not permit an analytic solution of the type (4.73)-(4.75), and we encounter a serious logical difficulty in using the logarithmic velocity profile as a substitute of the wall boundary condition. The actual numerical computation of the K-ε model with the logarithmic-velocity law as the boundary condition is not hindered by the difficulty such as the numerical divergence. The inconsistency is considered to be masked by the round-off numerical errors. This fact does not eliminate the mathematical unclearness concerning the use of the so-called log-law as the boundary condition.

In order to avoid the foregoing mathematical difficulty, we have two paths. One is to seek a different type of solution with the logarithmic-velocity law as the leading parts. Such a solution has not been found yet as far as the author is aware. Another path is to include the cross-diffusion or ε-diffusion effect in the K equation, as is done here. In case that the spatial distribution of ε is similar to the K counterpart, the cross-diffusion effect may be absorbed effectively into the K-diffusion term. However, there are various kinds of turbulent flows in which the spatial distribution of K and ε is not so similar. In such a case, the use of the cross-diffusion term becomes useful [for instance, we may mention a backward-facing step flow, as was pointed out by Kobayashi and Togashi (1996)].

B. Low-Friction-Velocity Solution

Let us consider a more general flow situation near a wall boundary, compared with channel turbulence. We take the x axis along the wall and the y axis to normal to it. In this case, the mean velocity U is also specified by Eq. (4.50), and obeys Eq. (4.51). We integrate it, and have

$$-R_{xy} + \nu \frac{dU}{dy} = \Pi y + u_W^2, \tag{4.84}$$

where $\Pi(= -G)$ is the pressure gradient itself, that is,

$$\Pi = \frac{dP}{dx} \tag{4.85}$$

[u_W is defined by the first relation of Eq. (4.54)]. As is seen from Eq. (4.59), the logarithmic-velocity law corresponds to the condition

$$u_W^2 \gg |\Pi y| . \tag{4.86}$$

In the case of positive Π or an adverse pressure gradient, the separation of turbulent boundary flow may occur. In this case, u_W may vanish locally or become small, and the condition (4.86) is invalidated. Then we need to consider the opposite case

$$\Pi y \gg u_W^2. \tag{4.87}$$

In the close vicinity of a separation point, the mean velocity normal to the wall, V, becomes large, and the mean velocity profile given by Eq. (4.50) is not proper. Therefore the following discussion is not applicable to such a region. The investigation based on Eq. (4.50), however, may capture some important flow properties in a region not very close to the separation point. In fact, the flow behaviors discussed below have been already detected observationally (Stratford 1959). Under the condition (4.87), we have

$$R_{xy} = -\Pi y, \tag{4.88}$$

from Eq. (4.84). Here the effect of ν was dropped since the close vicinity of the wall is excluded.

We use Eqs. (4.77) and (4.70) as the K and ε equations, respectively, and examine a solution of the modified K-ε model in the form

$$U = U_S + c_U \Pi^{1/2} y^{1/2}, \tag{4.89}$$

$$K = c_K \Pi y, \tag{4.90}$$

$$\varepsilon = c_\varepsilon \Pi^{3/2} y^{1/2}, \tag{4.91}$$

where c_U etc. are numerical factors (they may be regarded as weakly dependent on x in the sense that a local solution is sought). The relationship of this solution with the K-ε model was examined by Takemitsu (1989), who sought a substitute of the log-law as a numerical boundary condition near a separation point. From Eq. (4.8) for ν_T, we have

$$\nu_T = \left(C_\nu c_K^2 / c_\varepsilon \right) \Pi^{1/2} y^{3/2}. \tag{4.92}$$

We substitute Eqs. (4.89)-(4.92) into Eqs. (4.88), (4.77), and (4.70). The dependence on Π and y is satisfied identically, and the following relationship among the model constants are obtained:

$$2c_\varepsilon - c_K^2 c_U C_\nu = 0; \tag{4.93a}$$

$$2c_U c_\varepsilon - 4c_\varepsilon^2 + 3c_K^3 C_\nu \left(\frac{2}{\sigma_K} - \frac{1}{\sigma_{K\varepsilon}} \right) = 0; \tag{4.93b}$$

$$c_K^3 C_\nu + \left(c_U c_\varepsilon C_{\varepsilon 1} - 2c_\varepsilon^2 C_{\varepsilon 2} \right) \sigma_\varepsilon = 0. \tag{4.93c}$$

In the absence of U_S in Eq. (4.89), c_U was estimated by Stratford (1959) using a mixing-length theory as

$$c_U = 2/\kappa, \tag{4.94}$$

where κ is the Karman constant given by Eq. (4.69).

4.2.5. SHORTFALLS OF TURBULENT-VISCOSITY MODELING

A. Features of Model

A prominent feature of the turbulent-viscosity approximation to the Reynolds stress R_{ij}, Eq. (4.3), is the coincidence of the principal axis of the Reynolds-stress tensor with the mean-velocity-strain (S_{ij}) counterpart (recall that a symmetric matrix can be diagonalized under the rotation of the coordinate system). The property originates in the fact that Eq. (4.3) is generated by the interaction between the mean velocity and isotropic velocity fluctuation (this point will be clear from the statistical analyses in Chapters 6 and 7). Therefore the effect of anisotropy that Eq. (4.3) can represent is weak. In order to see the contribution of Eq. (4.3) to each turbulent intensity, we consider a channel flow discussed in Sec. 4.2.3.B. Under Eq. (4.3), we are led to the isotropic state

$$\langle u'^2 \rangle = \langle v'^2 \rangle = \langle w'^2 \rangle = (2/3)K. \tag{4.95}$$

The observation of channel flow, however, shows the strong anisotropy of intensities. In the application to the flow, Eq. (4.3) can reproduce U consistent with observations. This is due to the fortunate situation that only the shear-stress component of R_{ij}, R_{xy}, is linked with the U equation, as is seen from Eq. (4.51).

Another feature of Eq. (4.3) is the independence of mean rotational motion. We consider a more general expression for R_{ij}, in which the principal axis of R_{ij} deviates from that of S_{ij}. Namely, we write

$$R_{ij} = \frac{2}{3} K \delta_{ij} - \left[\nu_{ij\ell m} \frac{\partial U_m}{\partial x_\ell} \right]_D , \tag{4.96}$$

where $[A_{ij}]_D$ means the deviatoric part of A_{ij},

$$[A_{ij}]_D = A_{ij} - \frac{1}{3} A_{\ell\ell} \delta_{ij}, \tag{4.97}$$

and $\nu_{ij\ell m}$ obeys the constraint

$$\nu_{ij\ell m} = \nu_{ji\ell m}. \tag{4.98}$$

Using the mean counterpart of Eq. (2.51),

$$\frac{\partial U_j}{\partial x_i} = \frac{1}{2} \left(S_{ij} + \Omega_{ij} \right), \tag{4.99}$$

we rewrite Eq. (4.96) as

$$R_{ij} = \frac{2}{3} K \delta_{ij} - \frac{1}{2} \left[\nu_{ij\ell m} \left(S_{\ell m} + \Omega_{\ell m} \right) \right]_D . \tag{4.100}$$

Here S_{ij} is given by Eq. (4.4), and the mean vorticity tensor Ω_{ij} is defined by

$$\Omega_{ij} = \frac{\partial U_j}{\partial x_i} - \frac{\partial U_i}{\partial x_j}. \tag{4.101}$$

Any isotropic tensor can be written in terms of δ_{ij}, $\epsilon_{ij\ell}$ [the alternating tensor defined by Eq. (2.59)], and scalars. The isotropic form of $\nu_{ij\ell m}$ is given by

$$\nu_{ij\ell m} = \left(\nu_{ij\ell m} \right)_0 \equiv \nu_T \left(\delta_{i\ell} \delta_{jm} + \delta_{im} \delta_{j\ell} \right) + \nu'_T \delta_{ij} \delta_{\ell m}. \tag{4.102}$$

Then Eq. (4.100) results in Eq. (4.3), and the effect of mean rotational motion disappears. A method of retaining such an effect is to incorporate anisotropic effects into $\nu_{ij\ell m}$. This point will be discussed in the higher-order algebraic modeling of Sec. 4.3.

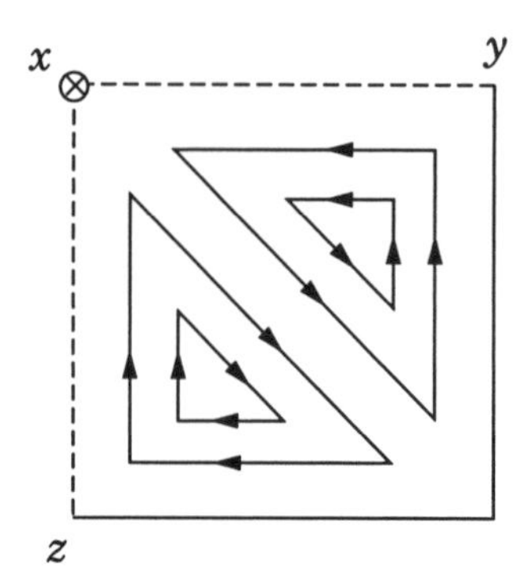

Figure 4.4. Secondary flows in the one-fourth cross section of a square duct.

B. Flows beyond the Scope of Turbulent-Viscosity Model
In order to see the shortfalls of the turbulent-viscosity model clearly, we consider a few typical flows whose qualitative behaviors are entirely beyond its reach.

B.1. Flow in a Square Duct. We consider a square-duct flow driven by a pressure gradient. Here we are concerned with its fully-developed state in which turbulence properties are homogeneous in the axial (x) direction. The most important feature of the flow is the appearance of secondary flows in the cross section, as is depicted schematically in Fig. 4.4. The magnitude of secondary flows is about one percent of the axial velocity. These flows, however, are important in the sense that they have great influence on the axial velocity profile.

The mean velocity in the fully-developed state is

$$\mathbf{U} = (U(y,z), V(y,z), W(y,z)) \,. \tag{4.103}$$

The occurrence of the secondary flows leads to nonvanishing Ω_x. From Eq. (2.100) for $\mathbf{U}$, Ω_x obeys

$$V\frac{\partial \Omega_x}{\partial y} + W\frac{\partial \Omega_x}{\partial z} = \frac{\partial^2}{\partial y \partial z}\left(R_{yy} - R_{zz}\right) - \left(\frac{\partial^2}{\partial y^2} - \frac{\partial^2}{\partial z^2}\right) R_{yz} + \nu \Delta \Omega_x. \tag{4.104}$$

Under Eq. (4.103), Eq. (4.3) gives

$$R_{yy} = \frac{2}{3}K - 2\nu_T \frac{\partial V}{\partial y}, \tag{4.105a}$$

$$R_{zz} = \frac{2}{3}K - 2\nu_T \frac{\partial W}{\partial z}, \tag{4.105b}$$

$$R_{yz} = -\nu_T \left(\frac{\partial W}{\partial y} + \frac{\partial V}{\partial z} \right). \tag{4.105c}$$

Then we have

$$R_{yy} - R_{zz} = -2\nu_T \left(\frac{\partial V}{\partial y} - \frac{\partial W}{\partial z} \right). \tag{4.106}$$

The state of no secondary flows is specified by

$$V = W = \Omega_x = 0, \tag{4.107}$$

which is consistent with Eq. (4.104) subject to Eqs. (4.105c) and (4.106). This conclusion is specifically related to Eq. (4.106). Under the condition (4.107), we have $R_{yy} = R_{zz}$, as is entirely similar to channel flow. Observationally, however, the turbulent intensities are anisotropic, resulting in $R_{yy} \neq R_{zz}$. This anisotropy does not allow the state given by Eq. (4.107) and generates secondary flows.

B.2. Swirling Flow in a Circular Pipe. The noticeable property of the turbulent-viscosity model (4.3) is the enhancement of the momentum diffusion. This point can be easily seen from the fact that in Eq. (2.100) for $\mathbf{U}$, ν is replaced with $\nu + \nu_T$. Flattening of the mean velocity profiles in channel and pipe flows, compared with the laminar counterparts, comes from this effect. Physically speaking, the turbulent viscosity breaks the specialty of flow properties at a location and leads to their spatial uniformity.

A typical instance in which a clear flow structure is maintained in turbulence is a swirling flow in a circular pipe (Kitoh 1991). The swirling motion whose rotational axis is in the axial or z direction is given at the entrance of the pipe. In the presence of swirling motion, the maximum-velocity point is off the central axis. The velocity profile is depicted schematically in Fig. 4.5, in the comparison with the no-swirling case. With the further increasing of entrance swirling, the flow direction reverses near the central axis. Weakening of the swirling motion in the downstream leads to the usual flat velocity profile. This fact indicates that swirling is crucial in keeping the velocity profile from flattening.

The turbulent-viscosity model cannot explain the foregoing feature of a swirling flow. In the numerical computation based on the model, the swirling given at the entrance is lost much more rapidly, compared with observations, resulting in the loss of the velocity profile with the local minimum at the central axis. The incapability of Eq. (4.3) to cope with a swirling flow arises from the fact that the model has no direct dependence on the mean vorticity $\mathbf{\Omega}$.

We assume that R_{ij} contains the terms linearly dependent on $\mathbf{\Omega}$; namely, we write

$$R_{ij} = (R_{ij})_{TV} + [\eta_i \Omega_j + \eta_j \Omega_i]_D, \tag{4.108}$$

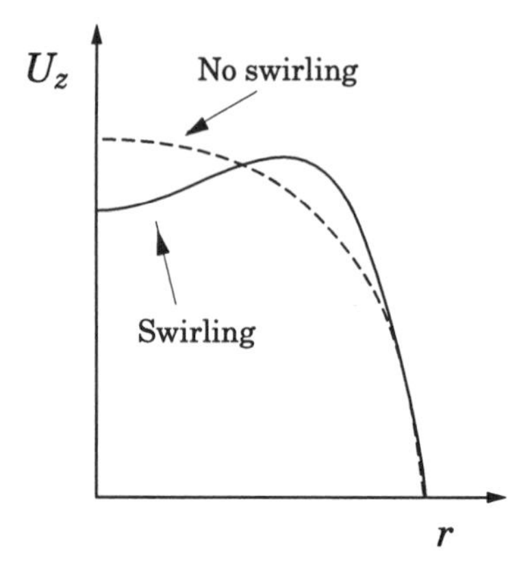

Figure 4.5. Axial velocity in a swirling pipe flow.

where $(R_{ij})_{TV}$ is the turbulent-viscosity part of R_{ij}, and $\boldsymbol{\eta}$ is an arbitrary vector. Under the reflection of the coordinate system, $\mathbf{x} \to -\mathbf{x}$, neither R_{ij} nor $\boldsymbol{\Omega}$ changes the sign. Therefore the sign of $\boldsymbol{\eta}$ should not change, unlike a usual or polar vector such as velocity. Such a vector is called an axial vector. For $\boldsymbol{\eta}$, we take a special form

$$\boldsymbol{\eta} = \nabla \psi, \tag{4.109}$$

where ψ is a scalar. The above requirement for $\boldsymbol{\eta}$ signifies that ψ is a pseudo-scalar, that is, the scalar changing its sign under the reflection. A typical example of pseudo-scalars is the helicity referred to in Secs. 2.5.3 and 2.6.2.C, specifically, its turbulent part $\langle \mathbf{u}' \cdot \boldsymbol{\omega}' \rangle$. The relationship with $\boldsymbol{\eta}$ will be examined in Chapter 6.

4.3. Higher-Order Algebraic Modeling

The simple, linear turbulent-viscosity model for the Reynolds stress can capture some of important properties of turbulent shear flow such as the logarithmic-velocity law. Its essence lies in the tendency to destroy locally distinct structures of a flow. As this result, the model fails in explaining the sustainment of ordered structures in turbulence. In order to alleviate the shortfall, we need to add some missing important effects into the model. The models rectified from this viewpoint are usually called higher-order models of the turbulent-viscosity type. In this section, we shall discuss this point.

4.3.1. LOWEST-ORDER NONLINEAR MODEL

We saw from Eqs. (4.100) and (4.102) that the turbulent-viscosity model, which is linear in S_{ij}, arises from the interaction of an isotropic turbulence state with the mean velocity gradient. One of the representative missing effects in the model is the nonlinear effect of the mean velocity gradient, that is, S_{ij} and Ω_{ij}. These two quantities are the primary factors generating and sustaining an anisotropic state of turbulence, and are appropriate for bringing anisotropic characteristics into the turbulent-viscosity model.

A. Simple Extension of Turbulent-Viscosity Model

We consider Eq. (4.100) from the standpoint of extending a linear model to a nonlinear one with the help of S_{ij} and Ω_{ij}. Up to the first order of S_{ij} and Ω_{ij}, $\nu_{ij\ell m}$ is given by

$$\nu_{ij\ell m} = (\nu_{ij\ell m})_0 + (\nu_{ij\ell m})_1, \tag{4.110}$$

where $(\nu_{ij\ell m})_0$ is defined by Eq. (4.102), and the first-order part $(\nu_{ij\ell m})_1$ is written as

$$\begin{aligned}(\nu_{ij\ell m})_1 = {} & \xi_1 (\delta_{i\ell} S_{mj} + \delta_{j\ell} S_{mi}) + \xi_2 (\delta_{im} S_{\ell j} + \delta_{jm} S_{\ell i}) \\ & + \xi_3 (\delta_{i\ell} \Omega_{mj} + \delta_{j\ell} \Omega_{mi}) + \xi_4 (\delta_{im} \Omega_{\ell j} + \delta_{jm} \Omega_{\ell i}) \\ & + \xi_5 \delta_{ij} S_{\ell m} + \xi_6 \delta_{ij} \Omega_{\ell m},\end{aligned} \tag{4.111}$$

where ξ_n's are the dimensional factors to be referred to below.

We substitute Eq. (4.111) into Eq. (4.100), and have

$$R_{ij} = (R_{ij})_1 + (R_{ij})_2 . \tag{4.112}$$

Here $(R_{ij})_1$ is the turbulent-viscosity model linear in S_{ij}, Eq. (4.3). On the other hand, $(R_{ij})_2$ expresses the lowest-order nonlinear effects concerning S_{ij} and Ω_{ij}, and is given by

$$\begin{aligned}(R_{ij})_2 = & - (\xi_1 + \xi_2) [S_{i\ell} S_{\ell j}]_D + \frac{1}{2} (\xi_1 - \xi_2 + \xi_3 - \xi_4) (S_{i\ell} \Omega_{\ell j} + S_{j\ell} \Omega_{\ell i}) \\ & - (\xi_3 - \xi_4) [\Omega_{i\ell} \Omega_{\ell j}]_D ,\end{aligned} \tag{4.113}$$

where the contributions from the last two terms in Eq. (4.111) have been absorbed into the first term in Eq. (4.100). More general nonlinear models were derived by Yoshizawa (1984) using a statistical method called a two-scale direct-interaction approximation (TSDIA) and by Speziale (1987, 1991) using a rational-mechanics approach. Equation (4.113) is written more simply as

$$(R_{ij})_2 = \zeta_1 [S_{i\ell} S_{\ell j}]_D + \zeta_2 (S_{i\ell} \Omega_{\ell j} + S_{j\ell} \Omega_{\ell i}) + \zeta_3 [\Omega_{i\ell} \Omega_{\ell j}]_D . \tag{4.114}$$

In the modeling of the K-ε type, the coefficients ζ_n's are

$$\zeta_n = C_{\zeta n}\left(K^3 \big/ \varepsilon^2\right), \tag{4.115}$$

where $C_{\zeta n}$'s are model constants. In a simplified version of Speziale's (1987) model corresponding to Eq. (4.114), they are

$$C_{\zeta 1} = 0.014,\ C_{\zeta 2} = 0.014,\ C_{\zeta 3} = 0. \tag{4.116}$$

The last of Eq. (4.116) means the disappearance of the Ω-Ω effect. This point will be explained in detail below.

B. Anisotropy Due to Nonlinear Effects

In what follows, we shall apply Eq. (4.114) to some fundamental turbulent flows and examine the anisotropic effects coming from it. In this course, some constraints on the model constants will be obtained.

B.1. Isotropic Turbulence in a Rotating Frame. In Sec. 2.5.2, we referred to the noninertial frame rotating with the angular velocity $\mathbf{\Omega}_F$. Equation (2.79b) indicates that the mean vorticity tensor Ω_{ij} should obey the transformation rule

$$\Omega_{ij} \to \Omega_{Tij} \equiv \Omega_{ij} + 2\epsilon_{ij\ell}\Omega_{F\ell}, \tag{4.117}$$

where Ω_{Tij} is called the total mean vorticity tensor. Then Eq. (4.114) may be rewritten as

$$\left(R_{ij}\right)_2 = \zeta_1 \left[S_{i\ell}S_{\ell j}\right]_D + \zeta_2 \left(S_{i\ell}\Omega_{T\ell j} + S_{j\ell}\Omega_{T\ell i}\right) + \zeta_3 \left[\Omega_{Ti\ell}\Omega_{T\ell j}\right]_D. \tag{4.118}$$

Speziale (1997) applied Eq. (4.118) to a decaying turbulence that is isotropic at the initial time, and drew the interesting conclusion that ζ_3 should vanish. First, we consider a decaying isotropic turbulence in an inertial frame. There K and ε obey Eqs. (4.39) and (4.40). We define the characteristic time τ_C

$$\tau_C = K/\varepsilon. \tag{4.119}$$

Then τ_C obeys the simple equation

$$\frac{d\tau_C}{dt} = C_{\varepsilon 2} - 1, \tag{4.120}$$

which leads to

$$\tau_C = \tau_{C0} + \left(C_{\varepsilon 2} - 1\right) t \tag{4.121}$$

(τ_{C0} is the initial value of τ_C). The characteristic time of decaying turbulence increases with time since $C_{\varepsilon 2} > 1$ [see Eq. (4.72)]. It is shown numerically (Bardina et al. 1987; Mansour et al. 1992; Shimomura 1997) and observationally (Jacquin et al. 1990) that ε tends to be suppressed in a rotating frame. As this result, τ_C becomes much larger, compared with the non-rotating case.

In the rotating frame with the angular velocity

$$\boldsymbol{\Omega}_F = (0, 0, \Omega_{F3}), \tag{4.122}$$

we have

$$S_{ij} = 0, \ \ \Omega_{Tij} = 2\epsilon_{ij3}\Omega_{F3}. \tag{4.123}$$

Equations (4.3) and (4.118) give

$$R_{ij} = \frac{2}{3}K\delta_{ij} + 4\zeta_3\left(\delta_{i3}\delta_{j3} - \frac{1}{3}\delta_{ij}\right)\Omega_{F3}^2, \tag{4.124}$$

where Eq. (2.60) has been used. As a quantity indicating the anisotropy of turbulence, we introduce

$$b_{ij} = \left(R_{ij} - \frac{2}{3}K\delta_{ij}\right)/K, \tag{4.125}$$

which vanishes in the isotropic case. Using Eqs. (4.115), (4.119), and (4.124), we have

$$b_{ij} = 4C_{\zeta 3}\left(\delta_{i3}\delta_{j3} - \frac{1}{3}\delta_{ij}\right)\Omega_{F3}^2\tau_C^2, \tag{4.126}$$

which gives

$$b_{11} = -(4/3)C_{\zeta 3}\Omega_{F3}^2\tau_C^2, \tag{4.127a}$$

$$b_{22} = -(4/3)C_{\zeta 3}\Omega_{F3}^2\tau_C^2, \tag{4.127b}$$

$$b_{33} = (8/3)C_{\zeta 3}\Omega_{F3}^2\tau_C^2. \tag{4.127c}$$

From the fact that τ_C in a rotating system becomes longer than Eq. (4.121), the three components in Eq. (4.127) do not remain finite as time elapses. It is shown, however, by Lumley (1978) that they should be between $-2/3$ and $4/3$. The Ω-Ω term in Eq. (4.118) violates this constraint and needs to be excluded.

B.2. Channel Turbulence. The mean velocity in channel flow is specified by Eq. (4.50). From Eqs. (4.3) and (4.114), we have

$$\langle u'^2 \rangle = \frac{2}{3}K + \left(\frac{1}{3}\zeta_1 + 2\zeta_2\right)\left(\frac{dU}{dy}\right)^2, \tag{4.128a}$$

$$\langle v'^2 \rangle = \frac{2}{3}K + \left(\frac{1}{3}\zeta_1 - 2\zeta_2 \right) \left(\frac{dU}{dy} \right)^2, \tag{4.128b}$$

$$\langle w'^2 \rangle = \frac{2}{3}K - \frac{2}{3}\zeta_1 \left(\frac{dU}{dy} \right)^2. \tag{4.128c}$$

It is known from observations that three turbulent intensities obey the relation

$$\langle u'^2 \rangle > \langle w'^2 \rangle > \langle v'^2 \rangle. \tag{4.129}$$

A prominent feature of wall-bounded flows is the suppression of the turbulent intensity normal to a wall. The condition (4.129) gives

$$C_{\zeta 1} + 2C_{\zeta 2} > 0, \ 2C_{\zeta 2} > C_{\zeta 1}, \tag{4.130}$$

from Eqs. (4.115) and (4.128). The choice of model constants by Speziale, Eq. (4.116), obeys this condition.

B.3. Square-Duct Flow. In Sec. 4.2.5.B, we noted that the occurrence of secondary flows in a square-duct flow is linked with the anisotropy of turbulent intensities in the cross section. Using Eq. (4.114) with ζ_3 dropped, we have

$$R_{yy} - R_{zz} = -2\nu_T \left(\frac{\partial V}{\partial y} - \frac{\partial Z}{\partial z} \right) + (\zeta_1 - 2\zeta_2) \left(\left(\frac{\partial U}{\partial y} \right)^2 - \left(\frac{\partial U}{\partial z} \right)^2 \right)$$
$$+4\zeta_1 \left(\left(\frac{\partial V}{\partial y} \right)^2 - \left(\frac{\partial W}{\partial z} \right)^2 \right) + 4\zeta_2 \left(\left(\frac{\partial V}{\partial z} \right)^2 - \left(\frac{\partial W}{\partial y} \right)^2 \right). \tag{4.131}$$

Owing to the second part of Eq. (4.131), $R_{yy} - R_{zz}$ does not vanish, irrespective of V and W. Therefore vanishing of Ω_x is not allowed in Eq. (4.104), resulting in the occurrence of secondary flows. The secondary-flow effects appear clearly on the mean velocity profile. The equi-velocity lines of the axial flow that are obtained by the K-ε model based on the linear turbulent-viscosity model are nearly circular. On the other hand, the counterparts obtained by experiments (Melling and Whitelaw 1976) and the models with the nonlinear correction (Speziale 1987; Nisizima 1990) tend to be rather square along the four walls. The origin of secondary flows was also discussed by Kajishima and Miyake (1992) using the large eddy simulation detailed in Chapter 5.

4.3.2. HIGHER-ORDER NONLINEAR MODELS

A. Third-Order Nonlinear Model

In the lowest-order nonlinear model (4.118), effects of frame rotation have been included through the total vorticity tensor Ω_{Tij}. A typical example

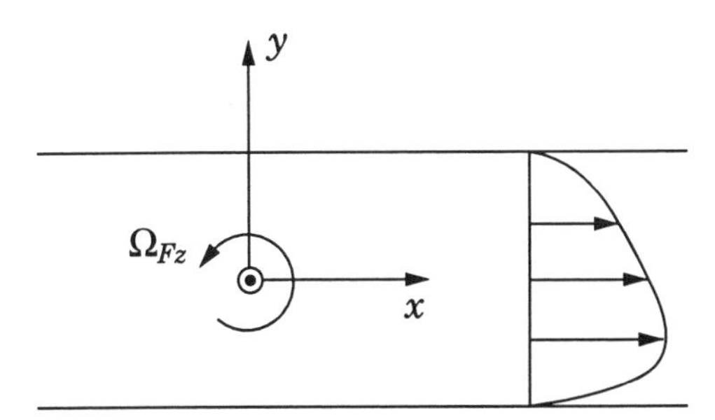

Figure 4.6. Rotating channel turbulence.

of flows beyond the scope of this model is a channel flow in a rotating frame (see Fig. 4.6). In this flow, the mean velocity $\mathbf{U}$ is unidirectional, as in Eq. (4.50), and the angular velocity of the frame rotation, $\mathbf{\Omega}_F$, is $(0, 0, \Omega_{Fz})$. The fluid near the lower wall is pushed towards the wall, whereas the upper counterpart tends to be pulled away. Then the velocity profile becomes asymmetric.

In the rotating frame, DU/Dt is replaced with $DU/Dt + 2(\mathbf{\Omega}_F \times \mathbf{U})_x$ [see Eq. (2.73)], which vanishes identically in channel flow. The rotation effect on R_{xy} occurs through the second term in Eq. (4.118) with ζ_3 dropped, but it also vanishes identically in this case. As a result, no effects of frame rotation appear in the lowest-order nonlinear K-ε model. We examine this situation on the basis of the Reynolds-stress transport equation (2.118). In the equation for R_{xy}, the second part on the left-hand side is reduced to

$$2\Omega_{F3}\left(R_{11} - R_{22}\right). \tag{4.132}$$

This quantity vanishes in the linear model for R_{ij}, Eq. (4.3), whereas the nonlinear model (4.114) gives the nonvanishing contribution to expression (4.132). Namely, the effects of frame rotation on channel flow appear in the combination of $\mathbf{\Omega}_F$ with the nonlinear effects of S_{ij} and Ω_{Tij}. In Eq. (4.118) with $\zeta_3 = 0$, effects of frame rotation occur through the linear coupling between S_{ij} and Ω_{ij}. In order to incorporate the frame-rotation effects on channel flow, we need to proceed to a higher-order nonlinear model.

A model for R_{ij} of the third order in S_{ij} and Ω_{ij} may be given without any specific difficulty. In this case, R_{ij} consists of the terms such as

$$\lambda_1 \left[S_{i\ell}S_{\ell m}S_{mj}\right]_D, \ \lambda_1' S_{ij}S_{\ell m}^2, \tag{4.133a}$$

$$\lambda_2 \left(S_{i\ell}S_{\ell m}\Omega_{mj} + S_{j\ell}S_{\ell m}\Omega_{mi}\right), \tag{4.133b}$$

$$\lambda_3 \left[S_{i\ell}\Omega_{\ell m}\Omega_{mj} + S_{j\ell}\Omega_{\ell m}\Omega_{mi}\right]_D, \ \lambda_3' S_{ij}\Omega_{\ell m}^2, \tag{4.133c}$$

where λ_n's are dimensional coefficients (we should note that $S_{\ell m}\Omega_{m\ell}$ and $\Omega_{i\ell}\Omega_{\ell m}\Omega_{mj}+\Omega_{j\ell}\Omega_{\ell m}\Omega_{mi}$ vanish). The S-Ω-Ω term in expression (4.133c) gives rise to no difficulty in a rotating isotropic turbulence, unlike the Ω-Ω term in Eq. (4.114), for S_{ij} vanishes there. In expression (4.133a), $S_{i\ell}S_{\ell m}S_{mj}$ is rewritten as

$$S_{i\ell}S_{\ell m}S_{mj}=\frac{1}{2}S_{ij}S_{\ell m}^2-\frac{1}{3}\delta_{ij}S_{\ell m}S_{mn}S_{n\ell}, \tag{4.134}$$

from Carley-Hamilton's formula referred to in the following *Note*. Then the $S_{i\ell}S_{\ell m}S_{mj}$-related part may be absorbed into the linear model (4.3). This point will be further discussed.

The terms such as expression (4.133b) have the same mathematical structure as expression (4.132) with the second-order nonlinear model as R_{ij}. Therefore they can explain the effects of frame rotation in channel turbulence. The coefficients λ_n's etc. are written as

$$(\lambda_n,\lambda_n')=(C_{\lambda_n},C_{\lambda_n}')\,K^4/\varepsilon^3, \tag{4.135}$$

from dimensional analysis (C_{λ_n}'s etc. are model constants).

Note: As a mathematical formula that is often used in turbulence modeling, we can mention Cayley-Hamilton's formula (Lumley 1978). We consider an arbitrary tensor of the second rank, A_{ij}. For this tensor, the formula says

$$A_{i\ell}A_{\ell m}A_{mj}=I_A A_{i\ell}A_{\ell j}-II_A A_{ij}+III_A\delta_{ij}, \tag{4.136}$$

where

$$I_A=A_{mm}, \tag{4.137a}$$

$$II_A=\left(A_{mm}A_{nn}-A_{mn}A_{nm}\right)/2\ , \tag{4.137b}$$

$$III_A=\left(A_{\ell\ell}A_{mm}A_{nn}-3A_{\ell\ell}A_{mn}A_{nm}+2A_{\ell m}A_{mn}A_{n\ell}\right)/6 \tag{4.137c}$$

(it should be noted that I_A etc. are invariant under the coordinate transformation). From Eq. (4.136), we can see that

$$A_{i\ell_1}A_{\ell_2\ell_3}\cdots A_{\ell_n j}=F\left\{\delta_{ij},A_{ij},A_{i\ell}A_{\ell j},I_A,II_A,III_A\right\}, \tag{4.138}$$

for $n\geq 3$. Specifically, for A_{ij} obeying $A_{ii}=0$, such as S_{ij}, we have

$$I_A=0,\ II_A=-A_{mn}A_{nm}/2,\ III_A=-A_{\ell m}A_{mn}A_{n\ell}/3. \tag{4.139}$$

B. Mathematical Structure of Nonlinear Models

In order to simply see the mathematical structure of the nonlinear model for R_{ij}, we summarize the results up to the third order:

$$\text{Linear: } C_\nu \frac{K^2}{\varepsilon} S_{ij}; \tag{4.140a}$$

$$\text{Second order: } C_{\zeta 1} \frac{K^3}{\varepsilon^2} \left[S_{i\ell} S_{\ell j} \right]_D, \; C_{\zeta 2} \frac{K^3}{\varepsilon^2} \left(S_{i\ell} \Omega_{\ell j} + S_{j\ell} \Omega_{\ell i} \right); \tag{4.140b}$$

$$\begin{aligned} \text{Third order: } & C_{\lambda 1} \frac{K^4}{\varepsilon^3} S_{ij} S_{\ell m}^2, \; C_{\lambda 2} \frac{K^4}{\varepsilon^3} S_{ij} \Omega_{\ell m}^2, \\ & C_{\lambda 3} \frac{K^4}{\varepsilon^3} \left(S_{i\ell} S_{\ell m} \Omega_{mj} + S_{j\ell} S_{\ell m} \Omega_{mi} \right), \\ & C_{\lambda 4} \frac{K^4}{\varepsilon^3} \left[S_{i\ell} \Omega_{\ell m} \Omega_{mj} + S_{j\ell} \Omega_{\ell m} \Omega_{mi} \right]_D, \end{aligned} \tag{4.140c}$$

where the constants C_{λ_n}'s have been reintroduced. A general method of constructing nonlinear models was discussed by Shih (1996) in detail.

We define the magnitude of the second-rank tensor A_{ij} by

$$\|A\| = \sqrt{(A_{ij})^2}. \tag{4.141}$$

From expressions (4.140a)-(4.140c), the nonlinear model may be regarded as an asymptotic series expansion whose successive ratio is characterized by the nondimensional quantities

$$\chi_U \equiv \frac{K}{\varepsilon} \|S\| \left(= \frac{K/\varepsilon}{1/\|S\|} \right) \tag{4.142}$$

and

$$\chi_\Omega \equiv \frac{K}{\varepsilon} \|\Omega\| \left(= \frac{K/\varepsilon}{1/\|\Omega\|} \right). \tag{4.143}$$

Here $1/\|S\|$ and $1/\|\Omega\|$ represent the time scales associated with the mean shearing and rotational motion, respectively. On the other hand, K/ε is the characteristic time scale of energy-containing eddies, as is shown by Eq. (3.50). Therefore this expansion is valid under the condition that the time scale of the mean flow is much longer than the counterpart of energy-containing eddies; namely, we require

$$\chi_U, \; \chi_\Omega \ll 1. \tag{4.144}$$

C. Functionalization of Model Constants
Some of the nonlinear terms may be absorbed into the lower-order ones, as was seen in Sec. 4.3.2.A, and the model constants in the latter are changed into the functionals concerning K, ε, S_{ij}, and Ω_{ij}. In the expression up to the third order in S_{ij} and Ω_{ij}, the first and second terms in expression (4.140c) are absorbed into expression (4.140a), and the model constant C_ν in the latter is replaced with

$$C_\nu \to C_\nu \left(1 + \frac{C_{\lambda 1}}{C_\nu} \frac{K^2}{\varepsilon^2} S_{\ell m}^2 + \frac{C_{\lambda 2}}{C_\nu} \frac{K^3}{\varepsilon^3} \Omega_{\ell m}^2 \right). \tag{4.145}$$

This indicates that C_ν is reduced to the functional of the second-order invariants

$$S_{\ell m}^2 \, (\equiv II_S) \, , \; \Omega_{\ell m}^2 \, (\equiv II_\Omega) \, . \tag{4.146}$$

From Carley-Hamilton's formula (4.138), the higher-order contributions to R_{ij} are also absorbed into the linear and the second-order parts, and the model constants C_ν, $C_{\zeta 2}$, etc. are reduced to functionals of the invariants

$$\begin{aligned} S_{\ell m} S_{mn} S_{n\ell} \, (\equiv III_S) \, , \; S_{\ell m} \Omega_{mn} \Omega_{n\ell} \, (\equiv III_{S\Omega\Omega}) \\ S_{\ell m} S_{mn} \Omega_{np} \Omega_{p\ell} \, (\equiv IV_{SS\Omega\Omega}) \, , \end{aligned} \tag{4.147}$$

in addition to Eq. (4.146) (we should note vanishing of $\Omega_{\ell m} \Omega_{mn} \Omega_{n\ell}$ and $S_{\ell m} S_{mn} \Omega_{n\ell}$). Third-order models with functional coefficients have been already proposed by Launder (1996) and Shih et al. (1997). Concerning the way how to functionalize model constants, we have no clear guiding principle. This point will be further discussed in Sec. 4.5.

D. Mathematical Meaning of Optimized Constants
The nonlinear algebraic model of the type (4.140) is an asymptotic series expansion characterized by the parameters χ_U and χ_Ω. In order to see the magnitude of these parameters in real flows, we consider a channel flow (see Fig. 4.2). In this case, they are reduced to

$$\chi_U = \chi_\Omega = \sqrt{2} \frac{K}{\varepsilon} \left| \frac{dU}{dy} \right| . \tag{4.148}$$

Using Eqs. (4.63), (4.64), and (4.67) for U, K, and ε, respectively, we have

$$\chi_U = \chi_\Omega = \sqrt{2/C_\nu}. \tag{4.149}$$

As C_ν, we adopt the value optimized in this flow or 0.09, which gives

$$\chi_U = \chi_\Omega = 4.7. \tag{4.150}$$

We consider functionalized model coefficients in the context of Eq. (4.150). For simplicity of discussion, we retain the effects of II_S, and functionalize C_ν in Eq. (4.8) as

$$C_\nu \to C_\nu F\left\{\left(K^2/\varepsilon^2\right) II_S\right\},\ F\{0\} = 1, \tag{4.151}$$

where $F\{A\}$ denotes a functional of A, and

$$\left(K^2/\varepsilon^2\right) II_S = \chi_U^2. \tag{4.152}$$

In Eq. (4.8), the choice of C_ν nearly equal to 0.09 is important for reproducing the logarithmic-velocity law (4.68) with the proper Karman constant $\kappa(\cong 0.4)$. This fact indicates that the functionalization (4.151) should be subject to either of the requirements:

(i) The functional $F\{(K^2/\varepsilon^2)II_S\}$ is nearly one in a channel flow, and the nonlinear effects are negligible;
(ii) The model constant C_ν takes a value different from the familiar one 0.09, and $C_\nu F\{(K^2/\varepsilon^2)II_S\}$ effectively becomes about 0.09 in a channel flow.

The quantity $(K^2/\varepsilon^2)II_S$ is estimated as about 20 from Eqs. (4.150) and (4.152). Considering this magnitude, the latter situation is much more probable. In reality, the counterpart of C_ν in the recent models using the functionalization of model coefficients is modified from 0.09 [Shih et al. (1994); Abe et al. (1997)]. Therefore we should consider that as far as the linear model (4.3) with Eq. (4.8) is adopted, the value 0.09 is an appropriate choice for C_ν in flows possessing the logarithmic velocity profile as one of the important properties. This point will be discussed from the theoretical viewpoint in Chapter 6.

4.3.3. NONEQUILIBRIUM EFFECTS ON TURBULENT VISCOSITY

The other typical flow that the linear model (4.3) with Eq. (4.8) cannot describe properly is the homogeneous-shear turbulence simply referred to in Secs. 3.1.1 and 3.4.2. In this flow, the velocity $\mathbf{U}$ is specified by Eq. (3.13). The equations for K and ε, (2.111) and (4.37), are reduced to

$$\frac{\partial K}{\partial t} = \varepsilon\left(C_\nu \frac{K^2}{\varepsilon^2} S^2 - 1\right), \tag{4.153}$$

$$\frac{\partial \varepsilon}{\partial t} = C_{\varepsilon 2}\frac{\varepsilon}{K^2}\left(C_\nu \frac{C_{\varepsilon 1}}{C_{\varepsilon 2}}\frac{K^2}{\varepsilon^2} S^2 - 1\right). \tag{4.154}$$

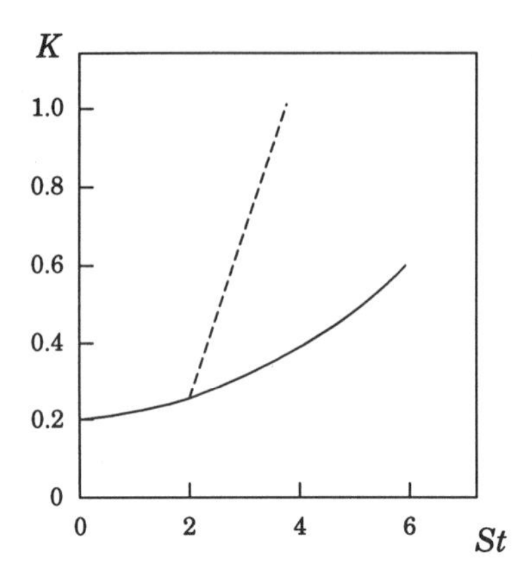

Figure 4.7. Evolution of K in homogeneous-shear turbulence (dashed line, K-ε model; solid line, DNS).

Owing to the absence of diffusion effects, the initial state that obeys

$$\frac{K}{\varepsilon}S > \sqrt{\frac{C_{\varepsilon 2}}{C_\nu C_{\varepsilon 1}}} (\cong 4) \tag{4.155}$$

is unstable, and both K and ε continue to increase indefinitely (note that $C_{\varepsilon 1}/C_{\varepsilon 2} \cong 0.7$).

Equation (4.8) for ν_T is derived from the combination of Eq. (4.9) with Eq. (3.48) for ℓ_C. This fact indicates that the equilibrium property of the inertial range is reflected strongly on ν_T. In homogeneous-shear turbulence, the injection rate of energy into the energy-containing range increases with time, and the energy dissipation process cannot catch up with the production one. In such a situation, the equilibrium assumption essential for the inertial range is invalidated. Then the K-ε model based on Eq. (4.8) is not applicable, and both K and ε are too overestimated, as is in Fig. 4.7 (initial KS/ε is about 28). In channel flow, the linear model for R_{ij} does not give rise to such a critical shortfall. Homogeneous-shear turbulence is similar to it in the sense that both the mean flows are unidirectional. Therefore the poor ability of the K-ε model arises not from the linear dependence of R_{ij} on S_{ij}, but from the lack of nonequilibrium effects on ν_T. A method of keeping the effects on ν_T is to incorporate $\partial\varepsilon/\partial t$ into it (Yoshizawa and Nisizima 1993; Yoshizawa 1994). This point will be discussed in Chapter 6.

4.4. Second-Order Modeling

In the second-order modeling based on the transport equation (2.105) for R_{ij}, we need to model the pressure-velocity-strain correlation Π_{ij}

[Eq. (2.107)], the destruction rate ε_{ij} [Eq. (2.108)], and the transport rate $T_{ij\ell}$ [Eq. (2.109)]. Of these three quantities, the modeling of Π_{ij} is the primary subject of the modeling. A weak point of the turbulent-viscosity-type modeling lies in the inability to describe strong anisotropic effects, as is exemplified through a rotating channel flow in Sec. 4.3.2.A. The second-order modeling directly dealing with R_{ij} makes us expect the high capability in the treatment of anisotropy.

4.4.1. PRESSURE-VELOCITY-STRAIN CORRELATION

In Sec. 2.6.2.A, it is remarked that Π_{ij} plays a role of governing the energy partition among three turbulence intensities, $\langle u_1'^2 \rangle$ etc. Whether or not the second-order modeling may meet the original expectation of the accurate description of anisotropy is heavily dependent on its proper modeling. The fluctuation of pressure, p', in constant-density fluid motion is given by Eq. (4.14). The difficulty in modeling p'-related correlations arises from the fact that p' is the aggregation of the contributions from the whole flow region; namely, p' is subject to long-distance effects through the solenoidal condition. In the K equation, we encountered the pressure-velocity correlation $\langle p'\mathbf{u}' \rangle$, but it is neglected in comparison with the triple velocity correlation $\langle (\mathbf{u}'^2/2)\mathbf{u}' \rangle$. Some supports for this approximation are given by observations and numerical simulations. In contrast with this, Π_{ij} is a critically important ingredient in the second-order modeling. The modeling has been studied intensively, but there is still much room for its further improvement in the application to complicated turbulent flows.

A. Slow and Rapid Terms

Using Eq. (4.14), Π_{ij} is given by

$$\begin{aligned} \Pi_{ij} = & \frac{1}{4\pi} \int_V \frac{1}{r} \left\langle \left(\frac{\partial u_j'(\mathbf{x})}{\partial x_i} + \frac{\partial u_i'(\mathbf{x})}{\partial x_j} \right) \frac{\partial^2 u_\ell'(\mathbf{x}') u_m'(\mathbf{x}')}{\partial x_\ell \partial x_m} \right\rangle dV \\ & + \frac{1}{2\pi} \int_V \frac{1}{r} \frac{\partial U_m(\mathbf{x}')}{\partial x_\ell} \left\langle \left(\frac{\partial u_j'(\mathbf{x})}{\partial x_i} + \frac{\partial u_i'(\mathbf{x})}{\partial x_j} \right) \frac{\partial u_\ell'(\mathbf{x}')}{\partial x_m} \right\rangle dV, \end{aligned} \tag{4.156}$$

where the contribution from the surface integral was dropped. We denote the first and second terms of Eq. (4.156) by $(\Pi_{ij})_S$ and $(\Pi_{ij})_R$, respectively:

$$(\Pi_{ij})_S = \frac{1}{4\pi} \int_V \frac{1}{r} \left\langle \left(\frac{\partial u_j'(\mathbf{x})}{\partial x_i} + \frac{\partial u_i'(\mathbf{x})}{\partial x_j} \right) \frac{\partial^2 u_\ell'(\mathbf{x}') u_m'(\mathbf{x}')}{\partial x_\ell \partial x_m} \right\rangle dV; \tag{4.157}$$

$$(\Pi_{ij})_R = \frac{1}{2\pi} \int_V \frac{1}{r} \frac{\partial U_m(\mathbf{x}')}{\partial x_\ell} \left\langle \left(\frac{\partial u_j'(\mathbf{x})}{\partial x_i} + \frac{\partial u_i'(\mathbf{x})}{\partial x_j} \right) \frac{\partial u_\ell'(\mathbf{x}')}{\partial x_m} \right\rangle dV. \tag{4.158}$$

Here $(\Pi_{ij})_S$ and $(\Pi_{ij})_R$ are often called the slow and rapid terms, respectively. This terminology comes from the difference between the responses of these two terms to external disturbances such as effects of the abrupt change of a boundary. In the process, the mean flow is affected firstly, resulting in the change of $\nabla\mathbf{U}$. This change generates that of velocity fluctuations, which affects Π_{ij}. This is the reason that Eq. (4.158) dependent on $\nabla\mathbf{U}$ is called the rapid term.

The response time scale to external disturbances such as boundary effects should not be confused with the time scale in the energy-cascade process (we should note that $\nabla\mathbf{U}$ is the dimension of the inverse of time). In the inertial range discussed in Sec. 3.2.1, Eq. (3.46) indicates that larger-scale fluctuations have a longer characteristic time or life time. This time scale is associated with the excitation of smaller-scale fluctuations through the energy cascade, but not with the response to external disturbances. The method of analyzing the rapid responses of fluctuations to $\nabla\mathbf{U}$ in a linear manner is called rapid distortion theory, and it is detailed in the work by Hunt and Carruthers (1990). In the current second-order modeling, little attention is paid to the long-distance effects of p'. The sole exception is the attempt by Durbin (1993) who modeled a Poisson-type equation for Π_{ij}.

Since Π_{ij} is traceless, that is, $\Pi_{ii} = 0$, we consider the traceless part of R_{ij}; namely, we write

$$B_{ij} = [R_{ij}]_D \equiv R_{ij} - (2/3)K\delta_{ij}. \tag{4.159}$$

The choice of B_{ij} for modeling Π_{ij} is also natural since the latter governs the anisotropy of R_{ij}, that is, the magnitude of B_{ij}. From Eq. (2.105), the equation for B_{ij} is given by

$$\frac{DB_{ij}}{Dt} = [P_{ij}]_D + \Pi_{ij} - [\varepsilon_{ij}]_D + \left[\frac{\partial T_{ij\ell}}{\partial x_\ell}\right]_D. \tag{4.160}$$

Specifically, $[P_{ij}]_D$ is written as

$$[P_{ij}]_D = -\left[B_{i\ell}\frac{\partial U_j}{\partial x_\ell} + B_{j\ell}\frac{\partial U_i}{\partial x_\ell}\right]_D - \frac{2}{3}KS_{ij}. \tag{4.161}$$

In the study of the second-order modeling, $(\Pi_{ij})_S$ and $(\Pi_{ij})_R$ have been modeled separately until recently. This separate treatment comes from the fact that $(\Pi_{ij})_S$ is not dependent on $\nabla\mathbf{U}$ explicitly, resulting in the survival in the latter absence. In what follows, we shall briefly refer to the method. In this context, we introduce the nondimensional counterpart of B_{ij} as

$$b_{ij} = B_{ij}/K \tag{4.162}$$

[b_{ij} has been already introduced by Eq. (4.125)].

A.1. Modeling of Slow Term. We assume that $(\Pi_{ij})_S$ is expressed in terms of b_{ij} in addition to K and ε; namely, we write

$$(\Pi_{ij})_S = (\Pi_{ij})_S \{b_{ij}, K, \varepsilon\}, \tag{4.163}$$

with the constraint

$$(\Pi_{ij})_S \{0, K, \varepsilon\} = 0 \tag{4.164}$$

(the isotropic case corresponds to vanishing Π_{ij}).

We expand $(\Pi_{ij})_S$ in b_{ij}, and retain the leading term, which is

$$(\Pi_{ij})_S = -C_{S1}\varepsilon b_{ij}, \tag{4.165}$$

where C_{S1} is a model constant. In case that turbulence is homogeneous but anisotropic, Eq. (4.160) is reduced to

$$\frac{\partial B_{ij}}{\partial t} = \Pi_{ij} - [\varepsilon_{ij}]_D . \tag{4.166}$$

In this equation, we have no terms generating and sustaining the anisotropy of turbulence since $\nabla \mathbf{U}$ is its primary cause. Therefore the anisotropy given at an initial time decays with time. In the combination of Eq. (4.165) with Eq. (4.166), this fact may be explained by choosing positive C_{S1}. Equation (4.165) is usually called Rotta's (1951) model.

We proceed to the second order in b_{ij}, and have

$$(\Pi_{ij})_S = -C_{S1}\varepsilon b_{ij} + C_{S2}\varepsilon [b_{i\ell}b_{\ell j}]_D , \tag{4.167}$$

where C_{S2} is a model constant. The third-order term is proportional to

$$\varepsilon [b_{i\ell}b_{\ell m}b_{mj}]_D . \tag{4.168}$$

This term is reduced to the form proportional to b_{ij} from Carley-Hamilton's formula (4.136). The same situation holds for the nth-order terms in $b_{ij} (n \geq 4)$. Finally, the model constants C_{S1} and C_{S2} are functionalized as

$$C_{S1} = C_{S1} \{II_b, III_b\} , \ C_{S2} = C_{S2} \{II_b, III_b\} , \tag{4.169}$$

where II_b and III_b are defined as

$$II_b = b_{\ell m}^2 , \ III_b = b_{\ell m}b_{mn}b_{n\ell}. \tag{4.170}$$

A.2. Modeling of Rapid Term. In Eq. (4.158) for the rapid term $(\Pi_{ij})_R$, the integrand is dependent explicitly on the gradients of both the mean and fluctuating velocity. In general, the spatial change of the fluctuating velocity is much more rapid, compared with the mean velocity. Then we expand

$$\frac{\partial U_i(\mathbf{x}')}{\partial x_j} = \frac{\partial U_i(\mathbf{x})}{\partial x_j} + \sum_{n=1}^{3} \frac{\partial^2 U_i(\mathbf{x})}{\partial x_j \partial x_n} r_n + \cdots, \tag{4.171}$$

where $\mathbf{r} = \mathbf{x} - \mathbf{x}'$. Retaining the first term, we write

$$(\Pi_{ij})_R = N_{ij\ell m} \frac{\partial U_m}{\partial x_\ell}, \tag{4.172}$$

where

$$N_{ij\ell m} = \frac{1}{2\pi} \int_V \frac{1}{r} \left\langle \left(\frac{\partial u_j'(\mathbf{x})}{\partial x_i} + \frac{\partial u_i'(\mathbf{x})}{\partial x_j} \right) \frac{\partial u_\ell'(\mathbf{x}')}{\partial x_m} \right\rangle dV \tag{4.173}$$

($r = |\mathbf{r}|$). The fourth-rank tensor $N_{ij\ell m}$ is subject to the constraint

$$N_{ii\ell m} = 0, \ N_{ij\ell m} = N_{ji\ell m}. \tag{4.174}$$

Equation (4.173) is not dependent on $\nabla \mathbf{U}$ explicitly, as is similar to the slow term $(\Pi_{ij})_S$, Eq. (4.157). Under the same modeling principle for the latter, $N_{ij\ell m}$ is modeled as

$$N_{ij\ell m} = N_{ij\ell m} \left\{ b_{ij}, K \right\}, \tag{4.175}$$

where the dependence on ε is not necessary, as will be seen below. We expand $N_{ij\ell m}$ in b_{ij}, and retain the terms up to its first order. As a result, we have

$$\begin{aligned} N_{ij\ell m} &= N_1 K \delta_{ij} \delta_{\ell m} + N_2 K \left(\delta_{i\ell} \delta_{jm} + \delta_{im} \delta_{j\ell} \right) + N_3 K b_{ij} \delta_{\ell m} + N_4 K b_{\ell m} \delta_{ij} \\ &+ N_5 K \left(b_{i\ell} \delta_{jm} + b_{j\ell} \delta_{im} \right) + N_6 K \left(b_{im} \delta_{j\ell} + b_{jm} \delta_{im} \right), \end{aligned} \tag{4.176}$$

where N_n's are nondimensional coefficients, and use has been made of the second constraint in Eq. (4.174). Further, the first constraint gives

$$3N_1 + 2N_2 = 0, \tag{4.177a}$$

$$3N_4 + 2N_5 + 2N_6 = 0. \tag{4.177b}$$

As a result, the number of the independent coefficients becomes four.

We can construct a higher-order expression for $(\Pi_{ij})_R$ by retaining the terms of the nth ($n \geq 2$) order in b_{ij}. The contributions from the nth-order

($n \geq 3$) terms are absorbed into the terms below the second order, owing to Cayley-Hamilton's formula (4.136). As a result, the coefficients N_n's are functionalized as

$$N_n = N_n \{II_b, III_b\}, \tag{4.178}$$

as is similar to Eq. (4.169). Such a general expression was discussed in details by Shih et al. (1992) and Shih (1996). Here we do not delve into their details since the interest of this book lies in the basic mathematical structures of second-order models.

B. Non-Separative Modeling of Π_{ij}

We have another choice of the modeling of Π_{ij}. The slow term $(\Pi_{ij})_S$ is not dependent explicitly on $\nabla \mathbf{U}$. The fact indicates that $(\Pi_{ij})_S$ has the terms surviving in the absence of $\nabla \mathbf{U}$ as far as turbulence is anisotropic, but not that the terms explicitly dependent on $\nabla \mathbf{U}$ are not allowed to enter a model for $(\Pi_{ij})_S$. This point will become clearer from an asymptotic solution for $\mathbf{u}'$ in Chapter 6.

One of the representative works based on the non-separative modeling of Π_{ij} is the one by Speziale et al. (1991). First, we assume a functional form

$$\Pi_{ij} = \Pi_{ij} \{b_{ij}, S_{ij}, \Omega_{ij}, K, \varepsilon\}. \tag{4.179}$$

Here the splitting of $\nabla \mathbf{U}$ into S_{ij} and Ω_{ij} is helpful to the clear understanding of its physical properties. In Eq. (4.179), no quantities intrinsic to the long-distance effects of p' are taken into account. We expand Π_{ij} in b_{ij}, S_{ij}, and Ω_{ij}, and retain the terms up to the first order in the latter two quantities. The choice comes from the linear dependence of Π_{ij} on $\nabla \mathbf{U}$ in Eq. (4.156), but the nonlinear dependence on S_{ij} and Ω_{ij} cannot be excluded from the original property of Π_{ij}.

Under the foregoing linear-dependence constraint, we have

$$\begin{aligned} \Pi_{ij} &= \Pi_1 \varepsilon b_{ij} + \Pi_2 \varepsilon \left[b_{i\ell} b_{\ell j} \right]_D + \Pi_3 K S_{ij} + \Pi_4 K \left[b_{i\ell} S_{\ell j} + b_{j\ell} S_{\ell i} \right]_D \\ &\quad + \Pi_5 K \left(b_{i\ell} \Omega_{\ell j} + b_{j\ell} \Omega_{\ell i} \right) + \Pi_6 K \left[b_{i\ell} b_{\ell m} S_{mj} + b_{j\ell} b_{\ell m} S_{mi} \right]_D \\ &\quad + \Pi_7 K \left(b_{i\ell} b_{\ell m} \Omega_{mj} + b_{j\ell} b_{\ell m} \Omega_{mi} \right). \end{aligned} \tag{4.180}$$

No explicit dependence on the terms of the third order in b_{ij} is clear from Carley-Hamilton's formula (4.136) and the related discussions in Sec. 4.3.2.A. Here Π_n's are the nondimensional coefficients and are written as functionals of

$$II_b,\ III_b,\ II_{bS} \left(\equiv \frac{K}{\varepsilon} b_{\ell m} S_{m\ell} \right),\ III_{bbS} \left(\equiv \frac{K}{\varepsilon} b_{\ell m} b_{mn} S_{n\ell} \right), \tag{4.181}$$

with Eq. (4.170) for II_b and III_b (we should note vanishing of $II_{b\Omega}$ and $III_{bb\Omega}$). Namely, we may write

$$\begin{aligned} \Pi_n &= \Pi_{n1}\{II_b, III_b\} + \Pi_{n2}\{II_b, III_b\} II_{bS} \\ &+ \Pi_{n3}\{II_b, III_b\} II_{bbS} \ (n = 1, 2), \end{aligned} \tag{4.182a}$$

$$\Pi_n = \Pi_n\{II_b, III_b\} \ (n = 3\text{-}7). \tag{4.182b}$$

No dependence of Eq. (4.182b) on II_{bS} and II_{bbS} comes from the assumption of the linear dependence of Π_{ij} on S_{ij} and Ω_{ij}. The frame-rotation effect may be incorporated into Eq. (4.179) through the replacement (4.117).

In case that only Π_{n1}'s for $n = 1$ and 3-5 are retained and are taken as constants, we have

$$\begin{aligned} \Pi_{ij} &= C_{\Pi 1}\varepsilon b_{ij} + C_{\Pi 3} K S_{ij} + C_{\Pi 4} K \left[b_{i\ell} S_{\ell j} + b_{j\ell} S_{\ell i} \right]_D \\ &+ C_{\Pi 5} K \left(b_{i\ell}\Omega_{\ell j} + b_{j\ell}\Omega_{\ell i} \right). \end{aligned} \tag{4.183}$$

Equation (4.183) corresponds to the model by Launder, Reece, and Rodi (1975), which is abbreviated as the LRR model and is the prototype of all the models for Π_{ij}. The first term is Rotta's model (4.165), and its model constant is usually chosen as

$$C_{\Pi 1} \equiv \lim_{b_{ij}\to 0} \Pi_1 = -C_{S1} \cong -1.8. \tag{4.184}$$

In the original version of the LRR model, Eq. (4.183) is written using the production rate of R_{ij}, P_{ij} [Eq. (2.106)], and P'_{ij} that is defined by

$$P'_{ij} = -R_{i\ell}\frac{\partial U_\ell}{\partial x_j} - R_{j\ell}\frac{\partial U_\ell}{\partial x_i}. \tag{4.185}$$

The physical meaning of P'_{ij} is not so clear as P_{ij}, but the use of P'_{ij} is necessary for the explicit manifestation of the Ω_{ij} effects on Π_{ij}.

Concerning $C_{\Pi 3}$, we have the important constraint

$$C_{\Pi 3} \equiv \lim_{b_{ij}\to 0} \Pi_3 = 2/5, \tag{4.186}$$

as will be shown in the following *Note*. In a simplified version of Eq. (4.180) by Speziale et al. (1991), the last two terms of Eq. (4.180) are dropped. Some of Π_n's are taken as constant, whereas the others are functionalized. The constant parts of Π_n's, $C_{\Pi n}$, are given by

$$C_{\Pi 2} \cong 1, \ C_{\Pi 4} \cong 0.3, \ C_{\Pi 5} \cong 0.1. \tag{4.187}$$

The foregoing simplified version was applied to some fundamental flows such as homogeneous-shear turbulence in non-rotating and rotating frames.

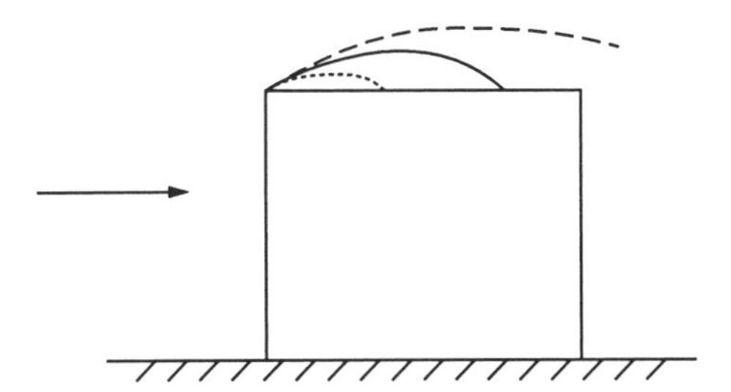

Figure 4.8. Separation in a flow past a cube (dashed line, second-order models; dotted line, turbulent-viscosity models; solid line, observations).

The linear turbulent-viscosity model based on Eq. (4.8) (the standard K-ε model) cannot cope with the nonequilibrium properties of homogeneous-shear turbulence and effects of frame rotation on turbulence, as was noted in Secs. 4.3.2 and 4.3.3. The simplified model overcomes these shortfalls, and effects of frame rotation on the time evolution of turbulence characteristics can be captured accurately. In the analysis of complex flows, however, there is room for the further improvement of second-order models. For instance, various second-order models fails in capturing some important behaviors of a flow passing over a cubic body (this is a model of wind flow impinging on a building), as was shown by Murakami et al. (1993). In observations, the flow separates at the front edge and reattaches at a location on the roof (see Fig. 4.8). In the current second-order models, this reattachment cannot be reproduced. In contrast, the linear turbulent-viscosity model leads to the estimate of too-short reattachment length.

Note: We consider the situation in which the steep mean velocity gradient is imposed impulsively, and pursue the time evolution of $\mathbf{u}'$. Then we retain the first part in Eq. (4.11) as

$$\Delta p' = -2\frac{\partial u'_j}{\partial x_i}\frac{\partial U_i}{\partial x_j}. \tag{4.188}$$

During this evolution, $\partial U_i/\partial x_j$ is assumed to be constant. Further, we assume the homogeneity of p' and $\mathbf{u}'$, and apply the Fourier representation (3.17) to Eq. (4.188). Then we have

$$p'(\mathbf{k}) = -2i\frac{k_\ell}{k^2}u'_m(\mathbf{k})\frac{\partial U_\ell}{\partial x_m}. \tag{4.189}$$

Using Eq. (4.189), we write

$$\left\langle p'\frac{\partial u'_j}{\partial x_i}\right\rangle = i\int k_i\frac{\left\langle p'(\mathbf{k})u'_j(\mathbf{k}')\right\rangle}{\delta\left(\mathbf{k}+\mathbf{k}'\right)}d\mathbf{k} = 2\frac{\partial U_\ell}{\partial x_m}\int\frac{k_i k_\ell}{k^2}\frac{\left\langle u'_m(\mathbf{k})u'_j(\mathbf{k}')\right\rangle}{\delta\left(\mathbf{k}+\mathbf{k}'\right)}d\mathbf{k}. \tag{4.190}$$

In Eq. (4.190), we assume the isotropy of $\mathbf{u}'$, which leads to vanishing of B_{ij} or b_{ij}. Under Eq. (3.35) for isotropic $\mathbf{u}'$, Eq. (4.190) is reduced to

$$\left\langle p' \frac{\partial u'_j}{\partial x_i} \right\rangle = 2 \frac{\partial U_\ell}{\partial x_m} \int \frac{k_i k_\ell}{k^2} D_{mj}(\mathbf{k}) Q(k) dk. \tag{4.191}$$

Here we note the following formulae:

$$\int \frac{k_i k_j}{k^2} d\mathbf{k} = \frac{1}{3} \delta_{ij} \int d\mathbf{k}; \tag{4.192a}$$

$$\int \frac{k_i k_j k_\ell k_m}{k^4} d\mathbf{k} = \frac{1}{15} \left(\delta_{ij}\delta_{\ell m} + \delta_{i\ell}\delta_{jm} + \delta_{im}\delta_{j\ell} \right) \int d\mathbf{k} \tag{4.192b}$$

(the latter can be easily confirmed by putting $i = j$ and $\ell = m$). Using Eq. (4.192), Eq. (4.191) is rewritten as

$$\left\langle p' \frac{\partial u'_j}{\partial x_i} \right\rangle = K \left(\frac{8}{15} \frac{\partial U_i}{\partial x_j} - \frac{2}{15} \frac{\partial U_j}{\partial x_i} \right), \tag{4.193}$$

where the turbulent energy K is related to its spectrum density $Q(k)$ as

$$K = \int Q(k) d\mathbf{k}. \tag{4.194}$$

From Eq. (4.193), we have

$$\Pi_{ij} = \frac{2}{5} K S_{ij}, \tag{4.195}$$

which leads to Eq. (4.186).

4.4.2. DESTRUCTION AND TRANSPORT EFFECTS

In Eq. (2.105), we have two terms yet to be modeled. One is the destruction term ε_{ij}, and another is the transport one $T_{ij\ell}$. We shall simply refer to their modeling.

A. Modeling of ε_{ij}

Most simply, ε_{ij} is assumed to be isotropic as

$$\varepsilon_{ij} = (1/3)\, \varepsilon \delta_{ij}. \tag{4.196}$$

The equation for each of three turbulent intensities, $\langle {u'_1}^2 \rangle$ etc., depends explicitly on ε_{11} etc. The anisotropy of ε_{ij} is expected to become important in the presence of the strong anisotropy of these intensities. This point may be confirmed through the direct numerical simulation (DNS) of a channel flow (Kim et al. 1987; Kasagi et al. 1992). The anisotropy of intensities is more dominant at the energy-containing scale, compared with the fine scales associated with the energy dissipation. This fact implies that the anisotropy of ε_{ij} is related to that of energy-containing quantities such as turbulent intensities.

We assume that ε_{ij} is modeled in the same manner as for Π_{ij}; namely, we write

$$[\varepsilon_{ij}]_D \equiv \varepsilon_{ij} - (1/3)\,\varepsilon\delta_{ij} = [\varepsilon_{ij}]_D \{b_{ij}, S_{ij}, \Omega_{ij}, K, \varepsilon\}. \tag{4.197}$$

On this basis, we can construct the model similar to Eq. (4.180) for Π_{ij}. The simplest anisotropic model is given by

$$[\varepsilon_{ij}]_D = C_D \varepsilon b_{ij}. \tag{4.198}$$

At this stage of modeling, the nondimensional coefficient C_D is left as unknown. Under the requirement that Eq. (4.198) is valid down to the close vicinity of a solid wall, we have

$$C_D = 2. \tag{4.199}$$

This point will be mentioned later in Sec. 4.7 on effects of low Reynolds number near a wall. A number of works have been done on more elaborate modeling of ε_{ij}, but we do not enter their details in this book (see Hallbäck et al. 1996).

B. Modeling of $T_{ij\ell}$

In Sec. 4.2.2.A, we performed the modeling of the transport term $\mathbf{T}_K$ [Eq. (2.114)] in the K equation (2.111). There we neglected the pressure-velocity correlation $\langle p'\mathbf{u}'\rangle$ with some observational supports, and focused attention on the triple velocity correlation $\langle(\mathbf{u}'^2/2)\mathbf{u}'\rangle$. This approximation gives rise to a critical shortfall in the close vicinity of a solid wall, as will be explained in Sec. 4.7.2. In the modeling of the transport term $T_{ij\ell}$ [Eq. (2.109)] in the R_{ij} equation (2.105), we also encounter the p'-related part. In the current second-order modeling, this part is also dropped.

The triple velocity correlation $\langle u'_i u'_j u'_\ell\rangle$ may be regarded as the rate of $u'_i u'_j$ that is transported by $\mathbf{u}'$ in the ℓ direction. In the scalar transport, the transport rate of a scalar θ' (for instance, the temperature fluctuation) by $\mathbf{u}'$ is modeled most simply as

$$\langle \theta' \mathbf{u}' \rangle = -\lambda_T \nabla \Theta, \tag{4.200}$$

where Θ is the mean of θ, and λ_T is the so-called turbulent diffusivity which corresponds to ν_T in the transport of momentum. Namely, Eq. (4.200) signifies that θ' is transported from the larger-Θ region to the smaller counterpart. This type of modeling is usually named the gradient-diffusion approximation and is used in the modeling of $\langle(\mathbf{u}'^2/2)\mathbf{u}'\rangle$ in Eq. (4.16). We apply Eq. (4.200) to $\langle u_i' u_j' u_\ell' \rangle$, and consider the symmetry concerning i, j, and ℓ. As a result, we may write

$$\left\langle u_i' u_j' u_\ell' \right\rangle = -\kappa_T \left(\frac{\partial R_{ij}}{\partial x_\ell} + \frac{\partial R_{j\ell}}{\partial x_i} + \frac{\partial R_{\ell i}}{\partial x_j} \right), \tag{4.201}$$

where κ_T is given by

$$\kappa_T = \nu_T / \sigma_R \tag{4.202}$$

(σ_R is a constant).

Using Eq. (4.8), we rewrite Eq. (4.201) as

$$\left\langle u_i' u_j' u_\ell' \right\rangle = -\frac{C_\nu}{\sigma_R} \frac{K}{\varepsilon} \left(K\delta_{\ell m} \frac{\partial R_{ij}}{\partial x_m} + K\delta_{im} \frac{\partial R_{j\ell}}{\partial x_m} + K\delta_{jm} \frac{\partial R_{\ell i}}{\partial x_m} \right). \tag{4.203}$$

Here we note that $(2/3)K\delta_{ij}$ is the leading isotropic part of R_{ij}, and make the replacement

$$\frac{2}{3} K\delta_{ij} \to R_{ij}. \tag{4.204}$$

As a result, Eq. (4.203) is reduced to

$$\left\langle u_i' u_j' u_\ell' \right\rangle = -\frac{3C_\nu}{2\sigma_R} \frac{K}{\varepsilon} \left(R_{\ell m} \frac{\partial R_{ij}}{\partial x_m} + R_{im} \frac{\partial R_{j\ell}}{\partial x_m} + R_{jm} \frac{\partial R_{\ell i}}{\partial x_m} \right). \tag{4.205}$$

From the viewpoint of the second-order modeling based on the direct use of R_{ij}, Eq. (4.205) seems to be superior to Eq. (4.203) and is believed so. In its practical use, however, we sometimes encounter numerical difficulties. For, R_{ij} may become negative, and the negativeness of diffusion coefficients is a primary seed of numerical instability. From the viewpoint of the cross-diffusion effect in Eq. (4.18), the ε counterpart may be included in Eq. (4.205). This point will be referred to in Chapter 6.

4.5. Relationship between Algebraic and Second-Order Models

In Secs. 4.2-4.4, the algebraic and second-order models are constructed in a rather independent manner. Through the comparison between Eqs. (4.118) and (4.183), however, we may see that the mathematical structures of these models are very similar each other. Provided that these two types of models may capture some important properties of turbulent shear flows, the algebraic model may be regarded as an approximate solution of the R_{ij} equation in the sense that R_{ij} is expressed explicitly using K, ε, S_{ij}, and Ω_{ij}. This viewpoint gives a clue to the understanding of the relationship between these two models. In what follows, we shall derive algebraic models from second-order ones. In Chapter 6, we shall consider the directly opposite procedure in a statistical theoretical manner.

4.5.1. DERIVATION OF LOWEST-ORDER NONLINEAR MODEL

As a second-order model, we consider Eq. (4.183) for Π_{ij} and Eq. (4.196) for ε_{ij}, (we shall refer to $T_{ij\ell}$ later). We combine these expressions with the R_{ij} equation, specifically, its B_{ij} counterpart (4.160). In Eq. (4.161), the deviatoric part of P_{ij}, $[P_{ij}]_D$, is rewritten as

$$[P_{ij}]_D = -\frac{1}{2}\left[B_{i\ell}S_{\ell j} + B_{j\ell}S_{\ell i}\right]_D - \frac{1}{2}\left(B_{i\ell}\Omega_{\ell j} + B_{j\ell}\Omega_{\ell i}\right) - \frac{2}{3}KS_{ij}. \quad (4.206)$$

Then Eq. (4.160) is reduced to

$$\begin{aligned}\frac{DB_{ij}}{Dt} &= C_{\Pi 1}\frac{\varepsilon}{K}B_{ij} + \left(C_{\Pi 3} - \frac{2}{3}\right)KS_{ij} + \left(C_{\Pi 4} - \frac{1}{2}\right)\left[B_{i\ell}S_{\ell j} + B_{j\ell}S_{\ell i}\right]_D \\ &+ \left(C_{\Pi 5} - \frac{1}{2}\right)\left(B_{i\ell}\Omega_{\ell j} + B_{j\ell}\Omega_{\ell i}\right) + \left[\frac{\partial T_{ij\ell}}{\partial x_\ell}\right]_D, \end{aligned} \quad (4.207)$$

where b_{ij} has been replaced with B_{ij} through Eq. (4.162).

We pay attention to the first term on the right-hand side of Eq. (4.207), and write

$$\begin{aligned} B_{ij} &= -\left(-\frac{2}{3C_{\Pi 1}} + \frac{C_{\Pi 3}}{C_{\Pi 1}}\right)\frac{K^2}{\varepsilon}S_{ij} + \left(\frac{1}{2C_{\Pi 1}} - \frac{C_{\Pi 4}}{C_{\Pi 1}}\right)\frac{K}{\varepsilon}\left[B_{i\ell}S_{\ell j} + B_{j\ell}S_{\ell i}\right]_D \\ &+ \left(\frac{1}{2C_{\Pi 1}} - \frac{C_{\Pi 4}}{C_{\Pi 1}}\right)\frac{K}{\varepsilon}\left(B_{i\ell}\Omega_{\ell j} + B_{j\ell}\Omega_{\ell i}\right) + \frac{1}{C_{\Pi 1}}\frac{K}{\varepsilon}\frac{DB_{ij}}{Dt} \\ &- \frac{1}{C_{\Pi 1}}\frac{K}{\varepsilon}\left[\frac{\partial T_{ij\ell}}{\partial x_\ell}\right]_D. \end{aligned} \quad (4.208)$$

Here the first term is nothing but the linear turbulent-viscosity model (4.3) with Eq. (4.8) for ν_T. The model constant in ν_T, C_ν, is given by

$$C_\nu = -\frac{(2/3) - C_{\Pi 3}}{C_{\Pi 1}}. \quad (4.209)$$

Using Eqs. (4.184) and (4.186), Eq. (4.209) is estimated as 0.14, which is a little larger than the widely-adopted value 0.09. This difference will be discussed in Sec. 4.5.2.

Using the linear model, we solve Eq. (4.208) by an iteration method. In the first iteration, we substitute the linear model into the right-hand side of Eq. (4.208). Then we have

$$\begin{aligned} B_{ij} = & -C_\nu \frac{K^2}{\varepsilon} S_{ij} + \left(\frac{2C_\nu C_{\Pi 4}}{C_{\Pi 1}} - \frac{C_\nu}{C_{\Pi 1}} \right) \frac{K^3}{\varepsilon^2} \left[S_{i\ell} S_{\ell j} \right]_D \\ & + \left(\frac{C_\nu C_{\Pi 4}}{C_{\Pi 1}} - \frac{C_\nu}{2C_{\Pi 1}} \right) \frac{K^3}{\varepsilon^2} \left(S_{i\ell} \Omega_{\ell j} + S_{j\ell} \Omega_{\ell i} \right) \\ & - \frac{C_\nu}{C_{\Pi 1}} \frac{K}{\varepsilon} \left(\frac{D}{Dt} \frac{K^2}{\varepsilon} \right) S_{ij} + R_{Bij}, \end{aligned} \tag{4.210}$$

where the remaining term R_{Bij} is given by

$$R_{Bij} = -\frac{C_\nu}{C_{\Pi 1}} \frac{K^3}{\varepsilon^2} \frac{DS_{ij}}{Dt} - \frac{1}{C_{\Pi 1}} \frac{K}{\varepsilon} \left[\frac{\partial T_{ij\ell}}{\partial x_\ell} \right]_D . \tag{4.211}$$

In the comparison with Eqs. (4.140a) and (4.140b), the first three parts of Eq. (4.210) correspond to the lowest- or second-order nonlinear model. Specifically, we should note that the Ω-Ω term in Eq. (4.118) does not occur, as was pointed out by Speziale (1996). Equation (4.211) generates the terms containing the second-order derivatives in space and time. This point makes contrast with the usual nonlinear models dependent on the first-order space derivatives only.

In Eq. (4.210), the fourth term may be absorbed into the first one. As a result, the linear model (4.3) is modified as

$$(R_{ij})_1 = \frac{2}{3} K \delta_{ij} - C_{\nu M} \frac{K^2}{\varepsilon} S_{ij}, \tag{4.212}$$

where the modified coefficient $C_{\nu M}$ is given by

$$\frac{C_{\nu M}}{C_\nu} = 1 + \frac{1}{C_{\Pi 1}} \frac{1}{K} \frac{D}{Dt} \frac{K^2}{\varepsilon}. \tag{4.213}$$

This is a kind of functionalization of model constant, similar to Eq. (4.145). In homogeneous-shear turbulence referred to in Sec. 4.3.3, D/Dt is replaced with $\partial/\partial t$. There K, ε, and K^2/ε increase with time, and the second part of Eq. (4.213) is negative, resulting in $C_{\nu M}/C_\nu < 1$. This effective decrease in the model coefficient tends to suppress the production rate of K and is helpful to rectify the overestimate of K and ε by Eq. (4.8) (see Fig. 4.7).

4.5.2. CONSOLIDATION OF INFINITE PERTURBATIONAL SOLUTION

In Sec. 4.5.1, we have seen that a second-order nonlinear model is obtained using a perturbational expansion of a second-order model. This procedure can be continued, in principle, to any orders in S_{ij}, Ω_{ij}, etc., although the mathematical manipulation is complicated. The parts of the resulting expression that are algebraic in S_{ij} and Ω_{ij} corresponds to an asymptotic expansion concerning small χ_U [Eq. (4.142)] and χ_Ω [Eq. (4.143)]. We also showed in Sec. 4.3.2.D that χ_U is estimated as about 5 in channel flow. For such χ_U, it is highly probable that the finite truncation of the asymptotic expansion does not always lead to significant results. The similar situation has been already encountered in the study of a statistical theory of small-scale turbulence (Chapter 3). There the Reynolds number, which is large in turbulence, is an expansion parameter. As a result, it is indispensable to summarize the terms up to an infinite order even if its summation is partial. This summation procedure may be called the consolidation of an infinite perturbational solution. The mathematical form of the resulting expression is dependent on each consolidation procedure and is not unique.

The methods of summarizing a perturbational solution of the R_{ij} equation up to an infinite order have been presented by Taulbee (1992) and Gatski and Speziale (1993). Both of the works originate from the pioneering work of Pope (1975) on the generalization of the turbulent-viscosity concept. Their starting point is the reduction of the R_{ij} equation into an algebraic form made by Rodi (1976). There we pick up the advection and diffusion terms, DR_{ij}/Dt and $\partial T_{ij\ell}/\partial x_\ell$, in Eq. (2.105) and make the approximation

$$\frac{DR_{ij}}{Dt} - \frac{\partial T_{ij\ell}}{\partial x_\ell} \cong \frac{R_{ij}}{K}\left(\frac{DK}{Dt} - \nabla \cdot \mathbf{T}_K\right), \tag{4.214a}$$

which leads to

$$\frac{DR_{ij}}{Dt} - \frac{\partial T_{ij\ell}}{\partial x_\ell} \cong \frac{R_{ij}}{K}\left(P_K - \varepsilon\right), \tag{4.214b}$$

from Eq. (2.111). The physical meaning of Eq. (4.214a) will be explained in the following *Note*. We substitute Eq. (4.214b) into Eq. (2.105), and have

$$\frac{P_K - \varepsilon}{K} R_{ij} = P_{ij} + \Pi_{ij} - \varepsilon_{ij} \tag{4.215a}$$

or

$$\frac{P_K - \varepsilon}{K} B_{ij} = [P_{ij}]_D + \Pi_{ij} - [\varepsilon_{ij}]_D . \tag{4.215b}$$

In case that Π_{ij} and $[\varepsilon_{ij}]_D$ are expressed in terms of B_{ij}, S_{ij}, and Ω_{ij}, Eq. (4.215b) is an algebraic equation concerning B_{ij}. For instance, we adopt the simple models for them, Eqs. (4.183) and (4.198). We substitute those models into Eq. (4.215b), and have

$$\left[P_K - \varepsilon + (C_D - C_{\Pi 1})\,\varepsilon\right] B_{ij} = -\left(\frac{2}{3} - C_{\Pi 3}\right) K^2 S_{ij}$$

$$+\left(C_{\Pi 4}-\frac{1}{2}\right)K\left[B_{i\ell}S_{\ell j}+B_{i\ell}S_{\ell j}\right]_D$$
$$+\left(C_{\Pi 5}-\frac{1}{2}\right)K\left(B_{i\ell}\Omega_{\ell j}+B_{i\ell}\Omega_{\ell j}\right), \qquad (4.216)$$

where use has been made of Eq. (4.206). A turbulent-viscosity model may be obtained by retaining the first term on the right-hand side of Eq. (4.216):

$$B_{ij}=-\frac{(2/3)-C_{\Pi 3}}{C_D-C_{\Pi 1}-1+(P_K/\varepsilon)}\frac{K^2}{\varepsilon}S_{ij}. \qquad (4.217)$$

In the context of Eq. (4.212), the model coefficient $C_{\nu M}$ is written as

$$C_{\nu M}=\frac{(2/3)-C_{\Pi 3}}{C_D-C_{\Pi 1}-1+(P_K/\varepsilon)}, \qquad (4.218a)$$

which is reduced to the constant coefficient

$$C_\nu=\frac{C_{\Pi 3}-(2/3)}{C_{\Pi 1}-C_D}\cong 0.07 \qquad (4.218b)$$

in case that $P_K=\varepsilon$, where use has been made of Eqs. (4.184), (4.186), and (4.199) for $C_{\Pi 1}$, $C_{\Pi 3}$, and C_D, respectively. Equation (4.218b) should be compared with Eq. (4.209) in the sense that both are the constant model coefficients attached to the linear turbulent-viscosity model. The condition that $P_K=\varepsilon$ is nearly satisfied in channel flow, and expression (4.218b) is closer to the widely-accepted choice of C_ν, 0.09. This difference lies in the modeling of ε_{ij}. In Eq. (4.207), it is modeled in an isotropic form, as in Eq. (4.196), whereas Eq. (4.198) is used in Eq. (4.216). From the viewpoint of retaining the terms up to the first order in B_{ij}, the latter treatment is more reasonable.

In a two-dimensional case, a general procedure for solving algebraic equations such as Eq. (4.216) was presented by Pope (1975). Taulbee (1992) and Gatski and Speziale (1993) extended this procedure to a three-dimensional case, and solved them. As a result, we have

$$B_{ij}=-L_1\frac{K^2}{\varepsilon}S_{ij}+L_2\frac{K^3}{\varepsilon^2}\left[S_{i\ell}S_{\ell j}\right]_D+L_3\frac{K^3}{\varepsilon^2}\left(S_{i\ell}\Omega_{\ell j}+S_{j\ell}\Omega_{\ell i}\right)+\cdots. \qquad (4.219)$$

Here nondimensional coefficients L_n's are functionals of II_S, II_Ω, III_S, $III_{SS\Omega}$, and $IV_{SS\Omega\Omega}$, besides K and ε, which are defined by expressions (4.146) and (4.147):

$$L_n=\frac{L_{nN}\left\{\tau_C^2 II_S,\tau_C^2 II_\Omega,\tau_C^3 III_S,\tau_C^3 III_{S\Omega\Omega},\tau_C^4 IV_{SS\Omega\Omega}\right\}}{L_{nD}\left\{\tau_C^2 II_S,\tau_C^2 II_\Omega,\tau_C^3 III_S,\tau_C^3 III_{S\Omega\Omega},\tau_C^4 IV_{SS\Omega\Omega}\right\}}, \qquad (4.220)$$

with $\tau_C = K/\varepsilon$. The functional form of L_{nN}'s and L_{nD}'s is dependent on the algebraic equations to be adopted and the method of solution. The difference between the works by Taulbee (1992) and Gatski and Speziale (1993) is attributed to the latter. In what follows, we shall omit these details and refer to the important mathematical structures of the resulting expressions.

In the linear turbulent-viscosity model, we have

$$\langle u'^2 \rangle = \frac{2}{3}K - 2\nu_T \frac{\partial U}{\partial x}, \tag{4.221}$$

where $\mathbf{U} = (U, V, W)$ and $\mathbf{u}' = (u', v', w')$. In the case of steep mean velocity change, positive $\langle u'^2 \rangle$ is not always guaranteed under Eq. (4.221). An entirely similar situation is encountered in the use of the second-order nonlinear model given in Sec. 4.3.1. Its nonlinear parts are useful in the explanation of some important turbulence properties, but they often become a cause of numerical instability in the application of the model. This instability is closely related to the point that the dominance of the leading term over the higher-order ones is not guaranteed.

A prominent feature of Eq. (4.219) is the occurrence of II_S, II_Ω, etc. in the denominator of Eq. (4.220), L_{nD}. The foregoing difficulty may be greatly alleviated by constructing L_{nD} that obeys

$$\lim_{\chi \to \infty} L_n = 0, \tag{4.222}$$

where χ is any of $\tau_C^2 II_S$, $\tau_C^2 II_\Omega$, etc. The construction of a numerically stable model is indispensable for the application to complex engineering flows. Such nonlinear models have been developed further by Shih et al. (1994) and Abe et al. (1997). Specifically, Abe et al. (1997) incorporated low-Reynolds-number effects important near a solid wall. In the sense of Eq. (4.222), the replacement of Eq. (4.213) with

$$\frac{C_{\nu M}}{C_\nu} = \left(1 - \frac{1}{C_{\Pi 1}} \frac{1}{K} \frac{D}{Dt} \frac{K^2}{\varepsilon} \right)^{-1} \tag{4.223}$$

is helpful in the analysis of homogeneous-shear turbulence with a high shear rate (Yoshizawa and Nisizima 1993), and the time evolution of K that is obtained by the standard K-ε model can be rectified (see Fig. 4.7).

Note: In the logarithmic-velocity region near a wall, we have Eq. (4.65), which is linked with the dominance of the energy-production and energy-dissipation effects, that is, $P_K \cong \varepsilon$. We extend Eq. (4.65) to a three-dimensional case, and assume

$$\frac{D}{Dt} \frac{R_{ij}}{K} \cong 0 \tag{4.224a}$$

or

$$\frac{DR_{ij}}{Dt} \cong \frac{R_{ij}}{K}\frac{DK}{Dt}. \tag{4.224b}$$

This relation corresponds to the DK/Dt-related part in Eq. (4.214a). The validity of this type of approximation is not guaranteed in case that the streamwise variation of flow properties, which is represented by D/Dt effects, is large.

4.6. Modeling of Scalar Diffusion

A scalar quantity θ, such as temperature and matter concentration, is called a passive scalar in case that it does not have any feedback effect on fluid motion. In this case, the scalar obeys Eq. (2.49) under a given velocity field $\mathbf{u}$. The temperature can exert force to fluid motion in the presence of buoyancy effects. In the Boussinesq approximation, θ also obeys Eq. (2.49), whereas $\mathbf{u}$ is governed by Eq. (2.96).

The modeling of passive-scalar diffusion may be done in a manner quite similar to the modeling of fluid motion that was discussed in detail in the previous sections. In the buoyancy case, its modeling is also similar, but the feedback effects on fluid motion bring some new aspects that are not seen in the passive-scalar case. The critical difference between two cases is related to the linearity. In what follows, we shall explain the modeling of passive-scalar diffusion with the aid of the knowledge about the modeling of fluid motion.

4.6.1. PASSIVE-SCALAR DIFFUSION

A. Linearity Principle

In the passive-scalar case, we have the important requirement called the linearity principle, which was first pointed out by Pope (1983) in a clear mathematical form. We consider two independent scalars, θ_1 and θ_2, both of which obey Eq. (2.49). In Eq. (2.102) for the mean scalars Θ_1 and Θ_2, we have two turbulent fluxes $\mathbf{H}_{\theta_1}$ and $\mathbf{H}_{\theta_2}$, which are defined as

$$\mathbf{H}_{\theta_n} = \left\langle \theta'_n \mathbf{u}' \right\rangle. \tag{4.225}$$

From the linearity of Eq. (2.49), any linear combination of θ_1 and θ_2,

$$\theta_{12} = C_1\theta_1 + C_2\theta_2, \tag{4.226}$$

also obeys Eq. (2.49). As a result, the scalar flux $\mathbf{H}_{\theta_{12}}$, which is defined by

$$\mathbf{H}_{\theta_{12}} = \left\langle \theta'_{12} \mathbf{u}' \right\rangle, \tag{4.227}$$

obeys the same relation as Eq. (4.226),

$$\mathbf{H}_{\theta_{12}} = C_1 \mathbf{H}_{\theta_1} + C_2 \mathbf{H}_{\theta_2}, \tag{4.228}$$

and any models for $\mathbf{H}_{\theta_{12}}$ should not violate Eq. (4.228).

In scalar diffusion, we have two typical time scales

$$\tau_u = K/\varepsilon, \ \tau_\theta = K_\theta / \varepsilon_\theta \ . \tag{4.229}$$

Here K_θ and ε_θ are the scalar variance and its destruction rate, respectively, which are defined by Eqs. (2.125) and (2.128). As is discussed in relation to Eq. (3.50), τ_u is a characteristic time scale of energy-containing eddies, whereas τ_θ is the scalar counterpart. The latter is not linear in θ', and it cannot enter a model for $\mathbf{H}_\theta$ in the passive-scalar case. Therefore the time scale affecting $\mathbf{H}_\theta$ is τ_u.

Let us consider the scalar variance K_θ in light of the linearity principle. Two independent scalars θ'_1 and θ'_2 lead to

$$K_{\theta_{12}} \equiv \left\langle \left(C_1 \theta'_1 + C_2 \theta'_2 \right)^2 \right\rangle = C_1^2 K_{\theta_1} + C_2^2 K_{\theta_2}. \tag{4.230}$$

Therefore the linearity principle in the sense of Eq. (4.228) does not hold in the $\langle \theta'^2 \rangle$-related modeling, and both the time scales τ_u and τ_θ may enter it. As an example supporting this point, we may mention the numerical study of Herring et al. (1982) based on the eddy-damped quasi-normal Markovian (EDQNM) approximation explained in Sec. 3.5.5.B. The study shows that the time-scale ratio γ, which is defined by

$$\gamma = \tau_\theta / \tau_u \ , \tag{4.231}$$

is closely related to the ratio of the peak-value wavenumber of the energy spectrum to the scalar counterpart. This fact also indicates that τ_θ is not determined simply by the macroscopic properties of fluid motion such as K and ε.

B. Algebraic Modeling of Scalar Flux

In the algebraic modeling of the turbulent scalar flux $\mathbf{H}_\theta$, we should note that $\mathbf{H}_\theta$ is linear in θ-related quantities such as the mean scalar Θ. Concerning the quantities expressing fluid motion, we adopt K, ε, S_{ij}, and Ω_{ij} in a manner quite similar to the algebraic modeling of the Reynolds stress R_{ij}.

A model containing the lowest-order correction to the linear turbulent-diffusivity approximation is written as

$$H_{\theta i} = -\frac{\nu_T}{Pr_T} \frac{\partial \Theta}{\partial x_i} + \left(\kappa_1 S_{ij} + \kappa_2 \Omega_{ij} \right) \frac{\partial \Theta}{\partial x_j}. \tag{4.232}$$

Here ν_T is the turbulent viscosity defined by Eq. (4.8), and Pr_T may be called the turbulent Prandtl number. The coefficients κ_n's ($n = 1, 2$) are written as

$$\kappa_n = C_{\kappa_n} \left(K^3 \big/ \varepsilon^2 \right), \tag{4.233}$$

where C_{κ_n}'s are constant. In a manner similar to the nonlinear modeling of R_{ij} in Sec. 4.3.2.A, we can functionalize C_{κ_n}'s with the aid of the invariants given by Eqs. (4.146) and (4.147). The turbulent Prandtl number Pr_T is dependent on the molecular counterpart Pr, but it is often chosen as constant.

We use Eq. (4.232) and examine effects of the mean velocity gradient on $\mathbf{H}_\theta$. We consider the situation in which both the mean velocity and temperature gradients are constant; namely, we write

$$\mathbf{U} = (Sy, 0, 0)\,, \ \Theta = S_\theta y. \tag{4.234}$$

Under Eq. (4.234), Eq. (4.232) is reduced to

$$H_{\theta y} = -\frac{\nu_T}{Pr_T} S_\theta, \tag{4.235a}$$

$$H_{\theta x} = (\kappa_1 - \kappa_2)\, S S_\theta. \tag{4.235b}$$

Equation (4.235a) is familiar in the sense that the heat transfer occurs in the direction opposite to the mean temperature gradient. On the other hand, Eq. (4.235b) represents the heat transfer normal to the mean temperature gradient in the combination with the mean velocity gradient. This type of transfer was examined in detail in the observation by Tavoularis and Corrsin (1981, 1985).

C. Modeling of Scalar Turbulence Equations

Equation (2.102) is combined with Eq. (4.232) to give a closed equation for Θ. In order to estimate the magnitude of the fluctuation of θ around Θ, we need the equation for the scalar variance K_θ, which is given by Eq. (2.126) and is very similar to the K equation (2.111). In it, the destruction and transport rates, ε_θ and $\mathbf{T}_\theta$, need to be modeled.

C.1. Modeling of $\mathbf{T}_\theta$. In Eq. (2.129) for $\mathbf{T}_\theta$, we pay attention to the turbulence transport term and neglect the molecular diffusion one. The former represents the transport rate of θ'^2 due to $\mathbf{u}'$. Under the gradient-diffusion approximation, we have

$$\langle \theta'^2 \mathbf{u}' \rangle = -\frac{\nu_T}{\sigma_\theta} \nabla K_\theta, \tag{4.236}$$

where σ_θ is a nondimensional coefficient and is usually chosen as constant.

Equation (4.236) should be compared with Eq. (4.16) for the energy transport rate $\langle (\mathbf{u}'^2/2)\mathbf{u}' \rangle$. In the latter case, the inhomogeneity of ε affects the transport rate in addition to that of K, and the significance of the ε-related part in Eq. (4.17) is discussed in Sec. 4.2.4.A. In this sense, the inhomogeneity of ε_θ exerts influence to Eq. (4.236). The scalar transport rate $\langle \theta'^2 \mathbf{u}' \rangle$ is generated originally by velocity fluctuations, and it is proper to consider that the rate is also dependent on the inhomogeneity of the quantities characterizing these fluctuations. The representative of such quantities is ε, as is seen from the inertial-range energy spectrum (3.45). Then Eq. (4.236) is replaced with

$$\langle \theta'^2 \mathbf{u}' \rangle = -\frac{\nu_T}{\sigma_\theta} \nabla K_\theta + \mu_1 \nabla \varepsilon_\theta - \mu_2 \nabla \varepsilon, \tag{4.237}$$

where the dimensional coefficients μ_n's are written as

$$\mu_1 = C_{\mu_1} \frac{K^2 K_\theta}{\varepsilon \varepsilon_\theta}, \quad \mu_2 = C_{\mu_2} \frac{K^2 K_\theta}{\varepsilon^2} \tag{4.238}$$

(C_{μ_n} are constants). Here C_{μ_1} is inferred to be positive from the correspondence between K and ε in Eq. (4.17). A larger-ε region corresponds to a larger-K region or the region accompanied by strong velocity fluctuations. Then θ'^2 is considered to be transported from a larger-ε region to a smaller-ε one, resulting in positive C_{μ_2}. A more general expression for $\langle (\mathbf{u}'^2/2)\mathbf{u}' \rangle$ will be discussed in Sec. 6.7.1.

C.2. Modeling of ε_θ equation. A model equation for the K dissipation rate ε is given by Eq. (4.37). There we emphasized that its mathematical basis is much weaker than the K equation (2.111). This situation arises from the fact that ε is not a quantity obeying a conservation property, unlike K. Then it is difficult to model the ε equation on the basis of its exact counterpart (4.26).

Entirely the same situation is encountered in the modeling of the equation for the K_θ destruction rate ε_θ. A representative model equation for ε_θ is

$$\frac{D\varepsilon_\theta}{Dt} = C_{\varepsilon\theta 1} \frac{\varepsilon_\theta}{K_\theta} P_\theta + C_{\varepsilon\theta 2} \frac{\varepsilon_\theta}{K} P_K - C_{\varepsilon\theta 3} \frac{\varepsilon_\theta^2}{K_\theta} - C_{\varepsilon\theta 4} \frac{\varepsilon_\theta \varepsilon}{K} + \nabla \cdot \left(\frac{\nu_T}{\sigma_{\varepsilon_\theta}} \nabla \varepsilon_\theta \right), \tag{4.239}$$

where $C_{\varepsilon\theta n}$'s ($n = 1-4$) and $\sigma_{\varepsilon\theta}$ are positive constants. The constants optimized through the application to various flows have the following interesting feature; the values of $C_{\varepsilon\theta 1}$ and $C_{\varepsilon\theta 2}$ are rather close to those of $C_{\varepsilon\theta 3}$ and $C_{\varepsilon\theta 4}$, respectively (for instance, see Nagano and Kim 1987).

We examine the above feature of $C_{\varepsilon_\theta n}$'s in the context of the ε equation (4.37). First, we write

$$\frac{D\varepsilon}{Dt} \cong C_{K\varepsilon} \frac{\varepsilon}{K} \frac{DK}{Dt}, \tag{4.240}$$

where $C_{K\varepsilon}$ is a positive constant. Its right-hand side is reduced to

$$C_{K\varepsilon} \frac{\varepsilon}{K} P_K - C_{K\varepsilon} \frac{\varepsilon^2}{K}, \tag{4.241}$$

except the diffusion term $\mathbf{T}_K$ in the K equation. In Eq. (4.37), we have two constants, $C_{\varepsilon 1}$ and $C_{\varepsilon 2}$, and they are rather close to each other, as is seen from Eq. (4.72). Therefore the primary mathematical properties of the ε equation can be found from the K equation. This point may be interpreted as follows. A representative quantity characterizing the turbulence state is the intensity of fluctuations, that is, K. In a region with larger K, the energy cascade tends to be more enhanced, resulting in larger ε. Then it is probable that some important properties of $D\varepsilon/Dt$ can be estimated from DK/Dt.

In light of the relationship between DK/Dt and $D\varepsilon/Dt$, $D\varepsilon_\theta/Dt$ is supposed to be related to DK_θ/Dt. Moreover we can add the effect linked with fluid motion. As such an effect, we choose $D\varepsilon/Dt$. Of course, we can use DK/Dt, resulting in the similar results under Eq. (4.240) (the choice of both is redundant). We write

$$\frac{D\varepsilon_\theta}{Dt} \cong C_{K_\theta \varepsilon_\theta} \frac{\varepsilon_\theta}{K_\theta} \frac{DK_\theta}{Dt} + C_{\varepsilon \varepsilon_\theta} \frac{\varepsilon_\theta}{\varepsilon} \frac{D\varepsilon}{Dt}, \tag{4.242}$$

where $C_{K_\theta \varepsilon_\theta}$ and $C_{\varepsilon \varepsilon_\theta}$ are model constants. As this consequence, we may see that $C_{K_\theta \varepsilon_\theta}$ corresponds to $C_{\varepsilon_\theta 1}$ and $C_{\varepsilon_\theta 3}$ in Eq. (4.239). This result supports the fact that the optimized values of $C_{\varepsilon_\theta 1}$ and $C_{\varepsilon_\theta 3}$ are rather close each other. The similar situation holds for $C_{\varepsilon \varepsilon_\theta}$. This point will be discussed from a statistical standpoint in Chapter 6.

A simple method of estimating ε_θ is the use of Eq. (4.231). From Eq. (4.229), it is reduced to

$$\varepsilon_\theta = \gamma \frac{\varepsilon}{K} K_\theta. \tag{4.243}$$

In general, γ is not a constant, and it depends on the properties of the energy and scalar-variance spectra, as was noted in Sec. 4.6.1.A. Equation (4.243) with the proper choice of γ, however, provides a simple, convenient model for ε_θ in engineering applications.

4.6.2. SECOND-ORDER SCALAR MODELING

The turbulent scalar flux $\mathbf{H}_\theta$ is governed by Eq. (2.120). Similar to Eq. (2.105) for R_{ij}, we need to model $\mathbf{\Pi}_\theta$ [Eq. (2.122)], ε_θ [Eq. (2.123)], and

$T_{\theta ij}$ [Eq. (2.124)]. Specifically, $\mathbf{\Pi}_\theta$, which corresponds to Π_{ij} [Eq. (2.107)], plays a central role in the second-order scalar modeling, just as the latter does in the modeling of fluid motion. In what follows, we shall construct a model for $\mathbf{\Pi}_\theta$ after the same modeling procedure as Π_{ij}.

What is important in the modeling of $\mathbf{H}_\theta$ is the linearity principle stressed by Pope (1983). From this principle, the characteristic time scale that enters $\mathbf{\Pi}_\theta$ is given by K/ε, but not by $K_\theta/\varepsilon_\theta$. Then we follow Eq. (4.179), and assume the functional form

$$\mathbf{\Pi}_\theta = \mathbf{\Pi}_\theta \left\{ \mathbf{H}_\theta, \Theta, b_{ij}, S_{ij}, \Omega_{ij}, K, \varepsilon \right\}. \tag{4.244}$$

The linearity principle suggests to us that $\mathbf{\Pi}_\theta$ consists of the two parts linearly dependent on $\mathbf{H}_\theta$ and $\nabla\Theta$, respectively. Then we write

$$\Pi_{\theta i} = \frac{\varepsilon}{K}\phi_{ij} H_{\theta j} + K\psi_{ij}\frac{\partial \Theta}{\partial x_j}. \tag{4.245}$$

Here ϕ_{ij} and ψ_{ij} should be written in terms of quantities related to fluid motion. In this sense, they are quite similar to Π_{ij} discussed in Sec. 4.4.1.B, except the fact that they are not symmetric tensors, unlike Π_{ij}. The modeling corresponding to Eq. (4.183) for Π_{ij} leads to

$$\begin{aligned}\phi_{ij} = {} & C_{\phi 1}\delta_{ij} + C_{\phi 2} b_{ij} + C_{\phi 3}\frac{K}{\varepsilon} S_{ij} + C_{\phi 4}\frac{K}{\varepsilon}\Omega_{ij} + C_{\phi 5}\frac{K}{\varepsilon} b_{i\ell} S_{\ell j} \\ & + C_{\phi 6}\frac{K}{\varepsilon} b_{j\ell} S_{\ell i} + C_{\phi 7}\frac{K}{\varepsilon} b_{i\ell}\Omega_{\ell j} + C_{\phi 8}\frac{K}{\varepsilon} b_{j\ell}\Omega_{\ell i},\end{aligned} \tag{4.246}$$

$$\begin{aligned}\psi_{ij} = {} & C_{\psi 1}\delta_{ij} + C_{\psi 2} b_{ij} + C_{\psi 3}\frac{K}{\varepsilon} S_{ij} + C_{\psi 4}\frac{K}{\varepsilon}\Omega_{ij} + C_{\psi 5}\frac{K}{\varepsilon} b_{i\ell} S_{\ell j} \\ & + C_{\psi 6}\frac{K}{\varepsilon} b_{j\ell} S_{\ell i} + C_{\psi 7}\frac{K}{\varepsilon} b_{i\ell}\Omega_{\ell j} + C_{\psi 8}\frac{K}{\varepsilon} b_{j\ell}\Omega_{\ell i},\end{aligned} \tag{4.247}$$

where $C_{\phi n}$'s $(n = 1-8)$ and $C_{\psi n}$'s $(n = 1-8)$ are model constants. The functionalization of these constants leads to a more elaborate model such as Eq. (4.180) with Eq. (4.182). As a model for $\mathbf{\Pi}_\theta$ that is faithful to the linearity principle, we may mention the work by Wakao and Kawamura (1996). In the model, the first term in Eq. (4.246) and the second one in Eq. (4.247) are retained.

The simplest model for the R_{ij} destruction rate ε_{ij} is given by Eq. (4.196), and represents the isotropic destruction. The counterpart of ε_θ vanishes since it is a vector and no preferred direction exists in the isotropic state. The leading contribution to $\boldsymbol{\varepsilon}_\theta$ is given by

$$\boldsymbol{\varepsilon}_\theta = C_{D\theta}\frac{\varepsilon}{K}\mathbf{H}_\theta, \tag{4.248}$$

where C_{D_θ} is a model constant. This should be compared with Eq. (4.198).

The remaining quantity to be modeled is the transport rate $T_{\theta ij}$, [Eq. (2.124)]. As is similar to the modeling of the R_{ij} counterpart $T_{ij\ell}$ in Sec. 4.4.2.B, we focus attention on its first part, and neglect the pressure-related one. The triple correlation $\langle u_i' u_j' \theta' \rangle$ may be regarded as the transport rate of $u_j'\theta'$ by $\mathbf{u}'$ in the i direction or that of $u_i'\theta'$ in the j direction. Under the gradient-diffusion approximation, we write

$$\left\langle u_i' u_j' \theta' \right\rangle = -\frac{\nu_T}{\sigma_{H_\theta}} \left(\frac{\partial H_{\theta j}}{\partial x_i} + \frac{\partial H_{\theta i}}{\partial x_j} \right), \tag{4.249}$$

where σ_{H_θ} is a constant, and the symmetry concerning i and j has been taken into account. After the manner reducing Eq. (4.201) to (4.205), we may rewrite Eq. (4.249) as

$$\left\langle u_i' u_j' \theta' \right\rangle = -\frac{3C_\nu}{2\sigma_{H_\theta}} \frac{K}{\varepsilon} \left(R_{i\ell} \frac{\partial H_{\theta j}}{\partial x_\ell} + R_{j\ell} \frac{\partial H_{\theta i}}{\partial x_\ell} \right). \tag{4.250}$$

Finally, let us simply refer to the relationship between the algebraic and second-order scalar modeling. As was seen in Sec. 4.5, the primary parts of an algebraic model for R_{ij} can be derived from the balance among the production, pressure-strain, and destruction terms. In the scalar case, we also assume such a balance in Eq. (2.120), and retain the terms linear in $\mathbf{H}_\theta$ and $\nabla\Theta$. As a result, we have

$$\mathbf{H}_\theta = -\frac{(2/3) - C_{\psi 1}}{C_{D_\theta} - C_{\phi 1}} \frac{K^2}{\varepsilon} \nabla\Theta, \tag{4.251}$$

which is the simplest turbulent-diffusivity model. By retaining the terms of the first order in $\nabla\mathbf{U}$, we have its extension, Eq. (4.232). The derivation of the scalar counterpart of Eq. (4.217) is also straightforward.

4.7. Low-Reynolds-Number Effects

All the turbulence models discussed in the previous sections belong to the category of the so-called high-Reynolds-number version. In these models, effects of the molecular viscosity (ν) do not appear, except through the implicit dependence of ε on it. Three-dimensional turbulence is subject to nonvanishing ε in the limit of vanishing ν, unlike two-dimensional turbulence, and the dependence of ε on ν is weak at high Reynolds number. This is the reason why the high-Reynolds-number version of models is significant in the application to real-world turbulent flows.

The foregoing situation does not deny the important role of ν in turbulence modeling. As an instance of important effects of ν, we may mention the situation in which a local Reynolds number becomes small, although the Reynolds number based on global quantities like the radius of a pipe is very large. In what follows, we shall consider effects of low Reynolds number from the viewpoint of a local Reynolds number.

4.7.1. LOCAL REYNOLDS NUMBER

As the representative local Reynolds number, we may mention the following two. One is the coordinate in wall units, y_W, which is defined as

$$y_W = u_W y/\nu, \tag{4.252}$$

where u_W is the friction velocity given by the first relation of Eq. (4.54). Equation (4.252) may be regarded as the Reynolds number based on this velocity and the distance from a solid wall. Another is

$$R_T = K^2/(\nu\varepsilon), \tag{4.253a}$$

which is also written as

$$R_T \propto \nu_T/\nu, \ \sqrt{K}\ell_C/\nu, \tag{4.253b}$$

where ν_T and ℓ_C (the characteristic spatial scale of energy-containing eddies) are defined by Eqs. (4.8) and (3.48), respectively. The second relation of Eq. (4.253b) signifies that R_T is the Reynolds number based on the length and velocity characterizing energy-containing eddies.

The wall-unit coordinate y_W plays an important role in expressing the velocity profile near a wall. For $y_W > 15$, we have the logarithmic velocity profile, Eq. (4.68) (see Fig. 4.3). It is derived as a local solution of the standard K-ε model. On the other hand, we have the entirely different velocity profile, Eq. (4.56), for $y_W < 10$. It comes from the balance of the pressure-gradient and viscous terms in the mean-velocity equation, and the nonlinear effect, which is an essential gradient in turbulence, does not occur at all. In this sense, the region of $y_W < 10$ is the low-Reynolds-number region. This near-wall region is the beyond the scope of the high-Reynolds-number version of turbulence models.

The inconsistency of high-Reynolds-number models with the noslip boundary condition can be seen clearly from the behaviors of ν_T. Near a solid wall, we have

$$u' = O(y), \ v' = O(y^2), \ w' = O(y), \tag{4.254}$$

where $\mathbf{u}' = (u', v', w')$, and the second relation comes from the solenoidal condition $\nabla \cdot \mathbf{u}' = 0$. Expression (4.254) gives

$$\langle u'v' \rangle = O(y^3), \tag{4.255}$$

resulting in

$$\nu_T = O(y^3), \tag{4.256}$$

under the turbulent-viscosity model. On the other hand, we have

$$K = O(y^2), \ \varepsilon = O(1), \tag{4.257}$$

which leads to

$$\nu_T = O(y^4). \tag{4.258}$$

This result contradicts Eq. (4.256). In order to rectify this inconsistency, we need to know the correct behaviors of various quantities near a solid boundary. This point will be referred to below.

Another local Reynolds number R_T [Eq. (4.253)] is associated with effects of ν twofold. In the close vicinity of a solid wall, K may become very small, resulting in small R_T. In this sense, R_T cannot be distinguished clearly from y_W. The latter, however, is related to u_W and loses its original meaning near a separation point, as is noted in Sec. 4.2.4.B. Then the use of both the local Reynolds numbers near a wall is not redundant.

The other aspect of R_T is not related directly to the noslip boundary condition, but it comes from the relationship with weakening of velocity fluctuations or relaminarization of turbulence. In real-world flow, we encounter various regions that possess different flow properties. For instance, velocity fluctuations are suppressed in a region subject to the rapid increase in the mean velocity. This phenomenon corresponds to the relaminarization of turbulent flow in the sense that kinetic energy is returned from small- to large-scale components of motion.

In the context of low-Reynolds-number effects on turbulence modeling, the study about the y_W and R_T effects linked with a solid-boundary is far ahead of the relaminarization counterpart. As the representative low-Reynolds-number algebraic models, we may mention the models by Myong and Kasagi (1990), Yang and Shih (1993), Abe et al. (1995), and Kawamura and Kawashima (1995). The second-order counterparts are represented by Launder and Shima (1989) and Shima (1993), and the critical review of various models is made in the work by So et al. (1991). The full use of DNS databases in turbulence modeling is detailed in the work by Kasagi and Shikazono (1995). Here we do not enter these details, but we shall simply refer to the near-wall asymptotic behaviors of some important statistics for the understanding of the guiding principle for constructing such models.

4.7.2. NEAR-WALL ASYMPTOTIC BEHAVIORS

The fluctuating velocity $\mathbf{u}'$ is expanded as

$$u' = a_1 y + a_2 y^2 + \cdots, \tag{4.259a}$$

$$v' = b_2 y^2 + \cdots, \tag{4.259b}$$

$$w' = c_1 y + c_2 y^2 + \cdots, \tag{4.259c}$$

near a solid wall. In general, the coefficients a_n's etc. are functions of x, z, and t. Corresponding to Eq. (4.259), we write the pressure fluctuation p' as

$$p' = d_0 + d_1 y + \cdots \tag{4.260}$$

(p' is defined as the original pressure fluctuation divided by constant density). From the $\mathbf{u}'$ equation at the wall,

$$\nabla p' = \nu \Delta \mathbf{u}', \tag{4.261}$$

we have the relation

$$\frac{\partial d_0}{\partial x} = 2\nu a_2, \; d_1 = 2\nu b_2, \; \frac{\partial d_0}{\partial z} = 2\nu c_2. \tag{4.262}$$

Equation (4.259) gives

$$\langle u'v' \rangle = \langle a_1 b_2 \rangle y^3 + \cdots, \tag{4.263}$$

$$K = \frac{1}{2}\langle a_1^2 + c_1^2 \rangle y^2 + \langle a_1 a_2 + c_1 c_2 \rangle y^3 + \cdots, \tag{4.264}$$

$$\varepsilon / \nu = \langle a_1^2 + c_1^2 \rangle + 4 \langle a_1 a_2 + c_1 c_2 \rangle y + \cdots. \tag{4.265}$$

From Eqs. (4.264) and (4.265), R_T [Eq. (4.253a)] is written as

$$R_T = \frac{\langle a_1^2 + c_1^2 \rangle}{4\nu^2} y^4 + \cdots. \tag{4.266}$$

In order to obtain the correct asymptotic behavior for ν_T, expression (4.256), we usually introduce the so-called wall-damping function f_W, which is a functional of y_W and R_T, as

$$f_W = f_W \{y_W, R_T\}, \tag{4.267}$$

where f_W is subject to the constraint

$$\lim_{y_W, R_T \to 0} f_W \{y_W, R_T\} = O(y^{-1}), \tag{4.268a}$$

$$\lim_{y_W, R_T \to \infty} f_W \{y_W, R_T\} = 1. \tag{4.268b}$$

By attaching this f_W to ν_T, we can recover the correct asymptotic behavior of $\langle u'v' \rangle$ in the close vicinity of a wall, Eq. (4.263). There is no method for uniquely introducing f_W. For example, f_W in the model by Myong an Kasagi (1990) is given by

$$f_W = \left(1 + D_1 R_T^{-1/2}\right)\left(1 - \exp\left(-\frac{y_W}{D_2}\right)\right). \tag{4.269}$$

Here D_1 and D_2 are model constants, and their optimized values are dependent on the magnitude of the constants in other wall-damping functions.

Let us examine the K equation (2.111). At the wall, we have

$$\nu \frac{\partial^2 K}{\partial y^2} - \varepsilon = \nabla \cdot \left(\left\langle \frac{\mathbf{u}'^2}{2}\mathbf{u}' \right\rangle\right) + \nabla \cdot \langle p'\mathbf{u}' \rangle. \tag{4.270}$$

From Eqs. (4.264) and (4.265), the left-hand side is written as

$$\nu \frac{\partial^2 K}{\partial y^2} - \varepsilon = 2\nu \langle a_1 a_2 + c_1 c_2 \rangle y + \cdots. \tag{4.271}$$

On the other hand, each term on the right-hand side gives

$$\nabla \cdot \left(\left\langle \left(\mathbf{u}'^2/2\right) \mathbf{u}' \right\rangle\right) = O(y^2), \tag{4.272}$$

$$\nabla \cdot \langle p'\mathbf{u}' \rangle = \langle \mathbf{u}' \cdot \nabla p' \rangle = 2\nu \langle a_1 a_2 + c_1 c_2 \rangle y + \cdots. \tag{4.273}$$

In deriving the latter, we used Eq. (4.262).

Here we should note that Eq. (4.271) balances with Eq. (4.273) but not with Eq. (4.272). In the turbulence modeling, however, we usually pay attention to the triple velocity correlation, and neglect the pressure-velocity correlation. The foregoing discussion shows clearly that in the close vicinity of a solid wall, the latter is more important and its modeling is indispensable for the proper description of near-wall behaviors of ε. The DNS of channel turbulence (Mansour et al. 1989) shows that ε is maximum at the wall. Most of turbulence models give maximum ε off the wall, owing to the neglect of the effect of Eq. (4.273). In order to rectify this shortfall, $\langle p'\mathbf{u}' \rangle$ is modeled by Kawamura and Kawashima (1995) as

$$\langle p'\mathbf{u}' \rangle = \frac{\nu}{2}\frac{K}{\varepsilon}\nabla\left(2\nu\left(\nabla\sqrt{K}\right)^2\right) \tag{4.274}$$

or in a more elaborate form, which satisfies the asymptotic behavior (4.273).

Finally, we refer to Eq. (4.198). Its left- and right-hand sides are

$$\varepsilon_{xy} = 4\nu \langle a_1 b_2 \rangle y + \cdots, \tag{4.275a}$$

$$C_D \varepsilon b_{xy} = 2C_D \nu \langle a_1 b_2 \rangle y, \tag{4.275b}$$

respectively. Therefore we are led to Eq. (4.199), that is, $C_D = 2$.

References

Abe, K., Kondoh, T., and Nagano, Y. (1995), Intern. J. Heat Mass Transfer **38**, 1467.

Abe, K., Kondoh, T., and Nagano, Y. (1997), Intern. J. Heat and Fluid Flow **18**, 266.

Bardina, J., Ferziger, J. H., and Rogallo, R. S. (1985), J. Fluid Mech. **154**, 321.

Comte-Bellot, G. and Corrsin, S. (1966), J. Fluid Mech. **25**, 657.

Durbin, P. A. (1993), J. Fluid Mech. **249**, 465.

Gatski, T. B. and Speziale, C. G. (1993), J. Fluid Mech. **254**, 59.

Hallbäck, M., Johansson, A. V., and Burden, A. D. (1996), in *Turbulence and Transition Modeling*, edited by M. Hallbäck, D. S. Henningson, A. V. Johansson, and P. H. Alfredson (Kluwer, Dordrecht), p. 81.

Herring, J. R., Schertzer, D., Lesieur, M., Newman, G. R., Chollet, J. P., and Larcheveque, H. (1982), J. Fluid Mech. **124**, 411.

Hunt, J. C. R. and Carruthers, D. J. (1990), J. Fluid Mech. **212**, 497.

Hussain, A. K. M. F. and Reynolds, W. C. (1975), J. Fluids Engn. **97**, 568.

Jacquin, L., Leuchter, O., Cambon, C., and Mathiew, J. (1990), J. Fluid Mech. **220**, 1.

Kajishima, T. and Miyake, Y. (1992), Computer and Fluids. **21**, 151.

Kasagi, N. and Shikazono, N. (1995), Proc. Roy. Soc. London A **451**, 257.

Kasagi, N., Tomita, Y., and Kuroda, A. (1992), J. Heat Transfer **114**, 598.

Kawamura, H. and Kawashima, N. (1995), in *Turbulence, Heat and Mass Transfer 1*, edited by K. Hanjalic and J. C. F. Pereira (Begell House, New York), p. 197.

Kim, J., Moin, P., and Moser, R. (1987), J. Fluid Mech. **177**, 133.

Kitoh, O. (1991), J. Fluid Mech. **225**, 445.

Kobayashi, T. and Togashi, S. (1996), JSME Intern. J. B **39**, 453.

Launder, B. E. (1996), in *Turbulence and Transition Modeling*, edited by M. Hallbäck, D. S. Henningson, A. V. Johansson, and P. H. Alfredson (Kluwer, Dordrecht), p. 193.

Launder, B. E., Reece, G., and Rodi, W. (1975), J. Fluid Mech. **68**, 537.

Launder, B. E. and Shima, N. (1989), AIAA J. **27**, 1319.

Lumley, J. L. (1978), Adv. in Appl. Mech. **18**, 123.

Mansour, N. N., Kim, J., and Moin, P. (1989), AIAA J. **27**, 1068.

Mansour, N. N., Cambon, C., and Speziale, C. G. (1992), in *Studies in Turbulence*, edited by G. Gatski, S. Sarkar, and C. G. Speziale (Springer, New York), p. 59.

Melling, A. and Whitelaw, J. H. (1976), J. Fluid Mech. **78**, 289.

Mellow, G. L. and Yamada, T. (1982), Rev. Geophys. Space Phys. **20**, 851.

Murakami, S., Mochida, A., and Ooka, R. (1993), in *Ninth Symposium on Turbulent Shear Flows*, 13-5.
Myong, H. K. and Kasagi, N. (1990), JSME Intern. J. **33**, 63.
Nagano, Y. and Kim, C. (1987), Trans. JSME B **53**, 1773.
Nisizima, S. (1990), Theor. Comput. Fluid Dyn. **2**, 61.
Pope, S. B. (1975), J. Fluid Mech. **72**, 331.
Pope, S. B. (1983), Phys. Fluids **26**, 404.
Rodi, W. (1976), Z. angew. Math. Mech. **56**, 219.
Rotta, J. C. (1951), Z. Phys. **129**, 547.
Shih, T. -H. (1996), in *Turbulence and Transition Modeling*, edited by M. Hallbäck, D. S. Henningson, A. V. Johansson, and P. H. Alfredson (Kluwer, Dordrecht), p. 155.
Shih, T. -H., Chen, J. -Y., and Lumley, J. L. (1992), AIAA J. **30**, 1553.
Shih, T. -H., Zhu, J., and Lumley, J. L. (1994), NASA TM 106644.
Shih, T. -H., Zhu, J., Liou, W., Chen, K. -H., Liu, N. -S., and Lumley, J. L. (1997), NASA TM 113112.
Shima, N (1993), J. Fluid Engn. **115**, 56.
Shimomura, Y. (1997), in *Eleventh Symposium on Turbulent Shear Flows*, 34-1.
So, R. M. C., Lai, Y. G., Zang, H. S., and Hwang, B. C. (1991), AIAA J. **29**, 1819.
Speziale, C. G. (1987), J. Fluid Mech. **178**, 459.
Speziale, C. G. (1991), Annu. Rev. Fluid Mech. **23**, 107.
Speziale, C. G. (1997), Intern. J. Nonlinear Mech. **33**, 579.
Speziale, C. G., Sarkar, S., and Gatski, T. (1991), J. Fluid Mech. **227**, 245.
Stratford, B. S. (1959), J. Fluid Mech. **5**, 1.
Takemitsu, N. (1987), Trans. JSME B **53**, 2928.
Takemitsu, N. (1989), Trans. JSME B **55**, 2684.
Takemitsu, N. (1990), J. Fluids Engn. **112**, 192.
Taulbee, D. (1992), Phys. Fluids A **4**, 2555.
Tavouralis, S. and Corrsin, S. (1981), J. Fluid Mech. **104**, 311.
Tavouralis, S. and Corrsin, S. (1985), Intern. J. Heat Mass Transfer **28**, 265.
Tennekes, H. and Lumley, J. L. (1972), *A First Course in Turbulence* (The MIT, Cambridge).
Wakao, Y. and Kawamura, H. (1996), Trans. JSME B **62**, 3934.
Yang, Z. and Shih, T. -H. (1993), NASA TM 106263.
Yoshizawa, A. (1984), Phys. Fluids **27**, 1377.
Yoshizawa, A. (1994), Phys. Rev. E **49**, 4065.
Yoshizawa, A. and Nisizima, S. (1993), Phys. Fluids A **5**, 3302.

CHAPTER 5

SUBGRID-SCALE MODELING

5.1. Large Eddy Simulation and Subgrid-Scale Modeling

In Chapters 3 and 4, we used the ensemble-mean procedure and divided a flow quantity into the large-scale part (mean field) and the fluctuation around it. A prominent feature of the resulting mean field is that its geometrical properties are dependent strongly on boundaries. For instance, the mean flow is one-dimensional in a flow between two parallel plates (channel turbulence) and two-dimensional in a square duct, whereas it is three-dimensional in a flow past a cubic body. In the first two examples, only the highly symmetric components of turbulent motion are retained as the mean field, and all the other components need to be modeled. The resulting system of equations for the mean field is simple, but the burden of the modeling of fluctuation effects becomes heavy, as is seen in Chapter 4. The energy-containing eddies possessing most of the energy of fluctuation, which were referred to in Sec. 3.2.2, are directly linked with the mean flow through the turbulent-energy production process represented by P_K [Eq. (2.112)]. This fact indicates that it is difficult to construct a universal ensemble-mean turbulence model applicable to various types of flows.

The advancement of the computer capability for these ten years is great. In the analyses of turbulent flows subject to simple geometrical conditions, such as channel and pipe flows, this merit cannot be utilized fully under the ensemble-mean modeling. On the other hand, the direct numerical simulation (DNS) of turbulent flows at Reynolds numbers higher than 10^4 is still beyond the capability of the present computer, as is noted in Chapter 1. A representative computational method bridging the gap between the DNS and the ensemble-mean modeling is large eddy simulation (LES). In LES, we use a spatial filtering and eliminate the small-scale components of motion contributing to the energy dissipation, and model their effects on the retained ones. The width of a filter is closely related to the size of computational grids. Therefore the spatial scale of the components to be modeled in LES is much smaller than the ensemble-mean counterpart. These components are much less affected by flow geometry than the large-scale ones. This type of modeling is called the subgrid-scale (SGS) modeling, and its universality is expected to be higher, compared with the ensemble-mean modeling. The SGS modeling, however, has various aspects different from the ensemble-mean one, leading to a cause of the difficulty of the modeling.

5.2. Filtering Procedures

5.2.1. INTRODUCTION OF FILTERING

In LES, we introduce a filter function $G(\mathbf{x}, \mathbf{x}')$ and eliminate small-scale components of a flow quantity f as

$$\overline{f}(\mathbf{x}) = \int G(\mathbf{x}, \mathbf{x}') f(\mathbf{x}') d\mathbf{x}' \tag{5.1}$$

(Leonard 1974). When two different filters are applied consecutively, we need to distinguish between them, and write the filtering based on a filter G_A as

$$\overline{f}^A(\mathbf{x}) = \int G_A(\mathbf{x}, \mathbf{x}') f(\mathbf{x}') d\mathbf{x}'. \tag{5.2}$$

On applying two filters, G_A and G_B, in this order, we use the definition

$$\overline{f}^{AB} = \int G_B(\mathbf{x}, \mathbf{x}') \overline{f}^A(\mathbf{x}') d\mathbf{x}'. \tag{5.3}$$

As this consequence, f is divided into two parts as

$$f = \overline{f} + f'. \tag{5.4}$$

Here $\overline{f}$ and f' are called the grid-scale (GS) and subgrid-scale (SGS) components, respectively.

In LES, the following two filters are often adopted. One is the top-hat filter, using which $\overline{f}$ is written as

$$\overline{f}(\mathbf{x}) = \frac{1}{\Delta^3} \int_{x-(\Delta/2)}^{x+(\Delta/2)} \int_{y-(\Delta/2)}^{y+(\Delta/2)} \int_{z-(\Delta/2)}^{z+(\Delta/2)} f(\mathbf{x}') dx' dy' dz'. \tag{5.5}$$

Another is the Gaussian filter, and the counterpart of Eq. (5.5) is

$$\overline{f}(\mathbf{x}) = \left(\frac{6}{\pi \Delta^2} \right)^{3/2} \int_{-\infty}^{\infty} \int_{-\infty}^{\infty} \int_{-\infty}^{\infty} \exp \left(- \frac{(\mathbf{x} - \mathbf{x}')^2}{\Delta^2/6} \right) f(\mathbf{x}') dx' dy' dz'. \tag{5.6}$$

In the application of Eqs. (5.5) and (5.6) to LES, the filter whose width is different in three directions as

$$\Delta = (\Delta_x, \Delta_y, \Delta_z) \text{ or } (\Delta_i; i = 1-3) \tag{5.7}$$

is often adopted.

In the Gaussian filter, the range of integration is $(-\infty, \infty)$, and its use in the direction normal to a boundary is not proper near it. Therefore in a channel flow, we use the top-hat filter in the (y) direction normal to two walls, and the Gaussian filter in the mean-flow (x) and spanwise (z) directions. In the SGS modeling that complements the effects of filtered-out components, we need to define one spatial scale characterizing three different Δ_i's, ℓ_F. As the typical examples of ℓ_F, we may mention

$$\ell_F = (\Delta_1 \Delta_2 \Delta_3)^{1/3} \tag{5.8a}$$

and

$$\ell_F = \sqrt{\left(\Delta_1^2 + \Delta_2^2 + \Delta_3^2\right)/3}. \tag{5.8b}$$

Both of them give rise to no critical difference in case that three Δ_i's do not differ so much one another. In the practical computation, however, the size of grids normal to a solid boundary is often much smaller than the other two. In the case, the difference of definition is not negligible.

The filters such as Eqs. (5.5) and (5.6) are difficult to apply to the study of SGS models using the statistical theories discussed in Chapter 3. For, these filters are not related to the ensemble-mean procedure, which is an essential ingredient of those theories. A filter applicable to the statistical SGS modeling is

$$\overline{f}(\mathbf{x}) = \int_{k<\pi/\Delta} f(\mathbf{k}) \exp\left(-i\mathbf{k}\cdot\mathbf{x}\right) d\mathbf{k} + \left\langle \int_{k>\pi/\Delta} f(\mathbf{k}) \exp\left(-i\mathbf{k}\cdot\mathbf{x}\right) d\mathbf{k} \right\rangle \tag{5.9}$$

(Yoshizawa 1982), where the Fourier component of $f(\mathbf{x})$, $f(\mathbf{k})$, is defined by Eq. (3.18), and π/Δ represents the wavenumber of the largest eddy in the interval Δ (we should note that the wavenumber $k(=|\mathbf{k}|)$ is related to the wavelength λ as $k = 2\pi/\lambda$ and that a pair of turbulent eddies of size Δ and opposite circulation corresponds to a single wave). The SGS component of f, f', is given by

$$f'(\mathbf{x}) = \int_{k>\pi/\Delta} f(\mathbf{k}) \exp\left(-i\mathbf{k}\cdot\mathbf{x}\right) d\mathbf{k} - \left\langle \int_{k>\pi/\Delta} f(\mathbf{k}) \exp\left(-i\mathbf{k}\cdot\mathbf{x}\right) d\mathbf{k} \right\rangle. \tag{5.10}$$

In the foregoing procedures, the filtering is made only spatially, but the temporal properties of f are also affected by the filtering. As Eq. (3.46) shows, each component of fluctuations has its intrinsic time scale. As a result, the spatial filtering-out of one component is equivalent to the temporal filtering-out of its intrinsic time scale.

5.2.2. FUNDAMENTAL PROPERTIES OF FILTERING

The filtering procedure brings some entirely new aspects that are not encountered in the ensemble-mean procedure. Such properties arise from the relation

$$\overline{\overline{f}} \neq \overline{f}, \ \overline{f'} \neq 0. \tag{5.11}$$

On applying the filtering to the product of f and g, we have

$$\overline{fg} = \overline{\overline{f}\,\overline{g}} + \overline{\overline{f}g'} + \overline{f'\overline{g}} + \overline{f'g'} \tag{5.12}$$

(Leonard 1974; Rogallo and Moin 1984). In Eq. (5.12), $\overline{\overline{f}g'}$ and $\overline{f'\overline{g}}$ express the degree of the correlation between the GS and SGS components of motion, and may be called the near-interaction effect. On the other hand, $\overline{f'g'}$ is the effect of the SGS-SGS interaction on the larger GS component, and mainly represents the distant-interaction effect in the scale hierarchy. In the ensemble-mean procedure, the counterparts of $\overline{\overline{f}g'}$ and $\overline{f'\overline{g}}$ vanish identically, and all the averaging effects occur in the form of the latter.

The so-called turbulent-viscosity approximation is proper for expressing the distant-interaction effect. In turbulence, there exists no clear-cut separation of scales between the mean and fluctuation, unlike the relationship between macroscopic and molecular motions through the molecular viscosity. In reality, the higher-order corrections to the turbulent-viscosity approximation become necessary for explaining various types of turbulent flows, as is noted in Chapter 4. These corrections originate in the near-interaction effects coming from the absence of the distinct scale separation. The role of the near interaction increases greatly in the filtering procedure, as is typically shown by $\overline{\overline{f}g'}$ and $\overline{f'\overline{g}}$.

In the decomposition by Eq. (5.4), $\overline{f}$ is the primary part of f from the viewpoint of energy-containing components. In this sense, the approximation

$$\overline{fg} \cong \overline{f}\,\overline{g} \tag{5.13}$$

is often useful in the estimate of filtered terms. For instance, $\overline{\overline{f}g'}$ may be approximated as

$$\overline{\overline{f}g'} \cong \overline{\overline{f}} \left(\overline{g} - \overline{\overline{g}} \right), \tag{5.14}$$

using Eq. (5.13).

In addition to Eq. (5.11), we should mention the important relation intrinsic to the filtering procedure pointed out first by Germano (1992), who paved the way for a new SGS modeling, that is, the dynamic one to be discussed later. We consider two filters with different filter width, G_A and G_B. We make the filtering of fg using G_A and define

$$\vartheta^A = \overline{fg}^A - \overline{f}^A \overline{g}^A. \tag{5.15}$$

For the consecutive application of G_A and G_B in this order, we write

$$\vartheta^{AB} = \overline{fg}^{AB} - \overline{f}^{AB}\overline{g}^{AB}. \tag{5.16}$$

The filtering of Eq. (5.15) using G_B leads to

$$\overline{\vartheta^A}^B = \overline{fg}^{AB} - \overline{\overline{f}^A\overline{g}^A}^B. \tag{5.17}$$

From Eqs. (5.16) and (5.17), we have

$$\vartheta^{AB} - \overline{\vartheta^A}^B = \overline{\overline{f}^A\overline{g}^A}^B - \overline{f}^{AB}\overline{g}^{AB}. \tag{5.18}$$

Namely, the right-hand side is written in terms of the GS quantities only. When some models are made for ϑ^A and ϑ^{AB}, they should obey Eq. (5.18). This relation is an invariance property of the filtering procedure and provides a constraint on model coefficients.

5.2.3. FILTERED EQUATIONS

A. Equations for GS Components

We apply the filtering procedure to the incompressible Navier-Stokes equation (2.30), and have the equation for the GS component $\overline{\mathbf{u}}$

$$\frac{\partial \overline{u}_i}{\partial t} + \frac{\partial}{\partial x_j}\overline{u}_i\overline{u}_j = -\frac{\partial \overline{p}}{\partial x_i} + \frac{\partial}{\partial x_j}\left(-L_{ij} - C_{ij} - \tau_{ij}\right) + \nu\Delta\overline{u}_i, \tag{5.19}$$

with the solenoidal condition

$$\nabla \cdot \overline{\mathbf{u}} = 0, \tag{5.20}$$

where $\overline{p}$ is the GS component of the original pressure divided by constant density, and

$$L_{ij} = \overline{\overline{u}_i\overline{u}_j} - \overline{u}_i\overline{u}_j, \tag{5.21}$$

$$C_{ij} = \overline{\overline{u}_i u'_j} + \overline{u'_i\overline{u}_j}, \tag{5.22}$$

$$\tau_{ij} = \overline{u'_i u'_i}. \tag{5.23}$$

In Eqs. (5.21)-(5.23), each term is usually named the Leonard term, the cross term, and the SGS Reynolds stress, respectively. The first two are the quantities specific to the filtering procedure, and the third is the SGS counterpart of the Reynolds stress in the ensemble-mean procedure.

Equations (5.21) and (5.22) correspond to the foregoing near-interaction effects on $\overline{\mathbf{u}}$ in the sense that $\overline{\mathbf{u}}$ itself is contained in the interaction. On the other hand, Eq. (5.23) not dependent directly on $\overline{\mathbf{u}}$ may be regarded as the term to which the distant parts in the scale hierarchy contribute much. From the viewpoint of modeling, L_{ij} is expressed in terms of the GS components only and may be computed in the course of LES, whereas C_{ij} and τ_{ij} are dependent on the SGS components and need to be modeled. This type of modeling is called the SGS modeling in the context of the ensemble-mean modeling of Chapter 4.

We perform the consecutive application of two filters G_A and G_B. First, the filtering based on G_A gives

$$\frac{\partial \overline{u}_i^A}{\partial t} + \frac{\partial}{\partial x_j}\overline{u}_i^A\overline{u}_j^A = -\frac{\partial \overline{p}^A}{\partial x_i} - \frac{\partial T_{ij}^A}{\partial x_j} + \nu\Delta\overline{u}_i^A, \tag{5.24}$$

where T_{ij}^A is defined as

$$T_{ij}^A = \overline{u_i u_j}^A - \overline{u}_i^A\overline{u}_j^A, \tag{5.25a}$$

which is also written as

$$T_{ij}^A = L_{ij}^A + C_{ij}^A + \tau_{ij}^A, \tag{5.25b}$$

where L_{ij}^A etc. are the counterparts of L_{ij} [Eq. (5.21)] etc. The consecutive application of G_A and G_B, which is defined by Eq. (5.3), leads to

$$\frac{\partial \overline{u}_i^{AB}}{\partial t} + \frac{\partial}{\partial x_j}\overline{u}_i^{AB}\overline{u}_j^{AB} = -\frac{\partial \overline{p}^{AB}}{\partial x_i} - \frac{\partial T_{ij}^{AB}}{\partial x_j} + \nu\Delta\overline{u}_i^{AB}, \tag{5.26}$$

with

$$T_{ij}^{AB} = \overline{u_i u_j}^{AB} - \overline{u}_i^{AB}\overline{u}_j^{AB}. \tag{5.27}$$

By Germano's identity (5.18), we have

$$T_{ij}^{AB} - \overline{T_{ij}^A}^B = \overline{\overline{u}_i^A\overline{u}_j^A}^B - \overline{u}_i^{AB}\overline{u}_j^{AB}. \tag{5.28}$$

B. Equations for SGS Components

We subtract Eq. (5.19) from Eq. (2.30), and have the equation for the SGS component $\mathbf{u}'$ as

$$\frac{\partial u_i'}{\partial t} + \overline{u}_j\frac{\partial u_i'}{\partial x_j} + u_j'\frac{\partial \overline{u}_i}{\partial x_j} + \frac{\partial}{\partial x_j}\left(u_i'u_j' - L_{ij} - C_{ij} - \tau_{ij}\right) = -\frac{\partial p'}{\partial x_i} + \nu\Delta u_i'. \tag{5.29}$$

We define the SGS energy K_S as

$$K_S = \overline{K_S'},\ \ K_S' = \mathbf{u}'^2/2 \tag{5.30}$$

(subscript S denotes the SGS counterpart of an ensemble-mean quantity). From Eq. (5.29), K_S obeys

$$\frac{\partial K_S}{\partial t} + \overline{\overline{u}_j \frac{\partial K'_S}{\partial x_j}} = -\overline{u'_i u'_j \frac{\partial \overline{u}_i}{\partial x_j}} - \nu \overline{\left(\frac{\partial u'_i}{\partial x_j} \right)^2}$$
$$+ \frac{\partial}{\partial x_i} \left(-\overline{K'_S u'_i} - \overline{p' u'_i} + \nu \frac{\partial K_S}{\partial x_i} \right) + \overline{u'_i \frac{\partial}{\partial x_j} (L_{ij} + C_{ij} + \tau_{ij})}. \tag{5.31}$$

Equation (5.31) has the mathematical structure similar to the ensemble-mean counterpart (2.111). The first three terms on the right-hand side of the former correspond to P_K [Eq. (2.112)], ε [Eq. (2.113)], and $\nabla \cdot \mathbf{T}_K$ [$\mathbf{T}_K$ is defined by Eq. (2.114)], respectively, although the fourth has no ensemble-mean counterpart. In the details of mathematical structures, however, Eq. (5.31) is much more complicated, as is seen from the second term on the left-hand side that is written using K'_S but not K_S.

In the ensemble-mean modeling, some transport equations such as the K equation (2.111) are adopted for describing the turbulence characteristics. The necessity of the complicated modeling based on their use is the return from the much reduction of computational burden. In LES, however, preference is given to the computational burden rather than the use of elaborate models, and the models not using SGS transport equations are widely used. A typical example of the SGS modeling in which the transport equations play an important role is thermally buoyant flow. In this case, explicit effects of buoyancy on K_S may be taken into account by adding the term

$$- \alpha \mathbf{g} \cdot \overline{\theta' \mathbf{u}'} \tag{5.32}$$

to the right-hand side of Eq. (5.31) under the Boussinesq approximation [see Eq. (2.96)].

A simple form of Eq. (5.31) is obtained using the approximation (5.13) and is written as

$$\frac{\partial K_S}{\partial t} + \overline{\overline{u}}_j \frac{\partial K_S}{\partial x_j} = -\overline{u'_i u'_j} \frac{\partial \overline{\overline{u}}_i}{\partial x_j} - \nu \overline{\left(\frac{\partial u'_i}{\partial x_j} \right)^2}$$
$$+ \frac{\partial}{\partial x_i} \left(-\overline{K'_S u'_i} - \overline{p' u'_i} + \nu \frac{\partial K_S}{\partial x_i} \right), \tag{5.33}$$

where the last term on the right-hand side of Eq. (5.31) was dropped. An equation simpler than Eq. (5.33) is given by

$$\frac{DK_S}{Dt_S} = -\overline{u'_i u'_j} \frac{\partial \overline{u}_i}{\partial x_j} - \nu \overline{\left(\frac{\partial u'_i}{\partial x_j} \right)^2} + \frac{\partial}{\partial x_i} \left(-\overline{K'_S u'_i} - \overline{p' u'_i} + \nu \frac{\partial K_S}{\partial x_i} \right), \tag{5.34}$$

with

$$\frac{D}{Dt_S} = \frac{\partial}{\partial t} + \overline{\mathbf{u}} \cdot \nabla. \tag{5.35}$$

Here $\overline{\mathbf{u}}$ has been used in place of $\overline{\overline{\mathbf{u}}}$. Equation (5.34) is of the same mathematical structure as the ensemble-mean counterpart (2.111), and is the most popular SGS-energy equation.

5.3. Classical Subgrid-Scale Modeling

5.3.1. SUBGRID-SCALE MODELING OF ONE-EQUATION TYPE

A. SGS Viscosity

In the simplest SGS modeling, we discard both the near-interaction terms, L_{ij} [Eq. (5.21)] and C_{ij} [Eq. (5.22)], and focus attention on τ_{ij} [Eq. (5.23)]. The resulting GS equation (5.19) without L_{ij} and C_{ij} is of the same mathematical form as the ensemble-mean equation for the mean velocity, (2.100), except the difference between τ_{ij} and R_{ij}. Considering that a turbulent-viscosity approximation is suitable for the description of the interaction among the distant parts in the scale hierarchy, we write

$$\tau_{ij} = \frac{2}{3} K_S \delta_{ij} - \nu_S \overline{s}_{ij}, \tag{5.36}$$

where $\overline{s}_{ij}$ is the GS velocity-strain tensor defined as

$$\overline{s}_{ij} = \frac{\partial \overline{u}_j}{\partial x_i} + \frac{\partial \overline{u}_i}{\partial x_j}, \tag{5.37}$$

and ν_S is the SGS viscosity.

In the ensemble-mean modeling, the turbulent viscosity ν_T is written in terms of two of K (turbulent energy), ε (dissipation rate), and ℓ_C (size of energy-containing eddies), as in Eqs. (4.8) and (4.9). A big difference between the ensemble-mean and SGS modeling lies in the characteristic spatial scale of the components to be modeled. In the former, ℓ_C cannot be specified in advance, but it is determined by Eq. (3.48). On the other hand, in the SGS modeling the filter width Δ is an important parameter characterizing the scale of modeled components. This comparison indicates that there exists the following correspondence between the two modeling:

$$K \leftrightarrow K_S, \ \varepsilon \leftrightarrow \varepsilon_S, \ \ell_C \leftrightarrow 2\Delta. \tag{5.38}$$

Here the SGS energy dissipation rate ε_S is given by

$$\varepsilon_S = \nu \overline{\left(\frac{\partial u'_j}{\partial x_i} \right)^2}, \tag{5.39}$$

which corresponds to the second term on the right-hand side of Eq. (5.33) or (5.34). In the last of expression (5.38), 2Δ is adopted from the reason that the wavelength of largest eddies or waves in the grid interval Δ is 2Δ, as was noted in relation to Eq. (5.10).

From Eq. (3.48) and the foregoing correspondence, we have

$$\varepsilon_S = C_\varepsilon^S K_S^{3/2} / \Delta \,, \tag{5.40}$$

where the constant C_ε^S is related to the Kolmogorov constant $K_O (\cong 1.5)$ as

$$C_\varepsilon^S = \pi \left(\frac{2}{3K_O} \right)^{3/2} (\cong 0.93) \,. \tag{5.41}$$

We replace ε in Eq. (4.8) for ν_T with ε_S, and have

$$\nu_S = C_\nu^S \sqrt{K_S} \Delta, \tag{5.42}$$

with

$$C_\nu^S = C_\nu \Big/ C_\varepsilon^S \,, \tag{5.43}$$

where C_ν^S is estimated as about 0.097 under Eq. (5.41) and $C_\nu = 0.09$ [Eq. (4.66)].

Summarizing Eqs. (5.36) and (5.42), we have reached a model for τ_{ij} with K_S adopted as a basic SGS quantity.

B. SGS-Energy Equation

We have two equations for K_S, (5.33) and (5.34), the difference between which lies in $\overline{\mathbf{u}}$ and $\overline{\overline{\mathbf{u}}}$. This difference depends on the type of filters, and becomes specifically important in relation to double-filtering quantities such as L_{ij} and C_{ij}. In the approximation dropping L_{ij} and C_{ij}, the difference is not so significant, and we adopt the simpler form (5.34). There we need to model the last diffusion term. As is in Sec. 4.2.2.A, we drop the p'-related effect, and write

$$\overline{K_S' \mathbf{u}'} = - \frac{\nu_S}{\sigma_K^S} \nabla K_S, \tag{5.44}$$

where σ_K^S is a model constant. Here we should note that it is not necessary to distinguish between Eqs. (4.16) and (4.17), under Eq. (5.40).

We substitute Eqs. (5.40) and (5.44) into Eq. (5.34), and have

$$\frac{DK_S}{Dt_S} = -\tau_{ij} \frac{\partial \overline{u}_i}{\partial x_j} - C_\varepsilon^S \frac{K_S^{3/2}}{\Delta} + \nabla \cdot \left(\frac{\nu_S}{\sigma_K^S} \nabla K_S \right) . \tag{5.45}$$

Here the ν effect in the last diffusion term is dropped. In order to retain this effect, the similar effect needs to be introduced into the SGS viscosity ν_S

and the second dissipation-related term. A model based on the SGS-energy equation was first used by Schumann (1975) in the LES of turbulent flows in a channel and a concentric annulus. In the study, the averaging over a grid volume is adopted as the filtering procedure. Equation (5.45) was also examined from the statistical viewpoint based on the statistical filter (5.9) (Yoshizawa 1982; Yoshizawa and Horiuti 1985).

5.3.2. SMAGORINSKY MODEL

A. Simplified SGS Model

The pioneering LES of turbulent flow was made by Deardorff (1970) for a channel flow. In this study, a model simpler than the foregoing one based on the SGS energy equation is adopted. In Eq. (5.45), we assume the quasi-balance between the first and second terms, which represent the production and dissipation processes of the SGS energy, respectively; namely, we write

$$\tau_{ij} \frac{\partial \overline{u}_i}{\partial x_j} + C_\varepsilon^S \frac{K_S^{3/2}}{\Delta} = 0. \tag{5.46}$$

This assumption corresponds to the equilibrium energy cascade in the inertial range, where the energy injection by GS eddies balances the dissipation by SGS ones. In the ensemble-mean sense, this type of assumption is reasonable in the logarithmic-law region, as is discussed in Sec. 4.2.3.B.

We substitute Eq. (5.36) with (5.42) into Eq. (5.46), and have

$$K_S = (C_K \Delta)^2 \overline{s}_{ij}^2, \tag{5.47}$$

where

$$C_K = \frac{1}{C_\varepsilon^S} \sqrt{\frac{C_\nu}{2}}. \tag{5.48}$$

From Eq. (5.47), Eq. (5.42) is rewritten as

$$\nu_S = (C_S \Delta)^2 \|\overline{s}\|, \tag{5.49}$$

where the magnitude of the SGS velocity-strain tensor $\overline{s}_{ij}$, $\|\overline{s}_{ij}\|$, is defined as

$$\|\overline{s}\| = \sqrt{\overline{s}_{ij}^2}, \tag{5.50}$$

and

$$C_S = \frac{1}{C_\varepsilon^S} \left(\frac{C_\nu^3}{2} \right)^{1/4} (\cong 0.15). \tag{5.51}$$

Equation (5.49) is called the Smagorinsky model after Smagorinsky (1963) introducing the concept of the SGS viscosity in the meteorological study, but it was derived in a little different manner, as is referred to in *Note*. Here we should emphasize that in the conventional definition, $\|\overline{s}_{ij}\|$ in Eq. (5.49) is replaced with $\|\overline{s}_{ij}\|/\sqrt{2}$.

The theoretical estimate of the Smagorinsky constant C_S was made first by Lilly (1966), who used the Kolmogorov spectrum in the inertial range and found $C_S = 0.15$. The foregoing estimate coincides with this value. In the LES of channel turbulence, however, Deardorff (1970) found that the choice of $C_S = 0.084$ leads to the results consistent with observations. On the other hand, the LES of decaying isotropic turbulence indicates that larger C_S is necessary for reproducing experimental data on the decay rate of the turbulent energy. This discrepancy is a great stumbling block for the universality of the LES method based on the Smagorinsky model with fixed C_S (this point will be later referred to in more detail).

Note: In the original derivation of the Smagorinsky model, we pay more attention to the dissipation rate of the SGS energy ε_S. Using ε_S and the filter width Δ, we may write

$$\nu_S = C'^S_\nu \Delta^{4/3} \varepsilon_S^{1/3}, \tag{5.52}$$

by dimensional analysis, where C'^S_ν is a constant. We substitute Eq. (5.36) with Eq. (5.52) into Eq. (5.46) with the second term replaced with ε_S. As a result, we have

$$\varepsilon_S = \left(C'^S_\nu/2\right)^{3/2} \Delta^2 \|s\|^3 . \tag{5.53}$$

From Eqs. (5.52) and (5.53), we reach Eq. (5.49), where C_S is replaced with

$$C_S = \left(C'^{S3}_\nu/2\right)^{1/4} . \tag{5.54}$$

Equation (5.53) is combined with the K_S counterpart

$$K_S = C^S_K \Delta^{2/3} \varepsilon^{2/3}, \tag{5.55}$$

resulting in Eq. (5.47). A merit of deriving SGS models using the K_S equation is that the knowledge obtained in the course of the study on the ensemble-mean modeling can be used fully, for instance, in the estimate of model constants.

B. Breakage of Universality of Smagorinsky Constant

The Smagorinsky model based on Eqs. (5.36) and (5.49) paved the way for computing a highly time-dependent and three-dimensional flow under the least approximation to the Navier-Stokes equation. This situation makes sharp contrast with the analysis by ensemble-mean turbulence modeling. Specifically, the simplicity of models is instrumental to the reduction of the computational burden. The pioneering works by Deardorff (1970) and Schumann (1975) made the LES seem a very promising method for simulating flows at high Reynolds number. These works were followed by the more elaborate simulation by Moin and Kim (1982) and Horiuti (1987).

LES has been further applied to other types of flows such as isotropic decaying flow and mixing-layer flow. Isotropic decaying flow is discussed in Sec. 4.2.3.A from the viewpoint of the K-ε model. Concerning mixing-layer flow, we have two types of flows. One is a temporally developing mixing layer. In this case, we virtually assume that two free streams with different speeds start to interact with each other at some instant, and persue the temporal development of the mixing process. Another is a spatially developing flow, in which two streams separated by a semi-infinite plate start to merge at the trailing edge. We pursue its spatial development of merging process. From the physical viewpoint, real-world mixing layers correspond to the latter. In LES, however, the former situation is preferred since the flow is homogeneous in two directions, that is, the streamwise and spanwise directions, and the computational burden is smaller.

In addition to the LES's of channel flows by Deardorff (1970), Schumann (1975), Moin and Kim (1982), and Horiuti (1987), isotropic turbulence was examined by Antonopoulos-Domis (1981), and mixing-layer flow by Hamba (1987). The most fundamental properties of isotropic, mixing-layer, and channel flows are the decaying law of turbulent energy, the similarity of temporally-developing mean velocity profile, and the logarithmic velocity law, respectively. In order to reproduce these properties consistent with observations, we need to choose the following different Smagorinsky constant C_S's:

$$\text{Isotropic flow: } C_S \cong 0.19; \tag{5.56a}$$

$$\text{Mixing-layer flow: } C_S \cong 0.13; \tag{5.56b}$$

$$\text{Channel flow: } C_S \cong 0.084. \tag{5.56c}$$

Here we should note that the foregoing Smagorinsky model corresponds to a high-Reynolds-number version of ensemble-mean models that were explained in Chapter 4. It cannot be extended straightforwardly to a solid wall where the noslip boundary condition is imposed. The simplest method of implementing the condition is to attach the wall-damping function of Van Driest's type

$$f_W^S = 1 - \exp\left(-y_W / A_W\right) \tag{5.57}$$

to the SGS viscosity ν_S, where the wall-unit coordinate y_w is defined by Eq. (4.55). The constant A_W is usually chosen as

$$A_W = 25. \tag{5.58}$$

This choice of A_W is significant for reproducing the proper logarithmic velocity profile, and f_W^S should be regarded as a part of the SGS model in channel flow.

The discrepancy of the Smagorinsky constant C_S between isotropic and channel flows is specifically large. Expression (5.56) clearly shows that C_S is not universal. This fact is a great stumbling block in the application of LES to real-world flows. They often consist of various types of flow region. For instance, we consider a flow passing over a cubic body on a solid wall, as is depicted schematically in Fig. 4.8. The flow far ahead of it is the near-wall flow along a solid boundary, whereas the mixing-layer region is observed behind the rear edge of the body.

5.3.3. IMPROVEMENT OF MODEL

A. Use of SGS-Energy Equation

As a cause of the discrepancy of the Smagorinsky constant C_S, we may mention the assumption of the balance of the SGS-energy production and dissipation rates, Eq. (5.46). The production rate is connected with the GS velocity gradient. In the ensemble-mean context, such balance holds best in channel flows, but not in isotropic and mixing-layer flows. Specifically, in mixing-layer flow the turbulent region continues to develop temporally or spatially. The increase in the turbulent energy K comes from their imbalance. This tendency is much more prominent in homogeneous-shear flow, as is discussed in Sec. 4.3.3. On the other hand, in channel flow two solid walls contribute to both the production and dissipation of energy, resulting in their balance. The difference of the energy production mechanism with the mean velocity gradient as a primary ingredient is supposed to be closely related to the difference of C_S. In reality, C_S becomes smaller with the increasing mean shear, as is seen from expression (5.56).

From the foregoing situation, a method of rectifying the Smagorinsky model is the direct use of the model based on the SGS-energy equation, which consists of Eqs. (5.36), (5.42), and (5.45). Following the pioneering work by Schumann, this type of model was applied by Horiuti (1985) to channel flow, and by Hamba (1987) to mixing-layer flow. Specifically, the SGS models of one-equation type have been used extensively in the meteorological field, in relation to effects of buoyancy (Nieuwstadt et al. 1991). Such models, however, have been hardly examined from the viewpoint of universality in the application to various types of flows. Recently, the test of the above one-equation model was made by Okamoto (1996), and it was concluded that the model with the fixed model constants cannot reproduce the results consistent with the observations in isotropic, mixing-layer, and channel flows. The finding indicates the necessity of improving the SGS-viscosity representation (5.42). This point will be referred to later.

B. Inclusion of Leonard and Cross Terms

In the Smagorinsky and foregoing one-equation models, we have dropped the Leonard and cross terms, L_{ij} [Eq. (5.21)] and C_{ij} [Eq. (5.22)]. It is meaningful to investigate whether or not their inclusion rectifies the shortfall of the Smagorinsky model. Of these two, L_{ij} is expressed in terms of the GS velocity only and may be taken into account without any modeling. The feature of an adopted filter appears explicitly through the term. In this case, however, we encounter the difficulty of the violation of the Galilean invariance. This point was first recognized by Speziale (1985). We move one frame to another with the relative constant velocity $\mathbf{V}$; namely, we write

$$\mathbf{u} \to \mathbf{u} + \mathbf{V}. \tag{5.59}$$

Then L_{ij} is subject to the transformation

$$L_{ij} \to L_{ij} + V_i \overline{\overline{u}}_j + \overline{\overline{u}}_i V_j - V_i \overline{u}_j - \overline{u}_i V_j. \tag{5.60}$$

The occurrence of the $\mathbf{V}$-related terms in Eq. (5.60) shows the violation of the Galilean invariance. Under Eq. (5.59), C_{ij} is transformed as

$$C_{ij} \to C_{ij} + V_i \left(\overline{u}_j - \overline{\overline{u}}_j \right) + \left(\overline{u}_i - \overline{\overline{u}}_i \right) V_j. \tag{5.61}$$

These $\mathbf{V}$-related terms cancel with the counterparts in Eq. (5.60). Therefore it is indispensable to add a proper model for C_{ij} as far as L_{ij} is retained.

A method of modeling C_{ij} is the use of the approximation (5.13). Under it, C_{ij} is written as

$$C_{ij} = \overline{\overline{u}}_i \left(\overline{u}_j - \overline{\overline{u}}_j \right) + \left(\overline{u}_i - \overline{\overline{u}}_i \right) \overline{\overline{u}}_j. \tag{5.62}$$

In the transformation (5.59), this model generates the $\mathbf{V}$-related terms canceling with the counterparts in Eq. (5.60), and is a promising candidate for the model of C_{ij}. Such a model was first proposed by Bardina (1983) and is usually called a scale-similarity model. This terminology is not so clear in the physical context, and it is more natural to consider the approximation (5.13) as the basis of the modeling.

In the test using the DNS database of channel flow, the high correlation between the original and modeled C_{ij}'s was shown by Horiuti (1989). There is, however, no report that the combination of the Smagorinsky model with L_{ij} and C_{ij}, Eqs. (5.21) and (5.62), leads to the elimination of the difficulty concerning the universality of C_S.

C. Inclusion of Shear Effects on SGS Viscosity

In Sec. 5.3.2.A, it is stated that the assumption of the balance between the SGS energy production and dissipation processes is closely related to the equilibrium state of the inertial range on which the Kolmogorov spectrum is founded. Therefore the abandonment of the assumption seems to contribute greatly to the recovery of nonequilibrium properties of the SGS state. This is not the case as far as Eq. (5.42) is adopted for the SGS viscosity. The representation has the same physical meaning as Eq. (4.9), except the difference between and Δ and ℓ_C. The latter is connected with the Kolmogorov spectrum, as is seen from Secs. 3.2.2 and 4.2.1. This fact signifies that dropping the foregoing assumption while using Eq. (5.42) is not sufficient for taking into full account nonequilibrium properties of SGS components.

In Eq. (4.223), we performed the modification of the model constant C_ν in the turbulent viscosity ν_T, through the inclusion of the D/Dt effect. In order to see the relationship of the effect with the energy production process, we consider

$$\frac{D}{Dt}\frac{K^2}{\varepsilon} = (2 - C_{\varepsilon 1})\frac{K}{\varepsilon}P_K - (2 - C_{\varepsilon 2})K + \frac{2K}{\varepsilon}\nabla\cdot\left(\frac{\nu_T}{\sigma_K}\nabla K\right) - \frac{K^2}{\varepsilon^2}\nabla\cdot\left(\frac{\nu_T}{\sigma_\varepsilon}\nabla\varepsilon\right), \tag{5.63}$$

where the K equation (2.111) [with Eq. (4.16)] and the ε equation (4.37) were used. On the right-hand side of Eq. (5.63), the first term comes from the production mechanisms of K and ε, and is reduced to

$$\frac{C_\nu (2 - C_{\varepsilon 1})}{2}\frac{K^3}{\varepsilon^2}\|S\|^2, \tag{5.64}$$

using Eq. (4.8), where $\|S\| = \sqrt{S_{ij}^2}$. The nonequilibrium properties of turbulence are associated with the dominance of the energy-production effect, resulting in the inability of the dissipation or destruction process to catch up with it (recall homogeneous-shear turbulence in Sec. 4.3.3). This fact suggests the importance of the energy-production effect on the SGS viscosity for incorporating nonequilibrium properties into its modeling.

From the foregoing discussion, we consider Eqs. (4.223) and (5.64), and modify Eq. (5.42) as

$$\nu_S = \frac{C_{\nu 1}^S}{1 + C'^S_{\nu 2}\left(K_S^2/\varepsilon_S^2\right)\|\overline{s}\|^2}\sqrt{K_S}\Delta, \tag{5.65}$$

where $C_{\nu 1}^{S}$ and $C'^{S}_{\nu 2}$ are constants [we should note that $C_{\Pi 1}$ in Eq. (4.223) is negative]. We use Eq. (5.40), and rewrite Eq. (5.65) as

$$\nu_S = \frac{C_{\nu 1}^{S}}{1 + C_{\nu 2}^{S} \left(\Delta^2 / K_S\right) \|\overline{s}\|^2} \sqrt{K_S} \Delta, \tag{5.66}$$

where $C_{\nu 2}^{S}$ is a model constant. Equation (5.66) was introduced by Okamoto (1996) using the statistical results in Chapter 6.

Finally, the nonequilibrium SGS model consists of Eqs. (5.36), (5.66), and the SGS-energy equation

$$\frac{DK_S}{Dt_S} = -\tau_{ij} \frac{\partial \overline{u}_i}{\partial x_j} - C_{\varepsilon}^{S} \frac{K_S^{3/2}}{\Delta} + \nabla \cdot \left(C_K^S \sqrt{K_S} \Delta \nabla K_S \right), \tag{5.67}$$

with the model constant C_K^S. This model was applied to isotropic, mixing-layer, and channel flows, and it was confirmed that the following set of model constants can really reproduce the primary properties of these flows:

$$C_{\nu 1}^{S} = 0.185, \; C_{\nu 2}^{S} = 0.025, \; C_{\varepsilon}^{S} = 0.835, \; C_K^S = 0.1 \tag{5.68}$$

[in channel flow, use is made of the wall-damping function (5.57)]. In Eq. (5.67), we assume the balance of the SGS energy production and dissipation rates, as is done in the derivation of the Smagorinsky model:

$$\frac{\left(\Delta^2 / K_S\right) \|\overline{s}\|^2}{1 + C_{\nu 2}^{S} \left(\Delta^2 / K_S\right) \|\overline{s}\|^2} = \frac{2 C_{\varepsilon}^{S}}{C_{\nu 1}^{S}}. \tag{5.69}$$

Under the set of model constants given by Eq. (5.68), we have

$$\left(\Delta^2 / K_S\right) \|\overline{s}\|^2 = 11.7. \tag{5.70}$$

We substitute Eq. (5.70) into Eq. (5.66), and have 0.14 as the resulting numerical factor. From Eqs. (5.43) and (5.51), this value gives $C_S = 0.2$ in the Smagorinsky model, which is much larger than the usually-adopted one in channel flow, 0.12, and is close to the counterpart for isotropic turbulence [see expression (5.56a)]. This fact indicates the complicated relationship among the model constants in SGS models.

The merit of the present SGS model of the one-equation type is not so large in the LES of engineering flows to which the dynamic SGS modeling to be explained later is applicable. In complicated flows such as meteorological ones, however, we have to include various effects represented by buoyancy ones. In those cases, the use of one-equation SGS models is indispensable, but it is difficult to use the dynamic version of one-equation SGS models in those phenomena. The classical SGS modeling of the one-equation type is expected to provide a useful numerical tool for such a purpose.

D. Algebraic SGS Model

The one-equation model based on the linear turbulent-viscosity model, Eq. (5.36), can overcome some of critical shortfalls of the classical Smagorinsky model. The model, however, is not sufficient for the treatment of anisotropic effects caused by external forces due to buoyancy, frame rotation, etc. In the presence of buoyancy force, no explicit buoyancy effect enters τ_{ij}, although parts of the effect can be taken into account through the equation for K_S, Eq. (5.67) with expression (5.32) added. In the case of frame rotation, its effect disappears entirely from the K_S equation. The shortfalls resulting from these properties become prominent in the LES, specifically, with coarse grid points.

A method of incorporating the anisotropic effects due to buoyancy, frame rotation, etc. is the algebraic second-order modeling of τ_{ij} (Shimomura 1994, 1997). The simplest of ensemble-mean algebraic second-order models is given by Eq. (4.216). We use the replacement (5.38) and construct the SGS version of Eq. (4.216). First, we define the deviatoric part of τ_{ij} as

$$[\tau_{ij}]_D = \tau_{ij} - (2/3)\, K_S \delta_{ij}. \tag{5.71}$$

Under the approximation of neglecting the effects of double filtering such as the cross term, we have

$$\begin{aligned}
&\left(-\tau_{ij}\frac{\partial \overline{u}_j}{\partial x_i} - \varepsilon_S + (C_D - C_{\Pi 1})\,\varepsilon_S\right)[\tau_{ij}]_D \\
&\quad = -\left(\frac{2}{3} - C_{\Pi 3}\right) K_S^2 \overline{s}_{ij} \\
&\quad + \left(C_{\Pi 4} - \frac{1}{2}\right) K_S \left[[\tau_{i\ell}]_D\, \overline{s}_{\ell j} + [\tau_{j\ell}]_D\, \overline{s}_{\ell i}\right]_D \\
&\quad + \left(C_{\Pi 5} - \frac{1}{2}\right) K_S \left([\tau_{i\ell}]_D\, \overline{\omega}_{\ell j} + [\tau_{j\ell}]_D\, \overline{\omega}_{\ell i}\right),
\end{aligned} \tag{5.72}$$

where Eq. (5.40) is used for ε_S, and the GS vorticity tensor $\overline{\omega}_{ij}$ is defined as

$$\overline{\omega}_{ij} = \frac{\partial \overline{u}_j}{\partial x_i} - \frac{\partial \overline{u}_i}{\partial x_j}. \tag{5.73}$$

The model constants C_D and $C_{\Pi n}$'s are given by Eqs. (4.199), (4.184), (4.186), and (4.187). Equation (5.72) is combined with Eq. (5.45) for K_S and leads to a closed SGS system of equations, which may be called an algebraic SGS model. In this model, effects of frame rotation are incorporated through the transformation rule

$$\overline{\omega}_{ij} \to \overline{\omega}_{ij} + 2\varepsilon_{ij\ell}\Omega_{F\ell}, \tag{5.74}$$

where $\boldsymbol{\Omega}_F$ is the angular velocity of a rotating frame. The usefulness of this type of algebraic model was confirmed by Shimomura (1997) through the application to the LES of homogeneous decaying turbulence in a rotating frame.

In this context, Canuto and Cheng (1997) adopted such a nonlinear model as is explained in Sec. 4.5.2, and demonstrated the nonuniversality of C_S in Eq. (5.49) from the viewpoint of effects of shear and stratification.

5.4. Dynamic Subgrid-Scale Modeling

In each of the ensemble-mean models in Chapter 4, we have several unknown constants. Some constraints are imposed on them from the viewpoint of model consistency, realizability, etc., but basically, these constants need to be optimized through the application to various types of turbulent flows. In addition, the near-wall correction to them is indispensable for the numerical computation subject to the noslip boundary condition. Entirely the same situation holds for the classical SGS models discussed in Sec. 5.3. In highly time-dependent, three-dimensional flows, however, it is difficult to make proper near-wall corrections to them.

In contrast with the ensemble-mean modeling, the number of model constants or coefficients is much smaller in the SGS modeling. Consequently, there occurs the possibility that the invariance properties intrinsic to the filtering procedure play a critical role in the estimate of these coefficients. In reality, Germano (1992), who discovered the identity (5.28), paved the way for an entirely new SGS modeling.

5.4.1. SMAGORINSKY-TYPE MODEL

A. Self-Consistent Determination of Smagorinsky Coefficient

We introduce two filters denoted by G_A and G_B, whose characteristic filter widths are Δ_A and Δ_B, respectively. Following the definition (5.2), the filtering of the Navier-Stoles equation using G_A leads to Eqs. (5.24) and (5.25). Next, we apply the filtering based on G_A and G_B consecutively to the Navier-Stokes equation, which is defined by Eq. (5.3). The resulting equation is given by Eq. (5.26) with Eq. (5.27). The latter obeys Germano's identity (5.28). In this identity, the right-hand side may be calculated explicitly in the course of LES, whereas the left-hand side to be modeled includes some unknown coefficients. The identity is instrumental to their estimate. This is the fundamental point of Germano's dynamic SGS modeling. The ensemble-mean procedure has no counterpart of Eq. (5.28) since $\overline{\overline{f}} = \overline{f}$ and $\overline{f'} = 0$.

Corresponding to the classical Smagorinsky model, we neglect L_{ij}^A and

C_{ij}^A in Eq. (5.25b), and retain τ_{ij}^A. A primary feature of τ_{ij}^A is the distant-interaction effect, as is noted in Sec. 5.2.2. Then we introduce an expression similar to the Smagorinsky model as

$$\left[T_{ij}^A\right]_D = -C_A \Delta_A^2 \left\|\overline{s}^A\right\| \overline{s}_{ij}^A, \tag{5.75}$$

where $[f_{ij}]_D$ denotes the deviatoric part of f_{ij} defined by Eq. (5.71), and C_A is a nondimensional coefficient. For T_{ij}^{AB}, we also assume the similar form

$$\left[T_{ij}^{AB}\right]_D = -C_{AB} \Delta_B^2 \left\|\overline{s}^{AB}\right\| \overline{s}_{ij}^{AB}, \tag{5.76}$$

with C_{AB} as another nondimensional coefficient. Here Δ_B has been used as a characteristic length scale in the consecutive filtering procedure since $\Delta_B > \Delta_A$.

We substitute Eqs. (5.75) and (5.76) into Eq. (5.28), and have

$$\Delta_A^2 \overline{C_A \left\|\overline{s}^A\right\| \overline{s}_{ij}^A}^B - \Delta_B^2 C_{AB} \left\|\overline{s}^{AB}\right\| \overline{s}_{ij}^{AB} = \left[\overline{\overline{u}_i^A \overline{u}_j^A}^B - \overline{u}_i^{AB} \overline{u}_j^{AB}\right]_D \equiv [L_{ij}]_D. \tag{5.77}$$

The simplest method of estimating C_A and C_B is to assume

$$C_A = C_{AB} = C_{DS}, \tag{5.78}$$

and neglect the dependence of C_A on $\mathbf{x}$ and t in the filtering operation. As a result, we reach

$$C_{DS} M_{ij} = [L_{ij}]_D, \tag{5.79}$$

where

$$M_{ij} = \Delta_A^2 \overline{\left\|\overline{s}^A\right\| \overline{s}_{ij}^A}^B - \Delta_B^2 \left\|\overline{s}^{AB}\right\| \overline{s}_{ij}^{AB}. \tag{5.80}$$

In general, there is no single C_{DS} that can exactly satisfy Eq. (5.79) that is the six-component relation. As the starting stage of the study of the dynamic SGS modeling by Germano et al. (1991), C_{DS} was estimated by multiplying Eq. (5.79) with $\overline{s}_{ij}^A$; namely, we have

$$C_{DS} = \overline{s}_{ij}^A [L_{ij}]_D \Big/ \left(\overline{s}_{ij}^A M_{ij}\right). \tag{5.81}$$

In the application of Eq. (5.81) to the LES of a channel flow, the denominator $\overline{s}_{ij}^A M_{ij}$ becomes very small or vanishes, in addition to its negativeness. The latter point will be referred to below. Then it was replaced with its average over the plane parallel to the wall.

The mathematical basis of the above replacement is not so clear. In place of the method, Lilly (1994) introduced an ingenious approach based on the least-square approximation. There we consider the square of the error in Eq. (5.79), which is given by

$$Q = \left(C_{DS}M_{ij} - [L_{ij}]_D\right)^2. \tag{5.82}$$

The minimization of Q or

$$\frac{\partial Q}{\partial C_{DS}} = 0 \tag{5.83}$$

leads to

$$C_{DS} = M_{ij}\,[L_{ij}]_D \Big/ M_{ij}^2\,. \tag{5.84}$$

Lilly's method can remove the difficulty related to vanishing of the denominator. In the application of Eq. (5.84), we still encounter negative C_{DS}, which is a primary cause of numerical instability. In order to alleviate this difficulty, we further average Eq. (5.84) over the plane in the case of a channel flow. Negative C_{DS}, which still survives after the averaging, is replaced with zero or some very small positive value.

In spite of the necessity of some numerical devices related to negative C_{DS}, the great merit of the dynamic SGS modeling lies in the fact that it does not need near-wall corrections such as Van Driest's damping function. It is difficult to construct such corrections in flows containing a separation point, a sharp edge, etc. The dynamic SGS modeling has made the straightforward application of LES to these flows feasible.

B. Lagrangian Averaging Procedure for Avoiding Negative Coefficients

In turbulent flow with one or two homogeneous directions, we can apply the line or plane averaging procedure and alleviate the difficulty related to the negative Smagorinsky coefficient. There are, however, a number of important flows with no homogeneous direction. As an alternative averaging procedure in such cases, a Lagrangian averaging procedure was presented by Meneveau et al. (1996). In this method, the foregoing plane averaging is replaced with the averaging along the trajectory of a fluid particle.

The positions of a fluid particle at time t and $t_1 (t' \le t_1 \le t)$, $\mathbf{x}$ and $\mathbf{x}'$, are related to each other as

$$\mathbf{x} = \mathbf{x}' + \int_{t'}^{t} \overline{\mathbf{u}}\left(\mathbf{x}\left(t_1\right), t_1\right) dt_1. \tag{5.85}$$

We write the error in Eq. (5.79) as

$$\gamma_{ij} = C_{DS}M_{ij} - [L_{ij}]_D\,. \tag{5.86}$$

Then the square of the error along the trajectory is given by

$$Q = \int_{t'}^{t} \gamma_{ij}\left(\mathbf{x}'\left(t_1\right), t_1\right)^2 W\left(\mathbf{x}(t), t; \mathbf{x}'\left(t_1\right), t_1\right) dt_1. \tag{5.87}$$

Here W is the weighting function obeying the constraint

$$\int_{t'}^{t} W\left(\mathbf{x}(t), t; \mathbf{x}'\left(t_1\right), t_1\right) dt_1 = 1, \tag{5.88}$$

and represents how far past events have influence on the present state (W is similar to the Green's or response function discussed in Sec. 3.5.2.D). The minimization of Q, which is expressed by Eq. (5.83), gives

$$C_{DS} = \Gamma_{ML} / \Gamma_{MM}, \tag{5.89}$$

where

$$\Gamma_{ML} = \int_{t'}^{t} M_{ij}\left(\mathbf{x}'\left(t_1\right), t_1\right) \left[L_{ij}\left(\mathbf{x}'\left(t_1\right), t_1\right)\right]_D W\left(\mathbf{x}(t), t; \mathbf{x}'\left(t_1\right), t_1\right) dt_1, \tag{5.90a}$$

$$\Gamma_{MM} = \int_{t'}^{t} M_{ij}\left(\mathbf{x}'\left(t_1\right), t_1\right)^2 W\left(\mathbf{x}(t), t; \mathbf{x}'\left(t_1\right), t_1\right) dt_1. \tag{5.90b}$$

In order to simplify Eq. (5.89), Meneveau et al. (1996) assumed that W is dependent on the time difference $t - t_1$, but not on the trajectory, and adopted the simplest form

$$W\left(\mathbf{x}(t), t; \mathbf{x}'\left(t_1\right), t_1\right) \equiv W\left(t, t_1\right) = \frac{1}{T_L} \exp\left(-\frac{t - t_1}{T_L}\right) H\left(t - t'\right), \tag{5.91}$$

where T_L is a time scale characterizing the duration time of past events, and $H(t)$ is the unit step function. Equation (5.91) obeys the response equation

$$\frac{DW\left(t, t'\right)}{Dt_S} + \frac{1}{T_L} W\left(t, t'\right) = \frac{1}{T_L} \delta\left(t - t'\right), \tag{5.92}$$

where D/Dt_S is defined by Eq. (5.35). Under Eqs. (5.91) and (5.92), Eqs. (5.90a) and (5.90b) are equivalent to the differential equations

$$\frac{D\Gamma_{ML}}{Dt_S} = -\frac{1}{T_L} \left[\Gamma_{ML} - M_{ij}\left(\mathbf{x}\left(t\right), t\right) L_{ij}\left(\mathbf{x}\left(t\right), t\right)\right], \tag{5.93a}$$

$$\frac{D\Gamma_{MM}}{Dt_S} = -\frac{1}{T_L} \left[\Gamma_{MM} - M_{ij}\left(\mathbf{x}\left(t\right), t\right)^2\right], \tag{5.93b}$$

respectively. Once T_L is designated, Eq. (5.93) may be calculated in the course of LES.

In the Lagrangian averaging procedure, the remaining important point is the proper choice of the time scale T_L, but there is no method of designating it uniquely. The typical examples of T_L are

$$T_L = C_{L1} \Delta \left(\Gamma_{ML} \Gamma_{MM}\right)^{-1/8}, \tag{5.94a}$$

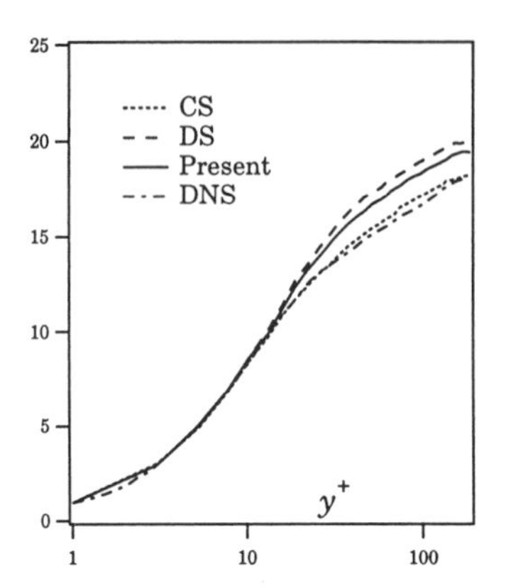

Figure 5.1. Mean velocity in a channel flow (CS, classical Smagorinsky model; DS, dynamical Smagorinsky model; Present, Eq. (5.108); DNS, direct numerical simulation).

$$T_L = C_{L2}\Delta \left(\Gamma_{ML}\right)^{-1/4}, \tag{5.94b}$$

where C_{L1} and C_{L2} are numerical coefficients. For C_{L1}, 1.5 was recommended by Meneveau et al. (1996) in the LES of a channel flow. It is highly probable, however, that the magnitude of these coefficients is dependent on each flow. It was shown by Murakami et al. (1997) that for C_{L2}, 0.2 is proper in the LES of a flow past a bluff body, although the counterpart of a channel flow is much larger (about 2). In this sense, the coefficient are an adjustable parameter, but this fact does not lower the value of the Lagrangian averaging method.

C. Corrections to Smagorinsky-Type Model

The dynamic SGS modeling paves the way for the LES that may determine model coefficients in a self-consistent manner. This fact, however, does not always guarantee that the results obtained from dynamic models can always provide better results, compared with the classical SGS models with carefully optimized model constants. Such an example is the LES of a channel flow using the dynamic Smagorinsky model, Eq. (5.84), that is combined with the plane averaging of C_{DS}. The resulting mean velocity profile deviates from the DNS counterpart, as is seen from Fig. 5.1. In reality, the classical Smagorinsky model, Eq. (5.49), with $C_S = 0.1$ and the near-wall correction, Eq. (5.57), gives much better results.

C.1. Near-Interaction Corrections. A method of improving the dynamic Smagorinsky model is the retention of the near-interaction effects expressed

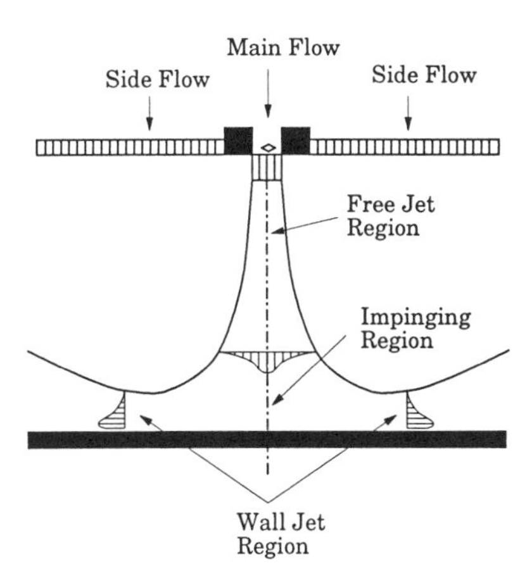

Figure 5.2. Plane impinging jet.

by the Leonard and cross terms, L_{ij} and C_{ij}. In this case, Eq. (5.62) is a candidate for the model of C_{ij}. The dynamic models, in which the Smagorinsky-type model is combined with such effects, have been proposed by Zang et al. (1993) and by Vreman et al. (1994), and are called mixed SGS models. In the LES's of channel flows, mixing-layer flows, etc., these models have been confirmed to improve the results of the simple Smagorinsky-type model.

Here we note the approximation (5.14) in the context of the modeling of near-interaction effects. In such a fine-grid system as may be adopted in a channel flow, it is a reasonable expression for modeling those effects. In turbulent flows encountered in real engineering phenomena, flow geometry is often very complicated, and the grid allocation becomes very sparse in some parts of the flow region. In the case, the high accuracy of Eq. (5.14) is not guaranteed. For instance, Tsubokura (1996) applied a mixed SGS model to a plane jet vertically impinging on a solid wall that is depicted schematically in Fig. 5.2, and found that the model is vulnerable to numerical instability. In this sense, the study on the modeling of near-interaction effects remains to be an important subject of the SGS modeling.

C.2. Localization of Model Coefficients. In the filtering procedure of Eq. (5.77), the spatial and temporal variation of the nondimensional coefficient $C_A(= C_{DS})$ was neglected. In the implementation of LES, the variation is a prominent feature of the dynamic modeling and is not small. Then it is desirable to retain this feature as far as possible. The determination of C_{DS} while keeping its spatial and temporal variations is called the localization.

When the spatial variation of C_{DS} is retained in Eq. (5.77), we are led to an integral equation concerning it. The numerical procedure solving such an equation was presented by Ghosal et al. (1995). The localization of the model coefficient from a little different viewpoint was proposed by Piomelli and Liu (1995). There the time dependence of the coefficient was retained by using C_{DS} one time step before in the time-marching method of solution.

5.4.2. NON-SMAGORINSKY-TYPE MODEL

In the current dynamic SGS modeling, the Smagorinsky-type expression (5.75) is adopted as a key element, which is instrumental to drain energy from the GS to SGS components. It is directly connected with the GS velocity gradient, and is founded on the assumption of the balance between the SGS energy production and dissipation processes, as is seen in the derivation of the classical Smagorinsky model. In the context of the ensemble-mean modeling in Chapter 4, the counterpart is called the mixing-length model and is not applicable to the analysis of a nonstationary state of turbulence. This shortfall may be attributed to the fact that the dissipation mechanism at small scales cannot catch up with the temporally-changing production one, resulting in the breakage of their balance. Such a typical instance is homogeneous-shear turbulence in Sec. 4.3.3.

In the filtering procedure, the GS field, which corresponds to the mean field in the ensemble-mean one, is always highly time-dependent even in case that ensemble-mean quantities are temporally constant. Therefore it is desirable to adopt a SGS model free from the assumption of the energy production and dissipation balance. In what follows, we shall make use of the knowledge about the ensemble mean modeling and derive a non-Smagorinsky-type SGS model (Yoshizawa et al. 1996).

From the $\mathbf{u}'$ equation (5.29), the deviatoric part of τ_{ij}^A, $[\tau_{ij}^A]_D$, obeys the equation similar to the ensemble-mean counterpart (2.105), which is written as

$$\frac{\partial}{\partial t}\left[\tau_{ij}^A\right]_D+\left[\overline{\overline{u}_\ell^A\frac{\partial}{\partial x_\ell}u_i'u_j'}^A\right]_D=\left[P_{ij}^A\right]_D+\Pi_{ij}^A-\left[\varepsilon_{ij}^A\right]_D+\frac{\partial}{\partial x_\ell}\left[T_{ij\ell}^A\right]_D+\left[F_{ij}^A\right]_D, \tag{5.95}$$

where

$$\left[P_{ij}^A\right]_D=\left[-\overline{u_i'u_\ell'\frac{\partial\overline{u}_j^A}{\partial x_\ell}}^A-\overline{u_j'u_\ell'\frac{\partial\overline{u}_i^A}{\partial x_\ell}}^A\right]_D, \tag{5.96}$$

$$\Pi_{ij}^A=\overline{p'\left(\frac{\partial u_j'}{\partial x_i}+\frac{\partial u_i'}{\partial x_j}\right)}^A, \tag{5.97}$$

$$\left[\varepsilon_{ij}^A\right]_D = \left[2\nu\overline{\frac{\partial u_i'}{\partial x_\ell}\frac{\partial u_j'}{\partial x_\ell}}^A\right]_D, \tag{5.98}$$

$$\left[T_{ij\ell}^A\right]_D = -\left[\overline{u_i'u_j'u_\ell'}^A + \overline{p'u_i'}^A\delta_{j\ell} + \overline{p'u_j'}^A\delta_{i\ell}\right]_D + \nu\frac{\partial}{\partial x_\ell}\left[\tau_{ij}^A\right]_D, \tag{5.99}$$

and $[F_{ij}^A]_D$ is the quantity intrinsic to the filtering procedure, whose details are omitted since they are not necessary here.

For the pressure-strain and destruction terms Π_{ij}^A and $[\varepsilon_{ij}^A]_D$, we adopt the simplest models

$$\Pi_{ij}^A = -C_{S1}^A\frac{\varepsilon_S^A}{K_S^A}\left[\tau_{ij}^A\right]_D, \tag{5.100}$$

$$\left[\varepsilon_{ij}^A\right]_D = C_D^A\frac{\varepsilon_S^A}{K_S^A}\left[\tau_{ij}^A\right]_D. \tag{5.101}$$

Here Eq. (5.100) is the SGS counterpart of the first term of Eq. (4.180) with Eq. (4.184), whereas Eq. (5.101) corresponds to Eq. (4.198). We substitute Eqs. (5.100) and (5.101) into Eq. (5.95), and have

$$\left(C_D^A + C_{S1}^A\right)\frac{\varepsilon_S^A}{K_S^A}\left[\tau_{ij}^A\right]_D = \left[P_{ij}^A\right]_D - \frac{\partial}{\partial t}\left[\tau_{ij}^A\right]_D - \left[\overline{\overline{u}_\ell^A\frac{\partial}{\partial x_\ell}u_i'u_j'}^A\right]_D$$
$$+ \frac{\partial}{\partial x_\ell}\left[T_{ij\ell}^A\right]_D + \left[F_{ij}^A\right]_D. \tag{5.102}$$

In order to construct a model as simple as possible, we retain only the first term on the right-hand side of Eq. (5.102). This is a procedure for deriving an algebraic second-order model from the transport equation for the Reynolds stress (see Sec. 4.5.2). Then we have

$$\left[\tau_{ij}^A\right]_D = -\frac{1}{C_\varepsilon^{SA}\left(C_D^A + C_{S1}^A\right)}\frac{\Delta_A}{\sqrt{K_S^A}}\left[\overline{u_i'u_\ell'\frac{\partial\overline{u}_j^A}{\partial x_\ell}}^A + \overline{u_j'u_\ell'\frac{\partial\overline{u}_i^A}{\partial x_\ell}}^A\right]_D, \tag{5.103}$$

where use has been made of Eq. (5.40) for ε_S^A. We apply the approximation (5.13) to Eq. (5.103), resulting in

$$\left[\tau_{ij}^A\right]_D = -\frac{1}{C_\varepsilon^{SA}\left(C_D^A + C_{S1}^A\right)}\frac{\Delta_A}{\sqrt{K_S^A}}\left[\overline{u_i'u_\ell'}^A\frac{\partial\overline{\overline{u}}_j^A}{\partial x_\ell} + \overline{u_j'u_\ell'}^A\frac{\partial\overline{\overline{u}}_i^A}{\partial x_\ell}\right]_D. \tag{5.104}$$

As the simplest approximation, we assume the isotropy of $\mathbf{u}'$ on the right-hand side of Eq. (5.104), and write

$$\overline{u_i'u_j'}^A = \frac{2}{3}K_S^A\delta_{ij}. \tag{5.105}$$

Then we have

$$\left[\tau_{ij}^{A}\right]_{D} = -\frac{2}{3C_{\varepsilon}^{SA}\left(C_{D}^{A}+C_{S1}^{A}\right)}\Delta_{A}\sqrt{K_{S}^{A}}\overline{\overline{s}}_{ij}^{A}. \tag{5.106}$$

For the estimate of K_S^A, we do not assume the balance between the SGS energy production and dissipation processes, but we use Eq. (5.13), that is,

$$K_{S}^{A} = \frac{1}{2}\overline{\mathbf{u}'^{2}}^{A} \cong \frac{1}{2}\left(\overline{\mathbf{u}}^{A} - \overline{\overline{\mathbf{u}}}^{A}\right)^{2}. \tag{5.107}$$

The combination of Eq. (5.106) with Eq. (5.107) leads to

$$\left[\tau_{ij}^{A}\right]_{D} = -C_{\tau}^{A}\Delta_{A}\sqrt{\left(\overline{\mathbf{u}}^{A} - \overline{\overline{\mathbf{u}}}^{A}\right)^{2}}\,\overline{\overline{s}}_{ij}^{A}, \tag{5.108}$$

where

$$C_{\tau}^{A} = \frac{\sqrt{2}}{3C_{\varepsilon}^{SA}\left(C_{D}^{A}+C_{S1}^{A}\right)}. \tag{5.109}$$

In Eq. (5.104), we may apply Eq. (5.13) to $\overline{u_i' u_j'}^A$ itself, as was done by Tsubokura (1996) and Tsubokura et al. (1997). In this case, we are led to an anisotropic turbulent-viscosity representation for $[\tau_{ij}^A]_D$. In the dynamic LES based on Eq. (5.108) and its anisotropic version, C_τ^A is determined in entirely the same manner as in Sec. 5.4.1.A.

The non-Smagorinsky-type model (5.108) was applied to a channel flow, and the resulting mean velocity profile is shown in Fig. 5.1. This model improves the results from the Smagorinsky-type model, but its degree of improvement is not so large. In addition, the classical Smagorinsky model gives better results. The merit of the non-Smagorinsky-type model, however, becomes large in complicated flow situations. Such an example is a plane jet that is ejected from a two-dimensional slit and impinges vertically on a solid wall (see Fig. 5.2). This flow consists mainly of three typical regions: a plane jet, an impinging flow, and a recirculating flow. An impinging flow is one of the representative flows that are difficult to analyze using ensemble-mean models. The non-Smagorinsky-type model (5.108) and its anisotropic version were applied by Tsubokura (1996) and Tsubokura et al. (1997) to the flow, and the detailed comparison with the observations was made. As a result, it was confirmed that the non-Smagorinsky-type models can provide much better results than the Smagorinsky-type model.

As the other non-Smagorinsky-type model, we may mention the one by Horiuti (1997). There τ_{ij} is modeled, in addition to C_{ij}, on the basis of the approximation similar to expression (5.13) and is not directly dependent on the GS velocity strain characterizing the SGS-viscosity approximation.

LES's of complex flows are usually performed under a sparse grid allocation. The validity of the model of the non-SGS-viscosity type in such a situation is an interesting subject of LES.

References

Antonopoulos-Domis, M. (1981), J. Fluid Mech. **104**, 55.

Bardina, J. (1983), Ph. D. Dissertation, Stanford University.

Canuto, V. M. and Cheng, Y. (1997), Phys. Fluids **9**, 1368.

Deardorff, J. W. (1970), J. Fluid Mech. **41**, 453.

Germano, M. (1992), J. Fluid Mech. **238**, 325.

Germano, M., Piomelli, U., Moin, P., and Cabot, W. H. (1991), Phys. Fluids A **3**, 1760.

Ghosal, S., Lund, T. S., Moin, P., and Akselvoll, K. (1995), J. Fluid Mech. **286**, 229.

Hamba, F. (1987), J. Phys. Soc. Jpn. **56**, 2721.

Horiuti, K. (1985), J. Phys. Soc. Jpn. **54**, 2855.

Horiuti, K. (1987), J. Comput. Phys. **71**, 343.

Horiuti, K. (1989), Phys. Fluids A **1**, 426.

Horiuti, K. (1997), Phys. Fluids **9**, 3443.

Leonard, A. (1974), Adv. in Geophys. **18**A, 237.

Lilly, D. K. (1966), in *Proceedings of IBM Scientific Computing Symposium on Environmental Sciences*, IBM Form No. 320-1951, p. 195.

Lilly, D. K. (1992), Phys. Fluids A **4**, 633.

Meneveau, C., Lund, T. S., and Cabot, W. H. (1996), J. Fluid Mech. **319**, 353.

Moin, P. and Kim, J. (1982), J. Fluid Mech. **118**, 341.

Murakami, S., Iizuka, S., Mochida, A., and Tominaga, Y., (1997), in *Direct and Large-Eddy Simulation II*, edited by J.-P. Chollet, P. R. Voke, and L. Kleiser (Kluwer, Dordrecht), p. 385.

Nieuwstadt, F. T. M., Mason, P. J., Moeng, C. H., and Schumann, U. (1991), in *Eighth Symposium on Turbulent Shear Flows*, 1-4-1.

Okamoto, M. (1996), in *Proceedings of Tenth Symposium on Computational Fluid Dynamics*, Japan Society of Computational Fluid Dynamics, p. 164.

Piomelli, U. and Liu, J. (1995), Phys. Fluids **7**, 839.

Rogallo, R. S. and Moin, P. (1984), Annu. Rev. Fluid Mech. **16**, 99.

Schumann, U. (1975), J. Comput. Phys. **18**, 376.

Shimomura, Y. (1994), J. Phys. Soc. Jpn. **63**, 5.

Shimomura, Y. (1997), in *Eleventh Symposium on Turbulent Shear Flows*, 34-1.

Smagorinsky, J. S. (1963), Mon. Weather Rev. **91**, 99.

Speziale, C. G. (1985), J. Fluid Mech. **156**, 55.
Tsubokura, M. (1996), Ph. D. Dissertation, University of Tokyo.
Tsubokura, M., Kobayashi, T., and Taniguchi, N. (1997), in *Eleventh Symposium on Turbulent Shear Flows*, 22-24.
Vreman, B., Geurts, B., and Kuerten, H. (1994), Phys. Fluids A **6**, 4057.
Yoshizawa, A. (1982), Phys. Fluids **25**, 1532.
Yoshizawa, A. and Horiuti, K. (1985), J. Phys. Soc. Jpn. **54**, 2834.
Yoshizawa, A., Tsubokura, M., Kobayashi, T., and Taniguchi, N. (1996), Phys. Fluids **8**, 2254.
Zang, Y., Street, R. L., and Koseff, J. R. (1993), Phys. Fluids A **5**, 3186.

CHAPTER 6

TWO-SCALE DIRECT-INTERACTION APPROXIMATION

6.1. Statistical Theory of Turbulent Shear Flow

A number of works have been done on statistical theoretical analyses of turbulence since the presentation of the direct-interaction approximation (DIA) method that is explained in Sec. 3.5. It is not too much, however, to say that most of them focus attention on homogeneous isotropic turbulence. In this course, one of a handful of theoretical works on turbulent shear flow was also done by Kraichnan (1964), the advocate of the DIA, and the formal expressions for the Reynolds stress and the turbulent scalar flux were derived with the aid of the DIA.

This interesting work also shows the difficulty in constructing a statistical theory of shear flow that is applicable to turbulence modeling. The primary cause of the difficulty is the coexistence of temporally and spatially slow variations of the mean field (F) and the fast ones of the fluctuating field (f'). This situation is depicted schematically in Fig. 6.1. In statistical-theoretical methods such as the DIA, the Fourier representation of flow field is a principal tool for analyzing turbulence properties. Its straightforward application to shear flow results in the wavenumber resolution of both the mean and fluctuating fields. The big difference between the variations of both the fields leads to the simultaneous treatment of a wide range of wavenumbers. Specifically, the mean field is directly affected by boundaries. The low-wavenumber components of the fluctuating field, which interact strongly with the mean field, are contaminated by the boundary effects. Therefore it is difficult to abstract universal properties of shear flow from the analytical expressions that are derived formally using the DIA.

A noticeable work following Kraichnan's attempt is Leslie's (1973) one. In the work, two space variables were introduced for distinguishing between different variations in shear flow. One is the centroid coordinate of two points $\mathbf{x}$ and $\mathbf{x}'$, $(\mathbf{x}+\mathbf{x}')/2$, and another is the difference coordinate, $\mathbf{x}-\mathbf{x}'$. The former is used for expressing the slow variation of two-point statistical quantities, whereas the latter is for their fast variation and the Fourier representation is applied to it. This interesting attempt, however, was not always advanced far enough to the theoretical derivation of various turbulence models discussed in Chapter 4. Its main cause is the mathemat-

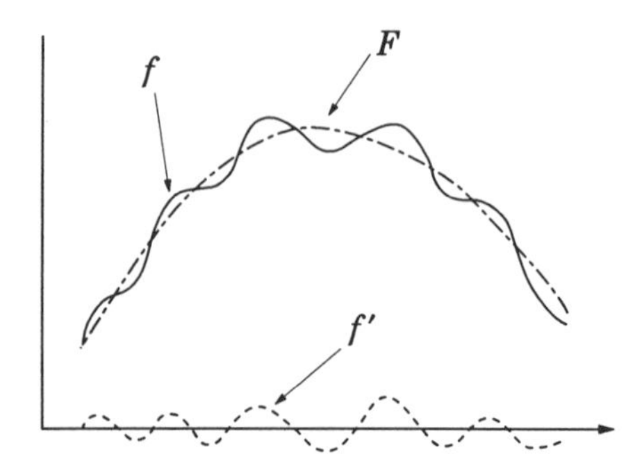

Figure 6.1. Coexistence of slow and fast variations.

ical complexity arising from the use of the centroid coordinate.

A method for analytically treating the state subject to two different temporal and spatial variations is the full use of their separation (here we should note that the separation does not mean the clear-cut one such as the separation between molecular and fluid motions). The method may be divided roughly into two categories. In one category, the effects of the mean field on the fluctuating one are incorporated in a perturbational manner. Afterwards, the feedback effects on the mean field are taken into account self-consistently through the mean-field equation with the former effects included. In another category, nonlinear effects of fast variations are directly incorporated into the slow-motion components in a form of renormalized coefficients. The former approach is a two-scale direct-interaction approximation (TSDIA) method (Yoshizawa 1984a), which will be detailed in this chapter. The latter corresponds to the renormalization group (RNG) method, which was highly developed in the study of solid-state physics and was applied by Yakhot and Orszag (1986) to turbulence analysis. This RNG method has been extended to shear turbulence by Rubinstein and Barton (1990, 1991) to derive the nonlinear models for the Reynolds stress and the turbulent scalar flux. Some problems in the application of the RNG method to turbulence have been scrutinized in several works, as is noted in Sec. 3.5.1 (see also Kraichnan 1987).

A different theoretical approach to shear turbulence was presented by Weinstock (1981, 1982) who focused attention on the pressure-strain correlation, Π_{ij} [Eq. (2.107)]. This quantity plays a central role of describing the anisotropic property of turbulent intensities, as is discussed in the second-order modeling of Chapter 4. In the work, the quasi-normal approximation method, which is referred to in Sec. 3.5.5.B, was applied, and the mathematical structure of Π_{ij} was examined in the wavenumber space. One of the interesting findings is the extension of Rotta's model (4.165) to an anisotropic form.

6.2. Multiple-Scale Method

In this chapter, we shall present the TSDIA method and discuss the turbulence modeling from the theoretical viewpoint. In this course, a statistical theoretical support will be given to the turbulence models constructed on the basis of conventional approaches. In the TSDIA, the concept of multiple time and space scales plays a key role. Therefore we shall give a brief account of the concept.

6.2.1. MODURATION OF FREQUENCY

A powerful mathematical tool for solving a differential equation subject to both of slow and fast variations is the multiple-scale method (Nayfeh 1973). For seeing the basic concept of the method, we consider an equation for a single pendulum with the nonlinear effects of amplitude included,

$$\frac{d^2\theta}{dt^2} + \theta = \delta\theta^3, \tag{6.1}$$

under the initial condition

$$\theta = 1, \ \frac{d\theta}{dt} = 0 \ \text{at} \ t = 0, \tag{6.2}$$

where δ is a small parameter. The linear part of Eq. (6.1) is called the unperturbed equation.

First we expand θ in δ as

$$\theta = \sum_{n=0}^{\infty} \delta^n \theta_n. \tag{6.3}$$

Then the first two terms, θ_0 and θ_1, obey

$$\frac{d^2\theta_0}{dt^2} + \theta_0 = 0, \tag{6.4}$$

$$\frac{d^2\theta_1}{dt^2} + \theta_1 = \theta_0^3. \tag{6.5}$$

The solution of Eq. (6.4) subject to the condition (6.2) is simply given by

$$\theta_0 = \cos t. \tag{6.6}$$

From Eqs. (6.5) and (6.6), we have

$$\frac{d^2\theta_1}{dt^2} + \theta_1 = \frac{3}{4}\cos t + \frac{1}{4}\cos 3t. \tag{6.7}$$

The solution of Eq. (6.7) is

$$\theta_1 = \frac{3}{8} t \sin t - \frac{1}{32} \cos 3t + C_1 \sin t + C_2 \cos t, \tag{6.8}$$

with the integral constants C_1 and C_2. From the condition (6.2) that leads to $\theta_1 = d\theta_1/dt = 0$, Eq. (6.8) is reduced to

$$\theta_1 = \frac{3}{8} t \sin t + \frac{1}{32} (\cos t - \cos 3t). \tag{6.9}$$

Summarizing Eqs. (6.6) and (6.9), we have

$$\theta = \cos t + \delta \left(\frac{3}{8} t \sin t + \frac{1}{32} (\cos t - \cos 3t) \right) + O\left(\delta^2\right). \tag{6.10}$$

In Eq. (6.10), we should note the first part in the $O(\delta)$ term. It signifies that this perturbational solution loses the mathematical meaning for $t \gg 1/|\delta|$ since the $O(\delta)$ term becomes larger than the leading one of $O(1)$. Namely, the expansion (6.10) is not uniformly valid for all t. The cause of this breakage lies in the first term on the right-hand side of Eq. (6.7). Its frequency is coincident with the normal frequency of Eq. (6.1), that of the unperturbed equation, resulting in the resonance of oscillation.

6.2.2. INTRODUCTION OF MULTIPLE SCALES

The cause of the breakage may be understood from a different viewpoint. Equation (6.10) is rewritten as

$$\theta = \cos \left(\left(1 - \frac{3}{8} \delta \right) t \right) + \delta \left(\frac{3}{32} (\cos t - \cos 3t) \right) + O\left(\delta^2\right). \tag{6.11}$$

On expanding Eq. (6.11) in δ, we may recover Eq. (6.10) up to $O(\delta)$. Physically speaking, the right-hand term of Eq. (6.1) changes or modulates the normal frequency by $O(\delta)$. Such a motion cannot be represented in terms of only the normal frequency of the unperturbed state and its higher harmonics. We rewrite the first term of Eq. (6.11) as

$$\cos \left(\tau - \frac{3}{8} T \right), \tag{6.12}$$

where

$$\tau = t, \; T = \delta t. \tag{6.13}$$

Equation (6.12) suggests the possibility that the use of two variables τ and T leads to a solution uniformly valid for t and T. These two variables are called the fast and slow variables, respectively.

Under Eq. (6.13), we have

$$\frac{d}{dt} = \frac{\partial}{\partial \tau} + \delta \frac{\partial}{\partial T}. \tag{6.14}$$

Then Eq. (6.1) is reduced to

$$\frac{\partial^2 \theta}{\partial \tau^2} + \theta = \delta \left(\theta^3 - 2 \frac{\partial^2 \theta}{\partial \tau \partial T} \right) + O\left(\delta^2\right). \tag{6.15}$$

The initial condition (6.2) is rewritten as

$$\theta = 1 \ \text{ at } \ \tau = T = 0, \tag{6.16a}$$

$$\frac{\partial \theta}{\partial \tau} = -\delta \frac{\partial \theta}{\partial T} \ \text{ at } \ \tau = T = 0. \tag{6.16b}$$

We substitute Eq. (6.3) into Eq. (6.15), and have

$$\frac{\partial^2 \theta_0}{\partial \tau^2} + \theta_0 = 0, \tag{6.17}$$

$$\frac{\partial^2 \theta_1}{\partial \tau^2} + \theta_1 = \theta_0^3 - 2 \frac{\partial^2 \theta_0}{\partial \tau \partial T}. \tag{6.18}$$

The initial condition (6.16) gives

$$\theta_0 = 1, \ \frac{\partial \theta_0}{\partial \tau} = 0 \ \text{ at } \ \tau = T = 0, \tag{6.19a}$$

$$\theta_1 = 0, \ \frac{\partial \theta_1}{\partial \tau} = -\frac{\partial \theta_0}{\partial T} \ \text{ at } \ \tau = T = 0. \tag{6.19b}$$

We write the solution of Eq. (6.17) in the form

$$\theta_0 = C e^{i\tau} + C^* e^{-i\tau}, \tag{6.20}$$

where C^* is the complex conjugate of C, and C is written as

$$C = \phi(T) \exp(i\psi(T)) \tag{6.21}$$

(we should note that ϕ and ψ are real functions of the slow variable T). From Eq. (6.19a), we have

$$\phi(0) = 1/2, \ \psi(0) = 0. \tag{6.22}$$

We substitute Eq. (6.20) into Eq. (6.18), resulting in

$$\begin{aligned} \frac{\partial^2 \theta_1}{\partial \tau^2} + \theta_1 &= \left(4\phi \frac{d\psi}{dT} + 6\phi^3 \right) \cos(\tau + \psi) + 4 \frac{d\phi}{dT} \sin(\tau + \psi) \\ &+ 2\phi^3 \cos(3(\tau + \psi)). \end{aligned} \tag{6.23}$$

As may be seen from Eqs. (6.7) and (6.8), we encounter the similar difficulty as far as the terms proportional to $\cos(\tau+\psi)$ and $\sin(\tau+\psi)$ are present. Then we require

$$\frac{d\phi}{dT}=0,\ \frac{d\psi}{dT}=-\frac{3}{2}\phi^2. \tag{6.24}$$

Under Eq. (6.22), we have

$$\phi=1/2,\ \psi=-(3/8)\,T. \tag{6.25}$$

The combination of Eq. (6.25) with Eqs. (6.20) and (6.21) leads to

$$\theta_0=\cos\left(\tau-\frac{3}{8}T\right). \tag{6.26}$$

This solution is coincident with the first term of Eq. (6.11).

The above method of solution is generally called the multiple-scale method. The $O(1)$ solution θ_0 has been determined in the analysis of the $O(\delta)$ equation. For finding the complete $O(\delta)$ solution, we need to introduce one more slow variable $\delta^2 t(=T')$ and investigate the $O(\delta^2)$ equation. In this case, C in Eq. (6.21) is a function of both T and T'.

6.3. Two-Scale Direct-Interaction Approximation

6.3.1. INTRODUCTION OF TWO SCALES

We use a small scale parameter δ and introduce two space and time variables:

$$\boldsymbol{\xi}\,(=\mathbf{x})\,,\ \mathbf{X}\,(=\delta\mathbf{x})\,;\ \ \tau\,(=t)\,,\ T\,(=\delta t)\,. \tag{6.27}$$

Here we should note that δ is not an actual parameter such as δ in Eq. (6.1), but it is an artificial one for incorporating the effects of slowly-varying quantities on the fast-varying ones in a perturbational manner. This parameter disappears automatically in the final results through the reverse replacement that $\mathbf{X}\to\delta\mathbf{x}$ and $T\to\delta t$, as will become clear later. Using Eq. (6.27), a flow quantity f, whose mean and fluctuating parts are given by F and f', respectively, is written as

$$f=F(\mathbf{X};T)+f'(\boldsymbol{\xi},\mathbf{X};\tau,T). \tag{6.28}$$

The smallness of δ indicates that the change of F is of $O(\delta)$ in the presence of the $O(1)$ change of original $\mathbf{x}$ and t. On the other hand, the change of f' is of $O(1)$ owing to the dependence on $\boldsymbol{\xi}$ and τ. These properties of Eq. (6.28) are consistent with the intrinsic features of the mean and fluctuating parts, F and f' (see Fig. 6.1). The dependence of f' on both $\mathbf{X}$ and T arises from the situation that f' is linked with F through the correlation functions of

f'. For instance, the fluctuating velocity $\mathbf{u}'$ is related to the mean velocity $\mathbf{U}$ through the Reynolds stress $\langle u'_i u'_j \rangle$. In the TSDIA formalism, we shall consider that f' is assumed to be statistically homogeneous and stationary in $\boldsymbol{\xi}$ and τ, and make full use of the Fourier transformation concerning $\boldsymbol{\xi}$.

Under Eq. (6.27), we have

$$\nabla = \nabla_\xi + \delta \nabla_X, \ \frac{\partial}{\partial t} = \frac{\partial}{\partial \tau} + \delta \frac{\partial}{\partial T}, \tag{6.29}$$

where $\nabla_\xi = (\partial/\partial \xi_i)$ and $\nabla_X = (\partial/\partial X_i)$. We apply Eqs. (6.28) and (6.29) to Eq. (2.104) for $\mathbf{u}'$, and have

$$\begin{aligned} &\frac{\partial u'_i}{\partial \tau} + U_j \frac{\partial u'_i}{\partial \xi_j} + \frac{\partial}{\partial \xi_j} u'_j u'_i + \frac{\partial p'}{\partial \xi_i} - \nu \Delta_\xi u'_i \\ &\quad = \delta \left(-u'_j \frac{\partial U_i}{\partial X_j} - \frac{D u'_i}{DT} - \frac{\partial p'}{\partial X_i} - \frac{\partial}{\partial X_j} \left(u'_j u'_i - R_{ij} \right) + 2\nu \frac{\partial u'_i}{\partial X_j \partial \xi_j} \right) \\ &\quad + \delta^2 \left(\nu \Delta_X u'_i \right), \end{aligned} \tag{6.30}$$

with the solenoidal condition

$$\frac{\partial u'_i}{\partial \xi_i} + \delta \frac{\partial u'_i}{\partial X_i} = 0, \tag{6.31}$$

where

$$\frac{D}{DT} = \frac{\partial}{\partial T} + \mathbf{U} \cdot \nabla_X. \tag{6.32}$$

In Eq. (6.30), it should be stressed that the Reynolds stress R_{ij} depends on $\mathbf{X}$ and T only since $\mathbf{u}'$ is assumed to be statistically homogeneous and stationary in $\boldsymbol{\xi}$ and τ (we should also note that R_{ij} is connected with the mean velocity $\mathbf{U}$ dependent on $\mathbf{X}$ and T only). In what follows, we shall drop the ν-related terms on the right hand side of Eq. (6.30) since their contributions to the Reynolds stress etc. are negligible in high-Reynolds-number turbulence.

6.3.2. FOURIER REPRESENTATION

We saw in Chapter 3 that the Fourier representation of the fluctuating field is instrumental to the understanding of homogeneous turbulence. In reality, almost all the statistical turbulence theories are founded on the Fourier representation. In the TSDIA, we assume the statistical homogeneity in $\boldsymbol{\xi}$ and apply the Fourier representation concerning $\boldsymbol{\xi}$ to Eqs. (6.30) and (6.31). On the other hand, the effects of the slow variation due to the mean field are dealt with in physical space. Through this procedure, we may alleviate the difficulty arising from the coexistence of slow and fast variations. This point is a crucial difference between Leslie's (1973) method and the TSDIA one (Yoshizawa 1984a).

What is important in the use of the Fourier representation in the TSDIA is that it is linked with the Galilean transformation based on the mean velocity $\mathbf{U}$; namely, we write

$$f(\boldsymbol{\xi}, \mathbf{X}; \tau, T) = \int f(\mathbf{k}, \mathbf{X}; \tau, T) \exp\left(-i\mathbf{k} \cdot (\boldsymbol{\xi} - \mathbf{U}\tau)\right) d\mathbf{k}. \tag{6.33}$$

The factor $\exp(-i\mathbf{k} \cdot (\boldsymbol{\xi} - \mathbf{U}\tau))$ signifies that the fast-varying motion of turbulent eddies is observed in the frame moving with the velocity $\mathbf{U}$. We apply Eq. (6.33) to Eqs. (6.30) and (6.31), and have

$$\begin{aligned} &\frac{\partial u'_i(\mathbf{k};\tau)}{\partial \tau} + \nu k^2 u'_i(\mathbf{k};\tau) - ik_i p'(\mathbf{k};\tau) \\ &\quad -ik_j \iint \delta(\mathbf{k} - \mathbf{p} - \mathbf{q}) d\mathbf{p} d\mathbf{q} u'_i(\mathbf{p};\tau) u'_j(\mathbf{q};\tau) \\ &\quad = \delta \left(-u'_j(\mathbf{k};\tau) \frac{\partial U_i}{\partial X_j} - \frac{D u'_i(\mathbf{k};\tau)}{DT_I} - \frac{\partial p'(\mathbf{k};\tau)}{\partial X_{Ii}} \right. \\ &\left. - \iint \delta(\mathbf{k} - \mathbf{p} - \mathbf{q}) d\mathbf{p} d\mathbf{q} \frac{\partial}{\partial X_{Ij}} \left(u'_i(\mathbf{p};\tau) u'_j(\mathbf{q};\tau) \right) + \delta(\mathbf{k}) \frac{\partial R_{ij}}{\partial X_j} \right), \end{aligned} \tag{6.34}$$

$$\mathbf{k} \cdot \mathbf{u}'(\mathbf{k};\tau) = \delta \left(-i \frac{\partial u'_i(\mathbf{k};\tau)}{\partial X_{Ii}} \right), \tag{6.35}$$

where

$$\left(\frac{D}{DT_I}, \nabla_{XI} \right) = \exp\left(-i\mathbf{k} \cdot \mathbf{U}\tau\right) \left(\frac{D}{DT}, \nabla_X \right) \exp\left(i\mathbf{k} \cdot \mathbf{U}\tau\right). \tag{6.36}$$

Here and hereafter, the dependence of f' on the slow variables $\mathbf{X}$ and T is not written explicitly, except when necessary. The new differential operators, D/DT_I and ∇_{XI}, arise from the Fourier representation linked with the Galilean transformation (6.33). Hereafter, the R_{ij}-related part in Eq. (6.34) will be dropped since it does not contribute to the following analyses.

A big difference between the TSDIA and the usual homogeneous turbulence theories explained in Chapter 3 lies in Eq. (6.35). The Fourier component $\mathbf{u}'(\mathbf{k};\tau)$ does not obey the solenoidal condition concerning $\mathbf{k}$, whereas the latter counterpart does. In the original TSDIA (Yoshizawa 1984a), the pressure p' was eliminated from Eq. (6.34) using Eq. (6.35), but the method made the following mathematical manipulation complicated and inaccurate. In order to avoid this shortfall, the following resolution of $\mathbf{u}'(\mathbf{k};\tau)$ was introduced by Hamba (1987a):

$$\mathbf{u}'(\mathbf{k};\tau) = \mathbf{u}'_S(\mathbf{k};\tau) + \delta \left(-i \frac{\mathbf{k}}{k^2} \frac{\partial u'_i(\mathbf{k};\tau)}{\partial X_{Ii}} \right). \tag{6.37}$$

Then $\mathbf{u'}_S(\mathbf{k};\tau)$ obeys the solenoidal condition concerning $\mathbf{k}$

$$\mathbf{k}\cdot\mathbf{u'}_S(\mathbf{k};\tau)=0, \tag{6.38}$$

as is the same as for homogeneous turbulence.

6.3.3. SCALE-PARAMETER EXPANSION

In Eq. (6.34), we may see that the explicit slowly-varying effects of $\mathbf{U}$ on $\mathbf{u'}$ as well as the implicit ones through $\mathbf{X}$ and T occur as the δ-related terms on the right-hand side. The left-hand side is free from such terms and is of the same form as for homogeneous turbulence, except the implicit dependence on $\mathbf{X}$ and T. This mathematical property paves the way for analyzing turbulent shear flow in the perturbational manner of making full use of the knowledge about homogeneous turbulence.

We expand $\mathbf{u'}$ in δ as

$$\mathbf{u'}(\mathbf{k};\tau)=\sum_{n=0}^{\infty}\delta^n\mathbf{u'}_n(\mathbf{k};\tau),\ \ \mathbf{u'}_S(\mathbf{k};\tau)=\sum_{n=0}^{\infty}\delta^n\mathbf{u'}_{Sn}(\mathbf{k};\tau), \tag{6.39a}$$

$$p'(\mathbf{k};\tau)=\sum_{n=0}^{\infty}\delta^n p'_n(\mathbf{k};\tau). \tag{6.39b}$$

We substitute Eq. (6.39) into Eqs. (6.34) and (6.37). Then we have

$$\begin{aligned}
&\frac{\partial u'_{ni}(\mathbf{k};\tau)}{\partial\tau}+\nu k^2u'_{ni}(\mathbf{k};\tau)-ik_ip'_n(\mathbf{k};\tau)\\
&-ik_j\iint\delta\left(\mathbf{k}-\mathbf{p}-\mathbf{q}\right)d\mathbf{p}d\mathbf{q}\left(u'_{ni}(\mathbf{p};\tau)u'_{0j}(\mathbf{q};\tau)+u'_{0i}(\mathbf{p};\tau)u'_{nj}(\mathbf{q};\tau)\right)\\
&=-u'_{n-1j}(\mathbf{k};\tau)\frac{\partial U_i}{\partial X_j}-\frac{Du'_{n-1i}(\mathbf{k};\tau)}{DT_I}-\frac{\partial p'_{n-1}(\mathbf{k};\tau)}{\partial X_{Ii}}\\
&-ik_j\sum_{m=1}^{n-1}\iint\delta\left(\mathbf{k}-\mathbf{p}-\mathbf{q}\right)d\mathbf{p}d\mathbf{q}u'_{mi}(\mathbf{p};\tau)u'_{n-mj}(\mathbf{q};\tau)\\
&-\sum_{m=0}^{n-1}\iint\delta\left(\mathbf{k}-\mathbf{p}-\mathbf{q}\right)d\mathbf{p}d\mathbf{q}\frac{\partial}{\partial X_{Ij}}\left(u'_{mi}(\mathbf{p};\tau)u'_{n-m-1j}(\mathbf{q};\tau)\right),
\end{aligned} \tag{6.40}$$

$$\mathbf{u'}_n(\mathbf{k};\tau)=\mathbf{u'}_{Sn}(\mathbf{k};\tau)-i\frac{\mathbf{k}}{k^2}\frac{\partial u'_{n-1i}(\mathbf{k};\tau)}{\partial X_{Ii}}, \tag{6.41}$$

where $\mathbf{u'}_{Sn}(\mathbf{k};\tau)$ obeys the solenoidal condition

$$\mathbf{k}\cdot\mathbf{u'}_{Sn}(\mathbf{k};\tau)=0, \tag{6.42}$$

from Eq. (6.38).

A. O(1) Field
The lowest-order field corresponding to $n = 0$ in Eqs. (6.40) and (6.41) obeys

$$\frac{\partial u'_{0i}(\mathbf{k};\tau)}{\partial \tau} + \nu k^2 u'_{0i}(\mathbf{k};\tau) - ik_i p'_0(\mathbf{k};\tau) \\ -ik_j \iint \delta\left(\mathbf{k}-\mathbf{p}-\mathbf{q}\right) d\mathbf{p} d\mathbf{q} u'_{0i}(\mathbf{p};\tau) u'_{0j}(\mathbf{q};\tau) = 0, \tag{6.43}$$

$$\mathbf{k}\cdot\mathbf{u}'_0(\mathbf{k};\tau) = 0, \tag{6.44}$$

where we should note that

$$\mathbf{u}'_0(\mathbf{k};\tau) = \mathbf{u}'_{S0}(\mathbf{k};\tau). \tag{6.45}$$

We apply Eq. (6.44) to Eq. (6.43), and have

$$\frac{\partial u'_{0i}(\mathbf{k};\tau)}{\partial \tau} + \nu k^2 u'_{0i}(\mathbf{k};\tau) \\ -iM_{ij\ell}(\mathbf{k}) \iint \delta(\mathbf{k}-\mathbf{p}-\mathbf{q}) d\mathbf{p} d\mathbf{q} u'_{0j}(\mathbf{p};\tau) u'_{0\ell}(\mathbf{q};\tau) = 0, \tag{6.46}$$

$$p'_0(\mathbf{k};\tau) = -\frac{k_i k_j}{k^2} \iint \delta(\mathbf{k}-\mathbf{p}-\mathbf{q}) d\mathbf{p} d\mathbf{q} u'_{0i}(\mathbf{p};\tau) u'_{0j}(\mathbf{q};\tau), \tag{6.47}$$

where

$$D_{ij}(\mathbf{k}) = \delta_{ij} - \left(k_i k_j / k^2\right), \tag{6.48}$$

$$M_{ij\ell}(\mathbf{k}) = (1/2)\left(k_j D_{i\ell}(\mathbf{k}) + k_\ell D_{ij}(\mathbf{k})\right). \tag{6.49}$$

Equation (6.46) is the same as Eq. (3.28) that is discussed in the context of homogeneous turbulence, except the implicit dependence on $\mathbf{X}$ and T.

The $O(\delta^n)$ field with $n \geq 1$ may be expressed in terms of $\mathbf{u}'_0(\mathbf{k};\tau)$, as is seen from Eq. (6.40). In order to evaluate various correlation functions in shear turbulence with the aid of this solution, we need to designate the statistical properties of the $O(1)$ field at high Reynolds number. One of those properties is the Kolmogorov law for the energy spectrum, Eq. (3.45). The DIA formalism in the usual Eulerian frame fails to reproduce the law, as is discussed in detail in Sec. 3.5.3.C. A method of rectifying this critical shortfall of the DIA formalism is to reformulate it in a Lagrangian frame. This method can reproduce the Kolmogorov law with the proper estimate of the numerical coefficient. The Lagrangian DIA formalism, however, is not applicable to turbulent shear flow owing to its complicated mathematical structure. As a simpler method for reproducing the law, we referred to the eddy-damped quasi-normal Markovianized (EDQNM) method in Sec. 3.5.5.B, but it is not equipped with Green's or response functions. The importance of Green's function in solving a linear system of equations such as Eq. (6.40) is clear. Therefore the EDQNM method is not suited, at least, for the following purpose.

In this context, the DIA formalism with the Green's function as a key ingredient is most superior. We fully use this merit, while making a big compromise in the derivation of the Kolmogorov spectrum. Namely, we resort to the artificial cut-off of the low-wavenumber components leading to the infra-red divergence of the response equation. This rather artificial theoretical device is not so unphysical from the viewpoint of turbulent shear flow. We divided $\mathbf{u}$ into $\mathbf{U}$ and $\mathbf{u}'$, but no clear-cut separation of scales exists between them. As a result, the very-low-wavenumber components of $\mathbf{u}$ should be considered to be included in $\mathbf{U}$. In reality, effects of $\mathbf{U}$ on $\mathbf{u}'_0(\mathbf{k};\tau)$ have been removed through the Galilean transformation, as in Eq. (6.33).

The response function associated with Eq. (6.46), $G'_{ij}(\mathbf{k};\tau,\tau')$, obeys

$$\frac{\partial G'_{ij}(\mathbf{k};\tau,\tau')}{\partial \tau} + \nu k^2 G'_{ij}(\mathbf{k};\tau,\tau') - D_{ij}(\mathbf{k})\delta(\tau-\tau') \\ = 2iM_{i\ell m}(\mathbf{k}) \iint \delta(\mathbf{k}-\mathbf{p}-\mathbf{q}) d\mathbf{p} d\mathbf{q} u'_{0\ell}(\mathbf{p};\tau) G'_{mj}(\mathbf{q};\tau,\tau'). \quad (6.50)$$

This type of equation has been already introduced as Eq. (3.100). Strictly speaking, G'_{ij} should be written as $G'_{ij}(\mathbf{k},\mathbf{X};\tau,\tau',T)$. As the simplest mathematical form for the statistical properties of the $O(1)$ field, we assume the isotropy, and write

$$Q_{ij}(\mathbf{k},\mathbf{X};\tau,\tau',T) = \frac{\left\langle u'_{0i}(\mathbf{k},\mathbf{X};\tau,T) u'_{0j}(\mathbf{k}',\mathbf{X};\tau',T) \right\rangle}{\delta(\mathbf{k}+\mathbf{k}')} \\ = D_{ij}(\mathbf{k}) \, Q(k,\mathbf{X};\tau,\tau',T), \quad (6.51)$$

$$G_{ij}(\mathbf{k},\mathbf{X};\tau,\tau',T) = \left\langle G'_{ij}(\mathbf{k},\mathbf{X};\tau,\tau',T) \right\rangle = D_{ij}(\mathbf{k}) \, G(k,\mathbf{X};\tau,\tau',T). \quad (6.52)$$

The assumption of the statistical isotropy is plausible since Eq. (6.46) is not directly dependent on the mean velocity gradient $\nabla \mathbf{U}$ that is a primary factor generating the anisotropy of turbulence.

Kolmogorov's $-5/3$ power spectrum in the inertial range is expressed in the form

$$Q(k,\mathbf{X};\tau,\tau',T) = \sigma(k,\mathbf{X};T) \exp\left(-\omega(k,\mathbf{X};T)\left|\tau-\tau'\right|\right), \quad (6.53)$$

$$G(k,\mathbf{X};\tau,\tau',T) = H(\tau-\tau') \exp\left(-\omega(k,\mathbf{X};T)(\tau-\tau')\right), \quad (6.54)$$

where

$$\sigma(k,\mathbf{X};T) = 0.12 \left(\varepsilon(\mathbf{X};T)\right)^{2/3} k^{-11/3}, \quad (6.55)$$

$$\omega\left(k, \mathbf{X}; T\right) = 0.42 \left(\varepsilon \left(\mathbf{X}; T\right)\right)^{1/3} k^{2/3}. \tag{6.56}$$

Here the numerical factors were determined with resort to the artificial cut-off of low-wavenumber components that is explained in Sec. 3.5.5.C. Specifically, the numerical factor in Eq. (6.55), 0.12, leads to the Kolmogorov constant 1.5.

B. $O(\delta)$ Field

The quantities such as the Reynolds stress, which are closely connected with the anisotropy of turbulence, play a central role in turbulent shear flow. In the present perturbational expansion (6.39), the $O(\delta)$ field gives the leading contribution to the Reynolds stress. From Eqs. (6.40) and (6.41), the $O(\delta)$ field obeys

$$\begin{aligned}
&\frac{\partial u'_{1i}(\mathbf{k};\tau)}{\partial \tau} + \nu k^2 u'_{1i}(\mathbf{k};\tau) - i k_i p'_1(\mathbf{k};\tau) \\
&\quad - i k_j \iint \delta\left(\mathbf{k} - \mathbf{p} - \mathbf{q}\right) d\mathbf{p} d\mathbf{q} \left(u'_{1i}(\mathbf{p};\tau) u'_{0j}(\mathbf{q};\tau) + u'_{0i}(\mathbf{p};\tau) u'_{1j}(\mathbf{q};\tau) \right) \\
&\quad = -u'_{0j}(\mathbf{k};\tau) \frac{\partial U_i}{\partial X_j} - \frac{D u'_{0i}(\mathbf{k};\tau)}{D T_I} - \frac{\partial p'_0(\mathbf{k};\tau)}{\partial X_{Ii}} \\
&\quad - \iint \delta\left(\mathbf{k} - \mathbf{p} - \mathbf{q}\right) d\mathbf{p} d\mathbf{q} \frac{\partial}{\partial X_{Ij}} \left(u'_{0i}(\mathbf{p};\tau) u'_{0j}(\mathbf{q};\tau) \right),
\end{aligned} \tag{6.57}$$

$$u'_1(\mathbf{k};\tau) = u'_{S1}(\mathbf{k};\tau) - i \frac{\mathbf{k}}{k^2} \frac{\partial u'_{0i}(\mathbf{k};\tau)}{\partial X_{Ii}}, \tag{6.58}$$

with

$$\mathbf{k} \cdot \mathbf{u}'_{S1}(\mathbf{k};\tau) = 0. \tag{6.59}$$

We substitute Eq. (6.58) into Eq. (6.57), and then apply Eq. (6.59). As a result, $p'_1(\mathbf{k};\tau)$ is given by

$$\begin{aligned}
p'_1(\mathbf{k};\tau) = & -2i \frac{k_j}{k^2} \frac{\partial U_j}{\partial X_i} u'_{0i}(\mathbf{k};\tau) \\
& -2 \frac{k_i k_j}{k^2} \iint \delta(\mathbf{k} - \mathbf{p} - \mathbf{q}) d\mathbf{p} d\mathbf{q} u'_{0i}(\mathbf{p};\tau) u'_{1j}(\mathbf{q};\tau) \\
& -2i \frac{1}{k^2} M_{\ell ij}(\mathbf{k}) \iint \delta(\mathbf{k} - \mathbf{p} - \mathbf{q}) d\mathbf{p} d\mathbf{q} \\
& \times \frac{\partial}{\partial X_{I\ell}} \left(u'_{0i}(\mathbf{p};\tau) u'_{0j}(\mathbf{q};\tau) \right).
\end{aligned} \tag{6.60}$$

We use Eq. (6.60) and eliminate $p'_1(\mathbf{k};\tau)$ from Eq. (6.57). Then we have

$$\frac{\partial u'_{S1i}(\mathbf{k};\tau)}{\partial \tau} + \nu k^2 u'_{S1i}(\mathbf{k};\tau)$$

$$
\begin{aligned}
&-2iM_{ij\ell}(\mathbf{k}) \iint \delta(\mathbf{k}-\mathbf{p}-\mathbf{q}) d\mathbf{p} d\mathbf{q} u'_{0j}(\mathbf{p};\tau) u'_{S1\ell}(\mathbf{q};\tau) \\
&= -D_{i\ell}(\mathbf{k}) u'_{0j}(\mathbf{k};\tau) \frac{\partial U_\ell}{\partial X_j} - D_{ij}(\mathbf{k}) \frac{D u'_{0j}(\mathbf{k};\tau)}{D T_I} \\
&+2M_{ij\ell}(\mathbf{k}) \iint \delta(\mathbf{k}-\mathbf{p}-\mathbf{q}) d\mathbf{p} d\mathbf{q} \frac{q_\ell}{q^2} u'_{0j}(\mathbf{p};\tau) \frac{\partial u'_{0m}(\mathbf{q};\tau)}{\partial X_{Im}} \\
&-D_{in}(\mathbf{k}) M_{j\ell mn}(\mathbf{k}) \iint \delta(\mathbf{k}-\mathbf{p}-\mathbf{q}) d\mathbf{p} d\mathbf{q} \\
&\qquad \times \frac{\partial}{\partial X_{Im}} \left(u'_{0j}(\mathbf{p};\tau) u'_{0\ell}(\mathbf{q};\tau) \right),
\end{aligned}
\tag{6.61}
$$

where

$$
M_{ij\ell m}(\mathbf{k}) = \frac{1}{2}\delta_{i\ell}\delta_{jm} + \frac{1}{2}\delta_{im}\delta_{j\ell} - \frac{k_i k_j}{k^2}\delta_{\ell m}. \tag{6.62}
$$

Equation (6.61) may be simply integrated using the Green's function $G'_{ij}(\mathbf{k};\tau,\tau')$ that is introduced in the context of the $O(1)$ field and is given by Eq. (6.50). Finally, we may write the solution of Eq. (6.61) as

$$
\mathbf{u}'_1(\mathbf{k};\tau) = \mathbf{u}'_{S1}(\mathbf{k};\tau) - i\frac{\mathbf{k}}{k^2}\frac{\partial u'_{0i}(\mathbf{k};\tau)}{\partial X_{Ii}}, \tag{6.63}
$$

where

$$
\begin{aligned}
u'_{S1i}(\mathbf{k};\tau) = &-\frac{\partial U_\ell}{\partial X_j} \int_{-\infty}^{\tau} d\tau_1 G'_{i\ell}(\mathbf{k};\tau,\tau_1) u'_{0j}(\mathbf{k};\tau_1) \\
&- \int_{-\infty}^{\tau} d\tau_1 G'_{ij}(\mathbf{k};\tau,\tau_1) \frac{D u'_{0j}(\mathbf{k};\tau_1)}{D T_I} \\
&+2M_{nj\ell}(\mathbf{k}) \iint \delta(\mathbf{k}-\mathbf{p}-\mathbf{q}) d\mathbf{p} d\mathbf{q} \int_{-\infty}^{\tau} d\tau_1 G'_{in}(\mathbf{k};\tau,\tau_1) \\
&\qquad \times \frac{q_\ell}{q^2} u'_{0j}(\mathbf{p};\tau_1) \frac{\partial u'_{0m}(\mathbf{q};\tau_1)}{\partial X_{Im}} \\
&-M_{j\ell mn}(\mathbf{k}) \iint \delta(\mathbf{k}-\mathbf{p}-\mathbf{q}) d\mathbf{p} d\mathbf{q} \int_{-\infty}^{\tau} d\tau_1 G'_{in}(\mathbf{k};\tau,\tau_1) \\
&\qquad \times \frac{\partial}{\partial X_{Im}} \left(u'_{0j}(\mathbf{p};\tau_1) u'_{0\ell}(\mathbf{q};\tau_1) \right).
\end{aligned}
\tag{6.64}
$$

This process of solving Eq. (6.61) clearly exhibits the merit of using the DIA with the Green's function as a key element. In the study of turbulent shear flow, such a merit can compensate for its shortfall associated with the inertial range.

C. Higher-Order Fields

The solution of $O(\delta^n)(n \geq 2)$ may be obtained in the manner entirely similar to the integration of the $O(1)$ system of equations. The actual mathematical manipulation is, however, complicated. The $O(\delta^2)$ solution was first given by Yoshizawa (1984a), but it was rectified by Okamoto (1994) on the basis of the exact fulfillment of the solenoidal condition by Hamba (1987), Eq. (6.37). In the latter, this calculation was further extended to the $O(\delta^3)$ field. In these works, attention was focused on the effects of the mean velocity $\mathbf{U}$ on $\mathbf{u}'$ to alleviate the mathematical complicatedness. Then the terms such as the third and fourth terms in Eq. (6.64) that are nonlinear in $\mathbf{u}'$ and are not dependent directly on $\mathbf{U}$ were dropped.

Under the foregoing constraint, the solution for $\mathbf{u}'_2$ is given by

$$\mathbf{u}'_2(\mathbf{k};\tau) = \mathbf{u}'_{S2}(\mathbf{k};\tau) - i\frac{\mathbf{k}}{k^2}\frac{\partial u'_{S1i}(\mathbf{k};\tau)}{\partial X_{Ii}} - \frac{\mathbf{k}k_i}{k^4}\frac{\partial^2 u'_{0j}(\mathbf{k};\tau)}{\partial X_{Ii}\partial X_{Ij}}, \tag{6.65}$$

where

$$\begin{aligned} u'_{S2i}(\mathbf{k};\tau) &= i\frac{k_m}{k^2}D_{\ell j}(\mathbf{k})\frac{\partial^2 U_m}{\partial X_j \partial X_n}\int_{-\infty}^{\tau} d\tau_1 G'_{i\ell}(\mathbf{k};\tau,\tau_1)u'_{0n}(\mathbf{k};\tau_1) \\ &- D_{\ell m}(\mathbf{k})\frac{\partial U_m}{\partial X_j}\int_{-\infty}^{\tau} d\tau_1 G'_{i\ell}(\mathbf{k};\tau,\tau_1)u'_{S1j}(\mathbf{k};\tau_1) \\ &+ i\frac{k_j}{k^2}D_{\ell m}(\mathbf{k})\frac{\partial U_j}{\partial X_n}\int_{-\infty}^{\tau} d\tau_1 G'_{i\ell}(\mathbf{k};\tau,\tau_1)\frac{\partial u'_{0n}(\mathbf{k};\tau_1)}{\partial X_{Im}} \\ &+ i\frac{k_m}{k^2}D_{\ell j}(\mathbf{k})\frac{\partial U_j}{\partial X_m}\int_{-\infty}^{\tau} d\tau_1 G'_{i\ell}(\mathbf{k};\tau,\tau_1)\frac{\partial u'_{0n}(\mathbf{k};\tau_1)}{\partial X_{In}} \\ &- \int_{-\infty}^{\tau} d\tau_1 G'_{ij}(\mathbf{k};\tau,\tau_1)\frac{Du'_{S1j}(\mathbf{k};\tau_1)}{DT_I} \\ &+ \frac{1}{k^2}D_{\ell j}(\mathbf{k})\int_{-\infty}^{\tau} d\tau_1 G'_{i\ell}(\mathbf{k};\tau,\tau_1)\frac{\partial^3 u'_{0m}(\mathbf{k};\tau_1)}{\partial \tau_1 \partial X_{Ij}\partial X_{Im}}. \end{aligned} \tag{6.66}$$

6.3.4. DERIVATION OF TURBULENT-VISCOSITY REPRESENTATION

As an example showing the mathematical procedure for calculating correlation functions by means of the TSDIA, we shall give the details of the derivation of the turbulent-viscosity representation for the Reynolds stress. In this course, the way how to apply the TSDIA to various types of shear flow will be understood.

A. Reynolds-Stress Spectrum
In the TSDIA, the Reynolds stress $\langle u'_i u'_j \rangle (= R_{ij})$ is written as

$$R_{ij} = \left\langle u'_i(\boldsymbol{\xi}, \mathbf{X}; \tau, T) u'_j(\boldsymbol{\xi}, \mathbf{X}; \tau, T) \right\rangle . \tag{6.67}$$

In the Fourier representation concerning $\boldsymbol{\xi}$, R_{ij} is given by

$$R_{ij} = \int R_{ij}(\mathbf{k}, \mathbf{X}; \tau, \tau, T) d\mathbf{k}, \tag{6.68}$$

where

$$R_{ij}(\mathbf{k}, \mathbf{X}; \tau, \tau, T) = \frac{\left\langle u'_i(\mathbf{k}, \mathbf{X}; \tau, T) u'_j(\mathbf{k}', \mathbf{X}; \tau, T) \right\rangle}{\delta\left(\mathbf{k} + \mathbf{k}'\right)} \tag{6.69}$$

[hereafter $R_{ij}(\mathbf{k}, \mathbf{X}; \tau, \tau, T)$ will be abbreviated as $R_{ij}(\mathbf{k})$]. We define the Reynolds-stress spectrum $\hat{R}_{ij}(k)$ by the average of $R_{ij}(\mathbf{k})$ on the spherical surface of radius k:

$$\hat{R}_{ij}(k) = \frac{1}{4\pi k^2} \int_{S(k)} R_{ij}(\mathbf{k}) dS \tag{6.70}$$

[$S(k)$ is the spherical surface with k as its radius].

Using Eq. (6.39), $R_{ij}(\mathbf{k})$ is expanded as

$$R_{ij}\left(\mathbf{k}\right) = \frac{\left\langle u'_{0i}\left(\mathbf{k}; \tau\right) u'_{0j}\left(\mathbf{k}'; \tau\right) \right\rangle}{\delta\left(\mathbf{k} + \mathbf{k}'\right)} + \delta \left(\frac{\left\langle u'_{1i}\left(\mathbf{k}; \tau\right) u'_{0j}\left(\mathbf{k}'; \tau\right) \right\rangle}{\delta\left(\mathbf{k} + \mathbf{k}'\right)} + \frac{\left\langle u'_{0i}\left(\mathbf{k}; \tau\right) u'_{1j}\left(\mathbf{k}'; \tau\right) \right\rangle}{\delta\left(\mathbf{k} + \mathbf{k}'\right)} \right) + O\left(\delta^2\right). \tag{6.71}$$

The first term in Eq. (6.71) is essentially the same as for homogeneous isotropic turbulence in Sec. 3.1.3. From Eq. (6.51), we have

$$\frac{\left\langle u'_{0i}(\mathbf{k}; \tau) u'_{0j}(\mathbf{k}'; \tau) \right\rangle}{\delta\left(\mathbf{k} + \mathbf{k}'\right)} = D_{ij}(\mathbf{k}) Q(k; \tau, \tau). \tag{6.72}$$

The first part of the $O(\delta)$ terms is written as

$$\begin{aligned} &\frac{\left\langle u'_{1i}(\mathbf{k}; \tau) u'_{0j}(\mathbf{k}'; \tau) \right\rangle}{\delta\left(\mathbf{k} + \mathbf{k}'\right)} \\ &= - \frac{\partial U_\ell}{\partial X_m} \int_{-\infty}^{\tau} d\tau_1 \frac{\left\langle G'_{i\ell}(\mathbf{k}; \tau, \tau_1) u'_{0m}(\mathbf{k}; \tau_1) u'_{0j}(\mathbf{k}'; \tau) \right\rangle}{\delta\left(\mathbf{k} + \mathbf{k}'\right)} \\ &- \int_{-\infty}^{\tau} d\tau_1 \frac{1}{\delta\left(\mathbf{k} + \mathbf{k}'\right)} \left\langle G'_{i\ell}(\mathbf{k}; \tau, \tau_1) \frac{D u'_{0\ell}(\mathbf{k}; \tau_1)}{D T_I} u'_{0j}(\mathbf{k}'; \tau) \right\rangle . \end{aligned} \tag{6.73}$$

Here only the first two terms of Eq. (6.64) contribute to Eq. (6.73), and the others vanish in the DIA formalism. This point will be referred to in *Note*.

In the DIA, the $O(1)$ equation (6.46) is integrated formally as Eq. (3.88); namely, we have

$$u'_{0i}(\mathbf{k};\tau) = v'_{0i}(\mathbf{k};\tau) + iM_{ij\ell}(\mathbf{k}) \iint \delta(\mathbf{k}-\mathbf{p}-\mathbf{q})d\mathbf{p}d\mathbf{q} \times \int_{-\infty}^{\tau} d\tau_1 g(k;\tau,\tau_1) u'_{0j}(\mathbf{p};\tau_1) u'_{0\ell}(\mathbf{q};\tau_1). \quad (6.74)$$

In the context of the TSDIA, the random quantity $\mathbf{v'}_0(\mathbf{k};\tau)$ should be interpreted as

$$\mathbf{v'}_0(\mathbf{k};\tau) = \mathbf{A}(\mathbf{k},\mathbf{X};T)\exp\left(-\nu k^2\tau\right), \quad (6.75)$$

whereas $g(k;\tau,\tau')$ is the same as Eq. (3.90), and is given by

$$g(k;\tau,\tau') = H(\tau-\tau')\exp\left(-\nu k^2(\tau-\tau')\right). \quad (6.76)$$

Similarly, Eq. (6.50) for the Green's function $G'_{ij}(\mathbf{k};\tau,\tau')$ is integrated as

$$G'_{ij}(\mathbf{k};\tau,\tau') = G_{Vij}(\mathbf{k};\tau,\tau') + 2iM_{n\ell m}(\mathbf{k}) \iint \delta(\mathbf{k}-\mathbf{p}-\mathbf{q})d\mathbf{p}d\mathbf{q} \times \int_{\tau'}^{\tau} d\tau_1 G_{Vin}(\mathbf{k};\tau,\tau_1) u'_{0\ell}(\mathbf{p};\tau_1) G'_{mj}(\mathbf{q};\tau_1,\tau'), \quad (6.77)$$

where

$$G_{Vij}(\mathbf{k};\tau,\tau') = D_{ij}(\mathbf{k})H(\tau-\tau')g(k;\tau,\tau') \quad (6.78)$$

[see Eq. (3.103)].

A.1. Calculation of $O(\delta)$ Terms. We substitute Eqs. (6.74) and (6.77) into the first term of Eq. (6.73). In the lowest-order contribution, we have

$$\frac{\left\langle G'_{i\ell}(\mathbf{k};\tau,\tau_1)u'_{0m}(\mathbf{k};\tau_1)u'_{0j}(\mathbf{k}';\tau)\right\rangle}{\delta(\mathbf{k}+\mathbf{k}')} = G_{Vi\ell}(\mathbf{k};\tau,\tau_1)Q_{Vmj}(\mathbf{k};\tau_1,\tau), \quad (6.79)$$

where

$$Q_{Vij}(\mathbf{k};\tau,\tau') = \frac{\left\langle v'_{0i}(\mathbf{k};\tau)v'_{0j}(\mathbf{k}';\tau')\right\rangle}{\delta(\mathbf{k}+\mathbf{k}')}, \quad (6.80)$$

and it should be noted that $G_{Vij}(\mathbf{k};\tau,\tau')$ is a deterministic quantity.

In Eq. (6.79), $Q_{Vij}(\mathbf{k};\tau,\tau')$ is the lowest-order component of $Q_{ij}(\mathbf{k};\tau,\tau')$. In the DIA formalism, $Q_{Vij}(\mathbf{k};\tau,\tau')$ and $G_{Vij}(\mathbf{k};\tau,\tau')$ are replaced with their exact counterparts, $Q_{ij}(\mathbf{k};\tau,\tau')$ and $G_{ij}(\mathbf{k};\tau,\tau')$, respectively. As a result, we have

$$\begin{aligned}&\frac{\partial U_\ell}{\partial X_m}\int_{-\infty}^{\tau} d\tau_1 \frac{\left\langle G'_{i\ell}(\mathbf{k};\tau,\tau_1)u'_{0m}(\mathbf{k};\tau_1)u'_{0j}(\mathbf{k}';\tau)\right\rangle}{\delta\left(\mathbf{k}+\mathbf{k}'\right)}\\&\quad=\frac{\partial U_\ell}{\partial X_m}\int_{-\infty}^{\tau} d\tau_1 G_{i\ell}(\mathbf{k};\tau,\tau_1)Q_{mj}(\mathbf{k};\tau_1,\tau).\end{aligned} \tag{6.81}$$

Under the isotropic form for $Q_{ij}(\mathbf{k};\tau,\tau')$ and $G_{ij}(\mathbf{k};\tau,\tau')$, Eqs. (6.51) and (6.52), Eq. (6.81) is reduced to

$$\begin{aligned}&\frac{\partial U_\ell}{\partial X_m}\int_{-\infty}^{\tau} d\tau_1 \frac{\left\langle G'_{i\ell}(\mathbf{k};\tau,\tau_1)u'_{0m}(\mathbf{k};\tau_1)u'_{0j}(\mathbf{k}';\tau)\right\rangle}{\delta\left(\mathbf{k}+\mathbf{k}'\right)}\\&=D_{i\ell}(\mathbf{k})D_{mj}(\mathbf{k})\left(\int_{-\infty}^{\tau} d\tau_1 G(k;\tau,\tau_1)Q(k;\tau,\tau_1)\right)\frac{\partial U_\ell}{\partial X_m}.\end{aligned} \tag{6.82}$$

In the second term of Eq. (6.73), the lowest-order parts of $\mathbf{u}'_0(\mathbf{k};\tau)$ and $G'_{ij}(\mathbf{k};\tau,\tau')$ give

$$\begin{aligned}&\frac{1}{\delta\left(\mathbf{k}+\mathbf{k}'\right)}\left\langle G'_{i\ell}(\mathbf{k};\tau,\tau_1)\frac{Du'_{0\ell}(\mathbf{k};\tau_1)}{DT_I}u'_{0j}(\mathbf{k}';\tau)\right\rangle\\&=\frac{1}{\delta\left(\mathbf{k}+\mathbf{k}'\right)}G_{Vi\ell}(\mathbf{k};\tau,\tau_1)\left\langle\frac{Dv'_{0\ell}(\mathbf{k};\tau_1)}{DT}v'_{0j}(\mathbf{k}';\tau)\right\rangle.\end{aligned} \tag{6.83}$$

Here and hereafter, we make the approximation to D/DT_I and ∇_{XI} [Eq. (6.36)] as

$$\left(\frac{D}{DT_I},\nabla_{XI}\right)\cong\left(\frac{D}{DT},\nabla_X\right). \tag{6.84}$$

The neglected part in Eq. (6.84) gives no contribution to Eq. (6.83) since it is an odd function of $\mathbf{k}$.

In Eq. (6.75), we impose the following constraint on $\mathbf{A}(\mathbf{k},\mathbf{X};T)$:

$$\mathbf{A}(\mathbf{k},\mathbf{X};T)=\mathbf{A}(\mathbf{k})Z(\mathbf{X};T). \tag{6.85}$$

Here the randomness of $\mathbf{A}(\mathbf{k},\mathbf{X};T)$ is assumed to come from that of $\mathbf{A}(\mathbf{k})$. This is a big constraint on parts of velocity fluctuation, $\mathbf{v}'_0$, that are dominant in the infinite past or $\tau\to-\infty$. The effect of the assumption, however, may be expected to become weak in a stationary state owing to strong nonlinear mixing effects of turbulence. Under Eq. (6.85), Eq. (6.83) is changed into

$$\frac{1}{\delta(\mathbf{k}+\mathbf{k}')}G_{Vi\ell}(\mathbf{k};\tau,\tau_1)\left\langle\frac{Dv'_{0\ell}(\mathbf{k};\tau_1)}{DT}v'_{0j}(\mathbf{k}';\tau)\right\rangle$$
$$=G_{Vi\ell}(\mathbf{k};\tau,\tau_1)\frac{D}{DT}\left(\frac{1}{2}\frac{\langle A_\ell(\mathbf{k})A_j(\mathbf{k}')\rangle}{\delta(\mathbf{k}+\mathbf{k}')}Z(\mathbf{X};T)^2\right.$$
$$\left.\times\exp\left(-\nu k^2\tau_1-\nu k'^2\tau\right)\right)$$
$$=\frac{1}{2}G_{Vi\ell}(\mathbf{k};\tau,\tau_1)\frac{DQ_{V\ell j}(\mathbf{k};\tau,\tau_1)}{DT}. \tag{6.86}$$

The renormalization of Eq. (6.86) leads to

$$\frac{1}{2}D_{ij}(\mathbf{k})\int_{-\infty}^{\tau}d\tau_1 G(k;\tau,\tau_1)\frac{DQ(k;\tau,\tau_1)}{DT}, \tag{6.87}$$

where use has been made of the second relation in Eq. (3.27).

Finally, we summarize Eqs. (6.72), (6.82), and (6.87), and obtain the expression for $R_{ij}(\mathbf{k})$ [Eq. (6.71)] up to $O(\delta)$ as

$$R_{ij}(\mathbf{k}) = D_{ij}(\mathbf{k})\left(Q(k;\tau,\tau)-\int_{-\infty}^{\tau}d\tau_1 G(k;\tau,\tau_1)\frac{DQ(k;\tau,\tau_1)}{Dt}\right)$$
$$-\ (D_{i\ell}(\mathbf{k})D_{mj}(\mathbf{k})+D_{j\ell}(\mathbf{k})D_{mi}(\mathbf{k}))$$
$$\times\left(\int_{-\infty}^{\tau}d\tau_1 G(k;\tau,\tau_1)Q(k;\tau,\tau_1)\right)\frac{\partial U_\ell}{\partial x_m}, \tag{6.88}$$

where it should be noted that the scale-expansion parameter δ has disappeared through the replacement of $\mathbf{X}\to\delta\mathbf{x}$ and $T\to\delta t$.

On substituting Eq. (6.88) into (6.70), we integrate the $D_{ij}(\mathbf{k})$-related parts on the spherical surface of radius k. In this integration, we use the formulae

$$\int\frac{k_ik_j}{k^2}d\mathbf{k}=\frac{1}{3}\delta_{ij}d\mathbf{k}, \tag{6.89a}$$

$$\int\frac{k_ik_jk_\ell k_m}{k^4}d\mathbf{k}=\frac{1}{15}\left(\delta_{ij}\delta_{\ell m}+\delta_{i\ell}\delta_{jm}+\delta_{im}\delta_{j\ell}\right)\int d\mathbf{k}. \tag{6.89b}$$

Then we have

$$\hat{R}_{ij}(k) = \frac{2}{3}\left(Q(k;\tau,\tau)-\int_{-\infty}^{\tau}d\tau_1 G(k;\tau,\tau_1)\frac{DQ(k;\tau,\tau_1)}{Dt}\right)\delta_{ij}$$
$$-\left(\frac{7}{15}\int_{-\infty}^{\tau}d\tau_1 G(k;\tau,\tau_1)Q(k;\tau,\tau_1)\right)S_{ij}, \tag{6.90}$$

where the mean velocity-strain tensor S_{ij} is defined by Eq. (4.4).

The Reynolds-stress spectrum (6.90) possesses the physical meaning a little different from the energy spectrum $E(k)$ in homogeneous isotropic turbulence that is discussed in Chapter 3. In Eq. (6.90), turbulence is regarded as nearly homogeneous anisotropic turbulence with the inhomogeneity represented locally through the slow variables. The anisotropy is averaged over the spherical surface with radius k. Substituting Eq. (6.90) to Eq. (6.68), we obtain the turbulent-viscosity approximation to the Reynolds stress R_{ij}. The relationship with the counterpart of the K-ε model, Eq. (4.3) with Eq. (4.8), will be discussed later.

Note: We confirm that the third term in Eq. (6.64) does not contribute to $\hat{R}_{ij}(\mathbf{k})$. Its contribution to Eq. (6.73) is written as

$$2M_{nh\ell}(\mathbf{k}) \iint \delta(\mathbf{k}-\mathbf{p}-\mathbf{q})d\mathbf{p}d\mathbf{q} \int_{-\infty}^{\tau} d\tau_1 \frac{q_\ell}{q^2} \times \frac{1}{\delta(\mathbf{k}+\mathbf{k}')} \left\langle G'_{in}(\mathbf{k};\tau,\tau_1) u'_{0j}(\mathbf{k}';\tau) u'_{0h}(\mathbf{p};\tau_1) \frac{\partial u'_{0m}(\mathbf{q};\tau_1)}{\partial X_{Im}} \right\rangle . \tag{6.91}$$

After the DIA procedure entirely similar to the one leading from Eq. (6.83) to Eq. (6.86), Eq. (6.91) is rewritten as

$$\frac{2}{3} M_{nh\ell}(\mathbf{k}) \iint \delta(\mathbf{k}-\mathbf{p}-\mathbf{q})d\mathbf{p}d\mathbf{q} \int_{-\infty}^{\tau} d\tau_1 \frac{q_\ell}{q^2} \times \frac{1}{\delta(\mathbf{k}+\mathbf{k}')} G_{in}(\mathbf{k};\tau,\tau_1) \frac{\partial}{\partial X_{Im}} \left\langle v'_{0j}(\mathbf{k}';\tau) v'_{0h}(\mathbf{p};\tau_1) v'_{0m}(\mathbf{q};\tau_1) \right\rangle , \tag{6.92}$$

which vanishes under the assumption of the Gaussianity of $\mathbf{v}'_0(\mathbf{k};\tau)$.

The nonvanishing contribution to Eq. (6.91) arises from the second part in Eq. (6.74) with $\mathbf{u}'_0(\mathbf{k};\tau)$ replaced with $\mathbf{v}'_0(\mathbf{k};\tau)$. This term is substituted into one of three $\mathbf{u}'_0(\mathbf{k};\tau)$ in Eq. (6.91). The resulting expression is odd concerning the wavenumber vector. Therefore it does not contribute to $\hat{R}_{ij}(\mathbf{k})$ owing to the integration on the spherical surface of radius k.

A.2. Deviation from Kolmogorov's Energy Spectrum. The turbulent energy K is related to R_{ij} as

$$K = (1/2)R_{ii} = (1/2) \int \hat{R}_{ii}(k) d\mathbf{k}, \tag{6.93}$$

from Eq. (6.70). The energy spectrum $E(k)$ is given by

$$E(k) = 4\pi k^2 \left(Q(k;\tau,\tau) - \int_{-\infty}^{\tau} d\tau_1 G(k;\tau,\tau_1) \frac{DQ(k;\tau,\tau_1)}{Dt} \right). \tag{6.94}$$

Here the second term vanishes in a stationary state of isotropic turbulence and represents a nonequilibrium property of $E(k)$. In order to see this point clearly, we substitute the inertial-range form for $Q(k;\tau,\tau')$ and $G(k;\tau,\tau')$, Eqs. (6.53)-(6.56), into Eq. (6.94). As a result, we have

$$E(k) = K_O \varepsilon^{2/3} k^{-5/3} - K_N \varepsilon^{-2/3} \frac{D\varepsilon}{Dt} k^{-7/3}, \tag{6.95}$$

with the numerical coefficients

$$K_O = 1.5, \ K_N = 0.90. \tag{6.96}$$

A typical nonequilibrium turbulence is homogeneous-shear turbulence referred to in Secs. 3.1.1 and 3.4.2. In the flow, Eq. (6.95) is replaced with

$$E(k) = K_O \varepsilon^{2/3} k^{-5/3} - K_N \varepsilon^{-2/3} \frac{\partial \varepsilon}{\partial t} k^{-7/3} \tag{6.97a}$$

$$= K_O \varepsilon_{\mathrm{eff}}^{2/3} k^{-5/3}. \tag{6.97b}$$

The ratio of the effective energy transfer rate $\varepsilon_{\mathrm{eff}}$ to the original one ε is

$$\frac{\varepsilon_{\mathrm{eff}}}{\varepsilon} = \left(1 - C_N \mathrm{sgn}\left(\frac{\partial \varepsilon}{\partial t}\right) (\ell_P k)^{-2/3}\right)^{3/2}, \tag{6.98}$$

as is given by Eq. (3.87), where the length scale ℓ_P is defined by Eq. (3.86) and $C_N = 0.6$.

The feature of the second part in Eq. (6.97a) is that the deviation from the Kolmogorov spectrum becomes more prominent as $k \to 0$. This situation makes sharp contrast with the so-called intermittency effect on $E(k)$ that is discussed in Sec. 3.4.1. The latter becomes important at large k. In turbulence modeling, the former type of deviation is usually more important. In homogeneous-shear turbulence in which ε as well as K continues to increase with time, Eq. (6.98) signifies that the transfer rate is smaller, compared with the equilibrium state with the same magnitude of ε. Physically, this fact indicates that the dissipation process cannot catch up with the production one instantaneously, resulting in the effective decrease in the energy transfer rate.

The spatial and temporal scales of energy-containing eddies in the equilibrium state are given by Eqs. (3.48) and (3.50). We simply redefine them as

$$\ell_C = K^{3/2}/\varepsilon, \ \tau_C = K/\varepsilon. \tag{6.99}$$

Then the ratio of ℓ_C to ℓ_P is

$$\ell_C/\ell_P = (\tau_C/\tau_P)^{3/2}, \tag{6.100}$$

where the time scale τ_P is introduced as

$$\tau_P = \varepsilon \left| \frac{\partial \varepsilon}{\partial t} \right|^{-1}. \tag{6.101}$$

Equation (6.101) is the time scale characterizing the change of the energy transfer rate that is closely connected with the change of the energy injection rate. Therefore ℓ_P may be called the energy-production scale. From the comparison with the direct numerical simulation (DNS) data, we may confirm that $\ell_P > \ell_C$ (Yoshizawa 1994). This fact supports that ℓ_P is a quantity linked more tightly with the mean field.

B. Relationship with K-ε Turbulent-Viscosity Model

From Eqs. (6.68), (6.70), and (6.90), we can obtain a turbulent-viscosity approximation to the Reynolds stress, Eq. (4.3), that is,

$$R_{ij} = \frac{2}{3} K \delta_{ij} - \nu_T S_{ij}, \tag{6.102}$$

where

$$K = \int Q(k;\tau,\tau) d\mathbf{k} - \int d\mathbf{k} \int_{-\infty}^{\tau} d\tau_1 G(k;\tau,\tau_1) \frac{DQ(k;\tau,\tau_1)}{Dt}, \tag{6.103}$$

$$\nu_T = \frac{7}{15} \int d\mathbf{k} \int_{-\infty}^{\tau} d\tau_1 G(k;\tau,\tau_1) Q(k;\tau,\tau_1). \tag{6.104}$$

When $Q(k;\tau,\tau)$ is approximated by the Kolmogorov spectrum (6.55), K does not converge in the limit of $k \to 0$, whereas it converges in the limit of $k \to \infty$. This fact indicates that K is heavily dependent on the properties of energy-containing eddies. These properties are connected with the mean flow that are greatly affected by external conditions. Therefore it is difficult to abstract an analytical expression applicable to various types of turbulent flows.

This situation may be seen from a little different viewpoint. The properties of the energy-containing range are partially reflected on the inertial-range ones. Then the inertial range is considered to be the intermediary that transfers part of turbulence information on the energy-containing range to the dissipation one. In hydrodynamics, it is rare to capture such intermediate characteristics in an analytical form, as is noted in Sec. 3.2.2. This point may be understood well by considering the difficulty of abstracting the analytical properties of laminar flow at intermediate Reynolds number (some prominent features of laminar flow at low and high Reynolds numbers can be found using singular perturbation methods).

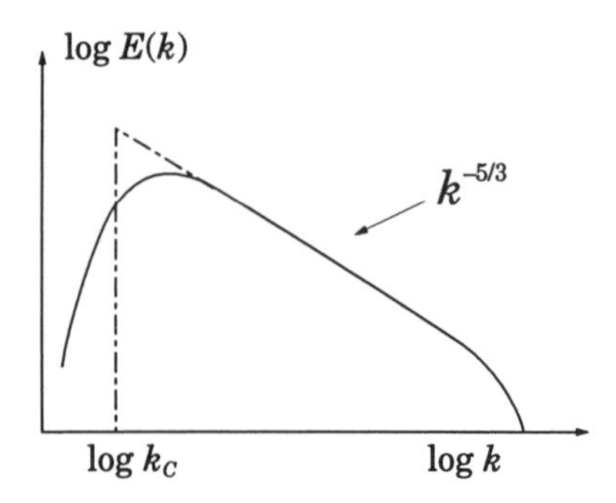

Figure 6.2. Inertial-range approximation to the energy spectrum.

We attempt to make full use of the properties of the inertial range and estimate K and ν_T. In the TSDIA, we consider that the very-low-wavenumber components of turbulent motion belong to the mean field and that the counterparts of the fluctuating field do not contribute much to the turbulence quantities such as the Reynolds stress. Then we approximate the integral as

$$\int d\mathbf{k} \to \int_{|\mathbf{k}| \geq k_C} d\mathbf{k}. \tag{6.105}$$

Here k_C is the wavenumber characterizing the largest energy-containing eddies (Fig. 6.2), and it is related to their size ℓ_C as

$$k_C(\mathbf{x};t) = 2\pi / \ell_C(\mathbf{x};t) , \tag{6.106}$$

where the dependence of k_C and ℓ_C on $\mathbf{x}$ and t (the slow variables in the TSDIA) is clear since the energy-containing eddies interact with the mean flow. It will be shown later that k_C or ℓ_C is not an arbitrary parameter, but that it is related to K and ε.

We substitute Eqs. (6.53) and (6.54) into Eq. (6.103), and have

$$\begin{aligned} K = 4\pi \int_{|\mathbf{k}| \geq k_C} & k^2 \sigma(k,\mathbf{x};t) dk \\ -2\pi \int_{|\mathbf{k}| \geq k_C} & k^2 \left(\frac{1}{\omega(k,\mathbf{x};t)} \frac{D\sigma(k,\mathbf{x};t)}{Dt} \right. \\ & \left. - \frac{\sigma(k,\mathbf{x};t)}{2\omega(k,\mathbf{x};t)^2} \frac{D\omega(k,\mathbf{x};t)}{Dt} \right) dk. \end{aligned} \tag{6.107}$$

From the inertial-range form, (6.55) and (6.56), Eq. (6.107) is reduced to

$$K = 4\pi\varepsilon(\mathbf{x};t)^{2/3} \int_{|\mathbf{k}| \geq k_C} k^{-5/3} dk$$

$$-2\pi \int_{|\mathbf{k}|\geq k_C} k^2 \left(\frac{\sigma}{\omega} \varepsilon(\mathbf{x};t)^{-1/3} k^{-2/3} \frac{D}{Dt} \left(\varepsilon(\mathbf{x};t)^{2/3} k^{-11/3} \right) \right.$$
$$\left. - \frac{\sigma}{2\omega} k^{-5} \frac{D}{Dt} \left(\varepsilon(\mathbf{x};t)^{1/3} k^{2/3} \right) \right) dk. \tag{6.108}$$

We apply the transformation

$$s = k/k_C \tag{6.109}$$

to Eq. (6.108), resulting in

$$K = 4\pi \varepsilon(\mathbf{x};T)^{2/3} k_C^{-2/3} \int_{s\geq 1} s^{-5/3} ds$$
$$-2\pi \left(\frac{\sigma}{\omega} \varepsilon(\mathbf{x};t)^{-1/3} k_C^{7/3} \frac{D}{Dt} \left(\varepsilon(\mathbf{x};t)^{2/3} k_C(\mathbf{x};t)^{-11/3} \right) \right.$$
$$\left. - \frac{\sigma}{2\omega} k_C^{-2} \frac{D}{Dt} \left(\varepsilon(\mathbf{x};t)^{1/3} k_C(\mathbf{x};t)^{2/3} \right) \right) \int_{s\geq 1} s^{-7/3} ds. \tag{6.110}$$

Here we should stress the following point. In the first part of the second term of Eq. (6.108), we do not perform the manipulation

$$\frac{D}{Dt} \left(\varepsilon(\mathbf{x};t)^{2/3} k^{-11/3} \right) = k^{-11/3} \frac{D}{Dt} \varepsilon(\mathbf{x};t)^{2/3}. \tag{6.111}$$

The left-hand side of Eq. (6.111) arises from $DQ(k, \mathbf{x}; \tau, \tau_1, t)/Dt$ in Eq. (6.103). The maximum value of $Q(k, \mathbf{x}; \tau, \tau_1, t)$ and the corresponding wavenumber are related to the mean field through $\mathbf{x}$ and t. We approximate $Q(k, \mathbf{x}; \tau, \tau_1, t)$ by its inertial-range form, and the maximum-value wavenumber is replaced by $k_C(\mathbf{x};t)$ (the lower limit of integral). Therefore we should interpret that the wavenumber on the left-hand side is scaled using $k_C(\mathbf{x};t)$. In this context, we may mention another method of cutting low-wavenumber components by Shimomura (1997b). The method will be referred to in *Note*.

Using Eq. (6.106), Eq. (6.110) is finally calculated as

$$K = C_{K1} \varepsilon^{2/3} \ell_C^{2/3} - C_{K2} \varepsilon^{-2/3} \ell_C^{4/3} \frac{D\varepsilon}{Dt} - C_{K3} \varepsilon^{1/3} \ell_C^{1/3} \frac{D\ell_C}{Dt}, \tag{6.112}$$

where

$$C_{K1} = 0.665,\ C_{K2} = 0.0581,\ C_{K3} = 0.465. \tag{6.113}$$

Entirely similarly, we have

$$\nu_T = C_{\nu\ell} \varepsilon^{1/3} \ell_C^{4/3}, \tag{6.114}$$

with

$$C_{\nu\ell} = 0.0542. \tag{6.115}$$

In order to see the relationship with the K-ε turbulent-viscosity model, we change Eq. (6.112) of the functional form $K = K\{\ell_C, \varepsilon\}$ into the form $\ell_C = \ell_C\{K, \varepsilon\}$. In Eq. (6.112), the first term comes from the $O(1)$ analysis, whereas the other two do from the $O(\delta)$ analysis. Taking this point into account, we solve Eq. (6.112) concerning ℓ_C in a perturbational manner. As a result, we have

$$\ell_C = C_{\ell 1} K^{3/2} \varepsilon^{-1} + C_{\ell 2} K^{3/2} \varepsilon^{-2} \frac{DK}{Dt} - C_{\ell 3} K^{5/2} \varepsilon^{-3} \frac{D\varepsilon}{Dt}, \tag{6.116}$$

where

$$C_{\ell 1} = 1.84, \ C_{\ell 2} = 4.36, \ C_{\ell 3} = 2.55. \tag{6.117}$$

As this result, k_C or ℓ_C, which was introduced as an unknown parameter at the stage of Eq. (6.105), has been expressed without any additional assumptions.

We substitute Eq. (6.116) into Eq. (6.114), and reach a K-ε turbulent-viscosity model

$$\nu_T = C_\nu K^2/\varepsilon, \tag{6.118}$$

with

$$C_\nu = 0.123. \tag{6.119}$$

In obtaining Eq. (6.118), we retained only the first term since the other two of $O(\delta)$ lead to the $O(\delta^2)$ contribution to R_{ij}. Such a contribution can also occur from the $O(\delta^2)$ analysis in Eq. (6.71), and these two contributions should be given in the combined form from the viewpoint of the order of expansion in the TSDIA. Equation (6.119) is compared with the optimized counterpart in Eq. (4.72), 0.09, and is rather larger. In Sec. 4.3.2.D, 0.09 is shown to be the value that the functionalized counterpart of C_ν takes under the situation that the energy production and dissipation processes balance each other. In the TSDIA formalism, the corrections to the turbulent-viscosity approximation (6.102) may occur in the higher-order analysis of R_{ij}. The validity of the theoretically-estimated value 0.123 should be judged in the context of these corrections. This point will be referred to in Secs. 6.4.2 and 6.8.

The foregoing result provides a statistical theoretical foundation to the K-ε turbulent-viscosity model for the Reynolds stress, which is the prototype of more elaborate turbulence models. This derivation process shows that Eq. (6.118) is constructed on a very simplified theoretical basis, and that there is much room for the improvement such as the inclusion of DK/Dt and $D\varepsilon/Dt$ effects.

Note: In the use of the inertial-range form, Eqs. (6.55) and (6.56), we introduced the low-wavenumber cut-off, as in Eq. (6.105). A method based on a little different cut-off was introduced by Shimomura (1997b) in the study about the transport rate $\mathbf{T}_K$ [Eq. (2.114)] (this work will be referred to later). There the unit step function $H(k-k_C)$ is attached to $Q(k;\tau,\tau')$. For instance, the second term in Eq. (6.103) is changed into

$$4\pi\int_0^\infty k^2dk\int_{-\infty}^{\tau} d\tau_1 G(k;\tau,\tau_1)\frac{D}{Dt}\left(Q\left(k;\tau,\tau_1\right)H\left(k-k_C\right)\right). \qquad (6.120)$$

As this consequence, $DQ(k;\tau,\tau_1)/Dt$ is rewritten as Eq. (6.111), and all the dependence on k_C or ℓ_C comes from $H(k-k_C)$. The resulting expressions are very similar to the foregoing counterparts, but the numerical coefficients change a little. The difference between those two methods, Eqs. (6.105) and (6.120), is the one in approximating the full spectral range by only the inertial one, and it is not so clear which approximation is better.

6.4. Derivation of Higher-Order Models for Reynolds Stress

In Sec. 6.3.4, we derived a turbulent-viscosity or linear model for the Reynolds stress R_{ij}. Some critical shortfalls of the model were pointed out in Sec. 4.2.5, and the necessity of higher-order models for rectifying them was discussed in Sec. 4.3 from the standpoint of conventional turbulence modeling. In what follows, we shall derive those models with the aid of the TSDIA.

6.4.1. SECOND-ORDER TSDIA ANALYSIS OF RETNOLDS STRESS

The $O(\delta^2)$ calculation of R_{ij} was made by Yoshizawa (1984a), which was improved by Okamoto (1994) following the exact treatment of the solenoidal condition by Hamba (1987a), Eqs. (6.37) and (6.38). In Eq. (6.71), the $O(\delta^2)$ terms are given by

$$\delta^2\left(\frac{\left\langle u'_{2i}\left(\mathbf{k};\tau\right)u'_{0j}\left(\mathbf{k}';\tau\right)\right\rangle}{\delta\left(\mathbf{k}+\mathbf{k}'\right)}+2\frac{\left\langle u'_{1i}\left(\mathbf{k};\tau\right)u'_{1j}\left(\mathbf{k}';\tau\right)\right\rangle}{\delta\left(\mathbf{k}+\mathbf{k}'\right)}\right.$$
$$\left.+\frac{\left\langle u'_{0i}\left(\mathbf{k};\tau\right)u'_{2j}\left(\mathbf{k}';\tau\right)\right\rangle}{\delta\left(\mathbf{k}+\mathbf{k}'\right)}\right). \qquad (6.121)$$

These terms may be calculated using Eqs. (6.63) and (6.65). This calculation is quite similar to the $O(\delta)$ counterpart, but the mathematical manipulation is much more complicated.

The final expression for R_{ij} may be rewritten as

$$R_{ij} = \frac{2}{3}K\delta_{ij} - \nu_T S_{ij} + \eta_1 \left[\frac{\partial U_i}{\partial x_\ell} \frac{\partial U_j}{\partial x_\ell} \right]_D + \eta_2 \left[\frac{\partial U_i}{\partial x_\ell} \frac{\partial U_\ell}{\partial x_j} + \frac{\partial U_\ell}{\partial x_i} \frac{\partial U_j}{\partial x_\ell} \right]_D$$
$$+ \eta_3 \left[\frac{\partial U_\ell}{\partial x_i} \frac{\partial U_\ell}{\partial x_j} \right]_D + \eta_4 \frac{DS_{ij}}{Dt}, \tag{6.122}$$

where $[A_{ij}]_D$ is the deviatoric part of A_{ij} defined by Eq. (4.97), and

$$K = C_{K1} \varepsilon^{2/3} \ell_C^{2/3} - C_{K2} \varepsilon^{-2/3} \ell_C^{4/3} \frac{D\varepsilon}{Dt} - C_{K3} \varepsilon^{1/3} \ell_C^{1/3} \frac{D\ell_C}{Dt}$$
$$+ C_{K4} \ell_C^2 S_{ij}^2, \tag{6.123}$$

$$\nu_T = C_{\nu\ell} \varepsilon^{1/3} \ell_C^{4/3} - C_{\nu\ell\varepsilon} \varepsilon^{-1} \ell_C^2 \frac{D\varepsilon}{Dt} - C_{\nu\ell\ell} \ell_C \frac{D\ell_C}{Dt}, \tag{6.124}$$

$$\eta_n = C_{\eta\ell n} \ell_C^2, \ (n = 1\text{-}4). \tag{6.125}$$

Here $C_{Kn} (n = 1\text{-}3)$ and $C_{\nu\ell}$ are given by Eqs. (6.113) and (6.115), respectively, and

$$C_{K4} = 0.00630, \tag{6.126a}$$

$$C_{\nu\ell\varepsilon} = 0.0105, \ C_{\nu\ell\ell} = 0.106, \tag{6.126b}$$

$$C_{\eta\ell1} = 0.0160, \ C_{\eta\ell2} = 0.00876, \ C_{\eta\ell3} = 0.00155, \ C_{\eta\ell4} = 0.0126. \tag{6.126c}$$

At the stage of the more primitive expression based on $Q(k; \tau, \tau')$ and $G(k; \tau, \tau')$, such as Eq. (6.104), we have the relationship

$$\eta_2 = (\eta_1 + \eta_3)/2 . \tag{6.127}$$

The importance of this relationship will be referred to later.

We solve Eq. (6.123) with respect to ℓ_C and have the expression for ℓ_C, which is the counterpart of Eq. (6.116), as

$$\ell_C = C_{\ell1} \frac{K^{3/2}}{\varepsilon} + C_{\ell2} \frac{K^{3/2}}{\varepsilon^2} \frac{DK}{Dt} - C_{\ell3} \frac{K^{5/2}}{\varepsilon^3} \frac{D\varepsilon}{Dt} - C_{\ell4} \frac{K^{7/2}}{\varepsilon^3} S_{ij}^2, \tag{6.128}$$

where $C_{\ell n}$'s $(n = 1\text{-}3)$ are given by Eq. (6.117), and

$$C_{\ell4} = 0.0595. \tag{6.129}$$

We substitute Eq. (6.128) into Eqs. (6.124) and (6.125), and retain the terms of the same order as the latter. As this result, we have

$$\nu_T = C_\nu \frac{K^2}{\varepsilon} - C_{\nu K} \frac{K^2}{\varepsilon^2} \frac{DK}{Dt} + C_{\nu\varepsilon} \frac{K^3}{\varepsilon^3} \frac{D\varepsilon}{Dt}, \tag{6.130}$$

$$\eta_n = C_{\eta n}\frac{K^3}{\varepsilon^2}, \ (n = 1\text{-}4), \tag{6.131}$$

with

$$C_\nu = 0.123, \ C_{\nu K} = 0.147, \ C_{\nu\varepsilon} = 0.0933, \tag{6.132a}$$

$$C_{\eta 1} = 0.0542, \ C_{\eta 2} = 0.0297, \ C_{\eta 3} = 0.00525, \ C_{\eta 4} = 0.0427. \tag{6.132b}$$

Equation (6.122) is a second-order nonlinear model for R_{ij}, and may be cast into the form (4.114), except the last η_4-related term; namely, we have

$$R_{ij} = \frac{2}{3}K\delta_{ij} - \nu_T S_{ij} + \zeta_1 \left[S_{i\ell}S_{\ell j}\right]_D + \zeta_2\left(S_{i\ell}\Omega_{\ell j} + S_{j\ell}\Omega_{\ell i}\right)$$
$$+\zeta_3\left[\Omega_{i\ell}\Omega_{\ell j}\right]_D + \zeta_4\frac{DS_{ij}}{Dt}, \tag{6.133}$$

where the mean vorticity tensor Ω_{ij} is defined by Eq. (4.101), and

$$\zeta_1 = \eta_2 = C_{\zeta 1}\frac{K^3}{\varepsilon^2}, \tag{6.134a}$$

$$\zeta_2 = \frac{1}{4}\left(\eta_1 - \eta_3\right) = C_{\zeta 2}\frac{K^3}{\varepsilon^2}, \tag{6.134b}$$

$$\zeta_3 = \frac{1}{2}\left(\eta_1 + \eta_3 - 2\eta_2\right) = C_{\zeta 3}\frac{K^3}{\varepsilon^2} = 0, \tag{6.134c}$$

$$\zeta_4 = \eta_4 = C_{\zeta 4}\frac{K^3}{\varepsilon^2}, \tag{6.134d}$$

with

$$C_{\zeta 1} = 0.0297, \ C_{\zeta 2} = 0.0122, \ C_{\zeta 4} = 0.0427. \tag{6.135}$$

Here we should emphasize vanishing of the coefficient connected with the Ω-Ω effects, ζ_3, which results from Eq. (6.127). In the original TSDIA presentation (Yoshizawa 1984a), ζ_3 did not vanish. The necessity of vanishing ζ_3 was first pointed out by Speziale (1987), and has been recently presented in a more clear form (Speziale 1997; see Sec. 4.3.1.B). Speziale's indication is supported by the TSDIA improved by the exact implementation of the solenoidal condition using Eqs. (6.37) and (6.38). Equation (6.135) should be compared with Speziale's optimized counterpart (4.116). The nonlinear model of the type (6.133) except the last term was also derived by Rubinstein and Barton (1990) using the renormalization group (RNG) method. Vanishing ζ_3, however, is not fulfilled in the work. The relation (6.127) was also derived by Rubinstein (1996) using the DIA in homogeneous-shear turbulence.

The primary properties of Eq. (6.133) are discussed in the context of second-order nonlinear model of Chapter 4. Specifically, the importance of the S-S and S-Ω terms is shown in relation to secondary flows in a square-duct flow in Sec. 4.3.1.B. Those flows were really reproduced by Speziale (1987) and Nisizima (1990) through the incorporation of the nonlinear parts into the K-ε model.

The implication of the D/Dt effects in ν_T [Eq. (6.130)] is discussed in Secs. 4.5.1 and 4.5.2 in close relation to homogeneous-shear turbulence. There Eq. (4.213) was derived by expanding a model transport equation for R_{ij} around the linear turbulent-viscosity representation. From Eq. (6.130), the counterpart of Eq. (4.213) is

$$\frac{C_{\nu M}}{C_\nu} = 1 - 0.759\left(1.57\frac{1}{\varepsilon}\frac{DK}{Dt} - \frac{K}{\varepsilon^2}\frac{D\varepsilon}{Dt}\right). \tag{6.136}$$

The numerical factor in Eq. (6.136), 1.57, is replaced with 2 in Eq. (4.213). The difference between them is rather small. In the actual use of (6.136), its renormalized form (4.223) may widen the range of applicability. Such a model is shown to mimic the DNS growth rate of K in homogeneous-shear flow in Fig. 4.7 (Yoshizawa and Nisizima 1993).

Using the K and ε equations (2.111) and (4.37), we may rewrite Eq. (4.223) as

$$\frac{C_{\nu M}}{C_\nu} = \left(1 - \frac{1}{C_{\Pi 1}}\left((2 - C_{\varepsilon 1})\frac{P_K}{\varepsilon} - (2 - C_{\varepsilon 1}) + 2\nabla\cdot\mathbf{T}_K - \nabla\cdot\mathbf{T}_\varepsilon\right)\right)^{-1}. \tag{6.137}$$

On dropping the ∇-related or diffusion effects, Eq. (6.137) is reduced to the model proposed first by Rodi (1976), and represents the modification of the turbulent-viscosity coefficient in the presence of the imbalance of energy production and dissipation rates.

6.4.2. RELATIONSHIP BETWEEN THEORETICALLY-ESTIMATED AND OPTIMIZED MODEL CONSTANTS

The mathematical structure of the TSDIA scale expansion may be understood from the comparison of Eq. (6.133) with expression (4.140), specifically, Eqs. (4.140a) and (4.140b). The nonlinear models in the conventional modeling consist of the terms related to $\nabla\mathbf{U}$, that is, S_{ij} and Ω_{ij}. As is stated in Sec. 4.3.2.B, they may be interpreted as a kind of expansion with χ_U [Eq. (4.142)] and χ_Ω [Eq. (4.143)] as the expansion parameters. From the D/Dt-related parts in ν_T [Eq. (6.130)] and the last term in Eq. (6.133), the TSDIA nonlinear model contains the terms concerning the first-order derivatives of K and ε, and the higher-order ones of $\mathbf{U}$, besides the S_{ij}- and Ω_{ij}-related parts. In the TSDIA formalism, the scale parameter δ is attached to the derivatives concerning the slow variables $\mathbf{X}$ and T [see Eq. (6.29)]. As this result, the second-order analysis of R_{ij} generates the terms dependent on second-order derivatives in addition to those quadratic in the first derivative. This point is an essential difference between the nonlinear models in the conventional and TSDIA modeling.

We can make entirely the same discussion on the $O(\delta^n)(n \geq 3)$ expression for R_{ij}. The $O(\delta^3)$ analysis of R_{ij}, which was carried out by Okamoto (1994), is quite complicated. The resulting expression consists of a lot of terms. Of these terms, we pick up the contributions to ν_T, and have

$$\nu_T = C_\nu \frac{K^2}{\varepsilon} - C_{\nu K} \frac{K^2}{\varepsilon^2} \frac{DK}{Dt} + C_{\nu\varepsilon} \frac{K^3}{\varepsilon^3} \frac{D\varepsilon}{Dt} - C_{\nu S} \frac{K^4}{\varepsilon^3} S_{ij}^2 - C_{\nu\Omega} \frac{K^4}{\varepsilon^3} \Omega_{ij}^2, \quad (6.138)$$

where the first three terms have been already given by Eq. (6.130), and

$$C_{\nu S} = 4.10 \times 10^{-3}, \ C_{\nu\Omega} = 8.63 \times 10^{-4}. \quad (6.139)$$

The last two terms were also pointed out by Horiuti (1990) on the basis of the TSDIA, except the numerical factors. There the S-S term was also derived using the expansion of an algebraic second-order model of Rodi's (1976) type around the linear turbulent-viscosity model. The Ω-Ω term, however, does not occur through such an expansion. This point will be referred to later.

In order to see the physical meaning of the S_{ij}-related term in Eq. (6.138), we retain the first and fourth (S-S) terms, and drop the Ω-Ω term since its numerical factor is much smaller than the S-S counterpart. As this consequence, we have

$$\nu_T = C_\nu \frac{K^2}{\varepsilon} \left(1 - \frac{C_{\nu S}}{C_\nu} \chi_U^2 \right), \quad (6.140)$$

where χ_U is defined by Eq. (4.142). The range of applicability of the straightforward expansion such as Eq. (6.140) is limited in the actual use. In order to widen the range, we apply a simple Padé approximation to it, resulting in

$$\nu_T = C_{\nu M} K^2 / \varepsilon, \quad (6.141)$$

where

$$C_{\nu M} = C_\nu \left(1 + \left(C_{\nu S} / C_\nu \right) \chi_U^2 \right)^{-1}. \quad (6.142)$$

In a channel flow, $C_{\nu M}$ needs to be about 0.09, and the theoretically-estimated value for C_ν [Eq. (6.119)], 0.123, is rather larger. From Eq. (4.150), χ_U is about 4 in a channel flow. Then $C_{\nu M}$ is effectively around 0.08. This fact indicates that the theoretically-estimated constant for C_ν does not need to be very close to the familiar value 0.09. What is most important is whether the correction terms to the widely-used turbulent viscosity or the first term of Eq. (6.138) are used or not. As far as only the first term is adopted in the study of near-wall flows, the choice of 0.09 is adequate. In a square-duct flow discussed in Sec. 4.2.5.B, the second-order nonlinear terms are essential for reproducing secondary flows, but their effects do not occur in the equation for the mean streamwise velocity. In such a case, the choice of 0.09 is appropriate for reproducing the logarithmic velocity law for it.

6.5. Derivation of Second-Order Models Using Renormalization

In Sec. 6.3, we performed the TSDIA analysis of the Reynolds stress R_{ij} to reveal its mathematical structure with the linear turbulent-viscosity model as the leading term. The transport equation for R_{ij}, (2.105), contains the unknown quantities such as Π_{ij} [Eq. (2.107)]. In the conventional second-order modeling, we construct model expressions for such quantities and close the equation. The resulting equation is complicated, as is seen in Sec. 4.4. It is usually solved by a numerical method, and attention has not been paid so much to the analytical property of the solution.

In the TSDIA formalism, we directly deal with the equation for $\mathbf{u}'$, (2.104), and construct a model for R_{ij}. This procedure is equivalent to finding a solution of the R_{ij} equation without solving the latter. Therefore it is significant to examine what type of equation is consistent with the model by the TSDIA. In Sec. 4.5, we explained the procedure reverse to it. There a nonlinear model is derived from a model transport equation for R_{ij} through the expansion around the linear turbulent-viscosity model.

In what follows, we shall first discuss on the derivation of a transport equation for R_{ij} with the aid of the TSDIA result about the second-order nonlinear model in Sec. 6.4 (Yoshizawa 1993a, b). Next, we shall expand the equation and derive the third-order nonlinear models for R_{ij}. Through this procedure, we can easily obtain the higher-order terms that are difficult to derive using the TSDIA because of its mathematical complicatedness.

6.5.1. RENORMALIZATION OF TSDIA SECOND-ORDER RESULTS

We, to begin with, rewrite the second-order nonlinear model, Eq. (6.133), as

$$\begin{aligned}
B_{ij} &\equiv R_{ij} - \frac{2}{3}K\delta_{ij} \\
&= -\nu_{TS}S_{ij} + C_{N1}\frac{K}{\varepsilon}\left[(\nu_{TS}S_{i\ell})S_{\ell j} + (\nu_{TS}S_{j\ell})S_{\ell i}\right]_D \\
&+ C_{N2}\frac{K}{\varepsilon}\left((\nu_{TS}S_{i\ell})\Omega_{\ell j} + (\nu_{TS}S_{j\ell})\Omega_{\ell i}\right) + C_A\frac{K}{\varepsilon}\frac{D}{Dt}(\nu_{TS}S_{ij}) \\
&+ \left(C_{NK}\frac{1}{\varepsilon}\frac{DK}{Dt} - C_{N\varepsilon}\frac{K}{\varepsilon^2}\frac{D\varepsilon}{Dt}\right)(\nu_{TS}S_{ij}),
\end{aligned} \tag{6.143}$$

where ν_{TS} is the static part of ν_T, which is defined as

$$\nu_{TS} = C_\nu K^2/\varepsilon, \tag{6.144}$$

and

$$C_{N1} = C_{\zeta 1}/(2C_\nu) = 0.121, \tag{6.145a}$$

$$C_{N2} = C_{\zeta 2}/C_\nu = 0.0992, \tag{6.145b}$$

$$C_A = C_{\zeta 4}/C_\nu = 0.347, \tag{6.145c}$$

$$C_{NK} = (C_{\nu K}/C_\nu) - 2C_A = 0.501, \tag{6.145d}$$

$$C_{N\varepsilon} = (C_{\nu\varepsilon}/C_\nu) - C_A = 0.411. \tag{6.145e}$$

In Eq. (6.143), we pay special attention to the point that the first term, $-\nu_{TS}S_{ij}$, is the first-order term in the TSDIA expansion. This term may be considered to correspond to the lowest-order terms in the DIA formalism for homogeneous turbulence, $Q_{Vij}(\mathbf{k};t,t')$ [Eq. (3.94)] and $G_{Vij}(\mathbf{k};t,t')$ [Eq. (3.104)]. In the latter, we performed the renormalization as

$$Q_{Vij}(\mathbf{k};t,t') \to Q_{ij}(\mathbf{k};t,t'), \; G_{Vij}(\mathbf{k};t,t') \to G_{ij}(\mathbf{k};t,t'), \tag{6.146a}$$

and obtained the DIA system of equations, (3.111) and (3.112), where $Q_{ij}(\mathbf{k};t,t')$ [Eq. (3.93)] and $G_{ij}(\mathbf{k};t,t')$ [Eq. (3.106)] are the exact counterparts of $Q_{Vij}(\mathbf{k};t,t')$ and $G_{Vij}(\mathbf{k};t,t')$, respectively. Following this renormalization procedure, we perform the replacement

$$-\nu_{TS}S_{ij} \to B_{ij} \tag{6.146b}$$

in Eq. (6.143), except the first term. As a result, we reach

$$B_{ij} = -\nu_{TS}S_{ij} - C_{N1}\frac{K}{\varepsilon}\left[B_{i\ell}S_{\ell j} + B_{j\ell}S_{\ell i}\right]_D - C_{N2}\frac{K}{\varepsilon}\left(B_{i\ell}\Omega_{\ell j} + B_{j\ell}\Omega_{\ell i}\right)$$
$$-C_A\frac{K}{\varepsilon}\frac{DB_{ij}}{Dt} - \left(C_{NK}\frac{1}{\varepsilon}\frac{DK}{Dt} - C_{N\varepsilon}\frac{K}{\varepsilon^2}\frac{D\varepsilon}{Dt}\right)B_{ij}. \tag{6.147}$$

Equation (6.147) is changed into a transport equation for B_{ij}; namely, it is given by

$$\frac{DB_{ij}}{Dt} = -\frac{1}{C_A}\left(1 + C_{NK}\frac{1}{\varepsilon}\frac{DK}{Dt} - C_{N\varepsilon}\frac{K}{\varepsilon^2}\frac{D\varepsilon}{Dt}\right)B_{ij} - \frac{1}{C_A}\frac{\varepsilon}{K}\nu_{TS}S_{ij}$$
$$-\frac{C_{N1}}{C_A}\left[B_{i\ell}S_{\ell j} + S_{i\ell}B_{\ell j}\right]_D - \frac{C_{N2}}{C_A}\left(B_{i\ell}\Omega_{\ell j} + B_{j\ell}\Omega_{\ell i}\right). \tag{6.148}$$

The exact equation for B_{ij}, which is derived from Eq. (2.105), is given by Eqs. (4.160) and (4.161). The comparison between Eq. (4.160) and Eq. (6.148) leads to the modeling

$$\Pi_{ij} - [\varepsilon_{ij}]_D = C_{\Pi 1}^{TS} \frac{\varepsilon}{K} B_{ij} + C_{\Pi 2}^{TS} K S_{ij} + C_{\Pi 3}^{TS} [B_{i\ell} S_{\ell j} + S_{j\ell} B_{\ell i}]_D + C_{\Pi 4}^{TS} (B_{i\ell} \Omega_{\ell j} + B_{j\ell} \Omega_{\ell i}), \quad (6.149)$$

where

$$C_{\Pi 1}^{TS} = -\frac{1}{C_A} \left(1 - C_{NK} \frac{1}{\varepsilon} \frac{DK}{Dt} + C_{N\varepsilon} \frac{K}{\varepsilon^2} \frac{D\varepsilon}{Dt} \right), \quad (6.150a)$$

$$C_{\Pi 2}^{TS} = \frac{2}{3} - \frac{C_\nu}{C_A} = 0.312, \quad (6.150b)$$

$$C_{\Pi 3}^{TS} = \frac{1}{2} - \frac{C_{N1}}{C_A} = 0.151, \quad (6.150c)$$

$$C_{\Pi 4}^{TS} = \frac{1}{2} - \frac{C_{N2}}{C_A} = 0.214 \quad (6.150d)$$

[the numerical coefficients in Eq. (6.150a) are given by Eq. (6.145)].

Equation (6.149) is the TSDIA counterpart of Eq. (4.183), except $[\varepsilon_{ij}]_D$. Specifically, the expression with the D/Dt effects dropped in Eq. (6.150a) corresponds to the model of Launder, Reece, and Rodi (1975), which is the prototype of all the models for Π_{ij}. Here we should note that Π_{ij} and $[\varepsilon_{ij}]_D$ cannot be separated from each other, as is seen from Eq. (4.198) of the same mathematical form as the first term in Eq. (6.149). The first term is the so-called Rotta's model, and Eq. (6.150a) expresses the nonequilibrium effect on its coefficient. The other work on the theoretical derivation of the equation for R_{ij} is the attempt by Rubinstein and Barton (1992). There Π_{ij} was examined directly using the RNG method.

In Eq. (6.148), we do not have the diffusion term corresponding to the last term in Eq. (4.160), which is written as Eq. (4.201) or (4.205) in the conventional turbulence modeling. The lack of diffusion effects comes from the approximation made in obtaining the $O(\delta^2)$ solution (6.66). In it, only the terms linear in $\mathbf{u}'_0$ and its $\mathbf{X}$ or T derivatives are retained, and the nonlinear counterparts are dropped. The latter are related to the diffusion effect, but such a calculation is very complicated and has not been done yet. In order to find a more general model for Π_{ij}, which corresponds to Eq. (4.180), we need to start from the nonlinear expression for R_{ij} that contains the terms of $O(\delta^n) (n \geq 3)$. The derivation of the diffusion term in the K equation (2.111) will be discussed in Sec. 6.6.

6.5.2. EXPANSION OF RENORMALIZED MODEL

The third-order TSDIA analysis of R_{ij} is very complicated, as is seen from the work of Okamoto (1994). On the other hand, the second-order nonlinear model of the type (6.133) is not still sufficient for the analyses of some kinds of basic turbulent shear flows. One typical instance of such flows is the frame rotation effect on a channel flow that is referred to in Sec. 4.3.2.A (see Fig. 4.6). In order to cope with the effect, a third-order nonlinear model for R_{ij} is necessary. A method for deriving such a high-order model without resort to the complicated TSDIA analysis is to use Eq. (6.147) and take the reverse step of going from it to nonlinear models such as Eq. (6.133) or (6.143).

The reverse step may be performed through the process of solving Eq. (6.147) by the iteration method with $-\nu_{TS}S_{ij}$ as the starting term. In the first iteration, we substitute $-\nu_{TS}S_{ij}$ into the B_{ij}-related terms on the right-hand side of Eq. (6.147), resulting naturally in the reproduction of Eq. (6.133) or (6.143). In such an iteration method, the first significant results arise from the second iteration. Namely, we substitute Eq. (6.133) into Eq. (6.147) and obtain various types of new terms.

A. Nonequilibrium Effects on Second-Order Nonlinear Model

A typical instance showing nonequilibrium properties of turbulence is homogeneous-shear turbulence that is referred to in Sec. 4.3.3. The temporal evolution of K and ε cannot be described correctly using the static turbulent-viscosity model (6.144). This shortfall may be rectified by incorporating the $\partial K/\partial t$ and $\partial \varepsilon/\partial t$ parts in the DK/Dt and $D\varepsilon/Dt$ effects into the turbulent viscosity, as in Eq. (6.136) or its extended form, Eq. (4.223). The latter is also obtained by equating the first two terms on the right-hand side of Eq. (6.148).

The other interesting nonequilibrium flow similar to it is the so-called spatially developing homogeneous-shear turbulence, which was experimentally examined in detail by Tavoularis and Corrsin (1981, 1985). This flow is generated by the shear-generation screen placed at one location, and the downstream turbulence properties are observed. In the flow, the streamwise change of the mean velocity is small and is approximated by Eq. (3.13), but the turbulence properties such as K and ε are spatially developing in the streamwise direction. The nonequilibrium effect in this flow is typically represented by $(\mathbf{U}\cdot\nabla)K$ and $(\mathbf{U}\cdot\nabla)\varepsilon$ in the DK/Dt and $D\varepsilon/Dt$ effects.

This flow is beyond the capability of the linear static turbulent-viscosity model, Eqs. (6.102) and (6.118), as is the case for temporally developing homogeneous-shear turbulence. A candidate for the models describing the flow is a second-order nonlinear model, Eqs. (6.133) and (6.134). The model, however, cannot properly explain the observed spatially developing turbulence characteristics. This deficiency is closely related to the lack of effects of the streamwise variation on the model coefficients ζ_n's.

The foregoing effects can be incorporated by performing the second iteration of Eq. (6.147). In the iteration, we substitute Eq. (6.133) into Eq. (6.147), as was stated above. Here we pay attention to the nonequilibrium effects on ν_T [Eq. (6.130)], and pick up their effects on the coefficients in the second-order nonlinear terms. As a result, we have

$$B_{ij} = -\nu_T S_{ij} - C_{\zeta 1}^N \frac{K^3}{\varepsilon^2} \left[S_{i\ell} S_{\ell j} + S_{j\ell} S_{\ell i} \right]_D - C_{\zeta 2}^N \frac{K^3}{\varepsilon^2} \left(S_{i\ell} \Omega_{\ell j} + S_{j\ell} \Omega_{\ell i} \right), \tag{6.151}$$

where

$$C_{\zeta n}^N = C_{\zeta n} \left(1 - \frac{C_{\nu K}}{C_\nu} \frac{1}{\varepsilon} \frac{DK}{Dt} + \frac{C_{\nu\varepsilon}}{C_\nu} \frac{K}{\varepsilon^2} \frac{D\varepsilon}{Dt} \right), \; (n = 1, 2). \tag{6.152}$$

This type of model was applied by Okamoto (1994) to a spatially developing homogeneous-shear flow and shown to give the results consistent with the observation by Tavoularis and Corrsin (1981, 1985).

B. Frame-Rotation Effects in Third-Order Nonlinear Model

In a rotating channel flow, the reverse of the frame rotation leads to that of the asymmetry of the mean velocity profile (the steep velocity gradient moves to the upper wall in Fig. 4.6). This fact indicates the importance of the terms linearly related to rotation effects. Then we pay attention to the terms linearly dependent on Ω in the second iteration (Yoshizawa 1993b). As a result, we have

$$B_{ij} = -\nu_T S_{ij} \left(B_{ij} \right)_{SN} - C_{N2} \frac{K}{\varepsilon} \left(\left(B_{i\ell} \right)_{SN} \Omega_{\ell j} + \left(B_{j\ell} \right)_{SN} \Omega_{\ell i} \right). \tag{6.153}$$

Here $(B_{ij})_{SN}$ expresses the nonlinear parts of Eq. (6.133) that are given by the third and fourth terms.

The frame-rotation form of Eq. (6.153) may be obtained through the replacement of the mean vorticity tensor Ω_{ij} with the total counterpart Ω_{Tij} defined by

$$\Omega_{Tij} = \Omega_{ij} + 2\epsilon_{ij\ell} \Omega_{F\ell}, \tag{6.154}$$

where $\mathbf{\Omega}_F$ is the angular velocity of frame rotation. As this consequence, B_{xy} is written as

$$B_{xy} = -C_\nu \frac{K^2}{\varepsilon} \frac{dU}{dy} - 2 C_{N2} \Omega_{F3} \frac{K}{\varepsilon} \left(\left(B_{xx} \right)_{SN} - \left(B_{yy} \right)_{SN} \right). \tag{6.155}$$

Equation (6.155) shows that the effect of frame rotation on a channel flow occurs in the combination with the effect of anisotropy, as is referred to in

Sec. 4.3.2 [see expression (4.132)]. The anisotropy arising from the nonlinear model is given by Eq. (4.128), and Eq. (6.155) is reduced to

$$B_{xy} = -C_\nu \frac{K^2}{\varepsilon}\frac{dU}{dy} - 8C_{N2}C_{\zeta 2}\Omega_{F3}\frac{K^4}{\varepsilon^3}\left(\frac{dU}{dy}\right)^2. \tag{6.156}$$

This equation indicates that the second term strengthens the effect of the first term at the lower wall, whereas the former weakens the latter at the upper wall. This is the reason of the asymmetry of the mean velocity profile in Fig. 4.6. The K-ε model with Eq. (6.156) incorporated was applied by Nisizima (1993) and was confirmed to reproduce such a velocity profile.

Finally, we should refer to the Ω_{ij}^2 effect on ν_T, which is given by the last term in Eq. (6.138). This effect arises from the $O(\delta^3)$ TSDIA analysis (Okamoto 1994), but such an effect does not occur from the iteration method based on Eq. (6.147). This point is related to the fact that the Ω-Ω term vanishes in the second-order nonlinear model (6.133). In the present iteration, the Ω effect on ν_T occurs in the combination of S_{ij}. In order to include the Ω_{ij}^2 effect, we need to renormalize a third-order nonlinear model, if partial, and derive the counterpart of Eq. (6.147).

6.6. Modeling of Turbulent Transport Equations

In the previous sections, we investigated the Reynolds stress in detail on the basis of the TSDIA. There we started from an algebraic nonlinear model, and renormalized it to reach a second-order model. In both types of models, K and ε play a key role in close relation to the characteristic time scale K/ε. The equation for K is given by Eq. (2.111), and the transport rate $\mathbf{T}_K$ [Eq. (2.114)] is the sole quantity to be modeled. Equation (4.16) is its simplest model in the conventional modeling. On the other hand, the counterpart of ε is Eq. (4.37), and is not founded on the mathematical basis so firm as for K.

In what follows, we shall examine these equations using the TSDIA. In the conventional modeling, the p'-related part of $\mathbf{T}_K$ [Eq. (2.114)], $\langle p'\mathbf{u}'\rangle$, is dropped, compared with the triple velocity correlation $\langle(\mathbf{u}'^2/2)\mathbf{u}'\rangle$. Through the following TSDIA analysis, we shall give a theoretical support to this approximation.

6.6.1. TURBULENT ENERGY EQUATION

The transport rate of K, $\mathbf{T}_K$, consists of two parts

$$\mathbf{J}_V = \left\langle \mathbf{u}'^2\mathbf{u}' \right\rangle, \tag{6.157}$$

$$\mathbf{J}_P = \langle p'\mathbf{u}' \rangle. \tag{6.158}$$

The TSDIA analysis of these parts was first done by Yoshizawa (1982), and afterwards, was refined by Shimomura (1997b) on the basis of the exact treatment of the solenoidal condition, Eqs. (6.37) and (6.38). In the same sense as Eqs. (6.67)-(6.69), Eqs. (6.157) and (6.158) are written as

$$\mathbf{J}_V = \iint \mathbf{J}_V(\mathbf{p},\mathbf{q},\mathbf{X};\tau,T)d\mathbf{p}d\mathbf{q}, \tag{6.159}$$

$$\mathbf{J}_P = \int \mathbf{J}_P(\mathbf{k},\mathbf{X};\tau,T)d\mathbf{k}, \tag{6.160}$$

where

$$J_{Vi}(\mathbf{p},\mathbf{q},\mathbf{X};\tau,T) = \frac{\left\langle u_i'(\mathbf{k},\mathbf{X};\tau,T)\,u_j'(\mathbf{p},\mathbf{X};\tau,T)\,u_j'(\mathbf{q},\mathbf{X};\tau,T)\right\rangle}{\delta(\mathbf{k}+\mathbf{p}+\mathbf{q})}, \tag{6.161}$$

$$J_{Pi}(\mathbf{k},\mathbf{X};\tau,T) = \frac{\langle p'(\mathbf{k},\mathbf{X};\tau,T)u_i'(\mathbf{k}',\mathbf{X};\tau,T)\rangle}{\delta(\mathbf{k}+\mathbf{k}')} \tag{6.162}$$

[$\mathbf{J}_V(\mathbf{p},\mathbf{q},\mathbf{X};\tau,T)$ and $\mathbf{J}_P(\mathbf{k},\mathbf{X};\tau,T)$ will be simply denoted as $\mathbf{J}_V(\mathbf{p},\mathbf{q};\tau)$ and $\mathbf{J}_P(\mathbf{k};\tau)$, respectively].

Using the expansion (6.39), we have

$$\begin{aligned}
J_{Vi}(\mathbf{p},\mathbf{q},\mathbf{X};\tau,T) &= \frac{\left\langle u_{0i}'(\mathbf{k},\mathbf{X};\tau,T)\,u_{0j}'(\mathbf{p},\mathbf{X};\tau,T)\,u_{0j}'(\mathbf{q},\mathbf{X};\tau,T)\right\rangle}{\delta(\mathbf{k}+\mathbf{p}+\mathbf{q})} \\
&+\delta\left(\frac{\left\langle u_{1i}'(\mathbf{k},\mathbf{X};\tau,T)\,u_{0j}'(\mathbf{p},\mathbf{X};\tau,T)\,u_{0j}'(\mathbf{q},\mathbf{X};\tau,T)\right\rangle}{\delta(\mathbf{k}+\mathbf{p}+\mathbf{q})}\right. \\
&\quad+\frac{\left\langle u_{0i}'(\mathbf{k},\mathbf{X};\tau,T)\,u_{1j}'(\mathbf{p},\mathbf{X};\tau,T)\,u_{0j}'(\mathbf{q},\mathbf{X};\tau,T)\right\rangle}{\delta(\mathbf{k}+\mathbf{p}+\mathbf{q})} \\
&\quad\left.+\frac{\left\langle u_{0i}'(\mathbf{k},\mathbf{X};\tau,T)\,u_{0j}'(\mathbf{p},\mathbf{X};\tau,T)\,u_{1j}'(\mathbf{q},\mathbf{X};\tau,T)\right\rangle}{\delta(\mathbf{k}+\mathbf{p}+\mathbf{q})}\right).
\end{aligned} \tag{6.163}$$

We apply the DIA formalism to the evaluation of Eq. (6.163). In the DIA evaluation, $\mathbf{u'}_0$ is replaced with $\mathbf{v'}_0$ in Eq. (6.74). The latter is assumed to be random or Gaussian, and the triple correlation concerning $\mathbf{v'}_0$ vanishes. Therefore the $O(1)$ part in Eq. (6.163) does not contribute to $\mathbf{J}_V$. Next, we substitute the second part of Eq. (6.74) with $\mathbf{u'}_0$ replaced with $\mathbf{v'}_0$ into one of $(\mathbf{u'}_0)$'s in the first term of Eq. (6.163). The resulting expression is of the forth order in $\mathbf{v'}_0$, and survives under the assumption of Gaussianity of $\mathbf{v'}_0$. This expression, however, is odd concerning the wavenumber owing to $M_{ij\ell}(\mathbf{k})$ in Eq. (6.74), and the renormalized counterpart vanishes after the wavenumber integration, owing to the statistical isotropy of $\mathbf{u'}_0$.

The nonvanishing contribution to $\mathbf{J}_V$ comes from the $O(\delta)$ terms. In Eq. (6.64), the first two terms are combined with the first part of the $O(\delta)$ terms in Eq. (6.163), and lead to the third-order correlation concerning $\mathbf{u}'_0$. These two terms vanish, as is similar to the $O(1)$ term in Eq. (6.163). On the other hand, the third and fourth terms in Eq. (6.64) may give nonvanishing contributions. We substitute the third term into the first part in Eq. (6.163). Then we have

$$2M_{nhm}(\mathbf{k}) \iint \delta\left(\mathbf{k}-\mathbf{p}'-\mathbf{q}'\right) d\mathbf{p}' d\mathbf{q}' \frac{q'_m}{q'^2} \int_{-\infty}^{\tau} d\tau_1 \\ \times \frac{1}{\delta\left(\mathbf{k}+\mathbf{p}+\mathbf{q}\right)} \left\langle G'_{in}\left(\mathbf{k};\tau,\tau_1\right) u'_{0h}\left(\mathbf{p}';\tau_1\right) \right. \\ \times \left. \frac{\partial u'_{0\ell}\left(\mathbf{q}';\tau_1\right)}{\partial X_\ell} u'_{0j}\left(\mathbf{p};\tau\right) u'_{0j}\left(\mathbf{q};\tau\right) \right\rangle . \tag{6.164}$$

For the treatment of the $\partial/\partial X_\ell$ operation on $\mathbf{u}'_0$, we make use of the procedure based on Eq. (6.85) that changes Eq. (6.83) into Eq. (6.86). Then we may rewrite Eq. (6.164) as

$$\frac{1}{2} M_{nhm}(\mathbf{k}) \iint \delta\left(\mathbf{k}-\mathbf{p}'-\mathbf{q}'\right) d\mathbf{p}' d\mathbf{q}' \frac{q'_m}{q'^2} \int_{-\infty}^{\tau} d\tau_1 \frac{1}{\delta\left(\mathbf{k}+\mathbf{p}+\mathbf{q}\right)} \\ \times G_{in}\left(\mathbf{k};\tau,\tau_1\right) \frac{\partial}{\partial X_\ell} \left\langle u'_{0h}\left(\mathbf{p}';\tau_1\right) u'_{0\ell}\left(\mathbf{q}';\tau_1\right) u'_{0j}\left(\mathbf{p};\tau\right) u'_{0j}\left(\mathbf{q};\tau\right) \right\rangle \\ = -\frac{1}{2} M_{nhm}(\mathbf{k}) \int_{-\infty}^{\tau} d\tau_1 G_{in}\left(\mathbf{k};\tau,\tau_1\right) \\ \times \left(\frac{q_m}{q^2} \frac{\partial}{\partial X_\ell} \left(Q_{jh}\left(\mathbf{p};\tau,\tau_1\right) Q_{j\ell}\left(\mathbf{q};\tau,\tau_1\right)\right) \right. \\ \left. + \frac{p_m}{p^2} \frac{\partial}{\partial X_\ell} \left(Q_{j\ell}\left(\mathbf{p};\tau,\tau_1\right) Q_{jh}\left(\mathbf{q};\tau,\tau_1\right)\right) \right), \tag{6.165}$$

where use has been made of Eqs. (6.51) and (6.52).

We summarize the contributions of the $O(\delta)$ terms, such as Eq. (6.165), to Eq. (6.163). This calculation is straightforward, but quite lengthy. The final expressions are given by

$$J_{Vi} = \iiint \delta\left(\mathbf{k}-\mathbf{p}-\mathbf{q}\right) d\mathbf{k} d\mathbf{p} d\mathbf{q} \int_{-\infty}^{\tau} d\tau_1 \\ \times \left(J_{1ij}\left(\mathbf{k},\mathbf{p},\mathbf{q}\right) G\left(k;\tau,\tau_1\right) \frac{\partial}{\partial x_j} \left(Q\left(p;\tau,\tau_1\right) Q\left(q;\tau,\tau_1\right)\right) \right. \\ \left. + J_{2ij}\left(\mathbf{k},\mathbf{p},\mathbf{q}\right) \frac{\partial}{\partial x_j} \left(G\left(k;\tau,\tau_1\right) Q\left(p;\tau,\tau_1\right) Q\left(q;\tau,\tau_1\right)\right) \right), \tag{6.166}$$

$$J_{Pi} = \iiint \delta(\mathbf{k}-\mathbf{p}-\mathbf{q}) d\mathbf{k} d\mathbf{p} d\mathbf{q} \int_{-\infty}^{\tau} d\tau_1$$
$$\times \left(J_{3ij}(\mathbf{k},\mathbf{p},\mathbf{q}) G(k;\tau,\tau_1) \frac{\partial}{\partial x_j} \left(Q(p;\tau,\tau_1) Q(q;\tau,\tau_1) \right) \right.$$
$$\left. + J_{4ij}(\mathbf{k},\mathbf{p},\mathbf{q}) \frac{\partial}{\partial x_j} \left(G(k;\tau,\tau_1) Q(p;\tau,\tau_1) Q(q;\tau,\tau_1) \right) \right), \quad (6.167)$$

where

$$J_{1ij}(\mathbf{k},\mathbf{p},\mathbf{q}) = \frac{p_n}{p^2} D_{\ell m}(\mathbf{q}) \left(M_{\ell mn}(\mathbf{k}) D_{ij}(\mathbf{p}) + M_{imn}(\mathbf{k}) D_{\ell j}(\mathbf{p}) \right)$$
$$+ \frac{q_n}{q^2} D_{\ell j}(\mathbf{q}) M_{\ell mn}(\mathbf{k}) D_{im}(\mathbf{p}) - M_{mnjr}(\mathbf{k}) D_{\ell n}(\mathbf{q}) D_{\ell r}(\mathbf{k}) D_{im}(\mathbf{p})$$
$$- M_{mnjr}(\mathbf{k}) D_{\ell m}(\mathbf{q}) \left(D_{\ell r}(\mathbf{k}) D_{in}(\mathbf{p}) + D_{ir}(\mathbf{k}) D_{\ell n}(\mathbf{p}) \right), \quad (6.168a)$$

$$J_{2ij}(\mathbf{k},\mathbf{p},\mathbf{q}) = \frac{2}{3k^2} M_{jmn}(\mathbf{k}) D_{\ell m}(\mathbf{p}) \left(2k_\ell D_{in}(\mathbf{q}) + k_i D_{\ell n}(\mathbf{q}) \right)$$
$$- \frac{4}{3p^2} M_{\ell mn}(\mathbf{k}) \left(p_\ell D_{jm}(\mathbf{p}) D_{in}(\mathbf{q}) + p_i D_{jn}(\mathbf{p}) D_{\ell m}(\mathbf{q}) \right)$$
$$- \frac{4}{3} \frac{p_\ell}{p^2} M_{imn}(\mathbf{k}) D_{jm}(\mathbf{p}) D_{\ell n}(\mathbf{q}), \quad (6.168b)$$

$$J_{3ij}(\mathbf{k},\mathbf{p},\mathbf{q})$$
$$= \frac{k_\ell k_m}{k^2} D_{\ell r}(\mathbf{p}) \left(M_{rsjn}(\mathbf{k}) D_{in}(\mathbf{k}) D_{ms}(\mathbf{q}) - \frac{q_s}{q^2} M_{irs}(\mathbf{k}) D_{jm}(\mathbf{q}) \right)$$
$$+ \frac{q_\ell q_m}{q^2} \left(2M_{srjn}(\mathbf{k}) D_{mn}(\mathbf{k}) D_{\ell s}(\mathbf{p}) D_{ir}(\mathbf{q}) \right.$$
$$\left. - M_{mrs}(\mathbf{k}) \left(\frac{q_s}{q^2} D_{\ell r}(\mathbf{p}) D_{ij}(\mathbf{q}) + \frac{p_s}{p^2} D_{j\ell}(\mathbf{p}) D_{ir}(\mathbf{q}) \right) \right), \quad (6.169a)$$

$$J_{4ij}(\mathbf{k},\mathbf{p},\mathbf{q}) = \frac{2}{3} \left(\frac{p_i p_\ell p_m}{p^4} M_{\ell nr}(\mathbf{k}) D_{jr}(\mathbf{p}) D_{mn}(\mathbf{q}) \right.$$
$$\left. + \frac{q_i q_\ell q_m}{q^4} M_{mnr}(\mathbf{k}) D_{in}(\mathbf{p}) D_{jr}(\mathbf{q}) - \frac{k_i k_\ell k_m}{p^4} M_{jnr}(\mathbf{k}) D_{\ell n}(\mathbf{p}) D_{mr}(\mathbf{q}) \right)$$
$$+ \frac{4}{3} \left(\left(\frac{2}{p^2} M_{j\ell m}(\mathbf{p}) + \frac{p_\ell p_s q_s}{p^2 q^2} \delta_{jm} \right) M_{\ell nr}(\mathbf{k}) D_{ir}(\mathbf{p}) D_{mn}(\mathbf{q}) \right.$$
$$+ \left(\frac{2}{q^2} M_{j\ell m}(\mathbf{q}) - \frac{q_\ell q_s k_s}{k^2 q^2} \delta_{jm} \right) M_{mnr}(\mathbf{k}) D_{\ell n}(\mathbf{p}) D_{ir}(\mathbf{q})$$
$$\left. - \left(\frac{2}{k^2} M_{j\ell m}(\mathbf{k}) - \frac{k_\ell k_s q_s}{k^2 q^2} \delta_{jm} \right) M_{inr}(\mathbf{k}) D_{\ell n}(\mathbf{p}) D_{mr}(\mathbf{q}) \right) \quad (6.169b)$$

[$M_{ij\ell m}(\mathbf{k})$ is defined by Eq. (6.62)].

We substitute the inertial range form for $Q(k;\tau,\tau')$ and $G(k;\tau,\tau')$, Eqs. (6.53) and (6.54). Here we need to cut the contributions of very-low-wavenumber components to the integrals, as is discussed in Sec. 6.3.4.B. In the original work of Yoshizawa (1982), the cut based on Eq. (6.105) was used. In the later Shimomura's (1997b) work, Eq. (6.53) was replaced as

$$Q\left(k;\tau,\tau'\right) \rightarrow Q\left(k;\tau,\tau'\right) H\left(k-k_C\right). \tag{6.170}$$

The difference between these two methods arises from

$$\begin{aligned}
&\int_{|\mathbf{k}|\geq k_C}\int_{|\mathbf{p}|\geq k_C}\int_{|\mathbf{q}|\geq k_C} \delta\left(\mathbf{k}-\mathbf{p}-\mathbf{q}\right) d\mathbf{k}d\mathbf{p}d\mathbf{q} \int_{-\infty}^{\tau} d\tau_1 \\
&\quad \times G\left(k;\tau,\tau_1\right)\frac{\partial}{\partial x_i}\left(Q\left(p;\tau,\tau_1\right)Q\left(q;\tau,\tau_1\right)\right) \\
&\quad \neq \iiint \delta\left(\mathbf{k}-\mathbf{p}-\mathbf{q}\right) d\mathbf{k}d\mathbf{p}d\mathbf{q} \int_{-\infty}^{\tau} d\tau_1 G\left(k;\tau,\tau_1\right) \\
&\quad \times \frac{\partial}{\partial x_i}\left(Q\left(p;\tau,\tau_1\right)H\left(p-k_C\right)Q\left(q;\tau,\tau_1\right)H\left(q-k_C\right)\right).
\end{aligned} \tag{6.171}$$

This point has been already referred to in Sec. 6.3 (*Note*). In Shimomura's (1997b) calculation, the metrics $J_{nij}(\mathbf{k},\mathbf{p},\mathbf{q})$'s were reduced to a more manageable form with the aid of REDUCE 3.

Equations (6.166) and (6.167) combined with Eq. (6.170) give

$$\mathbf{J}_V = -\left(C_{TVK}\left(K^2/\varepsilon\right)\nabla K - C_{TV\varepsilon}\left(K^3/\varepsilon^2\right)\nabla\varepsilon\right), \tag{6.172}$$

$$\mathbf{J}_P = C_{TPK}\left(K^2/\varepsilon\right)\nabla K - C_{TP\varepsilon}\left(K^3/\varepsilon^2\right)\nabla\varepsilon, \tag{6.173}$$

with

$$C_{TVK} = 0.331,\ C_{TV\varepsilon} = 0.0976, \tag{6.174a}$$

$$C_{TPK} = 0.0136,\ C_{TP\varepsilon} = 0.00467. \tag{6.174b}$$

Here $k_C (= 2\pi/\ell_C)$ was eliminated using the first relation of Eq. (6.128). As a result, the transport rate $\mathbf{T}_K$ [Eq. (2.114)], which is written as

$$\mathbf{T}_K = -\left(\left(1/2\right)\mathbf{J}_V + \mathbf{J}_P\right), \tag{6.175}$$

is expressed as

$$\mathbf{T}_K = C_{TK}\left(K^2/\varepsilon\right)\nabla K - C_{T\varepsilon}\left(K^3/\varepsilon^2\right)\nabla\varepsilon, \tag{6.176}$$

where

$$C_{TK} = 0.152,\ C_{T\varepsilon} = 0.0441. \tag{6.177}$$

From these results, we found the following two important points about $\mathbf{T}_K$. One is that a theoretical support has been given to the approximation of neglecting the p'-related correlation in the conventional modeling. Another is the occurrence of the $\nabla\varepsilon$-related diffusion effect, which may be called the cross-diffusion one. The occurrence of the effect is very natural since ε itself is a key term in the K equation (2.111) and its diffusion may affect K. In Eq. (6.176), positive C_{TK} and $C_{T\varepsilon}$ indicate that these two terms play an opposite role in the case of locally large K and ε. This situation may be understood from the fact that large ε tends to decrease K, resulting in the weakening of the K diffusion effect.

In general, the large-K region is accompanied by the large-ε one since large K and ε represent the state of strong turbulence. In case that the spatial profiles of K and ε are similar, the ε-related diffusion effect can be absorbed into the K counterpart. One of such examples is a channel flow in which the production and dissipation effects balance nearly with each other, and the diffusion effect is weak in addition to vanishing of the advection one. The latter two effects, however, may become comparable to the former two in more complicated turbulent flows. One instance is a backward facing-step flow. The importance of the ε diffusion effect in the flow was really confirmed in the work by Kobayashi and Togashi (1996) that is based on the algebraic second-order model. The theoretically-obtained values of C_{TK} and $C_{T\varepsilon}$ [Eq. (6.177)] are consistent with the counterparts optimized there.

6.6.2. ENERGY-DISSIPATION-RATE EQUATION

In Sec. 4.2.2.B, we discussed the equation governing the energy dissipation rate ε, which is given by Eq. (4.26). There we emphasized that the first two terms on the right-hand side are dominant in the limit of vanishing ν; namely, both of them are of $O(\nu^{-1/2})$, whereas the remaining terms remain finite in the limit. This fact gives rise to a serious difficulty in the modeling of the ε equation. Some phenomenological models have been proposed for explaining the observed form of the inertial- and dissipation-range energy spectra, but they cannot satisfy the exact cancellation of the foregoing two terms.

The energy dissipation at high Reynolds number occurs at fine scales. On the other hand, the energy is supplied at much larger scales, and the amount of energy to be dissipated is primarily determined by the energy-injection mechanism at large scales. The latter dominant role can be seen in nonequilibrium turbulent flows such as homogeneous-shear turbulence. In the flow, the energy-dissipation process cannot catch up with the energy-injection one. Such a time-lag effect reflects on the energy spectrum and plays a key role in the construction of a model properly describing the flow, as is discussed in Sec. 6.3.4.A. This fact suggests that it is appropriate to regard ε as the energy-injection rate from the mean to fluctuating field, rather than the traditional concept of the dissipation rate closely connected with fine scales. In what follows, we shall adopt this viewpoint and derive a model equation for ε.

In the K-ε model and its extended or nonlinear ones, we adopt K and ε as the most fundamental quantities characterizing turbulence state. This choice, however, is not so absolutely unique. In the TSDIA, the length scale of energy-containing eddies, ℓ_C, occurs as a primary characteristic turbulence quantity [see Eq. (6.112)]. From the standpoint of writing the turbulent viscosity ν_T in terms of two turbulence quantities, we can freely choose any two of K, ε, ℓ_C, and their combination.

In the TSDIA, the relationship among K, ε, and ℓ_C is given by Eq. (6.112) or (6.116) up to $O(\delta)$ and by Eq. (6.128) up to $O(\delta^2)$, respectively. In the sense that ℓ_C represents a characteristic spatial scale of energy-containing eddies, it is physically significant to perform the modeling based on ε and ℓ_C. We substitute Eq. (6.112) into the K equation, and eliminate $D\varepsilon/Dt$ using Eq. (4.37). The critical difference between the resulting equation and the usual K and ε ones arises from the diffusion term of the K equation. The $D\varepsilon/Dt$ and $D\ell_C/Dt$ terms in Eq. (6.112) give rise to the terms linearly dependent on $\Delta^2\varepsilon$ and $\Delta(D\ell_C/Dt)$ in the equation for ℓ_C; namely, the rank of the ℓ_C equation differs from that of the K and ε equations. From the viewpoint of the equal degree of importance of ε and ℓ_C, it is reasonable to assume the mathematical similarity between the ε and ℓ_C equations. Then we require the mathematical equivalence of the equations for K, ε, ℓ_C, and any combination of K, ε, and ℓ_C. This equivalence may be called the transferability of models.

The condition necessary for the transferability of models is the algebraic relationship among K, ε, and ℓ_C (Yoshizawa 1987). In the $O(\delta)$ expression for ℓ_C, (6.116), we require

$$\ell_C = C_{\ell 1} K^{3/2}/\varepsilon. \tag{6.178}$$

As a result, we have

$$\frac{D\varepsilon}{Dt} = \frac{C_{\ell 2}}{C_{\ell 3}}\frac{\varepsilon}{K}\frac{DK}{Dt} = C_{\varepsilon 1}^{TS}\frac{\varepsilon}{K}P_K - C_{\varepsilon 2}^{TS}\frac{\varepsilon^2}{K} + D_\varepsilon, \tag{6.179}$$

where

$$C_{\varepsilon 1}^{TS} = C_{\ell 2}/C_{\ell 3} = 1.71, \tag{6.180a}$$

$$C_{\varepsilon 2}^{TS} = C_{\ell 2}/C_{\ell 3} = 1.71. \tag{6.180b}$$

In Eq. (6.179), D_ε contains the contribution of the ∇-related diffusion effect in the K equation, but it cannot be written simply in a divergence form.

Therefore it is proper to call D_ε the diffusion-like term. This point will be referred to below.

In Eq. (6.179), we reproduced the primary parts of the ε equation used widely in the conventional turbulence modeling, Eq. (4.37). The counterparts of $C_{\varepsilon 1}^{TS}$ and $C_{\varepsilon 2}^{TS}$, $C_{\varepsilon 1}$ and $C_{\varepsilon 2}$, are optimized as about 1.4 and 1.9, respectively, and the theoretically-estimated values are close to them. In the approximation of dropping the diffusion-like term D_ε, we are led to the simple relation

$$\frac{D}{Dt}\log\left(\frac{K^{C_{\ell 2}/C_{\ell 3}}}{\varepsilon}\right) = 0. \tag{6.181}$$

Equation (6.181) expresses a kind of similarity between K and ε along the path of mean flow. This simple equation is shown to be a practically useful, simple model for analyzing some kinds of shear flows (Duranti and Pittaluga 1997).

In the $O(\delta)$ analysis of the TSDIA, we have the relation $C_{\varepsilon 1}^{TS} = C_{\varepsilon 2}^{TS}$. The equality of two coefficients is broken in the $O(\delta^2)$ analysis (Okamoto 1994). In Eq. (6.128), we also require Eq. (6.178), resulting in

$$\frac{D\varepsilon}{Dt} = C_{\varepsilon 1}^{TS}\frac{\varepsilon}{K}P_K - C_{\varepsilon 2}^{TS}\frac{\varepsilon^2}{K} - \frac{C_{\ell 4}}{C_{\ell 3}}KS_{ij}^2 + D_\varepsilon. \tag{6.182}$$

The third term on the right-hand side is rewritten as

$$\frac{C_{\ell 4}}{C_{\ell 3}}KS_{ij}^2 = \frac{2C_{\ell 4}}{C_{\ell 3}C_\nu}\frac{\varepsilon}{K}\left(\frac{\nu_T S_{ij}^2}{2}\right), \tag{6.183}$$

where ν_T is the turbulent viscosity expressed by Eq. (6.118), and C_ν is given by Eq. (6.119). Under the linear turbulent-viscosity approximation to the Reynolds stress R_{ij}, Eq. (6.183) is related to P_K as

$$\frac{C_{\ell 4}}{C_{\ell 3}}KS_{ij}^2 = \frac{2C_{\ell 4}}{C_{\ell 3}C_\nu}\frac{\varepsilon}{K}P_K. \tag{6.184}$$

We substitute Eq. (6.184) into Eq. (6.182), and have

$$\frac{D\varepsilon}{Dt} = C_{\varepsilon 1R}^{TS}\frac{\varepsilon}{K}P_K - C_{\varepsilon 2}^{TS}\frac{\varepsilon^2}{K} + D_\varepsilon, \tag{6.185}$$

where the renormalized counterpart of $C_{\varepsilon 1}^{TS}$, $C_{\varepsilon 1R}^{TS}$, is given by

$$C_{\varepsilon 1R}^{TS} = C_{\varepsilon 1}^{TS} - \frac{2C_{\ell 4}}{C_{\ell 3}C_\nu} = 1.33, \tag{6.186}$$

which should be compared with 1.4 in the conventional modeling.

The foregoing method of deriving a model equation for ε may be summarized as follows. We first express K in terms of ε and an appropriate turbulence quantity such as ℓ_C. In this course, various kinds of effects, for instance, the S_{ij} effect in Eq. (6.128) and buoyancy effects, may enter the relation. Next, we require the algebraic relation of the type (6.178) among these turbulence quantities, and obtain an equation for ε that consists of the contributions from the K equation itself and the foregoing effects. In the conventional turbulence modeling, D_ε in Eq. (6.179) is given by the last term in Eq. (4.37), as is similar to the counterpart in the K equation (2.111). The divergence form of the latter arises from the fact that the total amount of energy, $\int (\mathbf{u}^2/2)dV$, is conserved in the case of vanishing ν, as is discussed in Sec. 2.6.2.A. The ε-related quantities, however, do not possess such a property. Therefore the modeling of D_ε in the diffusion form should be regarded as a handy one, rather than the model based on a firm theoretical basis. With this reservation, the cross diffusion term in each of the K, ε, and ℓ_C equations is necessary for the transferability of models.

The other method of deriving a model equation for ε was presented by Hamba (1987b). In the method, the equation for the two-time velocity variance is evaluated using the TSDIA, and an equation for K^2/ε is derived. It is combined with the equation for K, resulting in the equation resembling Eq. (6.182). This method is similar to the foregoing one in the sense that no use is made of the exact equation for ε.

6.7. Scalar Transports and Body-Force Effects

In the previous sections, we gave a detailed account of the TSDIA formalism through the investigation of the Reynolds stress in incompressible turbulence. In turbulent flows in engineering and scientific fields, we encounter various types of effects. Some typical instances of them are the scalar transport, the frame-rotation effect, the buoyancy effect, etc. They have been referred to in Chapter 4 from the standpoint of conventional turbulence modeling. As is discussed there, a key element in scalar transport is the turbulent scalar flux $\mathbf{H}_\theta (= \langle \mathbf{u}'\theta' \rangle)$. We start from the analysis of $\mathbf{H}_\theta$ and examine some fundamental points in the modeling of scalar transport, with the aid of the TSDIA (Yoshizawa 1984b, 1985, 1988; Hamba 1987a; Okamoto 1996a).

6.7.1. SCALAR TRANSPORT

A. Turbulent Scalar Flux

The flux $\mathbf{H}_\theta$ may be evaluated in entirely the same manner as for the Reynolds stress R_{ij}. First, we write

$$\mathbf{H}_\theta = \left\langle \mathbf{u}'\left(\boldsymbol{\xi}, \mathbf{X}; \tau, T\right) \theta'\left(\boldsymbol{\xi}, \mathbf{X}; \tau, T\right) \right\rangle . \tag{6.187}$$

The Fourier resolution of $\mathbf{H}_\theta$ is given by

$$\mathbf{H}_\theta = \int \frac{\langle \mathbf{u}'(\mathbf{k},\mathbf{X};\tau,T)\,\theta'(\mathbf{k}',\mathbf{X};\tau,T)\rangle}{\delta(\mathbf{k}+\mathbf{k}')} d\mathbf{k}. \tag{6.188}$$

Hereafter the dependence of $\mathbf{u}'(\boldsymbol{\xi},\mathbf{X};\tau,T)$ and $\theta'(\boldsymbol{\xi},\mathbf{X};\tau,T)$ on $\mathbf{X}$ and T will not written explicitly, except when necessary.

A.1. Two-Scale Expansion. From Eq. (2.119), the two-scale version of the θ' equation is

$$\begin{aligned}
&\frac{\partial \theta'(\mathbf{k};\tau)}{\partial \tau} + \lambda k^2 \theta'(\mathbf{k};\tau) - ik_i \iint \delta(\mathbf{k}-\mathbf{p}-\mathbf{q}) d\mathbf{p}d\mathbf{q} u'_i(\mathbf{p};\tau)\theta'(\mathbf{q};\tau) \\
&\quad = \delta \left(-u'_i(\mathbf{k};\tau)\frac{\partial \Theta}{\partial X_i} - \frac{D\theta'(\mathbf{k};\tau)}{DT_I} - \iint \delta(\mathbf{k}-\mathbf{p}-\mathbf{q}) d\mathbf{p}d\mathbf{q} \right. \\
&\qquad \left. \times \frac{\partial}{\partial X_{Ii}} \left(u'_i(\mathbf{p};\tau)\theta'(\mathbf{q};\tau)\right) - \delta(\mathbf{k})\frac{\partial H_{\theta i}}{\partial X_i} \right).
\end{aligned} \tag{6.189}$$

We expand θ' as

$$\theta'(\mathbf{k};\tau) = \sum_{n=0}^{\infty} \delta^n \theta'_n(\mathbf{k};\tau), \tag{6.190}$$

which is the scalar counterpart of Eq. (6.39). We substitute Eqs. (6.39) and (6.190) into Eq. (6.189), and have

$$\begin{aligned}
&\frac{\partial \theta'_0(\mathbf{k};\tau)}{\partial \tau} + \lambda k^2 \theta'_0(\mathbf{k};\tau) \\
&\quad -ik_i \iint \delta(\mathbf{k}-\mathbf{p}-\mathbf{q}) d\mathbf{p}d\mathbf{q} u'_{0i}(\mathbf{p};\tau)\theta'_0(\mathbf{q};\tau) = 0,
\end{aligned} \tag{6.191}$$

$$\begin{aligned}
&\frac{\partial \theta'_1(\mathbf{k};\tau)}{\partial \tau} + \lambda k^2 \theta'_1(\mathbf{k};\tau) \\
&\quad -ik_i \iint \delta(\mathbf{k}-\mathbf{p}-\mathbf{q}) d\mathbf{p}d\mathbf{q} u'_{0i}(\mathbf{p};\tau)\theta'_1(\mathbf{q};\tau) \\
&\quad = -u'_{0i}(\mathbf{k};\tau)\frac{\partial \Theta}{\partial X_i} - \frac{D\theta'_0(\mathbf{k};\tau)}{DT_I} \\
&\quad - \iint \delta(\mathbf{k}-\mathbf{p}-\mathbf{q}) d\mathbf{p}d\mathbf{q} \frac{\partial}{\partial X_{Ii}} \left(u'_{0i}(\mathbf{p};\tau)\theta'_0(\mathbf{q};\tau)\right) \\
&\quad +ik_i \iint \delta(\mathbf{k}-\mathbf{p}-\mathbf{q}) d\mathbf{p}d\mathbf{q} u'_{1i}(\mathbf{p};\tau)\theta'_0(\mathbf{q};\tau),
\end{aligned} \tag{6.192}$$

and so on.

Corresponding to the Green's function for the velocity field, $G'_{ij}(\mathbf{k};\tau,\tau')$, we introduce the scalar Green's function $G'_\theta(\mathbf{k};\tau,\tau')$, which obeys

$$\frac{\partial G'_\theta(\mathbf{k};\tau,\tau')}{\partial \tau} + \lambda k^2 G'_\theta(\mathbf{k};\tau,\tau') - \delta(\tau-\tau') = ik_i \iint \delta(\mathbf{k}-\mathbf{p}-\mathbf{q}) d\mathbf{p} d\mathbf{q} u'_{0i}(\mathbf{p};\tau) G'_\theta(\mathbf{q};\tau,\tau'). \tag{6.193}$$

With the aid of $G'_\theta(\mathbf{k};\tau,\tau')$, Eq. (6.192) may be integrated formally as

$$\begin{aligned}
\theta'_1(\mathbf{k};\tau) = & -\frac{\partial \Theta}{\partial X_i} \int_{-\infty}^{\tau} d\tau_1 G'_\theta(\mathbf{k};\tau,\tau_1) u'_{0i}(\mathbf{k};\tau_1) \\
& - \int_{-\infty}^{\tau} d\tau_1 G'_\theta(\mathbf{k};\tau,\tau_1) \frac{D\theta'_0(\mathbf{k};\tau_1)}{DT_I} \\
& - \iint \delta(\mathbf{k}-\mathbf{p}-\mathbf{q}) d\mathbf{p} d\mathbf{q} \int_{-\infty}^{\tau} d\tau_1 G'_\theta(\mathbf{k};\tau,\tau_1) \\
& \quad \times \frac{\partial}{\partial X_{Ii}} \left(u'_{0i}(\mathbf{p};\tau_1) \theta'_0(\mathbf{q};\tau_1) \right) \\
& + ik_i \iint \delta(\mathbf{k}-\mathbf{p}-\mathbf{q}) d\mathbf{p} d\mathbf{q} \int_{-\infty}^{\tau} d\tau_1 \\
& \quad \times G'_\theta(\mathbf{k};\tau,\tau_1) u'_{1i}(\mathbf{p};\tau_1) \theta'_0(\mathbf{q};\tau_1).
\end{aligned} \tag{6.194}$$

A.2. First-Order Turbulent Scalar Flux. From Eqs. (6.39) and (6.190), we have

$$\begin{aligned}
\frac{\langle \mathbf{u}'(\mathbf{k};\tau)\theta'(\mathbf{k}';\tau)\rangle}{\delta(\mathbf{k}+\mathbf{k}')} = & \frac{\langle \mathbf{u}'_0(\mathbf{k};\tau)\theta'_0(\mathbf{k}';\tau)\rangle}{\delta(\mathbf{k}+\mathbf{k}')} \\
& + \delta \left(\frac{\langle \mathbf{u}'_1(\mathbf{k};\tau)\theta'_0(\mathbf{k}';\tau)\rangle}{\delta(\mathbf{k}+\mathbf{k}')} + \frac{\langle u'_0(\mathbf{k};\tau)\theta'_1(\mathbf{k}';\tau)\rangle}{\delta(\mathbf{k}+\mathbf{k}')} \right) + O\left(\delta^2\right).
\end{aligned} \tag{6.195}$$

For the statistical quantities characterizing the $O(1)$ velocity fluctuation $\mathbf{u}'_0(\mathbf{k};\tau)$, we use the isotropic expressions (6.51) and (6.52). As the scalar counterparts, we introduce

$$Q_\theta(k;\tau,\tau') = \frac{\langle \theta'_0(\mathbf{k};\tau)\theta'_0(\mathbf{k}';\tau')\rangle}{\delta(\mathbf{k}+\mathbf{k}')}, \tag{6.196}$$

$$G_\theta(k;\tau,\tau') = \left\langle G'_\theta(\mathbf{k};\tau,\tau')\right\rangle. \tag{6.197}$$

In the $O(1)$ field, we consider that there is no correlation between $\mathbf{u}'_0(\mathbf{k};\tau)$ and $\theta'_0(\mathbf{k};\tau)$ in the stationary state, that is,

$$\langle \mathbf{u}'_0(\mathbf{k};\tau)\theta'_0(\mathbf{k}';\tau')\rangle = 0. \tag{6.198}$$

This condition is reasonable since we have no preferred direction under the assumption of the isotropy of the $O(1)$ field.

In the analysis of the Reynolds stress R_{ij}, we approximated Eqs. (6.51) and (6.52) by the inertial-range form (6.53) and (6.54). In Sec. 3.3, it is mentioned that the scalar diffusion at high Peclet (temperature) or Schmidt (matter) number R_λ [Eq. (3.65)] may be characterized by three spectral ranges for the scalar variance $K_\theta (= \langle \theta'^2 \rangle)$, as is similar to the turbulent energy K. They are the variance-dominant, inertial, and destruction ranges, which correspond to the energy-containing, inertial, and dissipation ranges for K, respectively. The relative relationship between the ranges of the energy and scalar-variance spectra is dependent on the magnitude of the Prandtl number $P_r (= \nu/\lambda)$. In case that P_r is of $O(1)$, the inertial range of the energy spectrum is coincident with the scalar counterpart. In this case, $Q_\theta(\mathbf{k};\tau,\tau')$ and $G_\theta(\mathbf{k};\tau,\tau')$ may be written as

$$Q_\theta(k,\mathbf{X};\tau,\tau',T) = \sigma_\theta(k,\mathbf{X};T) \exp\left(-\omega(k,\mathbf{X};T)\left|\tau-\tau'\right|\right), \qquad (6.199)$$

$$G_\theta(k,\mathbf{X};\tau,\tau',T) = H\left(\tau-\tau'\right) \exp\left(-\omega_\theta(k,\mathbf{X};T)\left(\tau-\tau'\right)\right), \qquad (6.200)$$

with

$$\sigma_\theta(k,\mathbf{X};T) = 0.066\varepsilon_\theta(\mathbf{X};T)\left(\varepsilon(\mathbf{X};T)\right)^{-1/3} k^{-11/3}, \qquad (6.201)$$

$$\omega_\theta(k,\mathbf{X};T) = 1.60\omega(k,\mathbf{X};T). \qquad (6.202)$$

Here $\omega(k,\mathbf{X};T)$ is given by Eq. (6.56), and the destruction rate of K_θ, ε_θ, is defined by Eq. (2.128). Originally, $\omega(k,\mathbf{X};T)$ in $Q_\theta(\mathbf{k};\tau,\tau')$ differs from the velocity counterpart in Eq. (6.56) by some numerical factor, but the former is approximated by the latter for simplicity of analysis. The estimate of the numerical factors in Eqs. (6.201) and (6.202) is explained in Sec. 3.5.5.C (*Note*).

The calculation of Eq. (6.195) is quite similar to that of R_{ij}. We substitute Eqs. (6.63) and (6.194) into Eq. (6.195). From the condition (6.198), the nonvanishing contribution comes from the second part of the $O(\delta)$ terms in Eq. (6.195). Then $\mathbf{H}_\theta$ is expressed as

$$\mathbf{H}_\theta = -\lambda_T \nabla\Theta, \qquad (6.203)$$

where the turbulent diffusivity λ_T is given by

$$\lambda_T = \frac{2}{3}\int d\mathbf{k} \int_{-\infty}^{\tau} d\tau_1 G_\theta(k;\tau,\tau_1) Q(k;\tau,\tau_1). \qquad (6.204)$$

We substitute the inertial-range form for $Q(k;\tau,\tau')$ and $G_\theta(k;\tau,\tau')$, Eqs. (6.53) [with Eqs. (6.55) and (6.56)] and (6.200) [with (6.202)] into

Eq. (6.204). Here the lower limit of the integral concerning $\mathbf{k}$ is $k_C(= 2\pi/\ell_C)$, as is given by Eqs. (6.105) and (6.106) (we should note that G_θ is dependent on the statistics of $\mathbf{u}'_0$ only). As a result, we have

$$\lambda_T = C_{\lambda\ell}\varepsilon^{1/3}\ell_C^{4/3}, \tag{6.205}$$

where

$$C_{\lambda\ell} = 0.0596. \tag{6.206}$$

In the $O(\delta)$ analysis of $\mathbf{H}_\theta$, the first term of Eq. (6.116) should be retained for ℓ_C, resulting in

$$\lambda_T = C_\lambda K^2/\varepsilon, \tag{6.207}$$

where

$$C_\lambda = 0.134. \tag{6.208}$$

From Eqs. (6.118) and (6.207), the turbulent Prandtl number P_{rT} is

$$P_{rT} = \nu_T/\lambda_T = C_\nu/C_\lambda = 0.92. \tag{6.209}$$

For engineering purposes, the value 0.7-0.8 is often used as P_{rT}. In the case of high or low P_r, the inertial-range region of $G_\theta(k;\tau,\tau')$ does not coincide with that of $Q(k;\tau,\tau')$, and P_{rT} depends on P_r.

A.3. Second-Order Turbulent Scalar Flux. The foregoing $O(\delta)$ analysis may be extended to the $O(\delta^2)$ one in a straightforward manner, although it is rather tedious. This calculation was made by Yoshizawa (1985), and was later improved by Okamoto (1996a) using the exact treatment of the solenoidal condition due to Hamba (1987a). The turbulent scalar flux $\mathbf{H}_\theta$ is written as

$$H_{\theta i} = -\lambda_{TN}\frac{\partial\Theta}{\partial x_i} + \chi_1 S_{ij}\frac{\partial\Theta}{\partial x_j} + \chi_2\Omega_{ij}\frac{\partial\Theta}{\partial x_j} + \chi_3\frac{D}{Dt}\frac{\partial\Theta}{\partial x_i}, \tag{6.210}$$

where

$$\lambda_{TN} = C_\lambda\frac{K^2}{\varepsilon} - C_{\lambda K}\frac{K^2}{\varepsilon^2}\frac{DK}{Dt} + C_{\lambda\varepsilon}\frac{K^3}{\varepsilon^3}\frac{D\varepsilon}{Dt}, \tag{6.211}$$

$$\chi_n = C_{\chi n}\frac{K^3}{\varepsilon^2}. \tag{6.212}$$

Here the numerical factors $C_{\lambda K}$ etc. are

$$C_{\lambda K} = 0.0830,\ C_{\lambda\varepsilon} = 0.0534, \tag{6.213a}$$

$$C_{\chi 1} = 0.0434,\ C_{\chi 2} = 0.00915,\ C_{\chi 3} = 0.0362. \tag{6.213b}$$

Equation (6.210) has the same form as Eq. (4.232), except the last term of the former. The importance of S_{ij}- and Ω_{ij}-related terms is discussed in Sec. 4.6.1.B, in the context of the experiment by Tavoularis and Corrsin (1981, 1985) on the scalar diffusion in spatially developing homogeneous-shear flow. Equation (6.210) except the last term was also derived by Rubinstein and Barton (1991) using the RNG method. The D/Dt-related terms in Eq. (6.211) contribute to the suppression of the scalar-transport effect of the first term, as is similar to Eq. (6.136).

A.4. Transport Equation for Turbulent Scalar Flux. We use Eq. (6.210), and derive an equation for $\mathbf{H}_\theta$. For this purpose, we first rewrite it as

$$\begin{aligned} H_{\theta i} = & -\lambda_{TS}\frac{\partial \Theta}{\partial x_i} + C^H_{\lambda N1}\frac{K}{\varepsilon}S_{ij}\left(\lambda_{TS}\frac{\partial \Theta}{\partial x_j}\right) + C^B_{\lambda N1}\frac{K}{\varepsilon}\left(\nu_{TS}S_{ij}\right)\frac{\partial \Theta}{\partial x_j} \\ & + C_{\lambda N2}\frac{K}{\varepsilon}\Omega_{ij}\left(\lambda_{TS}\frac{\partial \Theta}{\partial x_j}\right) + C_{\lambda A}\frac{K}{\varepsilon}\frac{D}{Dt}\left(\lambda_{TS}\frac{\partial \Theta}{\partial x_i}\right) \\ & + \left(C_{\lambda NK}\frac{1}{\varepsilon}\frac{DK}{Dt} - C_{\lambda N\varepsilon}\frac{K}{\varepsilon^2}\frac{D\varepsilon}{Dt}\right)\left(\lambda_{TS}\frac{\partial \Theta}{\partial x_i}\right), \end{aligned} \tag{6.214}$$

where λ_{TS} is the static part of Eq. (6.211), that is,

$$\lambda_{TS} = C_\lambda K^2/\varepsilon, \tag{6.215}$$

the static turbulent viscosity ν_{TS} is given by Eq. (6.144), and

$$C_\lambda C^H_{\lambda N1} + C_\nu C^B_{\lambda N1} = C_{\chi 1}, \tag{6.216a}$$

$$C_{\lambda N2} = C_{\chi 2}/C_\lambda = 0.0687, \tag{6.216b}$$

$$C_{\lambda A} = C_{\chi 3}/C_\lambda = 0.270, \tag{6.216c}$$

$$C_{\lambda NK} = \frac{C_{\lambda K}}{C_\lambda} - 2C_{\chi 3} = 0.547, \tag{6.216d}$$

$$C_{\lambda N\varepsilon} = \frac{C_{\lambda\varepsilon}}{C_\lambda} - C_{\chi 3} = 0.362. \tag{6.216e}$$

Here we should note that $C^H_{\lambda N1}$ and $C^B_{\lambda N1}$ cannot be determined uniquely in the following renormalization procedure.

In Eq. (6.214), we perform the replacement

$$\lambda_{TS}\nabla\Theta \to -\mathbf{H}_\theta,\ \nu_{TS}S_{ij} \to -B_{ij}. \tag{6.217}$$

As a result, we have

$$H_{\theta i} = -\lambda_{TS}\frac{\partial \Theta}{\partial x_i} - C^H_{\lambda N1}\frac{K}{\varepsilon}S_{ij}H_{\theta j} - C^B_{\lambda N1}\frac{K}{\varepsilon}B_{ij}\frac{\partial \Theta}{\partial x_j} - C_{\lambda N2}\frac{K}{\varepsilon}\Omega_{ij}H_{\theta j}$$
$$-C_{\lambda A}\frac{K}{\varepsilon}\frac{DH_{\theta i}}{Dt} - \left(C_{\lambda NK}\frac{1}{\varepsilon}\frac{DK}{Dt} - C_{\lambda N\varepsilon}\frac{K}{\varepsilon^2}\frac{D\varepsilon}{Dt}\right)H_{\theta i}. \quad (6.218)$$

Equation (6.218) is equivalent to

$$\frac{DH_{\theta i}}{Dt} = -\frac{1}{C_{\lambda A}}\frac{\varepsilon}{K}\left(1 + C_{\lambda NK}\frac{1}{\varepsilon}\frac{DK}{Dt} - C_{\lambda N\varepsilon}\frac{K}{\varepsilon^2}\frac{D\varepsilon}{Dt}\right)H_{\theta i}$$
$$-\frac{1}{C_{\lambda A}}\frac{\varepsilon}{K}\lambda_{TS}\frac{\partial \Theta}{\partial x_i} - \left(\frac{C^H_{\lambda N1}}{C_{\lambda A}}S_{ij} + \frac{C_{\lambda N2}}{C_{\lambda A}}\Omega_{ij}\right)H_{\theta j} - \frac{C^B_{\lambda N1}}{C_{\lambda A}}B_{ij}\frac{\partial \Theta}{\partial x_j}. \quad (6.219)$$

The exact equation for $\mathbf{H}_\theta$, which is derived from Eq. (2.119) for θ', is given by Eq. (2.120). From the comparison with Eq. (6.219), we have

$$\Pi_{\theta i} - \varepsilon_{\theta i} = \left(C_{\theta\Pi 1}\delta_{ij} + C^S_{\theta\Pi 2}\frac{K}{\varepsilon}S_{ij} + C^\Omega_{\theta\Pi 2}\frac{K}{\varepsilon}\Omega_{ij}\right)\frac{\varepsilon}{K}H_{\theta j}$$
$$+ \left(C_{\theta\Pi 3}\delta_{ij} + C_{\theta\Pi 4}\frac{B_{ij}}{K}\right)K\frac{\partial \Theta}{\partial x_j}, \quad (6.220)$$

where

$$C_{\theta\Pi 1} = -\frac{1}{C_{\lambda A}}\left(1 + C_{\lambda NK}\frac{1}{\varepsilon}\frac{DK}{Dt} - C_{\lambda N\varepsilon}\frac{K}{\varepsilon^2}\frac{D\varepsilon}{Dt}\right), \quad (6.221a)$$

$$C^S_{\theta\Pi 2} = \frac{1}{2} - \frac{C^H_{\lambda N1}}{C_{\lambda A}}, \quad (6.221b)$$

$$C^\Omega_{\theta\Pi 2} = \frac{1}{2} - \frac{C_{\lambda N2}}{C_{\lambda A}}, \quad (6.221c)$$

$$C_{\theta\Pi 3} = \frac{2}{3} - \frac{C_\lambda}{C_{\lambda A}}, \quad (6.221d)$$

$$C_{\theta\Pi 4} = 1 - \frac{C^B_{\lambda N1}}{C_{\lambda A}}. \quad (6.221e)$$

The velocity counterpart of Eq. (6.220) is given by Eq. (6.149). In relation to the latter, we noted that Π_{ij} cannot be separated from ε_{ij} in the sense of turbulence modeling. The same situation holds for Eq. (6.220). This model should be compared with Eq. (4.245) by the conventional modeling and corresponds to the retention of the terms related to $C_{\phi n}(n = 1, 3, 4)$ in Eq. (4.246) and $C_{\psi n}(n = 1, 2)$ in Eq. (4.247).

A.5. Transport Equation for Scalar-Variance Destruction Rate. The scalar variance $K_\theta (= \langle \theta'^2 \rangle)$ obeys Eq. (2.126). The transport rate $\mathbf{T}_\theta$ [Eq. (2.129)] is the sole quantity to be modeled as long as the destruction rate of K_θ, ε_θ, is chosen as one of the fundamental turbulence quantities. First, we consider the modeling of the equation for ε_θ. From the same reason as for the modeling of the ε counterpart, we give up the modeling based on its exact equation. Instead, we shall follow the mathematical procedure adopted in the derivation of Eq. (6.179).

We use Eq. (6.194), and calculate K_θ on the basis of the TSDIA (Yoshizawa 1984b, 1988). To begin with, we have

$$K_\theta = \int Q_\theta (k;\tau,\tau) d\mathbf{k} - \int d\mathbf{k} \int_{-\infty}^{\tau} d\tau_1 G_\theta (k;\tau,\tau_1) \frac{DQ_\theta (k;\tau,\tau_1)}{Dt}, \quad (6.222)$$

which should be compared with the velocity counterpart (6.103). We apply the inertial-range form (6.199)-(6.202) to Eq. (6.222). At this time, we introduce the scalar length, ℓ_θ, that characterizes the low-wavenumber components of θ' fluctuation. This scale corresponds to that of energy-containing eddies, ℓ_C. In the integrals containing $Q_\theta(k;\tau,\tau')$, we cut the wavenumber integration as

$$\int d\mathbf{k} \to \int_{|\mathbf{k}| \geq 2\pi/\ell_\theta} d\mathbf{k} \quad (6.223)$$

[it should be recalled that Eq. (6.193) for $G'_\theta(k;\tau,\tau')$ is not dependent on θ']. Then we have

$$K_\theta = C_{K_\theta 1} \varepsilon_\theta \varepsilon^{-1/3} \ell_\theta^{2/3} - C_{K_\theta 2} \varepsilon^{-2/3} \ell_\theta^{4/3} \frac{D\varepsilon_\theta}{Dt} + C_{K_\theta 3} \varepsilon_\theta \varepsilon^{-5/3} \ell_\theta^{4/3} \frac{D\varepsilon}{Dt} - C_{K_\theta 4} \varepsilon_\theta \varepsilon^{-2/3} \ell_\theta^{1/3} \frac{D\ell_\theta}{Dt}, \quad (6.224)$$

where

$$C_{K_\theta 1} = 0.365,\ C_{K_\theta 2} = 0.0493,\ C_{K_\theta 3} = 0.0227,\ C_{K_\theta 4} = 0.193. \quad (6.225)$$

We solve Eq. (6.224) concerning ℓ_θ in a perturbational manner. The resulting expression is

$$\ell_\theta = C_{\ell_\theta 1} K_\theta^{3/2} \varepsilon_\theta^{-3/2} \varepsilon^{1/2} + C_{\ell_\theta 2} K_\theta^{3/2} \varepsilon_\theta^{-5/2} \varepsilon^{1/2} \frac{DK_\theta}{Dt} - C_{\ell_\theta 3} K_\theta^{5/2} \varepsilon_\theta^{-7/2} \varepsilon^{1/2} \frac{D\varepsilon_\theta}{Dt} + C_{\ell_\theta 4} K_\theta^{5/2} \varepsilon_\theta^{-5/2} \varepsilon^{-1/2} \frac{D\varepsilon}{Dt}, \quad (6.226)$$

where

$$C_{\ell_\theta 1} = 4.53,\ C_{\ell_\theta 2} = 14.7,\ C_{\ell_\theta 3} = 12.2,\ C_{\ell_\theta 4} = 3.76. \quad (6.227)$$

In Eq. (6.226), we require the algebraic relationship among ℓ_θ, K_θ, ε_θ, and ε from the transferability of models discussed in Sec. 6.6.2; namely, we put

$$\ell_\theta = C_{\ell\theta 1} K_\theta^{3/2} \varepsilon_\theta^{-3/2} \varepsilon^{1/2}, \tag{6.228}$$

which corresponds to Eq. (6.178). Vanishing of the remaining terms in Eq. (6.226) leads to

$$\frac{D\varepsilon_\theta}{Dt} = \frac{C_{\ell\theta 2}}{C_{\ell\theta 3}} \frac{\varepsilon_\theta}{K_\theta} \frac{DK_\theta}{Dt} + \frac{C_{\ell\theta 4}}{C_{\ell\theta 3}} \frac{\varepsilon_\theta}{\varepsilon} \frac{D\varepsilon}{Dt}. \tag{6.229}$$

From Eq. (2.126) for K_θ and Eq. (6.179) for ε, we have

$$\frac{D\varepsilon_\theta}{Dt} = C_{\varepsilon\theta 1} \frac{\varepsilon_\theta}{K_\theta} P_\theta + C_{\varepsilon\theta 2} \frac{\varepsilon_\theta}{K} P_K - C_{\varepsilon\theta 3} \frac{\varepsilon_\theta^2}{K_\theta} - C_{\varepsilon\theta 4} \frac{\varepsilon_\theta \varepsilon}{K_\theta}, \tag{6.230}$$

where

$$C_{\varepsilon\theta 1} = C_{\varepsilon\theta 3} = \frac{C_{\ell\theta 2}}{C_{\ell\theta 3}} = 1.20, \tag{6.231a}$$

$$C_{\varepsilon\theta 2} = \frac{C_{\ell\theta 4} C_{\varepsilon 1}^{TS}}{C_{\ell\theta 3}} = 0.527, \tag{6.231b}$$

$$C_{\varepsilon\theta 4} = \frac{C_{\ell\theta 4} C_{\varepsilon 2}^{TS}}{C_{\ell\theta 3}} = 0.527, \tag{6.231c}$$

and the terms related to effects of diffusion have been omitted.

In the conventional modeling referred to in Sec. 4.6.1.C, the model constants in Eq. (6.230) are often chosen as (Nagano and Kim 1987)

$$C_{\varepsilon\theta 1} \cong C_{\varepsilon\theta 3}, \; C_{\varepsilon\theta 2} \cong C_{\varepsilon\theta 4}. \tag{6.232}$$

The present TSDIA result (6.231) is consistent with expression (6.232), and the value of each constant is not far from the optimized counterpart. From the findings about both the model equations for ε and ε_θ, it may be concluded that the method based on the concept of transferability of models is a useful one for constructing model transport equations for the destruction rates to which direct theoretical approaches are difficult to apply.

Finally, we refer to the transport rate $\mathbf{T}_\theta$ [Eq. (2.129)], which is the sole quantity to be modeled in Eq. (2.126) for K_θ. It may be calculated using the TSDIA in a manner similar to the transport rate of K, $\mathbf{T}_K$, in Sec. 6.6.1. The original calculation by Yoshizawa (1984b) was improved by Shimomura (1997b) on the basis of the accurate treatment of the solenoidal condition (Hamba 1987). There we first derive the expression corresponding to Eq. (6.166), and then apply the inertial-range form, Eqs. (6.199)-(6.200) under the constraint (6.223). As a result, $\mathbf{T}_\theta$ with the molecular diffusivity λ dropped is written as the linear combination of

$$\varepsilon_\theta \ell_\theta \nabla \ell_\theta, \ \ell_\theta^2 \nabla \varepsilon_\theta, \ \varepsilon_\theta \varepsilon^{-1} \ell_\theta^2 \nabla \varepsilon. \tag{6.233}$$

Using the algebraic relationship (6.228), we express Eq. (6.233) in terms of K_θ, ε_θ, and ε. As this result, $\mathbf{T}_\theta$ is written as

$$\mathbf{T}_\theta = C_{TK_\theta} \frac{K_\theta^2 \varepsilon}{\varepsilon_\theta^2} \nabla K_\theta - C_{T\theta_\theta} \frac{K_\theta^3 \varepsilon}{\varepsilon_\theta^3} \nabla \varepsilon_\theta + C_{T\theta} \frac{K_\theta^3}{\varepsilon_\theta^2} \nabla \varepsilon, \tag{6.234}$$

where the model constants C_{TK_θ} etc. are positive.

In Eq. (6.234), the opposite contributions of the first two parts are similar to those in the turbulent-energy transport rate, Eq. (6.176). The third term expresses the effect of velocity fluctuations. In the conventional modeling of $\mathbf{T}_\theta$, only the first term of Eq. (6.234) is retained. In the logarithmic-velocity region, ε is inversely proportional to y (the distance from the wall), and the contribution from the third term is anticipated to be large.

6.7.2. FRAME-ROTATION EFFECTS

An example of frame-rotation effects is presented in Sec. 6.5.2.B, in the context of a rotating channel flow. Frame-rotation effects also become important in astro/geophysical flow phenomena, as is understood from the importance of the Coriolis force in meteorological flows. In the context of theoretical turbulence modeling, we may point out two interesting aspects of frame rotation. First, frame rotation brings a preferred direction along the axis of rotation, and the mirrorsymmetry of turbulence properties is lost even in isotropic turbulence. Then the isotropic form given by Eq. (6.51) is not sufficient, and effects of nonmirrorsymmetry need to be added to it. Their promising candidate is the helicity that is simply referred to in Secs. 2.5.3 and 2.6.2.C.

Another aspect is linked with the scale-parameter expansion in the TSDIA formalism. The mean vorticity $\mathbf{\Omega}$ or the mean vorticity tensor Ω_{ij} [Eq. (4.101)], which characterizes the global vortical motion of turbulent shear flow, is of $O(\delta)$ in the formalism since it is related to the first derivative of the mean velocity $\mathbf{U}$. In a frame rotating with the angular velocity $\mathbf{\Omega}_F$, the transformation rule

$$\mathbf{\Omega} \to \mathbf{\Omega} + 2\mathbf{\Omega}_F, \ \Omega_{ij} \to \Omega_{ij} + 2\epsilon_{ij\ell}\Omega_{F\ell}, \tag{6.235}$$

holds, as is discussed in Sec. 2.5.2. In the scale-parameter expansion of the TSDIA, $\mathbf{\Omega}_F$ is regarded as $O(1)$, and the analysis of $\mathbf{\Omega}_F$ effects is generally much simpler than that of the $\mathbf{\Omega}$ counterparts. Therefore the latter effects may be obtained from the former through the transformation rule (6.235).

A. Introduction of Helicity Effects

In a rotating frame, fluid motion is governed by Eq. (2.76). The counterpart of Eq. (6.34) for $\mathbf{u}'(\mathbf{k};\tau)$ is

$$\frac{\partial u_i'(\mathbf{k};\tau)}{\partial \tau} + \nu k^2 u_i'(\mathbf{k};\tau) - ik_i p'(\mathbf{k};\tau) + 2\epsilon_{ij\ell}\Omega_{Fj} u_\ell'(\mathbf{k};\tau)$$
$$-ik_j \iint \delta(\mathbf{k}-\mathbf{p}-\mathbf{q}) d\mathbf{p} d\mathbf{q} u_i'(\mathbf{p};\tau) u_j'(\mathbf{q};\tau)$$
$$= \delta\left(-u_j'(\mathbf{k};\tau)\frac{\partial U_i}{\partial X_j} - \frac{Du_i'(\mathbf{k};\tau)}{DT_I} - \frac{\partial p'(\mathbf{k};\tau)}{\partial X_{Ii}}\right.$$
$$\left. - \iint \delta(\mathbf{k}-\mathbf{p}-\mathbf{q}) d\mathbf{p} d\mathbf{q} \frac{\partial}{\partial X_{Ij}} \left(u_i'(\mathbf{p};\tau) u_j'(\mathbf{q};\tau)\right) + \delta(\mathbf{k})\frac{\partial R_{ij}}{\partial X_j}\right), \quad (6.236)$$

with the solenoidal condition (6.35).

We apply the scale-parameter expansion (6.39) to Eq. (6.236). Then the $O(1)$ equations (6.46) and (6.47) are replaced with

$$\frac{\partial u_{0i}'(\mathbf{k};\tau)}{\partial \tau} + \nu k^2 u_{0i}'(\mathbf{k};\tau) + 2D_{im}(\mathbf{k})\epsilon_{mj\ell}\Omega_{Fj} u_{0\ell}'(\mathbf{k};\tau)$$
$$-iM_{ij\ell}(\mathbf{k}) \iint \delta(\mathbf{k}-\mathbf{p}-\mathbf{q}) d\mathbf{p} d\mathbf{q} u_{0j}'(\mathbf{p};\tau) u_{0\ell}'(\mathbf{q};\tau) = 0, \quad (6.237)$$

$$p_0'(\mathbf{k};\tau) = -\frac{k_i k_j}{k^2} \iint \delta(\mathbf{k}-\mathbf{p}-\mathbf{q}) d\mathbf{p} d\mathbf{q} u_{0i}'(\mathbf{p};\tau) u_{0j}'(\mathbf{q};\tau)$$
$$-2i\frac{k_i}{k^2}\epsilon_{ij\ell}\Omega_{Fj} u_\ell'(\mathbf{k};\tau). \quad (6.238)$$

What is important here is that the reflectionally symmetric, isotropic form represented by Eq. (6.51) is not applicable to Eq. (6.237) containing a preferred direction through the Coriolis force. In order to make full use of the knowledge about the inertial-range properties, we further introduce the perturbational expansion in $\mathbf{\Omega}_F$. This approach is valid in the case of weak rotation, but a method of widening its range of applicability is to renormalize the result concerning $\mathbf{\Omega}_F$. The approach based on the expansion in an additional parameter is applicable to the study of other effects such as buoyancy effect. Therefore we shall give some of its main steps (Yokoi and Yoshizawa 1993).

A.1. Rotation-Parameter Expansion. In each of orders in Eq. (6.239), we further expand

$$\mathbf{u'}_n(\mathbf{k};\tau) = \sum_{m=0}^{\infty} \mathbf{u'}_{nm}(\mathbf{k};\tau). \quad (6.239)$$

For instance, the $O(1)$ field $\mathbf{u'}_0(\mathbf{k};\tau)$ is written as

$$\mathbf{u'}_0(\mathbf{k};\tau) = \sum_{m=0}^{\infty} \mathbf{u'}_{0m}(\mathbf{k};\tau). \tag{6.240}$$

Here $\mathbf{u'}_{00}(\mathbf{k};\tau)$ obeys the same equation as Eq. (6.46). The $O(|\boldsymbol{\Omega}_F|)$ part $\mathbf{u'}_{01}(\mathbf{k};\tau)$ is governed by

$$\begin{aligned}&\frac{\partial u'_{01i}(\mathbf{k};\tau)}{\partial \tau} + \nu k^2 u'_{01i}(\mathbf{k};\tau) \\ &-2iM_{ij\ell}(\mathbf{k}) \iint \delta(\mathbf{k}-\mathbf{p}-\mathbf{q}) d\mathbf{p} d\mathbf{q} u'_{00j}(\mathbf{p};\tau) u'_{01\ell}(\mathbf{q};\tau) \\ &= -2D_{im}(\mathbf{k}) \epsilon_{mj\ell} \Omega_{Fj} u'_{00\ell}(\mathbf{k};\tau).\end{aligned} \tag{6.241}$$

This equation may be integrated using the Green's function obeying

$$\begin{aligned}&\frac{\partial G'_{ij}(\mathbf{k};\tau,\tau')}{\partial \tau} + \nu k^2 G'_{ij}(\mathbf{k};\tau,\tau') - D_{ij}(\mathbf{k}) \delta(\tau-\tau') \\ &\quad = 2iM_{i\ell m}(\mathbf{k}) \iint \delta(\mathbf{k}-\mathbf{p}-\mathbf{q}) d\mathbf{p} d\mathbf{q} u'_{00\ell}(\mathbf{p};\tau) G'_{mj}(\mathbf{q};\tau,\tau');\end{aligned} \tag{6.242}$$

namely, we have

$$u'_{01i}(\mathbf{k};\tau) = -2\epsilon_{mj\ell} \Omega_{Fj} \int_{-\infty}^{\tau} d\tau_1 G'_{im}(\mathbf{k};\tau,\tau_1) u'_{00\ell}(\mathbf{k};\tau_1). \tag{6.243}$$

The $O(\delta)$ field $\mathbf{u'}_1(\mathbf{k};\tau)$ is given by Eq. (6.63), where $\mathbf{u'}_{S1}(\mathbf{k};\tau)$ obeys Eq. (6.61) with

$$2D_{im}(\mathbf{k}) \epsilon_{mj\ell} \Omega_{Fj} u'_{1j}(\mathbf{k};\tau) \tag{6.244}$$

added on the left-hand side. Then the $O(|\boldsymbol{\Omega}_F|^0)$ and $O(|\boldsymbol{\Omega}_F|)$ parts of $\mathbf{u'}_1(\mathbf{k};\tau)$, which are denoted by $\mathbf{u'}_{10}(\mathbf{k};\tau)$ and $\mathbf{u'}_{11}(\mathbf{k};\tau)$, are written as

$$\begin{aligned}u'_{10i}(\mathbf{k};\tau) &= -i\frac{k_i}{k^2}\frac{\partial u'_{00j}(\mathbf{k};\tau)}{\partial X_{Ij}} - \frac{\partial U_j}{\partial X_\ell}\int_{-\infty}^{\tau} d\tau_1 G'_{ij}(\mathbf{k};\tau,\tau_1) u'_{00\ell}(\mathbf{k};\tau_1) \\ &\quad - \int_{-\infty}^{\tau} d\tau_1 G'_{ij}(\mathbf{k};\tau,\tau_1)\frac{Du'_{00j}(\mathbf{k};\tau_1)}{DT_I} + N_{u'},\end{aligned} \tag{6.245}$$

$$\begin{aligned}u'_{11i}(\mathbf{k};\tau) &= -i\frac{k_i}{k^2}\frac{\partial u'_{01j}(\mathbf{k};\tau)}{\partial X_{Ij}} - \frac{\partial U_j}{\partial X_\ell}\int_{-\infty}^{\tau} d\tau_1 G'_{ij}(\mathbf{k};\tau,\tau_1) u'_{01\ell}(\mathbf{k};\tau_1) \\ &\quad - \int_{-\infty}^{\tau} d\tau_1 G'_{ij}(\mathbf{k};\tau,\tau_1)\frac{Du'_{01j}(\mathbf{k};\tau_1)}{DT_I} \\ &\quad + iM^{\Omega}_{j\ell mn}(\mathbf{k})\Omega_{Fj}\int_{-\infty}^{\tau} d\tau_1 G'_{in}(\mathbf{k};\tau,\tau_1)\frac{\partial u'_{00\ell}(\mathbf{k};\tau_1)}{\partial X_{Im}} + N_{u'},\end{aligned} \tag{6.246}$$

where

$$M^{\Omega}_{ij\ell m}(\mathbf{k}) = \epsilon_{ijn}\delta_{\ell m}\frac{k_n}{k^2} - \epsilon_{ijn}\delta_{\ell n}\frac{k_m}{k^2} - \epsilon_{imn}\delta_{j\ell}\frac{k_n}{k^2} - \epsilon_{ijn}\frac{k_\ell k_m k_n}{k^4}, \quad (6.247)$$

and $N_{u'}$ denotes the parts nonlinear in $\mathbf{u}'_{00}(\mathbf{k};\tau)$, which are omitted here since they do not contribute to the following calculation.

A.2. Introduction of Helicity Effect. The isotropic state of turbulence is assumed in Eq. (6.51). Strictly speaking, this state should be called isotropic and reflectionally symmetric. The mirrorsymmetry or reflectional symmetry is broken in the presence of frame rotation since it gives rise to a preferred direction through the axis of rotation.

The extension of Eq. (6.51) to the nonmirrorsymmetric isotropic form is made as

$$Q_{ij}(\mathbf{k},\mathbf{X};\tau,\tau',T) = \frac{\left\langle u'_{00i}(\mathbf{k},\mathbf{X};\tau,T)u'_{00j}(\mathbf{k}',\mathbf{X};\tau',T)\right\rangle}{\delta(\mathbf{k}+\mathbf{k}')}$$
$$= D_{ij}(\mathbf{k})Q(k,\mathbf{X};\tau,\tau',T) + \frac{i}{2}\frac{k_\ell}{k^2}\epsilon_{ij\ell}H(k,\mathbf{X};\tau,\tau',T). \quad (6.248)$$

The nonmirrorsymmetry of $Q_{ij}(\mathbf{k};\tau,\tau')$ is associated with the linear dependence of the second part on $\mathbf{k}$. It may be easily seen that Eq. (6.248) obeys the solenoidal condition

$$k_i Q_{ij}(\mathbf{k};\tau,\tau') = k_j Q_{ij}(\mathbf{k};\tau,\tau') = 0. \quad (6.249)$$

The energy of the $O(\delta^0|\mathbf{\Omega}_F|^0)$ field is written as

$$\left\langle \mathbf{u}'^2_{00}/2\right\rangle = \int Q(k;\tau,\tau)d\mathbf{k}. \quad (6.250)$$

Namely, the H-related part in Eq. (6.248) does not contribute to it.

Using Eq. (6.248), we have

$$\langle \mathbf{u}'_{00}\cdot\boldsymbol{\omega}'_{00}\rangle = \int H(k;\tau,\tau)d\mathbf{k}. \quad (6.251)$$

This is the $O(\delta^0|\mathbf{\Omega}_F|^0)$ part of the turbulent helicity $\langle \mathbf{u}'\cdot\boldsymbol{\omega}'\rangle$ that is referred to in Sec. 2.6.2.C, as a pseudo-scalar characterizing the nonmirrorsymmetry of turbulence state. The helicity spectrum $E_H(k)$, which corresponds to $E(k)$, is defined as

$$E_H(k) = 4\pi k^2 H(k;\tau,\tau). \quad (6.252)$$

A.3. Reynolds Stress. We examine the relationship of the frame-rotation effect on the Reynolds stress $R_{ij}(=\langle u'_i u'_j \rangle)$ with the turbulent helicity K_H, which is defined by

$$K_H = \langle \mathbf{u}' \cdot \boldsymbol{\omega}' \rangle. \tag{6.253}$$

First, we expand R_{ij} and retain the resulting expression up to $O(\delta|\boldsymbol{\Omega}_F|)$:

$$\begin{aligned} R_{ij} &= \left\langle u'_{00i} u'_{00j} \right\rangle + \left\langle u'_{00i} u'_{01j} \right\rangle + \left\langle u'_{01i} u'_{00j} \right\rangle \\ &+\delta \left(\left\langle u'_{00i} u'_{10j} \right\rangle + \left\langle u'_{10i} u'_{00j} \right\rangle + \left\langle u'_{00i} u'_{11j} \right\rangle + \left\langle u'_{11i} u'_{00j} \right\rangle \right. \\ &\left. + \left\langle u'_{01i} u'_{10j} \right\rangle + \left\langle u'_{10i} u'_{01j} \right\rangle \right). \end{aligned} \tag{6.254}$$

In the $O(\delta^0)$ part of R_{ij}, the second term is calculated as

$$\left\langle u'_{00i} u'_{01j} \right\rangle = -2\epsilon_{n\ell m} \Omega_{F\ell} \int d\mathbf{k} \int_{-\infty}^{\tau} d\tau_1 G_{jn}\left(\mathbf{k};\tau,\tau_1\right) Q_{im}\left(\mathbf{k};\tau,\tau_1\right), \tag{6.255}$$

from Eq. (6.243). We use the isotropic representations for $Q_{ij}(\mathbf{k};\tau,\tau')$ and $G_{ij}(\mathbf{k};\tau,\tau')$, Eqs. (6.248) and (6.52), and have

$$\left\langle u'_{00i} u'_{01j} \right\rangle = -2\epsilon_{n\ell m} \Omega_{F\ell} \int d\mathbf{k} D_{jn}(\mathbf{k}) D_{im}(\mathbf{k}) \int_{-\infty}^{\tau} d\tau_1 G\left(k;\tau,\tau_1\right) Q\left(k;\tau,\tau_1\right), \tag{6.256}$$

where the helicity part of Eq. (6.248) does not contribute since it leads to a function odd in $\mathbf{k}$. Equation (6.256) is antisymmetric concerning i and j, that is,

$$\left\langle u'_{00i} u'_{01j} \right\rangle = -\left\langle u'_{01i} u'_{00j} \right\rangle. \tag{6.257}$$

Then the second and third terms in the $O(\delta^0)$ part of Eq. (6.254) cancel with each other.

In the $O(\delta)$ part of R_{ij}, the first two terms give the so-called turbulent-viscosity representation, which is written in the form of the ν_T-related term in Eq. (6.102). The third term may be expressed as

$$\begin{aligned} \left\langle u'_{00i} u'_{11j} \right\rangle &= \frac{1}{4} \epsilon_{hmn} \epsilon_{inr} \Omega_{Fm} \int d\mathbf{k} D_{\ell h}(\mathbf{k}) \frac{k_j k_r}{k^4} \\ &\quad \times \int_{-\infty}^{\tau} d\tau_1 G\left(k;\tau,\tau_1\right) \frac{\partial H\left(k;\tau,\tau_1\right)}{\partial X_\ell} \\ &+\frac{1}{4} \epsilon_{imr} \Omega_{F\ell} \int d\mathbf{k} M^{\Omega}_{\ell mnh}(\mathbf{k}) D_{jh}(\mathbf{k}) \frac{k_r}{k^2} \\ &\quad \times \int_{-\infty}^{\tau} d\tau_1 G\left(k;\tau,\tau_1\right) \frac{\partial H\left(k;\tau,\tau_1\right)}{\partial X_n}. \end{aligned} \tag{6.258}$$

Here the first term comes from the first one of Eq. (6.246) combined with Eq. (6.243), whereas the second is the contribution from the fourth term of Eq. (6.246). We use Eq. (6.258) and have

$$\left\langle u'_{00i}u'_{11j}\right\rangle + \left\langle u'_{11i}u'_{00j}\right\rangle = \frac{17}{45}\Omega_{F\ell} I\left\{G, \frac{\partial H}{\partial X_\ell}\right\}\delta_{ij}$$
$$-\frac{1}{30}\left(\Omega_{Fi} I\left\{G, \frac{\partial H}{\partial X_j}\right\} + \Omega_{Fj} I\left\{G, \frac{\partial H}{\partial X_i}\right\} - \frac{2}{3}\Omega_{F\ell} I\left\{G, \frac{\partial H}{\partial X_\ell}\right\}\delta_{ij}\right), \quad (6.259)$$

where $I\{A,B\}$ is defined as

$$I\{A,B\} = \int k^{-2} d\mathbf{k} \int_{-\infty}^{\tau} d\tau_1 A(k;\tau,\tau_1) B(k;\tau,\tau_1). \quad (6.260)$$

The more details of the calculation of Eq. (6.259) are explained in *Note*. Entirely similarly, we have

$$\left\langle u'_{01i}u'_{10j}\right\rangle + \left\langle u'_{10i}u'_{01j}\right\rangle = -\frac{2}{3}\Omega_{F\ell} I\left\{G, \frac{\partial H}{\partial X_\ell}\right\}\delta_{ij}$$
$$+\frac{1}{10}\left(\Omega_{Fi} I\left\{G, \frac{\partial H}{\partial X_j}\right\} + \Omega_{Fj} I\left\{G, \frac{\partial H}{\partial X_i}\right\} - \frac{2}{3}\Omega_{F\ell} I\left\{G, \frac{\partial H}{\partial X_\ell}\right\}\delta_{ij}\right). \quad (6.261)$$

Summarizing the foregoing results, we have

$$R_{ij} = \frac{2}{3}K\delta_{ij} - \nu_T S_{ij} + \left[(2\Omega_{Fi} + \Omega_i)\gamma_j + (2\Omega_{Fj} + \Omega_j)\gamma_i\right]_D, \quad (6.262)$$

with

$$K = K_{NROT} + (1/3)(2\mathbf{\Omega}_F + \mathbf{\Omega})\cdot \mathbf{I}\{G, \nabla H\}, \quad (6.263)$$

$$\nu_T = \nu_{T,NROT}, \quad (6.264)$$

$$\gamma = (1/30)\,\mathbf{I}\{G, \nabla H\}, \quad (6.265)$$

where K_{NROT} and $\nu_{T,NROT}$ correspond to the nonrotating counterparts of K and ν_T, which are given by Eqs. (6.103) and (6.104), respectively. In Eqs. (6.262) and (6.263), we should note that the effects of the mean vorticity $\mathbf{\Omega}$ have been included on the basis of the transformation rule (6.235). In the TSDIA formalism, these added effects come originally from the $O(\delta^2|\mathbf{\Omega}_F|)$ calculation, which is rather tedious.

Note: In the first term of Eq. (6.258), the Green's function $G(k;\tau,\tau')$ is outside of the operator ∇_X. This point comes from the DIA procedure that the lowest-order or linear part of $G'_{ij}(\mathbf{k};\tau,\tau')$, $G_{Vij}(\mathbf{k};\tau,\tau')$, is independent of $\mathbf{X}$, as is seen from Eq. (6.78), and may be taken outside of $\nabla_{\mathbf{X}}$. In the second term of Eq. (6.258), the metric parts are calculated as follows. These parts are dependent on only k after the integration on the spherical surface with the radius k, $S(k)$. For instance, we write

$$\int_{S(k)} \epsilon_{imr} M^{\Omega}_{\ell mnh}(\mathbf{k}) D_{jh}(\mathbf{k}) \frac{k_r}{k^2} dS$$
$$= \frac{1}{k^2}\left(Z_1 \delta_{ij}\delta_{\ell n} + Z_2 \delta_{i\ell}\delta_{jn} + Z_3 \delta_{in}\delta_{j\ell}\right) \int_{S(k)} dS, \tag{6.266}$$

with unknown numerical coefficients Z_n's. Putting $i = j$ and $\ell = n$ in Eq. (6.266), we have

$$3Z_1 + Z_2 + Z_3 = \frac{4}{3}. \tag{6.267a}$$

Similarly, we have

$$Z_1 + 3Z_2 + Z_3 = \frac{2}{3}, \tag{6.267b}$$

$$Z_1 + Z_2 + 3Z_3 = \frac{2}{3}. \tag{6.267c}$$

As a result, Z_n's are

$$Z_1 = \frac{2}{5},\ Z_2 = \frac{1}{15},\ Z_3 = \frac{1}{15}. \tag{6.268}$$

We substitute Eq. (6.266) with Eq. (6.268) into Eq. (6.258), and finally obtain Eq. (6.259).

A.4. Modeling of Helicity Effects. In the case of vanishing H, we may apply the inertial-range form for $Q(k;\tau,\tau')$ and $G(k;\tau,\tau')$, Eqs. (6.55) and (6.56) to Eq. (6.104) for ν_T, and obtain Eq. (6.114). In the presence of H, however, any clear analytical expressions have not been so far found for $Q(k;\tau,\tau')$, $G(k;\tau,\tau')$, and $H(k;\tau,\tau')$. As a first step towards the examination of helicity effects, we adopt Eq. (6.114) for ν_T. For γ [Eq. (6.265)], we write $H(k;\tau,\tau')$ as

$$H(k,\mathbf{X};\tau,\tau',T) = K_H(\mathbf{X};T) f_H(k,\mathbf{X};\tau,\tau',T), \tag{6.269}$$

where f_H is normalized as

$$\int f_H(k,\mathbf{X},;\tau,\tau',T) d\mathbf{k} = 1, \tag{6.270}$$

using Eq. (6.253). Strictly speaking, K_H in Eq. (6.269) is the turbulent helicity of the $O(\delta^0|\mathbf{\Omega}_F|^0)$ field or

$$K_{H00}(\mathbf{X};T) = \left\langle \mathbf{u}'_{00} \cdot \boldsymbol{\omega}'_{00} \right\rangle, \tag{6.271}$$

but we have performed the renormalization

$$K_{H00}(\mathbf{X};T) \to K_H(\mathbf{X};T). \tag{6.272}$$

We substitute Eq. (6.269) into Eq. (6.265), and neglect the dependence of f_H on $\mathbf{X}$, resulting in

$$\gamma = \left(\frac{1}{30}\int k^{-2}d\mathbf{k}\int_{-\infty}^{\tau} d\tau_1 G(k;\tau,\tau_1) f_H(k;\tau,\tau_1)\right)\nabla K_H. \tag{6.273}$$

In Eq. (6.269), $H(k;\tau,\tau')$ may take both positive and negative signs. We assume that the sign of $H(k;\tau,\tau')$ is dependent on K_H. Namely, we consider that f_H is positive, and model the coefficient of ∇K_H in terms of K and ε. Then we have

$$\gamma = C_\gamma\left(K^4/\varepsilon^2\right)\nabla K_H, \tag{6.274}$$

using C_γ is a positive constant.

The turbulent helicity K_H obeys Eq. (2.132) with $\mathbf{\Omega}$ replaced with $\mathbf{\Omega} = \mathbf{\Omega} + 2\mathbf{\Omega}_F$. Here ε_H [Eq. (2.134)] and the second and third parts in $\mathbf{T}_H$ [Eq. (2.135)] need to be modeled. Equation (2.132) is simply modeled as

$$\begin{aligned}\frac{DK_H}{Dt} &= -R_{ij}\frac{\partial \Omega_j}{\partial x_i} + (\Omega_i + 2\Omega_{Fi})\frac{\partial R_{ij}}{\partial x_j} - C_H\frac{\varepsilon}{K}K_H \\ &+\nabla\cdot\left(\frac{\nu_T}{\sigma_H}\nabla K_H\right),\end{aligned} \tag{6.275}$$

with positive C_H and σ_H. Here the third term is the model for ε_H, and $K_H/(K/\varepsilon)$ represents the decay rate of K_H based on the turbulent characteristic time K/ε. The fourth term comes from the modeling of $\langle(\mathbf{u}'\cdot\boldsymbol{\omega}')\mathbf{u}'\rangle$ (the helicity transport rate by fluctuation) in Eq. (2.135), that is,

$$\langle(\mathbf{u}'\cdot\boldsymbol{\omega}')\,\mathbf{u}'\rangle = -\left(\nu_T/\sigma_H\right)\nabla K_H. \tag{6.276}$$

This modeling corresponds to the first term of Eq. (6.172).

A.5. Vortex Dynamo. As is noted in Sec. 2.5.3, nonvanishing $\mathbf{u}\cdot\boldsymbol{\omega}$ expresses a helical motion of fluid, and is related to the degree of duration of vortical motion. The foregoing analysis of R_{ij} indicates that the turbulent part of $\mathbf{u}\cdot\boldsymbol{\omega}$, K_H, is linked with the frame rotation and the mean vortical motion. Specifically, the inhomogeneity of turbulence plays a key role. In the context of homogeneous turbulence, it was pointed out by André and Lesieur (1977) that the helicity initially supplied to low-wavenumber components of motion tends to suppress the energy cascade. In the homogeneous case, however, there is no mechanism sustaining the helicity, and such a suppression effect eventually disappears. This point may be understood from the first two terms in Eq. (6.275).

In astro/geophysical flow phenomena, the rotation of a planet is one of the central elements governing them. As an interesting example showing the role of the helicity in a rotating planetary system, we consider the generation of mean or global vortical flow structures. This mechanism is called the vortex dynamo. The mean vorticity $\mathbf{\Omega}$ obeys

$$\frac{\partial \mathbf{\Omega}}{\partial t} = \nabla \times (\mathbf{U} \times (\mathbf{\Omega} + 2\mathbf{\Omega}_F) - \nu \nabla \times \mathbf{\Omega}) + \mathbf{I}_V, \tag{6.277}$$

where $\mathbf{I}_V$ expresses the induction rate of $\mathbf{\Omega}$ due to turbulence, and is given by

$$\mathbf{I}_V = \nabla \times \mathbf{V}_M, \tag{6.278}$$

with

$$\mathbf{V}_M = \langle \mathbf{u}' \times \boldsymbol{\omega}' \rangle . \tag{6.279}$$

Here $\mathbf{V}_M$, which is the vortical-motion counterpart of the Reynolds stress R_{ij}, is named the vortex-motive force in the vortex dynamo, and is related to R_{ij} as

$$V_{Mi} = -\frac{\partial R_{ij}}{\partial x_j} + \frac{\partial K}{\partial x_i} \tag{6.280}$$

[see Eq. (2.137)]. It should be noted that the second term in Eq. (6.280) does not contribute to the $\mathbf{\Omega}$ equation (6.278).

We substitute Eq. (6.262) into Eq. (6.280), and have

$$\mathbf{V}_M = -D_\gamma (\mathbf{\Omega} + 2\mathbf{\Omega}_F) - ((\mathbf{\Omega} + 2\mathbf{\Omega}_F) \cdot \nabla) \boldsymbol{\gamma} - \nu_T \nabla \times \mathbf{\Omega}, \tag{6.281}$$

with

$$D_\gamma = \nabla \cdot \boldsymbol{\gamma}. \tag{6.282}$$

Here we dropped the term that contains $\nabla \nu_T$ or vanish when the curl operator is applied. The contribution of the last term in Eq. (6.281) to the induction rate $\mathbf{I}_V$ is approximated as

$$\nabla \times (-\nu_T \nabla \times \mathbf{\Omega}) \cong \nu_T \Delta \mathbf{\Omega}, \tag{6.283}$$

which destroys global vortical structures. This point is clear from the original property of ν_T.

We shall consider vortical flow structures on a spherical surface. The coordinate system is depicted in Fig. 6.3. First, we examine the case in which the horizontal (x and y) spatial scales of a flow are much larger than the vertical (z) counterpart. Then we may put

$$\nabla = (0,\ 0,\ \partial/\partial z) . \tag{6.284}$$

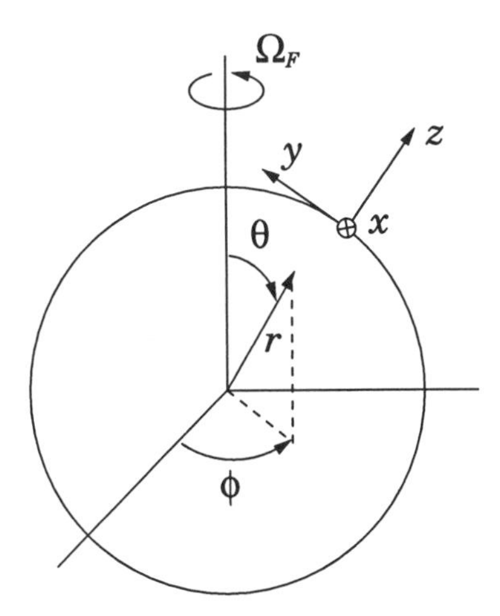

Figure 6.3. Coordinate system on the spherical surface.

At the early stage of vortex induction, the $\boldsymbol{\Omega}_F$-related terms in Eq. (6.281) are dominant, compared with the $\boldsymbol{\Omega}$ counterparts. Then the induction rate $\mathbf{I}_V$ is

$$\mathbf{I}_V \cong \nabla \times \left(-2D_\gamma \boldsymbol{\Omega}_F - 2\left(\boldsymbol{\Omega}_F \cdot \nabla\right)\gamma\right) = -2\left(\nabla D_\gamma\right) \times \boldsymbol{\Omega}_F, \tag{6.285}$$

where we should note that

$$\nabla \times \left(\left(\boldsymbol{\Omega}_F \cdot \nabla\right)\gamma\right) = 0, \tag{6.286}$$

under the approximation (6.284).

We write

$$\boldsymbol{\Omega}_F = \left(\Omega_{Fx},\ \Omega_{Fy},\ \Omega_{Fz}\right) = \left(0,\ |\boldsymbol{\Omega}_F|\sin\theta,\ |\boldsymbol{\Omega}_F|\cos\theta\right). \tag{6.287}$$

Using Eq. (6.287), Eq. (6.285) is reduced to

$$\mathbf{I}_V = \left(2\frac{\partial D_\gamma}{\partial z}\,|\boldsymbol{\Omega}_F|\sin\theta,\ 0,\ 0\right). \tag{6.288}$$

Equation (6.288) indicates that the vorticity with its axis along the x or zonal direction is induced in the low-latitude ($0 \ll \theta < \pi/2$) region at the early stage of vortex generation, through the interaction of frame rotation with the helicity inhomogeneity producing nonvanishing D_γ.

Once Ω_x is induced, the generation of the y component Ω_y follows. We substitute

$$\boldsymbol{\Omega} = \left(\Omega_x,\ 0,\ 0\right) \tag{6.289}$$

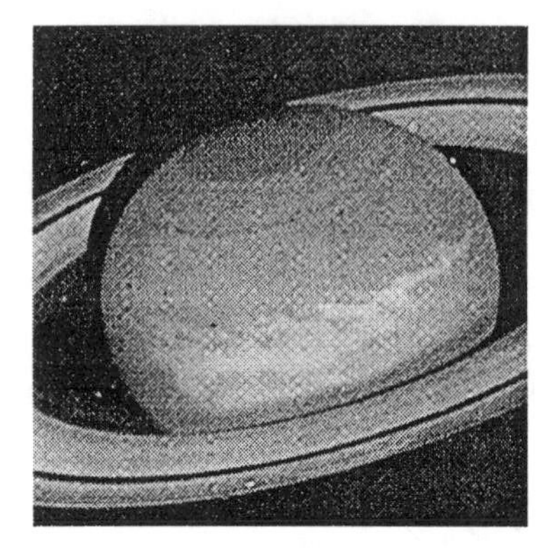

Figure 6.4. Saturn's white spot.

into the first term in Eq. (6.281), whose contribution to $\mathbf{I}_V$ is given by

$$(\nabla \times (-D_\gamma (\boldsymbol{\Omega} + 2\boldsymbol{\Omega}_F)))_y = -\frac{\partial}{\partial z} (D_\gamma \Omega_x) . \tag{6.290}$$

Equation (6.290) shows that induced Ω_x is combined with the helicity effects to further generate Ω_y. The vorticity having both the x and y components expresses a helical vortical structure. Such a structure is a promising candidate for Saturn's huge white spots shown in Fig. 6.4 (Yoshizawa and Yokoi 1991).

Next, we examine the case in which the horizontal scale becomes more important than the vertical one. In this case, we take

$$\nabla = (\partial/\partial x,\ \partial/\partial y,\ 0) . \tag{6.291}$$

Under the condition (6.291), we have

$$\boldsymbol{\gamma} = (\gamma_x,\ \gamma_y,\ 0) \propto \left(\frac{\partial K_H}{\partial x},\ \frac{\partial K_H}{\partial y},\ 0 \right), \tag{6.292}$$

from Eq. (6.274). At the early stage of vortex generation, we have

$$\begin{aligned} \mathbf{I}_V &\cong \nabla \times (-2D_\gamma \boldsymbol{\Omega}_F - 2(\boldsymbol{\Omega}_F \cdot \nabla) \boldsymbol{\gamma}) \\ &= \left(-2\frac{\partial D_\gamma}{\partial y} |\boldsymbol{\Omega}_F|_F \cos\theta,\ 2\frac{\partial D_\gamma}{\partial x} |\boldsymbol{\Omega}_F| \cos\theta,\ -2\frac{\partial D_\gamma}{\partial x} |\boldsymbol{\Omega}_F| \sin\theta \right) \\ &\quad + \left(0,\ 0,\ 2 |\boldsymbol{\Omega}_F| \sin\theta \left(\frac{\partial^2 \gamma_x}{\partial y^2} - \frac{\partial^2 \gamma_y}{\partial x \partial y} \right) \right), \end{aligned} \tag{6.293}$$

where the first and second parts in the second relation come from the counterparts in the first relation, respectively.

In the low-latitude ($0 \ll \theta < \pi/2$) region, Eq. (6.293) is approximated as

$$\mathbf{I}_V \cong \left(0,\ 0,\ 2\left|\mathbf{\Omega}_F\right|\sin\theta\left(-\frac{\partial D_\gamma}{\partial y}+\frac{\partial^2\gamma_x}{\partial y^2}-\frac{\partial^2\gamma_y}{\partial x\partial y}\right)\right). \tag{6.294}$$

Namely, the vortical motion with the rotation axis vertical to the surface is induced under the Coriolis force in such a region. This is an interesting mechanism in the context of the generation of tropical cyclones or typhoons in the low-latitude region.

A.6. Swirling Pipe Flow. The representative turbulent flow whose prominent feature lies in its vortical motion is a swirling flow in a circular pipe, which is referred to in Sec. 4.2.5.B. Its mean axial velocity has a profile with the local minimum velocity at the center as far as the swirling motion given at the entrance survives downstream (see Fig. 4.5; Kitoh 1991). In the K-ε model based on the simple turbulent-viscosity representation for the Reynolds stress R_{ij}, Eqs. (4.3) and (4.8), the swirling motion is lost too fast, compared with the observations. Since swirling is a kind of vortical motion, a model with effects of the mean vorticity $\mathbf{\Omega}$ included is expected to improve the shortfall of the simple turbulent-viscosity model.

Equations (6.262) (with Eq. (6.274) and $\mathbf{\Omega}_F = 0$) and (6.275) are combined with the familiar K-ε model to constitute a three-equation turbulence model with effects of helicity included. Here the model constants C_γ, C_H, and σ_H are chosen as

$$C_\gamma = 0.003,\ \ C_H = 1.5,\ \ \sigma_H = 1.6. \tag{6.295}$$

The model was applied to a swirling pipe flow, and the computed results were compared with the observation by Kitoh (Nisizima and Yokoi 1992; Yokoi and Yoshizawa 1993). Specifically, the $\mathbf{\Omega}$-related part in R_{ij} was confirmed to be instrumental to reproducing the velocity profile with the local minimum velocity at the central axis. This finding indicates the close linkage between frame-rotation and mean-vorticity effects.

B. Suppression of Energy Cascade

The other interesting example of effects of frame rotation on laboratory flows is the suppression of energy dissipation rate. In a frame rotating with the angular velocity $\mathbf{\Omega}_F$, the turbulent energy K obeys the same equation as Eq. (2.111). As a result, the effects of $\mathbf{\Omega}_F$ do not appear explicitly.

The simplest case showing the suppression effect due to $\mathbf{\Omega}_F$ is homogeneous isotropic turbulence in a rotating frame. In this case, K obeys

$$\frac{\partial K}{\partial t} = -\varepsilon. \tag{6.296}$$

From the large eddy simulation (LES) and DNS of the flow (Bardina et al. 1985; Shimomura 1997a), it is known that ε decreases with increasing $|\boldsymbol{\Omega}_F|$. This fact indicates that the energy spectrum $E(k)$ is greatly affected by the frame rotation, resulting in the change of ε.

In order to see the foregoing point, we seek the solution of Eq. (6.236) up to $O(\delta|\boldsymbol{\Omega}_F|^2)$ under the condition of isotropy. This calculation leads to

$$E(k) = K_O \varepsilon^{2/3} k^{-5/3} - K_N \varepsilon^{-2/3} \frac{\partial \varepsilon}{\partial t} k^{-7/3} - K_\Omega \varepsilon^{-4/3} \frac{\partial \varepsilon}{\partial t} \boldsymbol{\Omega}_F^2 k^{-11/3}, \quad (6.297)$$

where K_O and K_N have been already given by Eq. (6.96), and

$$K_\Omega = 2.83. \quad (6.298)$$

The last term in Eq. (6.297) shows that the frame-rotation effect occurs in the coupling with the unstationariness of turbulence (Okamoto 1995).

From Eq. (6.297), we have

$$K = C_{K1} \varepsilon^{2/3} \ell_C^{2/3} - C_{K2} \varepsilon^{-2/3} \ell_C^{4/3} \frac{\partial \varepsilon}{\partial t} - C_{K3} \varepsilon^{1/3} \ell_C^{1/3} \frac{\partial \ell_C}{\partial t}$$
$$- C_{K4} \varepsilon^{-4/3} \ell_C^{8/3} \frac{\partial \varepsilon}{\partial t} \boldsymbol{\Omega}_F^2 + C_{K5} \varepsilon^{-1/3} \ell_C^{5/3} \frac{\partial \ell_C}{\partial t} \boldsymbol{\Omega}_F^2, \quad (6.299)$$

which is the frame-rotation counterpart of Eq. (6.112), where $C_{Kn} (n = 1\text{-}3)$ are given by Eq. (6.113), and

$$C_{K4} = 0.00790, \quad C_{K5} = 0.0158. \quad (6.300)$$

We solve Eq. (6.299) concerning ℓ_C, and have

$$\ell_C = C_{\ell 1} K^{3/2} \varepsilon^{-1} + C_{\ell 2} K^{3/2} \varepsilon^{-2} \frac{\partial K}{\partial t} - C_{\ell 3} K^{5/2} \varepsilon^{-3} \frac{\partial \varepsilon}{\partial t}$$
$$- C_{\ell 4} K^{7/2} \varepsilon^{-4} \frac{\partial K}{\partial t} \boldsymbol{\Omega}_F^2 + C_{\ell 4} K^{9/2} \varepsilon^{-5} \frac{\partial \varepsilon}{\partial t} \boldsymbol{\Omega}_F^2, \quad (6.301)$$

where $C_{\ell n} (n = 1\text{-}3)$ are given by Eq. (6.117), and

$$C_{\ell 4} = 0.333, \quad C_{\ell 5} = 0.333. \quad (6.302)$$

We require the algebraic relationship, Eq. (6.178), and expand the resulting expression in $\boldsymbol{\Omega}_F^2$. As a result, we have

$$\frac{\partial \varepsilon}{\partial t} = -C_{\varepsilon 2}^{TS} \frac{\varepsilon^2}{K} - C_{\varepsilon\Omega} K \boldsymbol{\Omega}_F^2, \quad (6.303)$$

where $C_{\varepsilon 2}^{TS}$ is given by Eq. (6.180b), and

$$C_{\varepsilon\Omega} = 0.0931. \quad (6.304)$$

Equation (6.303) signifies that the frame rotation tends to suppress the dissipation rate, but it is not applicable for large $|\mathbf{\Omega}_F|$. In order to rectify this shortfall, the TSDIA calculation of $O(\delta|\mathbf{\Omega}_F|^4)$ was done by Okamoto (1995), and the result was combined with the Padé approximation to find a model equation for ε applicable for large $|\mathbf{\Omega}_F|$.

6.7.3. BUOYANCY EFFECTS

Effects of buoyancy are usually dealt with under the Boussinesq approximation, which is explained in Sec. 2.5.4. This approximation is originally valid for small temperature variation. It is often applied, however, to the case that seems to be beyond the theoretical constraint. The Navier-Stokes equation with the Boussinesq approximation included is given by Eq. (2.96). In what follows, we shall simply denote the gravity-included pressure p_g and the deviation of the temperature from the reference one θ_R, by p and θ, respectively. Then the counterpart of Eq (6.34) is

$$\frac{\partial u_i'(\mathbf{k};\tau)}{\partial \tau} + \nu k^2 u_i'(\mathbf{k};\tau) + i k_i p'(\mathbf{k};\tau) - \alpha g_i \theta'(\mathbf{k};\tau)$$
$$-ik_j \iint \delta(\mathbf{k}-\mathbf{p}-\mathbf{q}) d\mathbf{p} d\mathbf{q} u_i'(\mathbf{p};\tau) u_j'(\mathbf{q};\tau)$$
$$= \delta \left(-u_j'(\mathbf{k};\tau)\frac{\partial U_i}{\partial X_j} - \frac{D u_i'(\mathbf{k};\tau)}{D T_I} - \frac{\partial p'(\mathbf{k};\tau)}{\partial X_{Ii}} \right.$$
$$\left. - \iint \delta(\mathbf{k}-\mathbf{p}-\mathbf{q}) d\mathbf{p} d\mathbf{q} \frac{\partial}{\partial X_{Ij}} \left(u_i'(\mathbf{p};\tau) u_j'(\mathbf{q};\tau) \right) + \delta(\mathbf{k}) \frac{\partial R_{ij}}{\partial X_j} \right), \quad (6.305)$$

with the solenoidal condition (6.35). On the other hand, the temperature fluctuation $\theta'(\mathbf{k};\tau)$ obeys the same equation as Eq. (6.189).

We expand $\mathbf{u}'(\mathbf{k};\tau)$, $p'(\mathbf{k};\tau)$, and $\theta'(\mathbf{k};\tau)$ in the scale parameter δ as Eqs. (6.39) and (6.190). Moreover, we treat the θ'-related term on the left-hand side of Eq. (6.305) in entirely the same perturbational manner as for the case of the Coriolis force, Eq. (6.236). Namely, we expand $\mathbf{u}'(\mathbf{k};\tau)$ etc. as Eq. (6.239) etc. in α (the thermal expansion coefficient), and seek the solution up to $O(\delta\alpha)$. The details of the mathematical manipulation are quite similar to the case of the Coriolis-force effects (Yoshizawa 1983; Okamoto 1996a).

A. Reynolds Stress and Turbulent Heat Flux

The Reynolds stress R_{ij} is written as

$$R_{ij} = \frac{2}{3} K \delta_{ij} - \nu_T S_{ij} + \alpha \nu_{G\Theta} \left[g_i \frac{\partial \Theta}{\partial x_j} + g_j \frac{\partial \Theta}{\partial x_i} \right]_D, \quad (6.306)$$

where

$$K = \int Q(k;\tau,\tau)d\mathbf{k} - \int d\mathbf{k} \int_{-\infty}^{\tau} d\tau_1 G(k;\tau,\tau_1) \frac{DQ(k;\tau,\tau_1)}{Dt}$$
$$+\frac{2}{3}\alpha(\mathbf{g}\cdot\nabla\Theta)\int d\mathbf{k}\int_{-\infty}^{\tau} d\tau_1 \int_{-\infty}^{\tau_1} d\tau_2$$
$$\times G(k;\tau,\tau_1)G_\theta(k;\tau_1,\tau_2)Q(k;\tau,\tau_2), \quad (6.307)$$

$$\nu_{G\Theta} = \frac{7}{15}\int d\mathbf{k}\int_{-\infty}^{\tau} d\tau_1 \int_{-\infty}^{\tau_1} d\tau_2 G(k;\tau,\tau_1)G_\theta(k;\tau_1,\tau_2)Q(k;\tau,\tau_2), \quad (6.308)$$

and ν_T is the same as for the nonbuoyant case and is given by Eq. (6.104).

In the context of temperature, the turbulent heat flux $\mathbf{H}_\theta (= \langle \mathbf{u}'\theta' \rangle)$ and the temperature variance $K_\theta (= \langle \theta'^2 \rangle)$ are the most important quantities. The former is

$$\mathbf{H}_\theta = -\lambda_T \nabla\Theta - \lambda_G \alpha \mathbf{g} + \lambda_{GU}\alpha \left((\mathbf{g}\cdot\nabla)\mathbf{U} + \frac{1}{6}\nabla(\mathbf{g}\cdot\mathbf{U}) \right), \quad (6.309)$$

where the turbulent heat diffusivity λ_T is given by Eq. (6.204), and

$$\lambda_G = \frac{2}{3}\int d\mathbf{k}\int_{-\infty}^{\tau} d\tau_1 G(k;\tau,\tau_1)Q_\theta(k;\tau,\tau_1)$$
$$-\frac{1}{3}\int d\mathbf{k}\int_{-\infty}^{\tau} d\tau_1 \int_{-\infty}^{\tau} d\tau_2 G(k;\tau,\tau_1)G_\theta(k;\tau,\tau_2)\frac{DQ_\theta(k;\tau_1,\tau_2)}{Dt}$$
$$-\frac{1}{3}\int d\mathbf{k}\int_{-\infty}^{\tau} d\tau_1 \int_{-\infty}^{\tau_1} d\tau_2 G(k;\tau,\tau_1)$$
$$\times \left(G(k;\tau_1,\tau_2) + G_\theta(k;\tau_1,\tau_2)\right)\frac{DQ_\theta(k;\tau,\tau_2)}{Dt}, \quad (6.310)$$

$$\lambda_{GU} = \frac{2}{3}\int d\mathbf{k}\int_{-\infty}^{\tau} d\tau_1 \int_{-\infty}^{\tau_1} d\tau_2 G(k;\tau,\tau_1)G(k;\tau_1,\tau_2)Q_\theta(k;\tau,\tau_2). \quad (6.311)$$

The temperature variance K_θ is written as

$$K_\theta = \int Q_\theta(k;\tau,\tau)dk - \int d\mathbf{k}\int_{-\infty}^{\tau} d\tau_1 G_\theta(k;\tau,\tau_1)\frac{DQ_\theta(k;\tau,\tau_1)}{Dt}$$
$$+\frac{4}{3}\alpha(\mathbf{g}\cdot\nabla\Theta)\int d\mathbf{k}\int_{-\infty}^{\tau} d\tau_1 \int_{-\infty}^{\tau_1} d\tau_2$$
$$\times G_\theta(k;\tau,\tau_1)G(k;\tau_1,\tau_2)Q_\theta(k;\tau,\tau_2). \quad (6.312)$$

In Eq. (6.306), K, ν_T, and $\nu_{G\Theta}$ are dependent on $G_\theta(k;\tau,\tau')$, in addition to $G(k;\tau,\tau')$ and $Q(k;\tau,\tau')$, but not on $Q_\theta(k;\tau,\tau')$. As is noted in Sec. 6.7.1.A, $G_\theta(k;\tau,\tau')$ is determined by the statistical property of $u'_{00}(\mathbf{k};\tau)$, except the heat diffusivity λ [see Eq. (6.193)]. This fact indicates that the energy-containing eddies of velocity fluctuation contribute much to these quantities. Then we apply the $\mathbf{k}$-integration rule based on the velocity scale ℓ_C, Eq. (6.105). On the other hand, $\mathbf{H}_\theta$ and K_θ depend on $Q_\theta(k;\tau,\tau')$, apart from λ_T in $\mathbf{H}_\theta$. We adopt the $\mathbf{k}$-integration rule based on the scalar- or temperature-variance scale ℓ_θ, Eq. (6.223), for λ_G, $\lambda_{G\mathbf{U}}$, and K_θ, whereas λ_T obeys the former rule, Eq. (6.105). The calculation by Yoshizawa (1983) was improved by Okamoto (1996a) using the exact solenoidal condition, Eq. (6.37). In the latter, however, the definition of ε_θ is not consistent with the use of Eq. (6.201), and ε_θ needs to be replaced with $2\varepsilon_\theta$. Finally, we have

$$K = C_{K1}\varepsilon^{2/3}\ell_C^{2/3} - C_{K2}\varepsilon^{-2/3}\ell_C^{4/3}\frac{D\varepsilon}{Dt} - C_{K3}\varepsilon^{1/3}\ell_C^{1/3}\frac{D\ell_C}{Dt} + C_{K4}\ell_C^2\alpha\mathbf{g}\cdot\nabla\Theta, \tag{6.313}$$

$$\nu_{G\Theta} = C_{\nu_{G\Theta}\ell}\ell_C^2, \tag{6.314}$$

$$\lambda_G = C_{\lambda_G\ell1}\varepsilon_\theta\varepsilon^{-2/3}\ell_\theta^{4/3} - C_{\lambda_G\ell2}\varepsilon^{-1}\ell_\theta^2\frac{D\varepsilon_\theta}{Dt} + C_{\lambda_G\ell3}\varepsilon_\theta\varepsilon^{-2}\ell_\theta^2\frac{D\varepsilon}{Dt} - C_{\lambda_G\ell4}\varepsilon_\theta\varepsilon^{-1}\ell_\theta\frac{D\ell_\theta}{Dt}, \tag{6.315}$$

$$\lambda_{GU} = C_{\lambda_{GU}\ell}\varepsilon_\theta\varepsilon^{-1}\ell_\theta^2, \tag{6.316}$$

$$\begin{aligned} K_\theta &= C_{K_\theta1}\varepsilon_\theta\varepsilon^{-1/3}\ell_\theta^{2/3} - C_{K_\theta2}\varepsilon^{-2/3}\ell_\theta^{4/3}\frac{D\varepsilon_\theta}{Dt} \\ &+ C_{K_\theta3}\varepsilon_\theta\varepsilon^{-5/3}\ell_\theta^{4/3}\frac{D\varepsilon}{Dt} - C_{K_\theta4}\varepsilon_\theta\varepsilon^{-2/3}\ell_\theta^{1/3}\frac{D\ell_\theta}{Dt} \\ &+ C_{K_\theta5}\varepsilon_\theta\varepsilon^{-1}\ell_\theta^2\alpha\mathbf{g}\cdot\nabla\Theta, \end{aligned} \tag{6.317}$$

where $C_{Kn}(n = 1\text{-}3)$ and $C_{K_\theta n}(n = 1\text{-}4)$ are given by Eqs. (6.113) and (6.225), respectively, and

$$C_{K4} = 0.0139, \tag{6.318a}$$

$$C_{\nu_{G\Theta}\ell} = 0.00972, \tag{6.318b}$$

$$C_{\lambda_G\ell1} = 0.0426,\ C_{\lambda_G\ell2} = 0.0155,\ C_{\lambda_G\ell3} = 0.00896,\ C_{\lambda_G\ell4} = 0.0645, \tag{6.318c}$$

$$C_{\lambda_{GU}\ell} = 0.00597, \tag{6.318d}$$

$$C_{K_\theta5} = 0.0153. \tag{6.318e}$$

In the passive-scalar modeling, we chose K_θ and ε_θ as well as K and ε, as the fundamental turbulence quantities. In the present case, we also adopt these four and perform the modeling of buoyancy effects. For this purpose, we solve Eqs. (6.313) and (6.317) concerning ℓ_C and ℓ_θ, respectively, and have

$$\ell_C = C_{\ell 1} K^{3/2} \varepsilon^{-1} + C_{\ell 2} K^{3/2} \varepsilon^{-2} \frac{DK}{Dt} - C_{\ell 3} K^{5/2} \varepsilon^{-3} \frac{D\varepsilon}{Dt} - C_{\ell 4} K^{7/2} \varepsilon^{-3} \alpha \mathbf{g} \cdot \nabla \Theta, \tag{6.319}$$

$$\begin{aligned} \ell_\theta =& C_{\ell_\theta 1} K_\theta^{3/2} \varepsilon_\theta^{-3/2} \varepsilon^{1/2} + C_{\ell_\theta 2} K_\theta^{3/2} \varepsilon_\theta^{-5/2} \varepsilon^{1/2} \frac{DK_\theta}{Dt} \\ &- C_{\ell_\theta 3} K_\theta^{5/2} \varepsilon_\theta^{-7/2} \varepsilon^{1/2} \frac{D\varepsilon_\theta}{Dt} + C_{\ell_\theta 4} K_\theta^{5/2} \varepsilon_\theta^{-5/2} \varepsilon^{-1/2} \frac{D\varepsilon}{Dt} \\ &- C_{\ell_\theta 5} K_\theta^{7/2} \varepsilon_\theta^{-7/2} \varepsilon^{1/2} \alpha \mathbf{g} \cdot \nabla \Theta, \end{aligned} \tag{6.320}$$

where $C_{\ell n} (n = 1\text{-}3)$ and $C_{\ell_\theta n} (n = 1\text{-}4)$ are given by Eqs. (6.117) and (6.227), respectively, and

$$C_{\ell 4} = 0.131, \tag{6.321a}$$

$$C_{\ell_\theta 5} = 2.13. \tag{6.321b}$$

We substitute Eqs. (6.319) and (6.320) into Eqs. (6.314)-(6.316). Here we should note that the present calculation is up to $O(\alpha\delta)$. Then we have

$$\nu_{G\Theta} = C_{\nu_{G\Theta}} K^3 \varepsilon^{-2}, \tag{6.322}$$

$$\begin{aligned} \lambda_G =& C_{\lambda_G 1} K_\theta^2 \varepsilon_\theta^{-1} - C_{\lambda_G 2} K_\theta^2 \varepsilon_\theta^{-2} \frac{DK_\theta}{Dt} + C_{\lambda_G 3} K_\theta^3 \varepsilon_\theta^{-3} \frac{D\varepsilon_\theta}{Dt} \\ &- C_{\lambda_G 4} K_\theta^3 \varepsilon_\theta^{-2} \varepsilon^{-1} \frac{D\varepsilon}{Dt}, \end{aligned} \tag{6.323}$$

$$\lambda_{GU} = C_{\lambda_{GU}} K_\theta^3 \varepsilon_\theta^{-2}, \tag{6.324}$$

where

$$C_{\nu_{G\Theta}} = 0.0329, \tag{6.325a}$$

$$C_{\lambda_G 1} = 0.320, \ C_{\lambda_G 2} = 1.18, \ C_{\lambda_G 3} = 1.02, \ C_{\lambda_G 4} = 0.245, \tag{6.325b}$$

$$C_{\lambda_{GU}} = 0.243. \tag{6.325c}$$

B. Transport Equations for Turbulence Quantities

We consider the transport equations for the four fundamental turbulence quantities, K, ε, K_θ and ε_θ. The equation for K is

$$\frac{DK}{Dt} = P_K - \varepsilon + \nabla \cdot \mathbf{T}_K + B, \tag{6.326}$$

where B expresses the buoyancy effect and is given by

$$B = -\alpha \mathbf{g} \cdot \mathbf{H}_\theta, \tag{6.327}$$

whereas K_θ obeys Eq. (2.126). Effects of buoyancy occur explicitly as B in Eq. (6.326), compared with Eq. (2.111). Buoyancy effects on P_K and P_θ may be expressed through R_{ij} and $\mathbf{H}_\theta$. Under the approximation of neglecting the buoyancy effects on the transport rates $\mathbf{T}_K$ and $\mathbf{T}_\theta$, they are modeled as Eqs. (6.176) and (6.234), respectively (most simply, their first terms are retained). This approximation is plausible from the viewpoint of paying more attention to the low-order statistical quantities strongly affected by buoyancy effects.

In the absence of buoyancy effects, the model transport equations for ε and ε_θ are derived as Eqs. (6.185) and (6.230), respectively, on the basis of the TSDIA results combined with the principle of model transferability (Yoshizawa 1987). In the presence of buoyancy effects, we also require the same principle; namely, we impose the algebraic relationship on Eqs. (6.319) and (6.320), and find the effects of buoyancy on these equations (Okamoto 1996a). First, the counterpart of Eq. (6.185) is given by

$$\frac{D\varepsilon}{Dt} = C_{\varepsilon 1R}^{TS} \frac{\varepsilon}{K} P_K - C_{\varepsilon 2}^{TS} \frac{\varepsilon^2}{K} + C_{\varepsilon 3}^{TS} \frac{\varepsilon}{K} B - C_{\varepsilon 4}^{TS} K \alpha \mathbf{g} \cdot \nabla \Theta + D_\varepsilon, \tag{6.328}$$

where

$$C_{\varepsilon 3}^{TS} = \frac{C_{\ell 2}}{C_{\ell 3}} = 1.71, \tag{6.329a}$$

$$C_{\varepsilon 4}^{TS} = \frac{C_{\ell 4}}{C_{\ell 3}} = 0.0514, \tag{6.329b}$$

and D_ε is the diffusion-like term referred to in Sec. 6.6.2 [we should note that the model constant in the first term in Eq. (6.328) contains the contribution of $O(\delta^2)$]. The fourth term may be approximated as

$$C_{\varepsilon 4}^{TS} K \alpha \mathbf{g} \cdot \nabla \Theta \cong \frac{C_{\varepsilon 4}^{TS}}{C_\lambda} \frac{\varepsilon}{K} B, \tag{6.330}$$

using the leading term in Eq. (6.309), that is, Eqs. (6.203) and (6.207) in the nonbuoyant case. From Eq. (6.330), Eq. (6.328) is reduced to

$$\frac{D\varepsilon}{Dt} = C_{\varepsilon 1R}^{TS} \frac{\varepsilon}{K} P_K - C_{\varepsilon 2}^{TS} \frac{\varepsilon^2}{K} + C_{\varepsilon 3R}^{TS} \frac{\varepsilon}{K} B + D_\varepsilon, \tag{6.331}$$

where

$$C^{TS}_{\varepsilon 3R} = C^{TS}_{\varepsilon 3} - \frac{C^{TS}_{\varepsilon 4}}{C_\lambda} = 1.33. \tag{6.332}$$

Entirely similarly, we use Eq. (6.320), and have the buoyancy counterpart of Eq. (6.230) as

$$\begin{aligned}\frac{D\varepsilon_\theta}{Dt} &= C_{\varepsilon_\theta 1}\frac{\varepsilon_\theta}{K_\theta}P_\theta + C_{\varepsilon_\theta 2}\frac{\varepsilon_\theta}{K}P_K - C_{\varepsilon_\theta 3}\frac{\varepsilon_\theta^2}{K_\theta} - C_{\varepsilon_\theta 4}\frac{\varepsilon_\theta \varepsilon}{K_\theta} \\ &+ C_{\varepsilon_\theta 5}\frac{\varepsilon_\theta}{K}B - C_{\varepsilon_\theta 6}\frac{K_\theta \varepsilon}{K^2}B,\end{aligned} \tag{6.333}$$

where

$$C_{\varepsilon_\theta 5} = \frac{C_{\ell_\theta 4}C^{TS}_{\varepsilon 3R}}{C_{\ell_\theta 3}} = 0.403, \tag{6.334a}$$

$$C_{\varepsilon_\theta 6} = \frac{C_{\ell_\theta 5}}{C_{\ell_\theta 3}C_\lambda} = 1.30. \tag{6.334b}$$

Here we have dropped the diffusion-like terms.

C. Indications of Buoyancy Effects

In the TSDIA analysis of buoyancy effects under the Boussinesq approximation, they occur twofold. One is the effects on the momentum and heat transfer, R_{ij} and $\mathbf{H}_\theta$, and another is those on the transport equations for the fundamental turbulence quantities.

In order to simply see the effects of buoyancy on R_{ij} and $\mathbf{H}_\theta$, we consider the flow generated by a heated vertical wall, where the x and y directions are taken along and normal to the wall, respectively. We neglect the variation in the x direction, and write

$$\mathbf{U} = (U, 0, 0)\,,\ \mathbf{g} = (-g, 0, 0)\,. \tag{6.335}$$

The shear-stress component of R_{ij}, R_{xy}, and the x and y components of $\mathbf{H}_\theta$, $H_{\theta x}$ and $H_{\theta y}$, are expressed as

$$R_{xy} = -\nu_T\frac{dU}{dy} - \alpha\nu_{G\Theta}g\frac{d\Theta}{dy}, \tag{6.336}$$

$$H_{\theta x} = \lambda_G \alpha g, \tag{6.337a}$$

$$H_{\theta y} = -\lambda_T\frac{d\Theta}{dy} - \frac{1}{6}\nu_{GU}\alpha g\frac{dU}{dy}. \tag{6.337b}$$

In the presence of a rising flow due to a heated wall, dU/dy is positive near the wall, whereas $d\Theta/dy$ is negative. From Eq. (6.336), the contribution of the buoyancy part to R_{xy} is opposite to the velocity-shear counterpart. This fact may be interpreted as follows. The fluid near the wall, which has the lower velocity than the adjacent fluid, is dragged by the latter in the x direction. On the other hand, the buoyancy effect is highest near the wall, and the fluid near it, in turn, drags the adjacent one, resulting in the decrease in the shear stress. The similar situation holds in Eq. (6.337b) for the heat flux in the y direction or the direction along the temperature gradient. From Eq. (6.337a), the buoyancy effect gives rise to the heat flux in the x direction along which there is no mean temperature gradient. This situation is similar to the case in which the mean velocity gradient exists in the direction of mean temperature gradient (this point is discussed in Sec. 4.6.1.B).

Finally, we refer to the Boussinesq approximation (2.96). An approximate equation more faithful to the assumption of small temperature change is Eq. (2.95). The difference between these two equations lies in the pressure-gradient term in the latter. The TSDIA analysis based on Eq. (2.95) was made by Shimomura (1997c), and the pressure-gradient effect on the scalar flux was derived. This finding was discussed in the context of a natural convection due to a vertical heated wall, and the effect was shown not to be negligible.

6.8. Modeling Based on Renormalized Length Scales

In the previous sections discussing the turbulence modeling from the TSDIA results, the characteristic velocity and scalar length scales are required to obey the algebraic relationship with turbulence quantities such as K, ε, K_θ, and ε_θ, that is, Eqs. (6.178) and (6.228). This requirement, which comes from the transferability of models, leads to the modeling of the equations for kinetic-energy and scalar-variance destruction rates, without resort to the analysis of their exact counterparts. This merit, however, is accompanied by some shortfalls associated with the range of applicability of nonlinear expressions for the Reynolds stress, the scalar flux, etc.

A second-order nonlinear model for the Reynolds stress R_{ij} is given by Eq. (6.122) or (6.133) in the TSDIA analysis. We focus attention on the S-S-related nonlinear term. The ratio of its magnitude to that of the linear or second term is characterized by χ_U [Eq. (4.142)]. Such a nonlinear expression is vulnerable to large χ_U or the strong mean velocity shear that is encountered locally in real-world turbulent flows, and numerical instability often occurs in the computational implementation. A method of alleviating this difficulty is to abandon the requirement of the algebraic relationship concerning the length scales and renormalize them. In what follows, we shall discuss it from the viewpoint of the renormalization of length scales.

In the TSDIA analysis up to $O(\delta^2)$, the turbulent energy K is expressed in terms of its dissipation rate ε and the characteristic length scale ℓ_C as Eq. (6.123). In the present renormalization of ℓ_C, we pay special attention to the S effect, and write

$$K = C_{K1}\varepsilon^{2/3}\ell_C^{2/3} + C_{K4}\ell_C^2 S_{ij}^2. \tag{6.338}$$

We first rewrite

$$K = \ell_C^{8/3}\left(C_{K1}\varepsilon^{2/3}\ell_C^{-2} + C_{K4}\ell_C^{-2/3}S_{ij}^2\right). \tag{6.339}$$

The physical meaning of this rerwriting will be clear in the context of the turbulent-viscosity representation. In the absence of the S effect, we have Eq. (6.178) for ℓ_C. Using it, we approximate the inside of the parenthesis in Eq. (6.339). As this result, the renormalized counterpart of Eq. (6.178) is given by

$$\ell_C = C_{\ell 1}^N K^{3/2}/\varepsilon\ , \tag{6.340}$$

with

$$C_{\ell 1}^N = \frac{C_{\ell 1}}{(1 + C_{\ell 1S}\chi_U^2)^{3/8}}, \tag{6.341}$$

where $C_{\ell 1}$ is given by Eq. (6.117), and

$$C_{\ell 1S} = 0.0214. \tag{6.342}$$

We substitute Eq. (6.340) into Eqs. (6.124) and (6.125). Then we have

$$\nu_T = C_\nu^N \frac{K^2}{\varepsilon}, \tag{6.343}$$

$$\eta_n = C_{\eta n}^N \frac{K^3}{\varepsilon^2},\ (n = 1\text{-}4)\,, \tag{6.344}$$

where

$$C_\nu^N = \frac{C_\nu}{(1 + C_{\ell 1S}\chi_U^2)^{1/2}}, \tag{6.345}$$

$$C_{\eta n}^N = \frac{C_{\eta n}}{(1 + C_{\ell 1S}\chi_U^2)^{3/4}}. \tag{6.346}$$

Here C_ν and $C_{\eta n}(n = 1\text{-}4)$ are given by Eq. (6.132). In obtaining Eq. (6.343), we dropped the D/Dt-related effects that were also neglected in Eq. (6.339). The coefficients in the form of Eq. (6.133) may be obtained using Eq. (6.344) and the first relation in each of Eqs. (6.134a)-(6.134c).

In the limit of vanishing χ_U [Eq. (4.142)], C_ν^N [Eq. (6.345)] tends to C_ν, which is given as 0.123 from the TSDIA [see Eq. (6.132a)]. The latter is larger than the widely-adopted one 0.09 that is optimized through the comparison with wall-flow data. From Eq. (4.150), χ_U is 4.7 in channel flow. Then we have $C_\nu^N = 0.10$; namely, C_ν^N effectively takes the value close to the optimized counterpart in channel flow. On the other hand, the correction (6.345), which is based on Eq. (6.339), makes R_{ij} based on the turbulent-viscosity approximation remain finite in the limit of large mean velocity strain or χ_U [recall the discussion associated with Eq. (4.221)]. For instance, the effect of χ_U on ν_T becomes important in homogeneous-shear turbulence subject to a strong shear rate. In the flow, the static model (6.118) fails in reproducing the proper temporal evolution of K and ε, as is explained in Sec. 4.3.3. Equation (6.343) can properly describe this property in the comparison with the DNS database of homogeneous-shear turbulence (Nisizima 1997).

Another feature of the model based on the renormalized length scale lies in the strain effect on η_n [Eq. (6.344)]. The usual model, Eq. (6.122) or (6.133), is not valid for large χ_U, as was stated above. A typical example of flows in which the nonlinear parts of R_{ij} play a critical role is a square-duct flow accompanied by secondary flows in the cross section. From the foregoing property concerning χ_U, Eq. (6.133) is likely to encounter numerical instability in the region where the mean velocity strain becomes locally large, and the careful treatment is needed. On the other hand, the present model is numerically robust owing the presence of the χ_U effect in the denominator of Eq. (6.346), and can give better results, compared with Eq. (6.133) (Nisizima 1997). Such a theoretical device has also been made in the nonlinear models by Taulbee (1992), Gatski and Speziale (1993), Abe et al. (1995), and Shih et al. (1997). The merit of the present modeling based on the renormalized length scale is that effects of χ_U can be incorporated into higher-order nonlinear terms automatically. For instance, the third-order nonlinear model (6.153) becomes much more vulnerable to large χ_U. In it, the coefficients are proportional to

$$K^4 \Big/ \varepsilon^3 . \tag{6.347}$$

In the original TSDIA representation based on ℓ_C and ε, Eq. (6.347) is expressed as

$$\varepsilon^{-1/3} \ell_C^{8/3} . \tag{6.348}$$

We substitute Eq. (6.340) into Eq. (6.348), and have

$$\varepsilon^{-1/3} \ell_C^{8/3} \propto \frac{1}{1 + C_{\ell 1S} \chi_U^2} \frac{K^4}{\varepsilon^3} . \tag{6.349}$$

As a result, the robustness of the third-order nonlinear model concerning large χ_U much increases, compared with the original one.

The modeling based on the renormalization of length scales may be applied straightforwardly to that of scalar diffusion. Moreover, other effects can be included in an entirely similar manner. It should be noted, however, that the renormalization procedure of a length scale such as the one from Eqs. (6.338) to (6.340) is not unique, but which procedure is best should be judged from the comparison with various DNS and observational results.

6.9. Implications in Subgrid-Scale Modeling

A detailed account of the mathematical basis of the subgrid-scale (SGS) modeling in large eddy simulation (LES) is given in Chapter 5. In LES, simple models such as the Smagorinsky one of the mixing-length type are adopted, while making full use of computer resources. In the classical or nondynamic SGS modeling, the Smagorinsky model, which is the prototype of all the SGS models, cannot describe the quantitative properties of some representative turbulent flows without changing the model constant (see Sec. 5.3.2.B). A method of alleviating this deficiency is to use a transport equation for the SGS energy and lift the constraint on the SGS energy production and dissipation processes, resulting in the incorporation of grid-scale (GS) velocity-strain effects into the SGS viscosity. In the dynamic SGS modeling, the use of transport equations is too heavy from the viewpoint of computational burden. As far as attention is focused on the velocity field, the simple linear turbulent-viscosity SGS model may be improved through the combination with the proper modeling of the near-component interaction represented by the cross term (this point is explained in Sec. 5.3.3.B).

In the other fields represented by the meteorological one, there are a number of effects such as thermal ones. In the case, it is not possible to model all of them in the form of the turbulent viscosity that is closely related to the enhanced energy dissipation mechanism. In reality, some of those effects often play a role opposite to the dissipation mechanism. In order to cope with this situation, it is necessary to model them in a form different from the turbulent-viscosity one. In such a case, the use of suggestions from theoretical methods is helpful. In what follows, we shall give a procedure for deriving SGS models by making use of the TSDIA results about ensemble-mean statistical quantities.

6.9.1. NONDYNAMIC SUBGRID-SCALE MODELING

Under the filtering procedure (5.1), the GS velocity $\overline{\mathbf{u}}$ in a nonbuoyant flow is governed by Eqs. (5.19)-(5.23). The Smagorinsky model given by Eqs. (5.36) and (5.49) is founded on the assumption of the balance between the SGS energy production and dissipation processes. We focus attention on the SGS Reynolds stress τ_{ij}, and consider lifting of the constraint. For this purpose, we use Eqs. (6.141) and (6.142) for the ensemble mean turbulent viscosity ν_T, and write

$$\nu_T = \frac{C_\nu}{1 + (C_{\nu S}/C_\nu)(K/\varepsilon)^2 \|S\|^2} \frac{K^2}{\varepsilon}, \tag{6.350}$$

where $\|S\| = \sqrt{(S_{ij})^2}$.

The formal procedure for changing ensemble-mean expressions into the SGS counterparts is to replace the characteristic length scale of the velocity, ℓ_C, with 2Δ, and at the same time an ensemble-mean quantity Λ (for instance, the turbulent energy K) is changed into the SGS counterpart Λ_S; namely, we write

$$\ell_C \to 2\Delta, \ \Lambda \to \Lambda_S. \tag{6.351}$$

Here we should note that the wavelength of the largest eddy in the interval Δ is 2Δ. We apply the replacement of ℓ_C with 2Δ to the leading part of Eq. (6.116), resulting in

$$\varepsilon_S = C_\varepsilon^S \Delta^{-1} K_S^{3/2}, \tag{6.352}$$

where

$$C_\varepsilon^S = C_{\ell 1}/2. \tag{6.353}$$

The application of Eqs. (6.351) and (6.352) to Eq. (6.350) leads to

$$\nu_S = C_{SM} \Delta K_S^{1/2}, \tag{6.354}$$

with

$$C_{SM} = \frac{C_\nu \Big/ C_\varepsilon^S}{1 + \left(C_{\nu S} \Big/ \left(C_\nu C_\varepsilon^{S2}\right)\right)(\Delta^2/K_S)\|\bar{s}_{ij}\|^2}. \tag{6.355}$$

Equation (6.354) is essentially the same as Eq. (5.66), and contains two independent constants. The SGS energy K_S obeys Eq. (5.45). This type of one-equation model was applied by Okamoto (1996b) to isotropic, mixing-layer, and channel flows and was confirmed to possess the ability to explain their primary properties quantitatively, as was noted in Sec. 5.3.3.C.

In Sec. 5.3.3.D, we mentioned an algebraic SGS model by Shimomura (1997a). In the foregoing one-equation type model, the SGS Reynolds stress is still expressed in the turbulent-viscosity form. Such an expression is not sufficient for dealing with effects of rotation, buoyancy, etc. In the algebraic modeling, such effects may be included much more efficiently. From the TSDIA viewpoint, a second-order model is obtained through the renormalization of a nonlinear model for the Reynolds stress. Afterwards, an algebraic model is derived by approximating the advection and diffusion terms, as is done in the conventional modeling (see Sec. 4.5.2). The SGS counterpart is derived through the application of Eqs. (6.351) and (6.352) to such an algebraic model.

6.9.2. DYNAMIC SUBGRID-SCALE MODELING

In the dynamic SGS modeling, the model constants in nondynamic models are changed into temporally- and spatially-varying nondimensional coefficients and are determined in a self-consistent manner in the course of LES. In dynamic LES, models using transport equations for SGS quantities are too heavy because of the increasing number of coefficients to be determined. In order to perform the LES based on the SGS modeling with no transport equations, it is necessary to properly incorporate additional effects such as the buoyancy ones into model expressions for the SGS Reynolds stress etc. As an example, we shall discuss the dynamic modeling of buoyancy effects.

As is detailed in Sec. 5.4, we introduce two kinds of filter functions, G_A and G_B, which are characterized by the filter width Δ_A and $\Delta_B (> \Delta_A)$, respectively. The application of G_A gives

$$\frac{\partial \overline{u}_i^A}{\partial t} + \frac{\partial}{\partial x_j} \overline{u}_i^A \overline{u}_j^A = -\frac{\partial \overline{p}^A}{\partial x_i} - \frac{\partial T_{ij}^A}{\partial x_j} + \nu \Delta \overline{u}_i^A - \alpha g_i \overline{\theta}^A, \tag{6.356}$$

$$\frac{\partial \overline{\theta}^A}{\partial t} + \frac{\partial}{\partial x_i} \overline{u}_i^A \overline{\theta}^A = -\frac{\partial T_{\theta i}^A}{\partial x_i} + \lambda \Delta \overline{\theta}^A. \tag{6.357}$$

Here T_{ij}^A is defined by Eq. (5.25a), whereas $\mathbf{T}_\theta^A$ is given by

$$T_\theta^A = \overline{\mathbf{u}\theta}^A - \overline{\mathbf{u}}^A \overline{\theta}^A, \tag{6.358}$$

which is decomposed as

$$\mathbf{T}_\theta^A = \mathbf{L}_\theta^A + \mathbf{C}_\theta^A + \boldsymbol{\tau}_\theta^A, \tag{6.359}$$

with

$$\mathbf{L}_\theta^A = \overline{\overline{\mathbf{u}}^A \overline{\theta}^A}^A - \overline{\mathbf{u}}^A \overline{\theta}^A, \tag{6.360}$$

$$\mathbf{C}_\theta^A = \overline{\overline{\mathbf{u}}^A \theta'}^A + \overline{\mathbf{u}' \overline{\theta}^A}^A, \tag{6.361}$$

$$\boldsymbol{\tau}_\theta^A = \overline{\mathbf{u}'_i \theta'}^A. \tag{6.362}$$

The equations for the double-filtered velocity and temperature $\overline{\mathbf{u}}^{AB}$ and $\overline{\theta}^{AB}$ may be written similarly, and the counterparts of T_{ij}^A and $\mathbf{T}_\theta^A$, T_{ij}^{AB} and $\mathbf{T}_\theta^{AB}$, occur. Germano's (1992) identities are given by Eq. (5.28) and

$$\mathbf{T}_\theta^{AB} - \overline{\mathbf{T}_\theta^A}^B = \overline{\overline{\mathbf{u}}^A \overline{\theta}^A}^B - \overline{\mathbf{u}}^{AB} \overline{\theta}^{AB}. \tag{6.363}$$

In what follows, we shall focus attention on the buoyancy effects on τ_{ij}^{A} and $\boldsymbol{\tau}_{\theta}^{A}$ in T_{ij}^{AB} and $\mathbf{T}_{\theta}^{AB}$, respectively. In order to convert the ensemble-mean expressions to the SGS version based on the filter G_A, we make the replacement

$$\ell_C,\ \ell_\theta \to 2\Delta_A;\ \Lambda \to \Lambda_S^A, \tag{6.364}$$

following expression (6.351). We apply Eq. (6.364) to each first term of Eqs. (6.313) and (6.317), and have

$$\varepsilon_S^A \propto {K_S^A}^{3/2} \Delta_A^{-1}, \tag{6.365}$$

$$\varepsilon_{\theta S}^A \propto K_{\theta S}^A {K_S^A}^{1/2} \Delta_A^{-1}. \tag{6.366}$$

We combine Eqs. (6.365) and (6.366) with the SGS version of Eqs. (6.322)-(6.324), and have

$$\left[T_{ij}^A\right]_D = -C_\nu^A \Delta_A \sqrt{K_S^A}\,\overline{\overline{s}}_{ij}^A + \alpha C_{G\overline{\theta}}^A \Delta_A^2 \left[g_i \frac{\partial \overline{\overline{\theta}}^A}{\partial x_j} + g_j \frac{\partial \overline{\overline{\theta}}^A}{\partial x_i}\right]_D, \tag{6.367}$$

$$\left[\mathbf{T}_\theta^A\right]_D = -C_\lambda^A \Delta_A \sqrt{K_S^A}\,\nabla \overline{\overline{\theta}}^A - C_{\lambda_G}^A \Delta_A \frac{K_\theta^A}{\sqrt{K_S^A}} \alpha \mathbf{g}$$
$$+\alpha C_{G\overline{\mathbf{u}}}^A \Delta_A^2 \frac{K_\theta^A}{K_S^A}\left(\left(\mathbf{g}\cdot\nabla\right)\overline{\overline{\mathbf{u}}}^A + \frac{1}{6}\nabla\left(\mathbf{g}\cdot\overline{\overline{\mathbf{u}}}^A\right)\right), \tag{6.368}$$

where only the first term of the SGS version of Eq. (6.323) has been retained, and C_ν^A etc. are the nondimensional coefficients to be determined. The reason why the double-filtered velocity and temperature, $\overline{\overline{\mathbf{u}}}^A$ and $\overline{\overline{\theta}}^A$, occur is explained in Sec. 5.4.2 and is also simply referred to in the following *Note*. Similarly, $[\tau_{ij}^{AB}]_D$ and $[\boldsymbol{\tau}_\theta^{AB}]_D$, which result from the consecutive application of two filters G_A and G_B in this order, may be found from the replacement

$$\overline{\overline{\mathbf{u}}}^A \to \overline{\overline{\mathbf{u}}}^{AB},\ \overline{\overline{\theta}}^A \to \overline{\overline{\theta}}^{AB},\ K_S^A \to K_S^{AB},\ K_\theta^A \to K_\theta^{AB}, \tag{6.369a}$$

$$C_\nu^A \to C_\nu^{AB},\ C_{G\overline{\theta}}^A \to C_{G\overline{\theta}}^{AB},\ C_\lambda^A \to C_\lambda^{AB},\ \text{etc.}, \tag{6.369b}$$

$$\Delta_A \to \Delta_B. \tag{6.369c}$$

In relation to Eq. (6.369c), we should note that $\Delta_B > \Delta_A$.

In the usual dynamic SGS modeling, the first term of each of Eqs. (6.367) and (6.368) is written in the form of a mixing-length expression such as Eq. (5.75). The expression corresponds to

$$\sqrt{K_S^A} \propto \Delta_A \left\|\overline{\overline{s}}_{ij}^A\right\|, \tag{6.370}$$

which arises from the assumption of the balance between the SGS energy production and dissipation processes, as is stated in Sec. 5.3.2.A. In the presence of various effects such as thermal ones, however, such a balance cannot be expected to hold so accurately.

A method of modeling K_S^A and K_θ^A without resort to the foregoing assumption is to use Eq. (5.13), which leads to the so-called Bardina's (1983) model in the application to the cross term C_{ij}^A. Using the method, a non-Smagorinsky-type SGS model is proposed in Sec. 5.4.2 (Yoshizawa et al. 1996). From Eq. (5.13), we have

$$K_S^A = \frac{1}{2}\overline{\mathbf{u}'^2}^A \cong \frac{1}{2}\left(\overline{\mathbf{u}}^A - \overline{\overline{\mathbf{u}}}^A\right)^2, \tag{6.371}$$

$$K_\theta^A = \overline{\theta'^2}^A \cong \left(\overline{\theta}^A - \overline{\overline{\theta}}^A\right)^2. \tag{6.372}$$

In the dynamic determination of the nondimensional coefficients C_ν^A etc., we assume that they do not change so much in the two filtering based on G_A and G_B; namely, we write

$$C_\nu^A \cong C_\nu^{AB} \equiv C_\nu^D, \tag{6.373}$$

and so on. In order to show the procedure for determining them, we consider Eq. (6.367) related to the GS velocity. We apply the so-called Germano's identity (5.28) to T_{ij}^A and T_{ij}^{AB}, and have

$$C_\nu^D M_{Sij} + C_{G\overline{\theta}}^D M_{G\overline{\theta}ij} = [L_{ij}]_D, \tag{6.374}$$

where

$$M_{Sij} = \Delta_A \overline{\sqrt{K_S^A}\overline{\overline{s}}_{ij}^A}^B - \Delta_B \sqrt{K_S^{AB}}\overline{\overline{s}}_{ij}^{AB}, \tag{6.375}$$

$$M_{G\overline{\theta}ij} = \alpha\left(\Delta_B^2\left[g_i\frac{\partial\overline{\overline{\theta}}^{AB}}{\partial x_j} + g_j\frac{\partial\overline{\overline{\theta}}^{AB}}{\partial x_i}\right]_D - \Delta_A^2\overline{\left[g_i\frac{\partial\overline{\overline{\theta}}^A}{\partial x_j} + g_j\frac{\partial\overline{\overline{\theta}}^A}{\partial x_i}\right]_D}^B\right), \tag{6.376}$$

$$[L_{ij}]_D = \left[\overline{\overline{u}_i^A\overline{u}_B^A}^B - \overline{u}_i^{AB}\overline{u}_j^{AB}\right]_D. \tag{6.377}$$

Equation (6.374) is not mathematically well-posed since the number of coefficients is less than that of equations. In order to alleviate this difficulty, Lilly (1992) proposed an ingenious approach using the least-square method. Following the method, we define

$$Q = \left(C_\nu^D M_{Sij} + C_{G\overline{\theta}}^D M_{G\overline{\theta}ij} - [L_{ij}]_D\right)^2. \tag{6.378}$$

The coefficients C_ν^D and $C_{G\overline{\theta}}^D$ may be determined from the requirement

$$\frac{\partial Q}{\partial C_\nu^D} = \frac{\partial Q}{\partial C_{G\overline{\theta}}^D} = 0, \tag{6.379}$$

which leads to a simple linear system of equations

$$C_\nu^D M_{Sij}^2 + C_{G\overline{\theta}}^D M_{G\overline{\theta}ij} M_{Sij} = \left[L_{ij}\right]_D M_{Sij}, \tag{6.380a}$$

$$C_\nu^D M_{Sij} M_{G\overline{\theta}ij} + C_{G\overline{\theta}}^D M_{G\overline{\theta}ij}^2 = \left[L_{ij}\right]_D M_{G\overline{\theta}ij}. \tag{6.380b}$$

Solving this system, we have C_ν^D and $C_{G\overline{\theta}}^D$.

The nondimensional coefficients in Eq. (6.368) may be determined dynamically using the same procedure.

Note: The occurrence of the second term of Eq. (6.367) may be understood more simply as follows. We first consider the equation for the SGS velocity $\mathbf{u}'$ defined as

$$\mathbf{u}' = \mathbf{u} - \overline{\mathbf{u}}^A. \tag{6.381}$$

We pay attention to the buoyancy-related term in the equation. The contribution of the term to $\mathbf{u}'$, $\mathbf{u}'_B$, is given by

$$\mathbf{u}'_B = -\alpha T^A \mathbf{g}\theta', \tag{6.382}$$

where T^A is a proper characteristic time scale. Equation (6.382) contributes to $\overline{u'_i u'_j}^A$ as

$$\left[\overline{u'_i u'_j}^A\right]_B = -\alpha T^A \left[g_i \overline{u'_j \theta'} + g_j \overline{u'_i \theta'}\right]_D. \tag{6.383}$$

Using the first term of Eq. (6.368), we approximate Eq. (6.383). Since the time scale of SGS fluctuations is characterized by $\Delta_A/\sqrt{K_S^A}$, we finally reach the second part of Eq. (6.367).

References

Abe, K., Kondoh, T., and Nagano, Y. (1997), Intern. J. Heat and Fluid Flow **18**, 266.

André, J. C. and Lesieur, M. (1977), J. Fluid Mech. **81**, 187.

Bardina, J. (1983), Ph. D. Dissertation, Stanford University.

Bardina, J., Ferziger, J. H., and Rogallo, R. S. (1985), J. Fluid Mech. **154**, 321.

Duranti, S. and Pittaluga, F. (1997), private communication.

Gatski, T. B. and Speziale, C. G. (1993), J. Fluid Mech. **254**, 59.
Germano, M. (1992), J. Fluid Mech. **238**, 325.
Hamba, F. (1987a), J. Phys. Soc. Jpn. **56**, 79.
Hamba, F. (1987b), J. Phys. Soc. Jpn. **56**, 3771.
Horiuti, K. (1990), Phys. Fluids A **2**, 1708.
Kitoh, O. (1991), J. Fluid Mech. **225**, 445.
Kobayashi, T. and Togashi, S. (1996), JSME Intern. J. B **39**, 453.
Kraichnan, R. H. (1964), Phys. Fluids **7**, 1048.
Kraichnan, R. H. (1987), Phys. Fluids **30**, 2400.
Launder, B. E., Reece, G. J., and Rodi, W. (1975), J. Fluid Mech. **68**, 537.
Leslie, D. C. (1973), *Developments in the Theory of Turbulence* (Clarendon, Oxford).
Lilly, D. K. (1992), Phys. Fluids A **4**, 633.
Nagano, Y. and Kim, C. (1987), Trans. JSME B **53**, 1773.
Nayfeh, A. (1973), *Perturbation Methods* (Wiley, New York).
Nisizima, S. (1990), Theor. Comput. Fluid Dyn. **2**, 61.
Nisizima, S. (1993), Trans. JSME B **59**, 3032.
Nisizima, S. (1997), private communication.
Nisizima, S. and Yokoi, N. (1992), Trans. JSME B **58**, 2714.
Okamoto, M. (1994), J. Phys. Soc. Jpn. **63**, 2102.
Okamoto, M. (1995), J. Phys. Soc. Jpn. **64**, 2854.
Okamoto, M. (1996a), J. Phys. Soc. Jpn. **65**, 2044.
Okamoto, M. (1996b), in *Proceedings of Tenth Symposium on Computational Fluid Dynamics*, Japan Society of Computational Fluid Dynamics, p. 164.
Rodi, W. (1976), Z. angew. Math. and Mech. **56**, 219.
Rubinstein, R. (1996), Theor. Comput. Fluid Dyn. **8**, 377.
Rubinstein, R. and Barton, J. M. (1990), Phys. Fluids A **2**, 1472.
Rubinstein, R. and Barton, J. M. (1991), Phys. Fluids A **3**, 415.
Rubinstein, R. and Barton, J. M. (1992), Phys. Fluids A **4**, 1759.
Shih, T. -H., Zhu, J., Liou, W., Chen, K. -H., Liu, N. -S., and Lumley, J. L. (1997), NASA TM 113112.
Shimomura, Y. (1997a), in *Eleventh Symposium on Turbulent Shear Flows*, 3-34-1.
Shimomura, Y. (1997b), private communication (A theoretical study of the turbulent diffusion in incompressible shear flows and in passive scalars, submitted to Phys. Fluids).
Shimomura, Y. (1997c), private communication.
Speziale, C. G. (1987), J. Fluid Mech. **178**, 459.
Speziale, C. G. (1997), Intern. J. Nonlinear Mech. **33**, 579.
Taulbee, D. (1992), Phys. Fluids A **4**, 2555.

Tavouralis, S. and Corrsin, S. (1981), J. Fluid Mech. **104**, 311.
Tavouralis, S. and Corrsin, S. (1985), J. Heat Mass Transfer **28**, 265.
Weinstock, J. (1981), J. Fluid Mech. **105**, 369.
Weinstock, J. (1982), J. Fluid Mech. **116**, 1.
Yakhot, V. M. and Orszag, S. A. (1986), J. Sci. Comput. **1**, 3.
Yokoi, N. and Yoshizawa, A. (1993), Phys. Fluids A **5**, 464.
Yoshizawa, A. (1982), J. Phys. Soc. Jpn. **51**, 2326.
Yoshizawa, A. (1983), J. Phys. Soc. Jpn. **52**, 1194.
Yoshizawa, A. (1984a), Phys. Fluids **27**, 1377.
Yoshizawa, A. (1984b), J. Phys. Soc. Jpn. **53**, 1264.
Yoshizawa, A. (1985), Phys. Fluids **28**, 3226.
Yoshizawa, A. (1987), Phys. Fluids **30**, 628.
Yoshizawa, A. (1988), J. Fluid Mech. **195**, 541.
Yoshizawa, A. (1993a), Phys. Fluids A **5**, 707.
Yoshizawa, A. (1993b), Phys. Rev. E **48**, 273.
Yoshizawa, A. (1994), Phys. Rev. E **49**, 4065.
Yoshizawa, A. and Nisizima, S. (1993), Phys. Fluids A **5**, 3302 .
Yoshizawa, A. and Yokoi, N. (1991), J. Phys. Soc. Jpn. **60**, 2500.
Yoshizawa, A., Tsubokura, M., Kobayashi, T., and Taniguchi, N. (1996), Phys. Fluids **8**, 2254.

CHAPTER 7

MARKOVIANIZED ONE-POINT APPROACH

7.1. Necessity of Markovianized One-Point Approach

In Chapter 6, we presented the two-scale direct-interaction approximation (TSDIA) formalism for turbulent shear flow and discussed various kinds of turbulence models. In this course, some theoretical bases were given to the models that had been proposed by phenomenological approaches. A feature of the TSDIA in the derivation of turbulence models is the ability to estimate model constants. This ability is closely associated with the use of Kolmogorov's inertial-range law.

In complex turbulent shear flows such as compressible and magnetohydrodynamic ones, no analytical properties comparable to the Kolmogorov law have been yet established even for homogeneous turbulence. The situation has been already encountered in the study of frame-rotation or mean-vorticity effects on the Reynolds stress in Sec. 6.7.2. These effects are connected with the inhomogeneity of the turbulent helicity, as in Eq. (6.262). No definite information on the helicity spectrum, however, has been found in the current stage of the study on the helicity. The lack of the spectral information leads to the inability to estimate the model constant in the one-point expression (6.274), C_H.

In compressible and magnetohydrodynamic flows, the TSDIA analysis becomes quite tedious, and the number of undetermined constants in reduced one-point expressions increases. In these cases, other theoretical methods of constructing turbulence models while avoiding the mathematical complexity of the original TSDIA are necessary. One candidate for them is a two-scale method combined with a Markovianization procedure (Yoshizawa 1995), which will be called the Markovianized two-scale (MTS) method hereafter. Its feature lies in the point that all the mathematical manipulations are made in physical space. It becomes a useful tool in the calculation of correlation functions whose main properties comes from energy-containing or low-wavenumber eddies. This chapter is devoted to the explanation of the MTS method, which will be used in Chapters 8 and 9.

7.2. Markovianized Two-Scale Method

In order to see the key steps in the MTS method, we consider the case of constant-density turbulent shear flow, which is investigated in detail using the TSDIA in Chapter 6. In the TSDIA, we start from Eq. (6.30) for $\mathbf{u}'$, and introduce the Fourier representation (6.33). A great merit of the latter procedure is the simple elimination of the pressure p', leading to Eqs. (6.46) and (6.61). In physical space, p' obeys the equation of the Poisson type, (4.11), using whose solution (4.14), p' can be eliminated formally. The resulting system is a complicated differential-integral equation concerning time and space, and it is difficult to proceed further on its basis. This difficulty comes from the solenoidal condition that makes p' a long-distance force. A typical instance of flows to which the straightforward application of the TSDIA is difficult is compressible turbulent shear flow. In this case, the pressure is locally expressed using the density and the internal energy (temperature). As a result, the foregoing difficulty associated with long-distance effects is not so serious there. In the MTS method, we pay much less attention to the long-distance effects of pressure and treat the direct interaction between energy-containing eddies and the mean field.

7.2.1. SCALE-PARAMETER EXPANSION

We follow the TSDIA, and introduce two space and time scales, expression (6.27). The fluctuation of the velocity, $\mathbf{u}'$, in the solenoidal fluid motion obeys Eq. (6.30). In the MTS method, we perform the scale-parameter expansion in physical space, without proceeding to the wavenumber representation; namely, we write

$$\mathbf{u}'(\boldsymbol{\xi}, \mathbf{X}; \tau, T) = \sum_{n=0}^{\infty} \delta^n \mathbf{u}'_n(\boldsymbol{\xi}, \mathbf{X}; \tau, T), \tag{7.1a}$$

$$p'(\boldsymbol{\xi}, \mathbf{X}; \tau, T) = \sum_{n=0}^{\infty} \delta^n p'_n(\boldsymbol{\xi}, \mathbf{X}; \tau, T). \tag{7.1b}$$

We substitute Eq. (7.1) into Eq. (6.30). Then the $O(1)$ part of $\mathbf{u}'$, $\mathbf{u}'_0$, obeys

$$\frac{Du'_{0i}}{D\tau} + \frac{\partial}{\partial \xi_j} u'_{0j} u'_{0i} + \frac{\partial p'_0}{\partial \xi_i} - \nu \Delta_\xi u'_{0i} = 0, \tag{7.2}$$

where

$$\frac{D}{D\tau} = \frac{\partial}{\partial \tau} + \mathbf{U} \cdot \nabla_\xi. \tag{7.3}$$

The $O(\delta)$ counterpart of Eq. (7.2) is

$$L_u u'_{1i} = I_{ui} \equiv -u'_{0j}\frac{\partial U_i}{\partial X_j} - \frac{Du'_{0i}}{DT} - \frac{\partial p'_0}{\partial X_i} - \frac{\partial}{\partial X_j}\left(u'_{0j}u'_{0i} - \left\langle u'_{0j}u'_{0i}\right\rangle\right) + 2\nu\frac{\partial u'_{0i}}{\partial X_j \partial \xi_j}, \tag{7.4}$$

where the differential operator L_u is defined by

$$L_u u'_{1i} = \frac{Du'_{1i}}{D\tau} + \frac{\partial}{\partial \xi_j}\left(u'_{0j}u'_{1i} + u'_{1i}u'_{0j}\right) + \frac{\partial p'_1}{\partial \xi_i} - \nu\Delta_\xi u'_{1i}. \tag{7.5}$$

Equation (7.4) may be integrated formally as

$$\mathbf{u}'_1 = L_u^{-1}\mathbf{I}_u. \tag{7.6}$$

7.2.2. MARKOVIANIZATION BASED ON CHARACTERISTIC TIME SCALES

A. Markovianization in Isotropic Turbulence
In the formal solution (7.6), L_u^{-1} expresses the spatial and temporal integration using a Green's function, which corresponds to Eqs. (6.63) and (6.64) in the wavenumber space. For deriving turbulence models applicable to real-world turbulent flows with the aid of this solution, we need to reduce Eq. (7.6) to a one-point expression under some approximations.

The representative Markovianization method is the EDQNM (eddy-damped quasi-normal Markovianized) approximation in isotropic turbulence (Lesieur 1997) that is referred to in Sec. 3.5.5.B. In order to perform the Markovianization of Eq. (7.6), we briefly recall the key steps of the EDQNM method. We denote the Fourier representation of $\mathbf{u'}_0(\boldsymbol{\xi};\tau)$ by $\mathbf{u'}_0(\mathbf{k};\tau)$ (the dependence on $\mathbf{X}$ and T disappears in the homogeneous case), and the Green's function for $\mathbf{u'}_0(\mathbf{k};\tau)$ by $G_{ij}(\mathbf{k};\tau,\tau')$ [see Eq. (3.100) or (6.50)]. One of the important statistical quantities in isotropic turbulence is the two-time velocity covariance $Q(k;\tau,\tau')$, which is defined by

$$\left\langle u'_{0i}(\mathbf{k};\tau)u'_{0j}(\mathbf{k}';\tau')\right\rangle = D_{ij}(\mathbf{k})Q(k;\tau,\tau')\delta\left(\mathbf{k}+\mathbf{k}'\right), \tag{7.7}$$

with Eq. (3.25) for $D_{ij}(\mathbf{k})$. The one-time part of $Q(k;\tau,\tau')$, $Q(k;\tau,\tau)$, is related to the turbulent energy K_0 as

$$K_0 \equiv \left\langle \mathbf{u}'^2_0/2\right\rangle = \int Q(k;\tau)d\mathbf{k}, \tag{7.8}$$

with $Q(k;\tau) \equiv Q(k;\tau,\tau)$. Under the DIA (Kraichnan 1959; Leslie 1973), the equation for $Q(k;\tau)$ is written symbolically as

$$\frac{\partial Q(k;\tau)}{\partial \tau} + 2\nu k^2 Q(k;\tau) = F\left\{Q(k;\tau,\tau'), G(k;\tau,\tau')\right\}, \tag{7.9}$$

where $F\{A,B\}$ denotes a functional of A and B, and $G(k;\tau,\tau')$ is the counterpart of Eq. (7.7) that is given by

$$\left\langle G'_{ij}\left(\mathbf{k};\tau,\tau'\right)\right\rangle = D_{ij}\left(\mathbf{k}\right) G\left(k;\tau,\tau'\right), \tag{7.10}$$

and obeys the equation similar to Eq. (7.9).

Equation (7.9) shows that the two-time quantities, $Q(k;\tau,\tau')$ and $G(k;\tau,\tau')$, are still indispensable for knowing the one-time quantity $Q(k;\tau)$. In order to simplify Eq. (7.9) by reducing it to a one-time equation based on $Q(k;\tau)$, we first approximate

$$Q(k;\tau,\tau') = G(k;\tau,\tau')Q(k;\tau), \tag{7.11}$$

resulting in

$$\frac{\partial Q(k;\tau)}{\partial \tau} + 2\nu k^2 Q(k;\tau) = F\left\{Q(k;\tau), G(k;\tau,\tau')\right\}. \tag{7.12}$$

Next, we abandon the equation for $G(k;\tau,\tau')$ and introduce a model for the characteristic time scale $\tau_G(k)$ that is defined by

$$\tau_G(k) = \int_{-\infty}^{\tau} G(k;\tau,\tau')d\tau'. \tag{7.13}$$

This procedure is equivalent to the approximation

$$G(k;\tau,\tau') = H(\tau-\tau')\exp\left(-\frac{\tau-\tau'}{\tau_G(k)}\right) \tag{7.14}$$

[$H(\tau)$ is the unit step function]. Finally, we make a model for $\tau_G(k)$.

The choice of $\tau_G(k)$ is not unique, and the resulting expression inevitably contains adjustable parameters. In this sense, the EDQNM method is far from complete, compared with more elaborate theories such as the DIA in a Lagrangian framework (Kraichnan 1965; McComb 1990), as is stated Sec. 3.5.5. The merit of this method, however, lies in its simplicity, and it is applicable to flows much more complicated than isotropic turbulence, under the proper modeling of $\tau_G(k)$ (Cambon et al. 1981).

B. Introduction of Characteristic Time Scales
Considering that L_u^{-1} expresses the integration based on a Green's function G_u, we write Eq. (7.6) symbolically as

$$L_u^{-1}\mathbf{I}_u = \int_{-\infty}^{\tau} G_u(\tau, s)\mathbf{I}_u(s)ds, \tag{7.15}$$

where attention is focused on the time integral in L_u^{-1}. In order to construct a turbulence model applicable to real-world flows, we need to reduce Eq. (7.15) to a one-time representation. Considering that the contribution of $\mathbf{I}_u(s)$ near the latest time τ is most dominant, we make the Markovianization

$$L_u^{-1}\mathbf{I}_u = \tau_T \mathbf{I}_u(\tau), \tag{7.16}$$

where τ_T is the characteristic time scale given by

$$\tau_T = \int_{-\infty}^{\tau} G_u(\tau, s)ds. \tag{7.17}$$

Then Eq. (7.6) may be written simply as

$$\mathbf{u'}_1 = \tau_T \mathbf{I}_u. \tag{7.18}$$

Whether the MTS method works well or not is heavily dependent on the modeling of τ_T, as is entirely similar to the EDQNM method. In the TSDIA, the two-time properties of the inertial range, Eqs. (6.53)-(6.56), play a key role in the derivation of the one-point expressions for the Reynolds stress, the scalar flux, etc. and in the estimate of model constants. In complex flows encountered in engineering and sciences, such detailed information is usually not available. In the MTS method, we make full use of some limited information on the energy spectrum and estimate the characteristic time scale of turbulence.

We take the solenoidal turbulent motion as an example and discuss the modeling of the Reynolds stress. In the scale expansion (7.1), the statistical property of the $O(1)$ field $\mathbf{u'}_0$ at high Reynolds number is typically characterized by the inertial-range energy spectrum and time scale

$$E_0(k) \propto \varepsilon_0^{2/3} k^{-5/3}, \tag{7.19}$$

$$\tau(k) \propto \varepsilon_0^{-1/3} k^{-2/3}. \tag{7.20}$$

Of a wide range of spatial scales included in turbulent motion, the dominant contributions to the Reynolds stress, the scalar flux, etc. come from the so-called energy-containing eddies, whose scale is denoted by ℓ_C. Once ℓ_C is known, the corresponding time scale may be estimated from Eq. (7.20) with

$2\pi/\ell_C$ as k. In what follows, we shall estimate this time scale from $E_0(k)$ only, without resort to Eq. (7.20). This is a key step of the MTS method in applying it to more complex flows.

From the definition of ℓ_C, the turbulent energy of $\mathbf{u}'_0$, K_0, may be approximated as

$$K_0 = \int_{2\pi/\ell_C}^{\infty} E_0(k) dk. \tag{7.21}$$

Equation (7.19) is combined with Eq. (7.21) to lead to

$$K_0 \propto \varepsilon_0^{2/3} \ell_C^{2/3}, \tag{7.22}$$

which gives

$$\ell_C \propto K_0^{3/2} / \varepsilon_0 . \tag{7.23}$$

Therefore the time scale intrinsic to energy-containing eddies, τ_C, is

$$\tau_C = \ell_C \Big/ \sqrt{K_0} \propto K_0 / \varepsilon_0 . \tag{7.24}$$

As far as we are concerned with the energy-containing part of $\mathbf{u}'_0$, it is proper to adopt Eq. (7.24) as τ_T in Eq. (7.18); namely, we put

$$\tau_T = C_\tau K_0 / \varepsilon_0 . \tag{7.25}$$

Here the constant C_τ is unknown at this stage.

Equation (7.25) may be obtained by substituting Eq. (7.23) into Eq. (7.20) with $k = 2\pi/\ell_C$. This fact indicates that the characteristic time scale of energy-containing eddies may be estimated directly from the energy spectrum, without the help of the time-scale information in the wavenumber space such as Eq. (7.20). This point is not so important in simple solenoidal flows, but it becomes critical in case that no definite knowledge on the time scale has been yet established, as in compressible turbulence.

7.2.3. DERIVATION OF NONLINEAR MODEL FOR REYNOLDS STRESS

The Reynolds stress R_{ij} is expanded as

$$R_{ij} \equiv \left\langle u'_i u'_j \right\rangle = \left\langle u'_{0i} u'_{0j} \right\rangle + \delta \left(\left\langle u'_{0i} u'_{1j} \right\rangle + \left\langle u'_{1i} u'_{0j} \right\rangle \right) + O\left(\delta^2\right). \tag{7.26}$$

We substitute Eq. (7.18) into the $O(\delta)$ terms in Eq. (7.26), and use the characteristic time scale, Eq. (7.25). Here we focus attention on the interaction with the mean velocity field $\mathbf{U}$. Then we have

$$R_{ij} = \frac{2}{3} K \delta_{ij} - \tau_T \left[\langle u'_{0i} u'_{0\ell} \rangle \frac{\partial U_j}{\partial x_\ell} + \left\langle u'_{0j} u'_{0\ell} \right\rangle \frac{\partial U_i}{\partial x_\ell} \right]_D , \tag{7.27}$$

where subscript D means the deviatoric part of a second-rank tensor and is defined by Eq. (4.97), and the scale parameter δ has disappeared through the replacement $\mathbf{X} \to \delta\mathbf{x}$, as in the TSDIA.

We assume that the $O(1)$ statistics are isotropic; namely, we write

$$\left\langle u'_{0i} u'_{0j} \right\rangle = (2/3)\, K_0 \delta_{ij}. \tag{7.28}$$

We apply Eq. (7.28) to Eq. (7.27), and express the latter using $\mathbf{U}$ and the $O(1)$ turbulence quantities such as K_0. Finally, we resort to the renormalization procedure and replace these turbulence quantities with their exact counterparts. As a result, we reach the linear turbulent-viscosity model

$$R_{ij} = \frac{2}{3} K \delta_{ij} - \nu_T S_{ij}, \tag{7.29}$$

with the turbulent viscosity

$$\nu_T = (2/3)\, K \tau_T = C_\nu K^2 / \varepsilon, \tag{7.30a}$$

$$C_\nu = (2/3)\, C_\tau, \tag{7.30b}$$

where the renormalization of τ_T has been done. The numerical factor C_ν is often chosen as 0.09, which leads to

$$C_\tau = 0.14. \tag{7.31}$$

This procedure of estimating an unknown constant is one of key steps of the MTS method not using Green's functions.

A straightforward method of obtaining the nonlinear corrections to Eq. (7.29) is to perform the $O(\delta^2)$ calculation in Eq. (7.26). Such an analysis, however, leads to a rather complicated mathematical manipulation, as is seen from Eq. (6.30). One method that is much simpler but may capture some of the $O(\delta^2)$ results is the iterative use of Eq. (7.27). For this purpose, we see Eq. (7.27) from a different viewpoint. Equations (7.2) and (7.4) may be regarded as the first iteration procedure for solving Eq. (6.30) with the right-hand side as a perturbational part. In the forgoing derivation of the linear model (7.29), we started from the isotropic $O(1)$ statistics, Eq. (7.28), and obtained the simplest anisotropic effects generated by the mean velocity gradient. Now we proceed to the second iteration procedure; namely, we substitute this first iteration result, Eq. (7.29), into Eq. (7.27). As a result, we have

$$\begin{aligned} R_{ij} &= \frac{2}{3} K \delta_{ij} - \nu_T S_{ij} + \tau_T \nu_T \left[S_{i\ell} S_{\ell j} \right]_D \\ &+ \frac{1}{2} \tau_T \nu_T \left(S_{i\ell} \Omega_{\ell j} + S_{j\ell} \Omega_{\ell i} \right). \end{aligned} \tag{7.32}$$

Equation (7.32) is the simplest of the so-called nonlinear models that are discussed in Secs. 4.3 and 6.4, and corresponds to Eq. (4.114) or (6.133). The coefficient of the nonlinear terms is

$$\tau_T \nu_T = C_\tau C_\nu \frac{K^3}{\varepsilon^2} = 0.012 \frac{K^3}{\varepsilon^2}. \tag{7.33}$$

The numerical factors in Eq. (7.32) are smaller than the counterparts of Eqs. (4.116) and (6.135). As its main reason, we may mention the lack of the pressure effect. In the MTS method, we focus attention on the effect of the mean velocity gradient, that is, the first term on the right-hand side of Eq. (6.30), and neglect the contribution of the pressure In the solenoidal motion, the mean velocity is connected directly with the pressure neglected here. With this reservation in mind, the merit of the MTS method lies in the fact that the primary mathematical structures of the one-point expressions for correlation functions may be derived without heavy mathematical manipulations.

In Eq. (7.4), the second term on the right-hand side contributes to Eq. (7.27) as

$$-\tau_T \left[\frac{D}{Dt} \left\langle u'_{0i} u'_{0j} \right\rangle \right]_D . \tag{7.34}$$

In the second iteration, Eq. (7.34) gives

$$\left(0.026 \frac{K^2}{\varepsilon^2} \frac{DK}{Dt} - 0.013 \frac{K^3}{\varepsilon^3} \frac{D\varepsilon}{Dt} \right) S_{ij} + 0.013 \frac{K^3}{\varepsilon^2} \frac{DS_{ij}}{Dt}. \tag{7.35}$$

Here the coefficient of S_{ij} should be compared with the second and third terms of Eq. (6.130), while the second part corresponds to the last of Eq. (6.133).

The entirely similar analysis can be performed for other correlation functions in turbulent shear flows, such as the turbulent scalar flux. The MTS method is a useful approach, at least, in case that two-point closure methods such as the TSDIA are difficult to apply because of mathematical complicatedness. It will be fully utilized in the turbulence modeling of compressible and magnetohydrodynamic flows in Chapters 8 and 9.

References

Cambon, C., Jeandel, D., and Mathieu, J. (1981), J. Fluid Mech. **104**, 247.

Kraichnan, R. H. (1959), J. Fluid Mech. **5**, 497.

Kraichnan, R. H. (1965), Phys. Fluids **8**, 575.

Lesieur, M. (1997), *Turbulence in Fluids* (Kluwer, Dordrecht).
Leslie, D. C. (1973), *Developments in the Theory of Turbulence* (Clarendon, Oxford).
McComb, M. D. (1990), *The Physics of Fluid Turbulence* (Clarendon, Oxford).
Yoshizawa, A. (1995), Phys. Fluids **7**, 3105.

CHAPTER 8

COMPRESSIBLE TURBULENCE MODELING

8.1. Features of Compressible Turbulence

In Chapters 3-7, the solenoidal motion of turbulent flow is studied exclusively, except buoyancy effects of weak density change through the Boussinesq approximation. In aeronautical, aerospace, and astrophysical fields, however, flows with high Mach number are observed frequently. Their prominent feature is the occurrence of a shock wave across which flow changes from the supersonic to subsonic state. There fluid is highly compressed, and the steep streamwise gradient of density generates large density fluctuations.

Distinct effects of fluid compressibility appear even in the absence of a shock-wave region. Such a typical example is a turbulent free-shear layer (see Fig. 8.1). In this flow, two parallel streams with different speeds start to merge behind a splitting plate. The width of the mixing region grows nearly linearly with the distance from the trailing edge of the plate. Compressibility effects on the flow are characterized by the Mach number of the difference between the two speeds, which is usually named the convective Mach number. A primary feature of this flow is the drastic decrease in the growth rate of the layer with the increasing convective Mach number. The similar flow situation may be observed in a turbulent jet; the jet growth rate is suppressed with the increasing Mach number based on the difference between the central jet velocity and that of the surrounding flow.

The quantitative observation of high-speed flows is difficult in general. A free-shear layer flow is the representative one whose properties are examined in detail by experiments. The advancement of a computer is also making possible the study of compressible turbulence by the direct numerical simulation (DNS) of the primitive fluid equations, although it is limited to the special cases with simple flow geometry (Lele 1994). A great merit of the DNS is the ability to elucidate the properties of compressibility-related turbulence quantities that are difficult to be measured directly, such as the dilatational energy dissipation rate and the pressure-dilatation correlation. The DNS has been already performed of isotropic and homogeneous-shear turbulence (Kida and Orszag 1990; Sarkar et al. 1991a; Sarkar 1992; Fujiwara and Arakawa 1993; Blaisdell et al. 1993; Sarkar 1995), the interaction

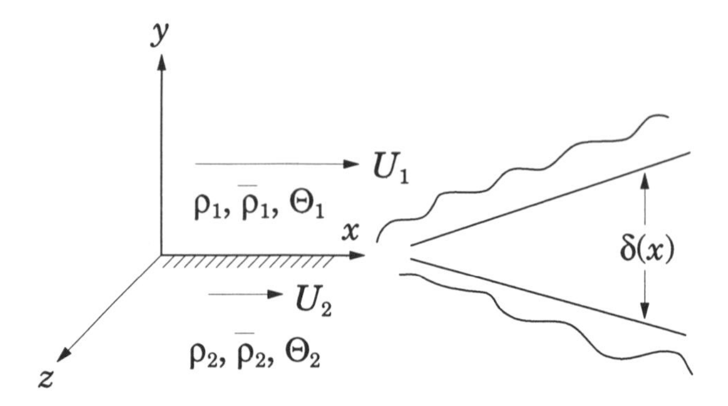

Figure 8.1. Free-shear layer.

of isotropic turbulence with a weak shock wave (Lee et al. 1993, 1997), and a free-shear layer (Vreman et al. 1996), etc.

Compressible turbulence modeling based on the ensemble-mean and filtering procedures is in a premature stage, compared with the incompressible counterpart. Fluid compressibility makes the momentum and temperature (internal-energy) equations inseparable, and leads to a complicated system of equations in a turbulent state. In this chapter, we shall focus attention on the ensemble-mean modeling of compressible flow. In the modeling, we have two types of averaging procedures. One is the conventional ensemble averaging procedure, and another is the mass-weighted one, in which the averaging is made in the combination with the density. The mass-weighted ensemble-averaging procedure is widely used in the current study of aeronautical turbulence modeling. Under it, the resulting system of equations is very similar to the low-Mach-number counterpart, and the solenoidal modeling is extended straightforwardly. There effects of compressibility are taken into account mainly through the mean density change. On the other hand, the application of the conventional ensemble averaging to the compressible system of equations generates a number of density-related correlation functions. This procedure reveals effects of compressibility explicitly, but their proper modeling is a difficult task.

The method based on the combination of the mass-weighted averaging procedure with the solenoidal turbulence modeling may reproduce primary properties of supersonic flows in wall-bounded flows such as a channel one. The fact indicates that the compressibility effects in those flows can be explained through the change of the mean density associated with that of the mean temperature. This point is confirmed by the DNS of a supersonic channel flow (Coleman et al. 1995; Huang et al. 1995). Such a method, however, fails to reproduce the suppression of the growth rate of a free-shear layer with the increasing convective Mach number. This finding suggests that compressibility effects on the layer are subject to the mechanism entirely different from that in a channel flow. The velocity and density fluctuations enhanced in front of a shock wave are greatly attenuated across it (Lee et al. 1993). In this context, a free-shear layer is supposed to share more features with a flow interacting with a shock wave.

In this chapter, we shall give a brief account of the difference between the conventional and mass-weighted ensemble-mean procedures, and discuss the modeling of compressible turbulence from the theoretical viewpoint of the two-scale direct-interaction approximation (TSDIA) and the Markovianized two-scale (MTS) method.

8.2. Conventional and Mass-Weighted Ensemble Averaging

8.2.1. BASIC PROPERTIES OF MASS-WEIGHTED AVERAGING

The conventional and mass-weighted ensemble averages of a flow quantity f are written as

$$\bar{f} = F \equiv \langle f \rangle , \tag{8.1a}$$

$$\hat{f} = \{f\} \equiv \langle \rho f \rangle / \bar{\rho} , \tag{8.1b}$$

respectively. As is the same as in Chapters 4, 6, and 7, the conventional ensemble average of f is denoted by the corresponding capital letter F, except ρ. We distinguish between the fluctuations related to two averaging procedures as

$$f' = f - F, \tag{8.2a}$$

$$f'' = f - \hat{f}. \tag{8.2b}$$

These two fluctuations obey

$$\langle f' \rangle = \{f''\} = 0, \tag{8.3a}$$

$$\{f'\} = - \langle f'' \rangle = \hat{f} - F = \langle \rho' f' \rangle / \bar{\rho} . \tag{8.3b}$$

From Eq. (8.3b), F and $\hat{f}$ are related to each other as

$$\hat{f} = F + \frac{\langle \rho' f' \rangle}{\bar{\rho}}, \tag{8.4a}$$

$$F = \hat{f} + \langle f'' \rangle . \tag{8.4b}$$

In the momentum and internal-energy (e) equations (2.29a) and (2.44a), we encounter the nonlinearity of the third order in ρ, $\mathbf{u}$, and e. We write it symbolically as $\rho f g$. Its conventional ensemble averaging leads to

$$\langle \rho f g \rangle = \bar{\rho} F G + \bar{\rho} \langle f' g' \rangle + F \langle \rho' g' \rangle + G \langle \rho' f' \rangle + \langle \rho' f' g' \rangle . \tag{8.5}$$

The left-hand side of Eq. (8.5) is regarded as the mass-weighted averaging of fg as

$$\langle \rho f g \rangle = \bar{\rho} \{fg\}, \tag{8.6}$$

which is reduced to

$$\langle \rho f g \rangle = \bar{\rho} \left(\hat{f} \hat{g} + \{f'' g''\} \right). \tag{8.7}$$

The comparison between Eqs. (8.5) and (8.7) shows that the mathematical form of a third-order quantity dependent on ρ is greatly simplified under the mass-weighted averaging. All the advection parts in the compressible system of equations are linearly dependent on ρ. This is the reason why the mass-weighted averaging is used exclusively in the turbulence modeling of high-speed aeronautical flows.

8.2.2. CONVENTIONAL AVERAGING OF FLUID EQUATIONS

The compressible system of fluid equations consists of the mass, momentum, and internal-energy conservation laws, Eqs. (2.10), (2.29), and (2.44). This system is supplemented with the thermodynamic relation for a perfect gas, Eq. (2.47). In this chapter, we newly define

$$s_{ij} = \frac{\partial u_j}{\partial x_i} + \frac{\partial u_i}{\partial x_j} - \frac{2}{3} \nabla \cdot \mathbf{u} \delta_{ij}, \tag{8.8}$$

which is reduced to the older one, Eq. (2.52), in the solenoidal limit. Using this definition, the dissipation function ϕ [Eq. (2.45)] is rewritten as

$$\phi = \mu s_{ij} \frac{\partial u_i}{\partial x_j}. \tag{8.9}$$

A. Mean-Field Equations

The application of the conventional averaging to these equations leads to

$$\frac{\partial \bar{\rho}}{\partial t} + \nabla \cdot (\bar{\rho} \mathbf{U}) = -\nabla \cdot \langle \rho' \mathbf{u}' \rangle, \tag{8.10}$$

$$\begin{aligned} \frac{\partial}{\partial t} \bar{\rho} U_i + \frac{\partial}{\partial x_j} \bar{\rho} U_j U_i = &-\frac{\partial P}{\partial x_i} + \frac{\partial}{\partial x_j} \left(-\bar{\rho} \left\langle u'_j u'_i \right\rangle - U_j \left\langle \rho u'_i \right\rangle - U_i \left\langle \rho' u'_j \right\rangle \right) \\ &+ \frac{\partial}{\partial x_j} \bar{\mu} S_{ji} + R_{PUi}, \end{aligned} \tag{8.11}$$

$$\begin{aligned} \frac{\partial}{\partial t} \bar{\rho} E + \nabla \cdot (\bar{\rho} \mathbf{U} E) = &\nabla \cdot (-\bar{\rho} \langle \mathbf{u}' e' \rangle - E \langle \rho' \mathbf{u}' \rangle - \mathbf{U} \langle \rho' e' \rangle) \\ &- P \nabla \cdot \mathbf{U} + \nabla \cdot \left(\bar{\kappa} \nabla \frac{E}{C_V} \right) + \Phi + R_{PE}, \end{aligned} \tag{8.12}$$

where the fluctuating parts of μ and κ were neglected, and

$$R_{PUi} = \frac{\partial}{\partial x_j}\left(-\left\langle \rho' u_j' u_i' \right\rangle\right) - \frac{\partial}{\partial t}\left\langle \rho' u_i' \right\rangle, \tag{8.13}$$

$$\Phi = \bar{\mu} S_{ij} \frac{\partial U_i}{\partial x_j} + \bar{\mu}\left\langle s_{ij}' \frac{\partial u_i'}{\partial x_j} \right\rangle, \tag{8.14}$$

$$R_{PE} = \nabla \cdot \left(-\left\langle \rho' e' \mathbf{u}' \right\rangle\right) - \frac{\partial}{\partial t}\left\langle \rho' e' \right\rangle - \left\langle p' \nabla \cdot \mathbf{u}' \right\rangle \tag{8.15}$$

(R_P means residual parts).

The residual terms $\mathbf{R}_{PU}$ and R_{PE} may be neglected in general, compared with the terms in Eqs. (8.11) and (8.12). For instance, the DNS's of some compressible flows show

$$\left|\bar{\rho}\left\langle u_i' u_j' \right\rangle\right| \gg \left|\left\langle \rho' u_i' u_j' \right\rangle\right|, \tag{8.16a}$$

$$\left|P \nabla \cdot \mathbf{U}\right| \gg \left|\left\langle p' \nabla \cdot \mathbf{u}' \right\rangle\right|. \tag{8.16b}$$

Equation (8.16b) indicates that the pressure-dilatation correlation $\langle p' \nabla \cdot \mathbf{u}' \rangle$ has small influence on the change of the mean internal energy E. This quantity also occurs in the equation for the velocity variance, but it is not always negligible there. In Eq. (8.14) for the mean dissipation function Φ, the second part expresses the conversion of the kinetic energy to the thermodynamic one at the fluctuation level.

The mean pressure P is given by

$$P = (\gamma - 1)\left(\bar{\rho} E + \left\langle \rho' e' \right\rangle\right), \tag{8.17}$$

which is approximated as

$$P \cong (\gamma - 1)\bar{\rho} E. \tag{8.18}$$

B. Turbulence Equations

Corresponding to Eq. (8.17), the fluctuation of p, p', is written as

$$p' = (\gamma - 1)\left(\rho' E + \bar{\rho} e' + \rho' e' - \left\langle \rho' e' \right\rangle\right) \cong (\gamma - 1)\left(\rho' E + \bar{\rho} e'\right). \tag{8.19}$$

The fluctuating field ρ', $\mathbf{u}'$, and e' obey

$$\frac{D\rho'}{Dt} + \nabla \cdot \left(\rho' \mathbf{u}'\right) + \bar{\rho} \nabla \cdot \mathbf{u}' = -\left(\mathbf{u}' \cdot \nabla\right)\bar{\rho} - \rho' \nabla \cdot \mathbf{U}, \tag{8.20}$$

$$\frac{Du'_i}{Dt} + (\mathbf{u}' \cdot \nabla) u'_i - \frac{1}{\bar{\rho}} \frac{\partial}{\partial x_j} \bar{\mu} s'_{ji} + (\gamma - 1) \frac{\partial e'}{\partial x_i} + (\gamma - 1) \frac{E}{\bar{\rho}} \frac{\partial \rho'}{\partial x_i}$$
$$= -(\mathbf{u}' \cdot \nabla) U_i - (\gamma - 1) \frac{\rho'}{\bar{\rho}} \frac{\partial E}{\partial x_i} - (\gamma - 1) \frac{e'}{\bar{\rho}} \frac{\partial \bar{\rho}}{\partial x_i} - \frac{\rho'}{\bar{\rho}} \frac{DU_i}{Dt}, \tag{8.21}$$

$$\frac{De'}{Dt} + (\mathbf{u}' \cdot \nabla) e' - \frac{1}{\bar{\rho}} \nabla \cdot \left(\bar{\kappa} \nabla \frac{e'}{C_V} \right) + (\gamma - 1) E \nabla \cdot \mathbf{u}'$$
$$= -(\mathbf{u}' \cdot \nabla) E - \frac{\rho'}{\bar{\rho}} \frac{DE}{Dt} - (\gamma - 1) e' \nabla \cdot \mathbf{U} - (\gamma - 1) \frac{\rho'}{\bar{\rho}} E \nabla \cdot \mathbf{U}, \tag{8.22}$$

where the Lagrange derivative based on the conventional mean velocity $\mathbf{U}$ is defined by

$$\frac{D}{Dt} = \frac{\partial}{\partial t} + \mathbf{U} \cdot \nabla. \tag{8.23}$$

In the full mathematical form, the following terms should be included in the above system of equations:

$$\nabla \cdot \langle \rho' \mathbf{u}' \rangle, \ \langle (\mathbf{u}' \cdot \nabla) \mathbf{u}' \rangle, \ \langle (\mathbf{u}' \cdot \nabla) e' \rangle, \text{ etc.}; \tag{8.24a}$$

$$\frac{\rho'}{\bar{\rho}} \frac{D\mathbf{u}'}{Dt}, \ \frac{\rho'}{\bar{\rho}} (\mathbf{u}' \cdot \nabla) \mathbf{U}, \ \frac{\rho'}{\bar{\rho}} (\mathbf{u}' \cdot \nabla) \mathbf{u}', \text{ etc.}; \tag{8.24b}$$

$$\frac{\rho'}{\bar{\rho}} \frac{De'}{Dt}, \ \frac{\rho'}{\bar{\rho}} (\mathbf{u}' \cdot \nabla) E, \ \frac{\rho'}{\bar{\rho}} (\mathbf{u}' \cdot \nabla) e', \text{ etc.} \tag{8.24c}$$

In Eq. (8.24a), these averaged quantities do not contribute to the equations for the velocity and density variances. Each term in Eqs. (8.24b) and (8.24c) can be neglected, compared with the retained counterpart in Eqs. (8.21) and (8.22), since we usually have

$$|\rho'| / \bar{\rho} = O\left(10^{-1}\right). \tag{8.25}$$

For instance, the first term in Eq. (8.24b) is dropped, compared with the first one on the left-hand side of Eq. (8.21).

As the representative quantities characterizing compressible turbulence, we may mention the velocity variance K, which is simply called the turbulent energy, and the density variance K_ρ:

$$K = \left\langle \mathbf{u}'^2 / 2 \right\rangle; \tag{8.26}$$

$$K_\rho = \left\langle \rho'^2 \right\rangle. \tag{8.27}$$

Strictly speaking, it is not accurate to call K the turbulent energy, unlike the solenoidal case. The kinetic energy is written as

$$\left\langle \rho \frac{\mathbf{u}^2}{2} \right\rangle = \bar{\rho} \left(\frac{\mathbf{U}^2}{2} \right) + \bar{\rho} \left\langle \frac{\mathbf{u}'^2}{2} \right\rangle + \mathbf{U} \cdot \langle \rho' \mathbf{u}' \rangle + \left\langle \rho' \frac{\mathbf{u}'^2}{2} \right\rangle . \tag{8.28}$$

In the original sense of turbulence parts, the sum of the second through fourth terms should be called the turbulent energy, although $\bar{\rho}K$ usually gives a dominant contribution.

From Eqs. (8.20) and (8.21), K and K_ρ obey

$$\bar{\rho} \frac{DK}{Dt} = -\bar{\rho} \left\langle u_i' u_j' \right\rangle \frac{\partial U_i}{\partial x_j} - \bar{\rho}\varepsilon + \langle p' \nabla \cdot \mathbf{u}' \rangle - \langle \rho' \mathbf{u}' \rangle \cdot \frac{D\mathbf{U}}{Dt} + \nabla \cdot \mathbf{T}_K , \tag{8.29}$$

$$\frac{DK_\rho}{Dt} = -2 \langle \rho' \mathbf{u}' \rangle \cdot \nabla \bar{\rho} - 2K_\rho \nabla \cdot \mathbf{U} - 2\bar{\rho} \langle \rho' \nabla \cdot \mathbf{u}' \rangle - \left\langle \rho'^2 \nabla \cdot \mathbf{u}' \right\rangle - \nabla \cdot \left\langle \rho'^2 \mathbf{u}' \right\rangle . \tag{8.30}$$

In Eq. (8.29), ε is the dissipation rate of the turbulent energy in compressible motion and is defined by

$$\varepsilon = \bar{\nu} \left\langle s_{ij}' \frac{\partial u_i'}{\partial x_j} \right\rangle , \tag{8.31}$$

with $\bar{\nu} = \bar{\mu}/\bar{\rho}$, and $\mathbf{T}_K$ is the counterpart of Eq. (2.114) (the transport rate) and is given by

$$T_{Ki} = -\langle p' u_i' \rangle - \bar{\rho} \left\langle \frac{\mathbf{u}'^2}{2} u_i' \right\rangle + \frac{\partial}{\partial x_j} \bar{\mu} \left\langle u_i' s_{ji}' \right\rangle . \tag{8.32}$$

Let us simply refer to the mathematical structures of Eqs. (8.29) and (8.30) in the context of the solenoidal turbulent-energy equation (2.111). In the latter case, the mean velocity gradient, specifically, the mean velocity strain plays a key role in the generation of the turbulent energy. In Eq. (8.29), the first term on the right-hand side fills this role, and its mathematical structure is the same as the solenoidal counterpart, except implicit effects of compressibility through $\langle u_i' u_j' \rangle$.

Equation (8.31) for ε may be written as

$$\varepsilon = \frac{1}{2} \bar{\nu} \left\langle {\omega_{ij}'}^2 \right\rangle + 2\bar{\nu} \left\langle \frac{\partial u_j'}{\partial x_i} \frac{\partial u_i'}{\partial x_j} \right\rangle - \frac{2}{3} \bar{\nu} \left\langle (\nabla \cdot \mathbf{u}')^2 \right\rangle , \tag{8.33}$$

where

$$\omega'_{ij} = \frac{\partial u'_j}{\partial x_i} - \frac{\partial u'_i}{\partial x_j} = \epsilon_{ij\ell}\omega'_\ell \tag{8.34}$$

[see Eq. (2.61)]. Using Eq. (2.60), we have

$$\varepsilon = \bar{\nu}\left\langle \boldsymbol{\omega}'^2 \right\rangle + 2\bar{\nu}\left\langle \frac{\partial u'_j}{\partial x_i}\frac{\partial u'_i}{\partial x_j} \right\rangle - \frac{2}{3}\bar{\nu}\left\langle (\nabla \cdot \mathbf{u}')^2 \right\rangle. \tag{8.35}$$

In the homogeneous case, we note

$$\frac{\partial}{\partial x_i}\left(u'_j \frac{\partial u'_i}{\partial x_j} \right) = 0, \tag{8.36}$$

and have

$$\left\langle \frac{\partial u'_j}{\partial x_i}\frac{\partial u'_i}{\partial x_j} \right\rangle = -\left\langle u'_j \frac{\partial^2 u'_i}{\partial x_i \partial x_j} \right\rangle = \left\langle (\nabla \cdot \mathbf{u}')^2 \right\rangle. \tag{8.37}$$

As a result, Eq. (8.35) is reduced to

$$\varepsilon = \varepsilon_S + \varepsilon_D, \tag{8.38}$$

where

$$\varepsilon_S = \bar{\nu}\left\langle \boldsymbol{\omega}'^2 \right\rangle, \tag{8.39}$$

$$\varepsilon_D = (4/3)\,\bar{\nu}\left\langle (\nabla \cdot \mathbf{u}')^2 \right\rangle \tag{8.40}$$

(subscripts S and D denote solenoidal and dilatational parts, respectively). This splitting is exact in homogeneous flow only, but it shows the enhancement of energy dissipation due to fluid compressibility. Equation (8.40) is usually called the dilatational dissipation rate.

The third term in Eq. (8.29), $\langle p' \nabla \cdot \mathbf{u}' \rangle$ (the pressure-dilatation correlation), expresses the work done by the surrounding fluid. This quantity can take both positive and negative signs. It is positive in isotropic decaying turbulence (Sarkar et al. 1991a; Fujiwara and Arakawa 1993), whereas it behaves much more complicatedly in homogeneous-shear turbulence; namely, it becomes both positive and negative in time, but it is negative on the average (Blaisdell et al. 1993; Sarkar 1995). The contribution of $\langle p' \nabla \cdot \mathbf{u}' \rangle$ to Eq. (8.29) is not so large in homogeneous-shear and free-shear flows (Sarkar 1995; Fujiwara 1996; Vreman et al. 1996).

The fourth term may become important in the presence of large streamwise variation of turbulent state, as in the vicinity of a shock wave. In such a region, we may approximate

$$\frac{D\mathbf{U}}{Dt} = -\frac{1}{\bar{\rho}}\nabla P, \tag{8.41}$$

and the term is rewritten as

$$-\langle \rho' \mathbf{u}' \rangle \cdot \frac{D\mathbf{U}}{Dt} = \frac{1}{\bar{\rho}} \langle \rho' \mathbf{u}' \rangle \cdot \nabla P. \tag{8.42}$$

As will be shown later, we may model

$$\langle \rho' \mathbf{u}' \rangle = -\nu_\rho \nabla \bar{\rho}, \tag{8.43}$$

with the positive coefficient ν_ρ, resulting in

$$-\langle \rho' \mathbf{u}' \rangle \cdot \frac{D\mathbf{U}}{Dt} = -\frac{\nu_\rho}{\bar{\rho}} \nabla \bar{\rho} \cdot \nabla P. \tag{8.44}$$

Across a shock wave, both $\nabla \bar{\rho}$ and ∇P are positive. Therefore Eq. (8.44) contributes to the suppression of turbulence.

In the context of the transport rate $\mathbf{T}_K$ [Eq. (8.32)], we should pay attention to the first p'-related term. In the solenoidal case, it is dropped, compared with the second term, except the close vicinity of a solid wall. This approximation is supported by the TSDIA, as is shown in Sec. 6.6.1. In the compressible case, $\langle p' \mathbf{u}' \rangle$ is rewritten as

$$\langle p' \mathbf{u}' \rangle = (\gamma - 1) \left(\langle \rho' \mathbf{u}' \rangle E + \langle e' \mathbf{u}' \rangle \bar{\rho} \right), \tag{8.45}$$

from Eq. (8.19) and occurs in the form of $\nabla \cdot \langle p' \mathbf{u}' \rangle$ in the K equation. In the region with large $\nabla \bar{\rho}$ and ∇E, for instance, in the vicinity of a shock wave, these quantities may become large, and $\nabla \cdot \langle p' \mathbf{u}' \rangle$ is not negligible there. This point may be confirmed by the DNS (Lee et al. 1993).

In Eq. (8.30) for K_ρ, the first term has the mathematical structure similar to the first term or production rate in the solenoidal turbulent-energy equation (2.111). It is related to the mean density gradient $\nabla \bar{\rho}$. As far as Eq. (8.43) is a valid approximation, the first term is always positive and contributes to the generation of density fluctuations. The second term, which is dependent on $\nabla \cdot \mathbf{U}$, behaves a little differently. In the case of negative $\nabla \cdot \mathbf{U}$ or contraction, density fluctuations are enhanced, whereas they are suppressed in the case of positive $\nabla \cdot \mathbf{U}$ or expansion. In the context of a shock wave, the term contributes to the enhancement and suppression of density fluctuations ahead of and behind it, respectively.

The last term in Eq. (8.30) expresses the diffusion effect of ρ' due to turbulence, and its role may be compared with $\nabla \cdot \mathbf{T}_K$ in Eq. (2.111). The remaining terms, that is, the third and fourth ones, are not so clear in their physical meaning. This point is deeply associated with the fact that the total amount of ρ^2, $\int \rho^2 dV$, is not conserved, unlike the kinetic energy in the solenoidal motion with molecular effects dropped. In compressible flow, the total amount of kinetic energy itself is not conserved in the absence of molecular effects, but its combination with the thermodynamic energy is conserved. As a result, Eq. (8.29) still possesses the mathematical properties similar to the solenoidal counterpart. Of those two terms, the third term is considered dominant under the condition (8.25). It contributes to the generation (or suppression) of density fluctuation in the case of anti-correlation (or normal correlation) between ρ' and $\nabla \cdot \mathbf{u}'$.

8.2.3. MASS-WEIGHTED AVERAGING OF FLUID EQUATIONS

A. Mean-Field Equations

We apply the ensemble averaging to Eqs. (2.10), (2.29), and (2.44). We use Eq. (8.7) and write the advection parts in terms of the mass-weighted quantities. The resulting equations are

$$\frac{\partial \bar{\rho}}{\partial t} + \nabla \cdot (\bar{\rho}\hat{\mathbf{u}}) = 0, \tag{8.46}$$

$$\frac{\partial}{\partial t}\bar{\rho}\hat{u}_i + \frac{\partial}{\partial x_j}\bar{\rho}\hat{u}_j\hat{u}_i = -\frac{\partial P}{\partial x_i} + \frac{\partial}{\partial x_j}\left(-\bar{\rho}\left\{u_j''u_i''\right\}\right) + \frac{\partial}{\partial x_j}\left\langle \mu s_{ji}\right\rangle, \tag{8.47}$$

$$\frac{\partial}{\partial t}\bar{\rho}\hat{e} + \nabla\cdot(\bar{\rho}\hat{e}\hat{\mathbf{u}}) = \nabla\cdot\left(-\bar{\rho}\left\{e''\mathbf{u}''\right\}\right) - \left\langle p\nabla\cdot\mathbf{u}\right\rangle + \nabla\cdot\left\langle \kappa\nabla\frac{e}{C_V}\right\rangle + \left\langle\phi\right\rangle. \tag{8.48}$$

On the right-hand sides of Eqs. (8.47) and (8.48), we should note the mixture of the conventional and mass-weighted averaged quantities. The terms except the Reynolds stress $\{u_j''u_i''\}$ and the turbulent internal-energy flux $\{e''\mathbf{u}''\}$ are yet to be expressed in terms of mass-weighted quantities. Of these terms, the mean pressure P may be simply rewritten as

$$P = \left\langle (\gamma - 1)\rho e\right\rangle = (\gamma - 1)\bar{\rho}\hat{e}. \tag{8.49}$$

On the other hand, $\langle p\nabla\cdot\mathbf{u}\rangle$ is reduced to

$$\left\langle p\nabla\cdot\mathbf{u}\right\rangle = \left\langle (\gamma-1)\rho e\nabla\cdot\mathbf{u}\right\rangle = (\gamma-1)\bar{\rho}\left(\hat{e}\left\{\nabla\cdot\mathbf{u}\right\} + \left\{e''\left(\nabla\cdot\mathbf{u}\right)''\right\}\right). \tag{8.50}$$

Concerning the $\hat{e}$-related part, we use Eq. (8.3b), and have

$$\left\{\nabla\cdot\mathbf{u}\right\} - \nabla\cdot\hat{\mathbf{u}} = \left\{\nabla\cdot\mathbf{u}''\right\} = \frac{1}{\bar{\rho}}\left\langle\rho\nabla\cdot\mathbf{u}''\right\rangle = -\frac{1}{\bar{\rho}}\left\langle\mathbf{u}''\cdot\nabla\rho\right\rangle \tag{8.51a}$$

$$\cong -\frac{1}{\bar{\rho}}\left\langle\mathbf{u}''\right\rangle\cdot\nabla\bar{\rho} = \frac{1}{\bar{\rho}^2}\left\langle\rho'\mathbf{u}'\right\rangle\cdot\nabla\bar{\rho}. \tag{8.51b}$$

In order to express $\{\nabla\cdot\mathbf{u}\}$ in terms of mass-weighted quantities, we need to write $\langle\rho'\mathbf{u}'\rangle$ using those quantities. Such a situation is also encountered in the molecular-effect-related quantities, the last term in Eq. (8.47) and the last two terms in Eq. (8.48).

B. Turbulence Equations

At the level of the mean-field equations, the mass-weighted averaging leads to a simpler system of equations, specifically, concerning the advection parts. On the other hand, it is not always the case at the level of the turbulence equations. In order to see this point, we consider the equation for ρ'. Subtracting Eq. (8.46) from Eq. (2.10), we have

$$\frac{\partial \rho'}{\partial t} + \nabla \cdot \left(\bar{\rho} \mathbf{u}'' + \rho' \hat{\mathbf{u}} + \rho' \mathbf{u}'' \right) = 0. \tag{8.52}$$

Then the equation for K_ρ is written as

$$\frac{\hat{D} K_\rho}{\hat{D} t} = -2 \left\langle \rho' \mathbf{u}'' \right\rangle \cdot \nabla \bar{\rho} - 2 K_\rho \nabla \cdot \hat{\mathbf{u}} - 2 \bar{\rho} \left\langle \rho' \nabla \cdot \mathbf{u}'' \right\rangle$$
$$- \left\langle \rho'^2 \nabla \cdot \mathbf{u}'' \right\rangle - \nabla \cdot \left\langle \rho'^2 \mathbf{u}'' \right\rangle, \tag{8.53}$$

where $\hat{D}/\hat{D}t$ is the Lagrange derivative based on $\hat{\mathbf{u}}$, which is defined by Eq. (8.23) with $\mathbf{U}$ replaced with $\hat{\mathbf{u}}$. On the right-hand side, the first and third terms are written as

$$- 2 \left\langle \rho' \mathbf{u}'' \right\rangle \cdot \nabla \bar{\rho} = 2 \bar{\rho} \left\langle \mathbf{u}'' \right\rangle \cdot \nabla \bar{\rho} = -2 \left\langle \rho' \mathbf{u}' \right\rangle \cdot \nabla \bar{\rho}, \tag{8.54}$$

$$- 2 \bar{\rho} \left\langle \rho' \nabla \cdot \mathbf{u}'' \right\rangle = -2 \bar{\rho}^2 \left(\left\{ \nabla \cdot \mathbf{u}'' \right\} - \nabla \cdot \left\langle \mathbf{u}'' \right\rangle \right). \tag{8.55}$$

From Eqs. (8.3b) and (8.51), Eq. (8.55) is also related to $\langle \rho' \mathbf{u}' \rangle$. Therefore it is necessary to express the conventional averaged quantity $\langle \rho' \mathbf{u}' \rangle$ in terms of mass-weighted ones. The similar situation is encountered in the equation for the mass-weighted turbulent energy $\hat{K}$, which is defined as

$$\hat{K} = \left\{ (1/2)\, \mathbf{u}^2 \right\} - (1/2)\, \hat{\mathbf{u}}^2 = \left\{ (1/2)\, \mathbf{u}''^2 \right\}. \tag{8.56}$$

The foregoing discussion indicates that the mass-weighted averaging procedure does not always lead to a system of turbulence equations mathematically much simpler than the ensemble-mean counterpart. Specifically, it is necessary to express the ensemble-mean correlation functions such as $\langle \rho' \mathbf{u}' \rangle$ in terms of mass-weighted ones. Statistical turbulence theories for shear flow such as the TSDIA are founded on the ensemble-averaging basis, and we have no mass-weighted statistical theories for examining those quantities at least at present. This situation becomes a big stumbling block for performing the compressible turbulence modeling from the theoretical standpoint comparable to the incompressible one.

8.2.4. CURRENT COMPRESSIBLE TURBULENCE MODELING

Compressible turbulence models are classified into two categories. In the first category, full use is made of the mathematical similarity of the mass-weighted mean equation to the solenoidal counterparts, and the conventional algebraic and second-order models in solenoidal turbulent flow are extended to compressible ones (Rubesin 1989; Gatski 1997). In this case, effects of fluid compressibility are taken into account mainly through the mean density change. Such an extension cannot always cope with highly compressible flows satisfactorily. This situation is reflected on the fact that the higher-order compressible turbulence models do not always lead to the absolute superiority to the lower-order ones. Its typical example is the model by Baldwin and Lomax (1978). It is the so-called mixing-length model using no transport equations for turbulence quantities, and the turbulent viscosity is prescribed in an algebraic form. This model is still used widely in the analysis of high speed flows, although mixing-length models have been already expelled in the solenoidal turbulence modeling, except the subgrid-scale one explained in Chapter 5.

The foregoing situation indicates that it is necessary to incorporate compressibility effects more distinctly into the mass-weighted mean equations. In the second category aiming at such modeling, attention is paid to the dilatational dissipation ε_D and the pressure-dilatation correlation $\langle p'\nabla \cdot \mathbf{u}'\rangle$ in the equation for $\hat{K}$ that is essentially the same form as the conventional one (8.29) (Zeman 1990; Sarkar et al. 1991b; Sarkar 1992). Specifically, ε_D is modeled in terms of the solenoidal dissipation rate ε_S [Eq. (8.39)] and the turbulent Mach number M_T, as will be given by Eq. (8.140). The enhanced energy dissipation is instrumental to the explanation of the drastic suppression of the growth rate of a free-shear flow that is beyond the reach of the models in the first category. This enhancement effect also plays a key role in the multiple-scale model by Liou et al. (1995), which may reproduce the suppression of the growth rate.

The compressible models in the second category still do not contain the turbulence quantities related directly to fluid compressibility, either. It was pointed out using the TSDIA explained later that compressibility effects in turbulent shear flow are tightly linked with fluctuations of density, and that density variance K_ρ is a key quantity in compressible turbulence (Yoshizawa 1990, 1992). This point may be understood easily from the frequent occurrence of the mass flux $\langle \rho'\mathbf{u}'\rangle$ in the conventional mean equations. The simplest model in which the K_ρ equation is combined with the K and ε equations leads to a turbulence model of the three-equation type. Such a model based on the modeling of ε_D and $\langle p'\nabla \cdot \mathbf{u}'\rangle$ in terms of K_ρ was presented by Taulbee and VanOsdol (1991) and may explain the suppression of the growth of a free-shear flow. A similar model is the one by Fujiwara and Arakawa (1993) who adopted the combination of K_ρ and the variance of compressible velocity fluctuation as the quantity characterizing direct effects of compressibility.

In the DNS's of homogeneous-shear and free-shear flows, effects of compressibility through ε_D and $\langle p' \nabla \cdot \mathbf{u}' \rangle$ on the turbulent-energy equation are not dominant enough to guarantee the suppression of turbulence (Sarkar 1995, Fujiwara 1996, and Vreman et al. 1996). The compressibility effects occur most clearly through the change of the Reynolds stress, resulting in the decrease in the turbulent-energy production. In the context of the algebraic modeling, this finding indicates the importance of explicitly including compressibility effects on the turbulent viscosity. Such modeling will be discussed later.

In the current study of compressible turbulence modeling, attention has been focused on the phenomena in aeronautical and mechanical-engineering fields. An interesting example of flows in entirely different fields is astrophysical flow, specifically, bipolar jets in accretion disks. Around a high-mass stellar object such as neutron stars and active galactic nuclei, highly electrically conducting gases or plasmas exist in the form of a circular disk called accretion ones. The surrounding gases accrete onto the central object under the action of gravity force. One of the prominent features of accretion disks is bipolar jets normal to the disks. These jets are highly collimated. The mechanism of collimation has not been clarified yet, but the confinement of plasma due to magnetic fields has been inferred to be its major mechanism. It is known well from laboratory experiments on thermo-nuclear fusion that plasma cannot be confined for long by magnetic fields only. The drastic suppression of the growth of turbulence in a free-shear layer under effects of compressibility is a promising mechanism for explaining the jet collimation, in addition to magnetic-field effects (some phenomena associated with accretion disks are discussed in Chapter 10 in the context of magnetohydrodynamic turbulence).

8.3. TSDIA Analysis of Compressible Flow

In the solenoidal case, we may apply the TSDIA to the equations for velocity and scalar fluctuations and abstract the information useful in turbulence modeling. In the case, the Kolmogorov's inertial-range properties play a key role, as is seen in Chapter 6. In compressible flow, however, such definite properties have not been established yet. As a result, the range of applicability of the TSDIA findings to compressible modeling is narrow, compared with the solenoidal counterpart. Nevertheless, those findings are helpful to understanding of the fundamentals in modeling compressible shear flow. Specifically, they are expected to give an important clue to the mass-weighted modeling with little support from statistical-theoretical approaches.

8.3.1. KEY MATHEMATICAL PROCEDURES

A. Two-Scale Expansion

We follow the TSDIA procedures in the solenoidal motion and introduce two space and time variables (6.27), leading to the two-scale expression (6.28). We apply these procedures to a compressible system, Eqs. (8.20)-(8.22). Then we have

$$\frac{D\rho'}{D\tau} + \nabla_\xi \cdot (\rho' \mathbf{u}') + \bar{\rho} \nabla_\xi \cdot \mathbf{u}' = \delta \left(- (\mathbf{u}' \cdot \nabla_X) \bar{\rho} - \rho' \nabla_X \cdot \mathbf{U} - \frac{D\rho'}{DT} - \bar{\rho} \nabla_X \cdot \mathbf{u}' + N_\rho \right), \tag{8.57}$$

$$\begin{aligned} &\frac{Du'_i}{D\tau} + (\mathbf{u}' \cdot \nabla_\xi)\, u'_i - \bar{\nu} \frac{\partial s'_{ji}}{\partial \xi_j} + (\gamma - 1) \frac{\partial e'}{\partial \xi_i} + (\gamma - 1) \frac{E}{\bar{\rho}} \frac{\partial \rho'}{\partial \xi_i} \\ &\quad = \delta \left(- (\mathbf{u}' \cdot \nabla_X)\, U_i - (\gamma - 1) \frac{\rho'}{\bar{\rho}} \frac{\partial E}{\partial X_i} - (\gamma - 1) \frac{e'}{\bar{\rho}} \frac{\partial \bar{\rho}}{\partial X_i} \right. \\ &\qquad \left. - \frac{\rho'}{\bar{\rho}} \frac{DU_i}{DT} - \frac{Du'_i}{DT} - (\gamma - 1) \frac{\partial e'}{\partial X_i} - (\gamma - 1) \frac{E}{\bar{\rho}} \frac{\partial \rho'}{\partial X_i} + N_{ui} \right), \end{aligned} \tag{8.58}$$

$$\begin{aligned} &\frac{De'}{D\tau} + (\mathbf{u}' \cdot \nabla_\xi)\, e' - \bar{\lambda} \Delta_\xi e' + (\gamma - 1) E \nabla_\xi \cdot \mathbf{u}' \\ &\quad = \delta \left(- (\mathbf{u}' \cdot \nabla_X)\, E - \frac{\rho'}{\bar{\rho}} \frac{DE}{DT} - (\gamma - 1)\, e' \nabla_X \cdot \mathbf{U} \right. \\ &\qquad \left. - (\gamma - 1)\, \rho' \frac{E}{\bar{\rho}} \nabla_X \cdot \mathbf{U} - \frac{De'}{DT} - (\gamma - 1) E \nabla_X \cdot \mathbf{u}' + N_e \right), \end{aligned} \tag{8.59}$$

where $\bar{\nu} = \bar{\mu}/\bar{\rho}$ and $\bar{\lambda} = \bar{\kappa}/(\bar{\rho} C_V)$, and

$$\frac{D}{D\tau} = \frac{\partial}{\partial \tau} + \mathbf{U} \cdot \nabla_\xi, \ \nabla_\xi = \left(\frac{\partial}{\partial \xi_i} \right), \tag{8.60a}$$

$$\frac{D}{DT} = \frac{\partial}{\partial T} + \mathbf{U} \cdot \nabla_X, \ \nabla_X = \left(\frac{\partial}{\partial X_i} \right), \tag{8.60b}$$

and N_ρ, $\mathbf{N}_u$, and N_e denote the terms nonlinear in ρ, $\mathbf{u}'$, and e'. We focus attention on the interaction between the mean field and fluctuations, and neglect N_ρ etc. hereafter.

We use the Fourier representation concerning the fast variables $\boldsymbol{\xi}$, Eq. (6.33) (in what follows, the dependence on the slow variables $\mathbf{X}$ and T will not be written explicitly). Then we have

$$\frac{\partial \rho'(\mathbf{k};\tau)}{\partial \tau} - ik_i \iint \delta(\mathbf{k}-\mathbf{p}-\mathbf{q}) d\mathbf{p} d\mathbf{q} u_i'(\mathbf{p};\tau)\rho'(\mathbf{q};\tau) - ik_i\bar{\rho}u_i'(\mathbf{k};\tau)$$
$$= \delta\left(-u_i'(\mathbf{k};\tau)\frac{\partial\bar{\rho}}{\partial X_i} - \rho'(\mathbf{k};\tau)\nabla_X\cdot\mathbf{U} - \frac{D\rho'(\mathbf{k};\tau)}{DT_I} - \bar{\rho}\frac{\partial u_i'(\mathbf{k};\tau)}{\partial X_{Ii}}\right), \quad (8.61)$$

$$\frac{\partial u_i'(\mathbf{k};\tau)}{\partial \tau} - i\iint \delta(\mathbf{k}-\mathbf{p}-\mathbf{q}) d\mathbf{p} d\mathbf{q} M_{Cij\ell}(\mathbf{p},\mathbf{q}) u_j'(\mathbf{p};\tau)u_\ell'(\mathbf{q};\tau)$$
$$+\bar{\nu}k^2 u_i'(\mathbf{k};\tau) + \frac{1}{3}\bar{\nu}k_i k_j u_j'(\mathbf{k};\tau)$$
$$-i(\gamma-1)k_i e'(\mathbf{k};\tau) - i(\gamma-1)k_i E\frac{\rho'(\mathbf{k};\tau)}{\bar{\rho}}$$
$$= \delta\left(-u_j'(\mathbf{k};\tau)\frac{\partial U_i}{\partial X_j} - (\gamma-1)\frac{\rho'(\mathbf{k};\tau)}{\bar{\rho}}\frac{\partial E}{\partial X_i}\right.$$
$$-(\gamma-1)\frac{e'(\mathbf{k};\tau)}{\bar{\rho}}\frac{\partial\bar{\rho}}{\partial X_i} - \frac{\rho'(\mathbf{k};\tau)}{\bar{\rho}}\frac{DU_i}{DT} - \frac{Du_i'(\mathbf{k};\tau)}{DT_I}$$
$$\left.-(\gamma-1)\frac{\partial e'(\mathbf{k};\tau)}{\partial X_{Ii}} - (\gamma-1)\frac{E}{\bar{\rho}}\frac{\partial\rho'(\mathbf{k};\tau)}{\partial X_{Ii}}\right), \quad (8.62)$$

$$\frac{\partial e'(\mathbf{k};\tau)}{\partial \tau} - i\iint \delta(\mathbf{k}-\mathbf{p}-\mathbf{q}) d\mathbf{p} d\mathbf{q} q_i u_i'(\mathbf{p};\tau)e'(\mathbf{q};\tau)$$
$$+\bar{\lambda}k^2 e'(\mathbf{k};\tau) - i(\gamma-1)k_i E u_i'(\mathbf{k};\tau)$$
$$= \delta\left(-u_i'(\mathbf{k};\tau)\frac{\partial E}{\partial X_i} - \frac{\rho'(\mathbf{k};\tau)}{\bar{\rho}}\frac{DE}{DT}\right.$$
$$-(\gamma-1)e'(\mathbf{k};\tau)\nabla_X\cdot\mathbf{U} - (\gamma-1)\frac{\rho'(\mathbf{k};\tau)}{\bar{\rho}}E_X\nabla\cdot\mathbf{U}$$
$$\left.-\frac{De'(\mathbf{k};\tau)}{DT_I} - (\gamma-1)E\frac{\partial u_i'(\mathbf{k};\tau)}{\partial X_{Ii}}\right), \quad (8.63)$$

where

$$M_{Cij\ell}(\mathbf{p},\mathbf{q}) = (1/2)(q_j\delta_{i\ell} + p_\ell\delta_{ij}), \quad (8.64)$$

and ∇_{XI} and D/DT_I are defined by Eq. (6.36).

We apply the scale-parameter expansion (6.39) to Eqs. (8.61)-(8.63). The $O(1)$ system of equations consists of

$$\frac{\partial \rho_0'(\mathbf{k};\tau)}{\partial \tau} - ik_i\iint \delta(\mathbf{k}-\mathbf{p}-\mathbf{q}) d\mathbf{p} d\mathbf{q} u_{0i}'(\mathbf{p};\tau)\rho_0'(\mathbf{q};\tau)$$
$$= ik_i\bar{\rho}u_{0i}'(\mathbf{k};\tau), \quad (8.65)$$

$$\frac{\partial u'_{0i}(\mathbf{k};\tau)}{\partial \tau} - i\iint \delta\left(\mathbf{k}-\mathbf{p}-\mathbf{q}\right)d\mathbf{p}d\mathbf{q}M_{Cij\ell}\left(\mathbf{p},\mathbf{q}\right)u'_{0j}(\mathbf{p};\tau)u'_{0\ell}(\mathbf{q};\tau)$$
$$+\bar{\nu}k^2u'_{0i}(\mathbf{k};\tau) + \frac{1}{3}\bar{\nu}k_ik_ju'_{0j}(\mathbf{k};\tau)$$
$$= i\left(\gamma-1\right)k_ie'_0(\mathbf{k};\tau) + i\left(\gamma-1\right)k_i\frac{E}{\bar{\rho}}\rho'_0(\mathbf{k};\tau), \tag{8.66}$$

$$\frac{\partial e'_0(\mathbf{k};\tau)}{\partial \tau} - i\iint \delta\left(\mathbf{k}-\mathbf{p}-\mathbf{q}\right)d\mathbf{p}d\mathbf{q}q_iu'_{0i}(\mathbf{p};\tau)e'_0(\mathbf{q};\tau) + \bar{\lambda}k^2e'_0(\mathbf{k};\tau)$$
$$= i\left(\gamma-1\right)k_iEu'_{0i}(\mathbf{k};\tau). \tag{8.67}$$

On the other hand, the $O(\delta)$ system is given by

$$\frac{\partial \rho'_1(\mathbf{k};\tau)}{\partial \tau} - ik_i\iint \delta\left(\mathbf{k}-\mathbf{p}-\mathbf{q}\right)d\mathbf{p}d\mathbf{q}u'_{0i}(\mathbf{p};\tau)\rho'_1(\mathbf{q};\tau) - ik_i\bar{\rho}u'_{1i}(\mathbf{k};\tau)$$
$$= -u'_{0i}(\mathbf{k};\tau)\frac{\partial\bar{\rho}}{\partial X_i} - \rho'_0(\mathbf{k};\tau)\nabla_X\cdot\mathbf{U} - \frac{D\rho'_0(\mathbf{k};\tau)}{DT_I} - \bar{\rho}\frac{\partial u'_{0i}(\mathbf{k};\tau)}{\partial X_{Ii}}, \tag{8.68}$$

$$\frac{\partial u'_{1i}(\mathbf{k};\tau)}{\partial \tau} - 2i\iint \delta\left(\mathbf{k}-\mathbf{p}-\mathbf{q}\right)d\mathbf{p}d\mathbf{q}M_{Cij\ell}\left(\mathbf{p},\mathbf{q}\right)u'_{0j}(\mathbf{p};\tau)u'_{1\ell}(\mathbf{q};\tau)$$
$$+\bar{\nu}k^2u'_{1i}(\mathbf{k};\tau) + \frac{1}{3}\bar{\nu}k_ik_ju'_{1j}(\mathbf{k};\tau)$$
$$-i\left(\gamma-1\right)k_ie'_1(\mathbf{k};\tau) - i\left(\gamma-1\right)k_i\frac{E}{\bar{\rho}}\rho'_1(\mathbf{k};\tau)$$
$$= -u'_{0j}(\mathbf{k};\tau)\frac{\partial U_i}{\partial X_j} - \left(\gamma-1\right)\frac{\rho'_0(\mathbf{k};\tau)}{\bar{\rho}}\frac{\partial E}{\partial X_i}$$
$$-\left(\gamma-1\right)\frac{e'_0(\mathbf{k};\tau)}{\bar{\rho}}\frac{\partial\bar{\rho}}{\partial X_i} - \frac{\rho'_0(\mathbf{k};\tau)}{\bar{\rho}}\frac{DU_i}{DT} - \frac{Du'_{0i}(\mathbf{k};\tau)}{DT_I}$$
$$-\left(\gamma-1\right)\frac{\partial e'_0(\mathbf{k};\tau)}{\partial X_{Ii}} - \left(\gamma-1\right)\frac{E}{\bar{\rho}}\frac{\partial\rho'_0(\mathbf{k};\tau)}{\partial X_{Ii}}, \tag{8.69}$$

$$\frac{\partial e'_1(\mathbf{k};\tau)}{\partial \tau} - i\iint \delta\left(\mathbf{k}-\mathbf{p}-\mathbf{q}\right)d\mathbf{p}d\mathbf{q}q_iu'_{0i}(\mathbf{p};\tau)e'_1(\mathbf{q};\tau)$$
$$+\bar{\lambda}k^2e'_1(\mathbf{k};\tau) - i\left(\gamma-1\right)k_iEu'_{1i}(\mathbf{k};\tau)$$
$$= -u'_{0i}(\mathbf{k};\tau)\frac{\partial E}{\partial X_i} - \frac{\rho'_0(\mathbf{k};\tau)}{\bar{\rho}}\frac{DE}{DT}$$
$$-\left(\gamma-1\right)e'_0(\mathbf{k};\tau)\nabla_X\cdot\mathbf{U} - \left(\gamma-1\right)\frac{\rho'_0(\mathbf{k};\tau)}{\bar{\rho}}E\nabla_X\cdot\mathbf{U}$$
$$-\frac{De'_0(\mathbf{k};\tau)}{DT_I} - \left(\gamma-1\right)E\frac{\partial u'_{0i}(\mathbf{k};\tau)}{\partial X_{Ii}}, \tag{8.70}$$

where we neglected the terms nonlinear in ρ, $\mathbf{u}'$, and e' that occur newly through the scale-parameter expansion.

Here we have two methods of dealing with the above system of equations. In one method, we assume weak compressibility and examine its effects on the solenoidal turbulent motion (Yoshizawa 1986). In this case, the Kolmogorov law may be fully utilized in analyzing various correlation functions. In another method, we consider a highly compressible flow state through the direct use of Eqs. (8.65)-(8.67) (Yoshizawa 1992). In what follows, we shall adopt the latter method.

B. Basic-Field Expansion

In the TSDIA, we formally integrate the $O(1)$ system of equations by using the Green's functions associated with it. In the present case, the $O(1)$ system consisting of Eqs. (8.65)-(8.67) leads to three coupled Green's functions. In order to reduce this complexity, we shall introduce the expansion based on the concept of the basic field, resulting in three uncoupled Green's functions for the density, velocity and internal energy.

We define the basic field $(\rho'_B, \mathbf{u}'_B, e'_B)$ by

$$\frac{\partial \rho'_B(\mathbf{k};\tau)}{\partial \tau} - ik_i \iint \delta(\mathbf{k}-\mathbf{p}-\mathbf{q}) d\mathbf{p} d\mathbf{q} u'_{Bi}(\mathbf{p};\tau)\rho'_B(\mathbf{q};\tau) = 0, \tag{8.71}$$

$$\frac{\partial u'_{Bi}(\mathbf{k};\tau)}{\partial \tau} - i \iint \delta(\mathbf{k}-\mathbf{p}-\mathbf{q}) d\mathbf{p} d\mathbf{q} M_{Cij\ell}(\mathbf{p},\mathbf{q})\, u'_{Bj}(\mathbf{p};\tau) u'_{B\ell}(\mathbf{q};\tau)$$
$$+\bar{\nu}k^2 u'_{Bi}(\mathbf{k};\tau) + \frac{1}{3}\bar{\nu}k_i k_j u'_{Bj}(\mathbf{k};\tau) = 0, \tag{8.72}$$

$$\frac{\partial e'_B(\mathbf{k};\tau)}{\partial \tau} - i \iint \delta(\mathbf{k}-\mathbf{p}-\mathbf{q}) d\mathbf{p} d\mathbf{q} q_i u'_{Bi}(\mathbf{p};\tau) e'_B(\mathbf{q};\tau)$$
$$+ \bar{\lambda}k^2 e'_B(\mathbf{k};\tau) = 0. \tag{8.73}$$

For this basic field, we introduce three Green's functions, G'_ρ, G'_u, and G'_e, as

$$\frac{\partial G'_\rho(\mathbf{k};\tau,\tau')}{\partial \tau} - ik_i \iint \delta(\mathbf{k}-\mathbf{p}-\mathbf{q}) d\mathbf{p} d\mathbf{q} u'_{Bi}(\mathbf{p};\tau) G'_\rho(\mathbf{q};\tau,\tau')$$
$$= \delta(\tau-\tau'), \tag{8.74}$$

$$\frac{\partial G'_{uij}(\mathbf{k};\tau,\tau')}{\partial \tau} + \bar{\nu}k^2 G'_{uij}(\mathbf{k};\tau,\tau') + \frac{1}{3}\bar{\nu}k_i k_\ell G'_{u\ell j}(\mathbf{k};\tau,\tau')$$
$$-2i \iint \delta(\mathbf{k}-\mathbf{p}-\mathbf{q}) d\mathbf{p} d\mathbf{q} M_{Ci\ell m}(\mathbf{p},\mathbf{q})\, u'_{B\ell}(\mathbf{p};\tau) G'_{umj}(\mathbf{q};\tau,\tau')$$
$$= \delta_{ij}\delta(\tau-\tau'), \tag{8.75}$$

$$\frac{\partial G'_e(\mathbf{k};\tau,\tau')}{\partial \tau} - i\iint \delta(\mathbf{k}-\mathbf{p}-\mathbf{q})d\mathbf{p}d\mathbf{q}q_i u'_{Bi}(\mathbf{p};\tau)G'_e(\mathbf{q};\tau,\tau')$$
$$+\bar{\lambda}k^2 G'_e(\mathbf{k};\tau,\tau') = \delta(\tau-\tau'). \tag{8.76}$$

We expand the $O(1)$ field $(\rho'_0, \mathbf{u}'_0, e'_0)$ around the basic one $(\rho'_B, \mathbf{u}'_B, e'_B)$ as

$$f'_0 = f'_B + \sum_{m=1}^{\infty} f'_{0m}, \tag{8.77}$$

and solve Eqs. (8.65)-(8.67) in an iterative manner (subscript m indicates the contributions from the mth iteration). Then the first iteration gives

$$\rho'_0(\mathbf{k};\tau) = \rho'_B(\mathbf{k};\tau) + ik_i\bar{\rho}\int_{-\infty}^{\tau} d\tau_1 G'_\rho(\mathbf{k};\tau,\tau_1)u'_{Bi}(\mathbf{k};\tau_1), \tag{8.78}$$

$$u'_{0i}(\mathbf{k};\tau) = u'_{Bi}(\mathbf{k};\tau) + i(\gamma-1)k_j\int_{-\infty}^{\tau} d\tau_1 G'_{uij}(\mathbf{k};\tau,\tau_1)e'_B(\mathbf{k};\tau_1)$$
$$+i(\gamma-1)k_j(E/\bar{\rho})\int_{-\infty}^{\tau} d\tau_1 G'_{uij}(\mathbf{k};\tau,\tau_1)\rho'_B(\mathbf{k};\tau_1), \tag{8.79}$$

$$e'_0(\mathbf{k};\tau) = e'_B(\mathbf{k};\tau) + ik_i(\gamma-1)E\int_{-\infty}^{\tau} d\tau_1 G'_e(\mathbf{k};\tau,\tau_1)u'_{Bi}(\mathbf{k};\tau_1). \tag{8.80}$$

We substitute Eqs. (8.78)-(8.80) into the $O(\delta)$ system, Eqs. (8.68)-(8.70), and may solve it in entirely the same manner. The leading contributions of the $O(1)$ field $(\rho'_0, \mathbf{u}'_0, e'_0)$ to these equations are obtained from the replacement of $(\rho'_0, \mathbf{u}'_0, e'_0)$ with the basic field $(\rho'_B, \mathbf{u}'_B, e'_B)$. Then Eqs. (8.68)-(8.70) may be simply integrated using the Green's functions G'_ρ, G'_u, and G'_e. For example, we have

$$\rho'_1(\mathbf{k};\tau) = \int_{-\infty}^{\tau} d\tau_1 G'_\rho(\mathbf{k};\tau,\tau_1)\left(-u'_{Bi}(\mathbf{k};\tau_1)\frac{\partial\bar{\rho}}{\partial X_i} - \rho'_B(\mathbf{k};\tau_1)\nabla_X\cdot\mathbf{U}\right.$$
$$\left. - \frac{D\rho'_B(\mathbf{k};\tau_1)}{DT_I} - \bar{\rho}\frac{\partial u'_{Bi}(\mathbf{k};\tau_1)}{\partial X_{Ii}} + ik_i\bar{\rho}u'_{Bi}(\mathbf{k};\tau_1)\right). \tag{8.81}$$

C. Statistics of Basic Field and Renormalization

Equation (8.72) is not dependent on any quantities giving rise to and maintaining the anisotropic state of turbulence. Then we assume that $\mathbf{u}'_B$ is statistically isotropic, and that ρ'_B and e'_B are the fluctuations generated by it. Then we write

$$\frac{\langle\rho'_B(\mathbf{k};\tau)\rho'_B(\mathbf{k}';\tau')\rangle}{\delta(\mathbf{k}+\mathbf{k}')} = Q_\rho(k;\tau,\tau'), \tag{8.82}$$

$$\frac{\left\langle u'_{Bi}(\mathbf{k};\tau)u'_{Bj}(\mathbf{k}';\tau')\right\rangle}{\delta\left(\mathbf{k}+\mathbf{k}'\right)} = Q_{uij}(\mathbf{k};\tau,\tau')$$

$$= D_{ij}(\mathbf{k})Q_{uS}(k;\tau,\tau') + \Pi_{ij}(\mathbf{k})Q_{uC}(k;\tau,\tau'), \tag{8.83}$$

$$\frac{\langle e'_B(\mathbf{k};\tau)e'_B(\mathbf{k}';\tau')\rangle}{\delta\left(\mathbf{k}+\mathbf{k}'\right)} = Q_e(k;\tau,\tau'), \tag{8.84}$$

$$\left\langle G'_\rho(\mathbf{k};\tau,\tau')\right\rangle = G_\rho(k;\tau,\tau'), \tag{8.85}$$

$$\left\langle G'_{uij}(\mathbf{k};\tau,\tau')\right\rangle = G_{uij}(\mathbf{k};\tau,\tau')$$

$$= D_{ij}(\mathbf{k})G_{uS}(k;\tau,\tau') + \Pi_{ij}(\mathbf{k})G_{uC}(k;\tau,\tau'), \tag{8.86}$$

$$\langle G'_e(\mathbf{k};\tau,\tau')\rangle = G_e(k;\tau,\tau'), \tag{8.87}$$

where the solenoidal projection operator $D_{ij}(\mathbf{k})$ is given by Eq. (6.48), and the compressible counterpart $\Pi_{ij}(\mathbf{k})$ is defined as

$$\Pi_{ij}(\mathbf{k}) = k_i k_j \Big/ k^2 \tag{8.88}$$

(subscripts S and C denote solenoidal and compressible, respectively). Strictly speaking, $\rho'_B(\mathbf{k};\tau)$, $Q_\rho(k;\tau,\tau')$, etc. should be written as $\rho_B(\mathbf{k},\mathbf{X};\tau,T)$, $Q_\rho(k,\mathbf{X};\tau,\tau',T)$, etc., respectively.

Moreover, we assume

$$\langle\rho'_B(\mathbf{k};\tau)\mathbf{u}'_B(\mathbf{k};\tau')\rangle = \langle e'_B(\mathbf{k};\tau)\mathbf{u}'_B(\mathbf{k};\tau')\rangle = \langle\rho'_B(\mathbf{k};\tau)e'_B(\mathbf{k};\tau')\rangle = 0. \tag{8.89}$$

Vanishing of the first two correlation functions, the mass and internal-energy fluxes, is natural, for the basic field is isotropic and has no preferred direction. On the other hand, ρ'_B and e'_B do not interact directly with each other, as is seen from Eqs. (8.71) and (8.73), and it is plausible to neglect their correlation.

We combine Eqs. (8.78)-(8.81), etc. with the basic-field statistics, Eqs. (8.82)-(8.87) and (8.89), and calculate some important correlation functions with the aid of the DIA renormalization procedure such as Eq. (6.86) to Eq. (6.87). The final results will be given and discussed in Sec. 8.3.2.

8.3.2. MATHEMATICAL STRUCTURES OF CORRELATION FUNCTIONS

From the mean-field equations (8.10)-(8.12) and the turbulence equations (8.29) and (8.30), we need to know the mathematical structures of the following correlation functions:

$$\text{Mass flux}: \langle \rho' \mathbf{u}' \rangle; \tag{8.90a}$$

$$\text{Velocity variance}: \left\langle u_i' u_j' \right\rangle; \tag{8.90b}$$

$$\text{Internal-energy flux}: \langle e' \mathbf{u}' \rangle; \tag{8.90c}$$

$$\text{Density/internal-energy correlation}: \langle \rho' e' \rangle; \tag{8.90d}$$

$$\text{Dilatational dissipation rate}: \varepsilon_D; \tag{8.90e}$$

$$\text{Pressure-dilatation correlation}: \langle p' \nabla \cdot \mathbf{u}' \rangle, \tag{8.90f}$$

and so on.

A. TSDIA Results

In order to show the results compactly, we introduce the following abbreviation form for time and wavenumber integrals:

$$I_n \{A\} = \int k^{2n} A(k, \mathbf{x}; \tau, \tau, t) d\mathbf{k}, \tag{8.91a}$$

$$I_n \{A, B\} = \int k^{2n} d\mathbf{k} \int_{-\infty}^{\tau} d\tau_1 A(k, \mathbf{x}; \tau, \tau_1, t) B(k, \mathbf{x}; \tau, \tau_1, t). \tag{8.91b}$$

Here we should note that the replacement

$$\mathbf{X} \to \delta \mathbf{x}, \ T \to \delta t \tag{8.92}$$

is made in the final step of the TSDIA analysis, resulting in the automatic disappearance of the scale parameter δ. In what follows, we shall focus attention on the relationship with the inhomogeneity of the mean field and drop the contributions from that of turbulence quantities only.

The correlation functions (8.90a)-(8.90f) may be written as

$$\begin{aligned} \langle \rho' \mathbf{u}' \rangle = & -\frac{1}{3} \left(2I_0 \{G_\rho, Q_{uS}\} + I_0 \{G_\rho, Q_{uC}\} \right) \nabla \bar{\rho} \\ & -\frac{1}{3} (\gamma - 1) \left(2I_0 \{G_{uS}, Q_\rho\} + I_0 \{G_{uC}, Q_\rho\} \right) \frac{\nabla E}{\bar{\rho}} \\ & -\frac{1}{3} \left(2I_0 \{G_{uS}, Q_\rho\} + I_0 \{G_{uC}, Q_\rho\} \right) \frac{1}{\bar{\rho}} \frac{D\mathbf{U}}{Dt}, \end{aligned} \tag{8.93}$$

$$\begin{aligned} \left\langle u_i' u_j' \right\rangle = \frac{2}{3} K \delta_{ij} - \frac{1}{15} \Big(& 7I_0 \{G_{uS}, Q_{uS}\} + 3I_0 \{G_{uS}, Q_{uC}\} \\ & + 3I_0 \{G_{uC}, Q_{uS}\} + 2I_0 \{G_{uC}, Q_{uC}\} \Big) S_{ij}, \end{aligned} \tag{8.94}$$

$$\langle e'\mathbf{u}'\rangle = -\frac{1}{3}\left(2I_0\left\{G_e, Q_{uS}\right\} + I_0\left\{G_e, Q_{uC}\right\}\right)\nabla E$$
$$-\frac{1}{3}(\gamma-1)\left(2I_0\left\{G_{uS}, Q_e\right\} + I_0\left\{G_{uC}, Q_e\right\}\right)\frac{\nabla\bar{\rho}}{\bar{\rho}}, \tag{8.95}$$

$$\langle \rho' e'\rangle = I_0\left\{G_e, Q_\rho\right\}\frac{1}{\bar{\rho}}\left((\gamma-1)E\nabla\cdot\mathbf{U} + \frac{DE}{Dt}\right), \tag{8.96}$$

$$\varepsilon_D = \frac{4}{3}\nu I_1\left\{Q_{uC}\right\}, \tag{8.97}$$

$$\langle p'\nabla\cdot\mathbf{u}'\rangle = -(\gamma-1)I_1\left\{G_{uS}, Q_{uC}\right\}\bar{\rho}E, \tag{8.98}$$

$$\langle \rho'\nabla\cdot\mathbf{u}'\rangle = (\gamma-1)I_1\left\{G_{uC}, Q_\rho\right\}\frac{E}{\bar{\rho}} - I_1\left\{G_\rho, Q_{uC}\right\}\bar{\rho}, \tag{8.99}$$

where

$$K = I_0\left\{Q_{uS}\right\} + \frac{1}{2}I_0\left\{Q_{uC}\right\} - I_0\left\{G_{uS}, \frac{DQ_{uS}}{Dt}\right\} - \frac{1}{2}I_0\left\{G_{uC}, \frac{DQ_{uC}}{Dt}\right\}$$
$$-\frac{1}{3}\left(2I_0\left\{G_{uS}, Q_{uS}\right\} + \frac{1}{2}I_0\left\{G_{uC}, Q_{uC}\right\}\right)\nabla\cdot\mathbf{U}. \tag{8.100}$$

The pressure-velocity correlation $\langle p'\mathbf{u}'\rangle$ may be calculated using Eq. (8.45).

B. Suggestions to Turbulence Modeling

Equation (8.94) is a turbulent-viscosity expression for $\langle u_i'u_j'\rangle$. Here the turbulent viscosity ν_T is given by

$$\nu_T = (1/15)\left(7I_0\left\{G_{uS}, Q_{uS}\right\} + 3I_0\left\{G_{uS}, Q_{uC}\right\}\right.$$
$$\left. + 3I_0\left\{G_{uC}, Q_{uS}\right\} + 2I_0\left\{G_{uC}, Q_{uC}\right\}\right). \tag{8.101}$$

In the current compressible turbulence modeling, a turbulent-viscosity model for $\langle u_i'u_j'\rangle$ is widely used, and ν_T is written in the same form as the solenoidal counterpart, Eq. (4.8), which corresponds to the first part of Eq. (8.101). As a result, the compressibility effect on $\langle u_i'u_j'\rangle$ occurs implicitly through K and ε. This is one of the reasons why such a model combined with the K and ε equations without explicit compressibility effects fails to reproduce the drastic suppression of the growth rate of a compressible free-shear layer.

Equation (8.101) shows the necessity of the explicit inclusion of compressibility effects. For this purpose, we need the knowledge about the compressible Green's function G_{uC}, the spectrum of the compressible velocity variance, Q_{uC}, etc., but we do not have any definite knowledge about them.

As a method of obtaining the results applicable to one-point compressible turbulence modeling, we may mention two approaches. One is to derive one-point expressions for $\langle u_i' u_j' \rangle$ etc. by assuming some proper spectral forms for Q_{uC}, G_{uC}, etc. Another is to resort to the Markovianized two-scale method explained in Chapter 7.

From the mean-field equations and turbulence equations, (8.10)-(8.12), (8.29), and (8.30), we can see that $\langle \rho' \mathbf{u}' \rangle$ is one of the primary quantities characterizing effects of compressibility. In Eq. (8.93), the first term corresponds to the familiar gradient-diffusion expression (8.43). The second ∇E-related part seems rather unusual, compared with the $\nabla \bar{\rho}$-related one. The former expresses the contribution of Eq. (8.69) to $\langle \rho' \mathbf{u}' \rangle$, whereas the latter comes from Eq. (8.68) for ρ_1'. The similar situation holds for Eq. (8.95) for $\langle e' \mathbf{u}' \rangle$, which is also dependent on $\nabla \bar{\rho}$ in addition to ∇E. The $\nabla \bar{\rho}$ dependence arises from Eq. (8.69).

In the K equation (8.29), the typical compressibility-related quantities are ε_D, $\langle p' \nabla \cdot \mathbf{u}' \rangle$, and $\langle p' \mathbf{u}' \rangle$, in addition to $\langle \rho' \mathbf{u}' \rangle$. The leading contribution to $\langle p' \nabla \cdot \mathbf{u}' \rangle$ is given by Eq. (8.98), which is always negative. The DNS of compressible turbulence shows that it is positive in decaying isotropic flow, whereas it tends to be negative in homogeneous-shear flow. The present finding is not consistent with the DNS of isotropic flow. This discrepancy is supposed to come from the fact that in the present analysis, we focus attention on the contributions of the mean field generating inhomogeneity of turbulence.

The pressure-velocity correlation $\langle p' \mathbf{u}' \rangle$ is calculated using Eq. (8.45). In the solenoidal motion, p' is obtained by the solution of Eq. (4.11). The TSDIA analysis of $\langle p' \mathbf{u}' \rangle$ in the case, which is explained in detail in Sec. 6.6.1, indicates that it is less important than the triple velocity correlation $\langle (\mathbf{u}'^2/2) \mathbf{u}' \rangle$. In a high-speed flow containing a shock wave, both $\nabla \bar{\rho}$ and ∇E become large, and the contribution of Eq. (8.45) to $\langle p' \mathbf{u}' \rangle$ may become important.

In the K_ρ equation (8.30), the third term is dominant, compared with the fourth, under the condition of Eq. (8.25). Equation (8.99) consists of two parts, and the sign of the first part is opposite to that of the second. In case that the first term is dominant, it plays the role similar to ε in the K equation (8.29). The last term $\langle \rho'^2 \mathbf{u}' \rangle$ corresponds to $\bar{\rho} \langle (\mathbf{u}'^2/2) \mathbf{u}' \rangle$ in $\mathbf{T}_K$ [Eq. (8.32)]. Its TSDIA analysis is very complicated, as is inferred from Sec. 6.6.1. The finding about $\langle (\mathbf{u}'^2/2) \mathbf{u}' \rangle$ in the solenoidal case suggests that a plausible modeling of $\langle \rho'^2 \mathbf{u}' \rangle$ is

$$\left\langle \rho'^2 \mathbf{u}' \right\rangle = -\nu_{\rho\rho} \nabla K_\rho, \tag{8.102}$$

with the turbulent diffusivity $\nu_{\rho\rho}$ as

$$\nu_{\rho\rho} \propto \nu_T. \tag{8.103}$$

C. Alternatives of TSDIA Approaches

The TSDIA formalism is based on the combination of the scale-parameter expansion and renormalization procedures, and its analysis based on one set of fundamental variables is not always equivalent to the counterpart based on the other set, even in a similar physical state. In the foregoing TSDIA analysis, emphasis is put on the effects of density fluctuation. This fact does not rule out the importance of pressure fluctuation under a different choice of fundamental variables. In reality, the importance was pointed out on the basis of the DNS of compressible homogeneous-shear turbulence (Fujiwara 1996). The TSDIA calculation based on the choice of $(\rho, \mathbf{u}, p)$ was done by Hamba and Blaisdell (1997). There proper spectral forms are assumed for G_{uC}, Q_{uC}, etc., and one-point expressions for various correlation functions were derived.

An adiabatic state of compressible flow is expressed in terms of ρ and p [see Eq. (8.148) below]. A quantity appropriate for describing the deviation from the state is the entropy s. Under the choice of $(\rho, \mathbf{u}, s)$ and the condition of weak compressibility, the TSDIA analysis was done by Rubinstein and Erlebacher (1997) using the expansion around the solenoidal motion, as is similar to the work by Yoshizawa (1986), and effects of turbulence on the sound speed and transport coefficients were discussed. In this approach, a coupled system of Green's functions is directly treated, unlike the present one. This work was extended by Hamba (1997) to a highly compressible flow state, and the model of the four-equation type, which consists of the transport equations for K, ε, $\langle s'^2 \rangle$, and K_C, was proposed (K_C is the compressible part of K).

8.4. Markovianized Two-Scale Approach to Compressible Flows

One of the typical flows in which effects of compressibility appear clearly is a free-shear layer, as was referred to in Sec. 8.1. Its drastic decrease in the growth rate of the layer is associated with the convective Mach number M_C that characterizes the magnitude of the velocity difference of two streams. Compressibility effects occur at the stage of the transition from the laminar to turbulent state, which corresponds to low M_C (Sandham and Reynolds 1991). It is not clear, however, how long these effects survive at the stage of high M_C.

In the context of turbulence modeling, this phenomenon has been explained by the enhancement of energy dissipation through ε_D and $\langle p' \nabla \cdot \mathbf{u}' \rangle$ in the turbulent-energy equation (8.29). The DNS of a free-shear layer, however, puts more emphasis on the decrease in the production rate that is given by its first term (Vreman et al. 1996). A method of incorporating compressibility effects on the production rate is to include those on the turbulent viscosity ν_T. For this purpose, we need the knowledge about the compressible part of the velocity covariance, as is seen from Eq. (8.101). In what follows, we shall start from a simple assumption on such a part, and examine compressibility effects on ν_T with resort to the Markovianized two-scale (MTS) method that is detailed in Chapter 7 (Yoshizawa 1995)

8.4.1. MODELING BASED ON MTS METHOD

A. Formal Integration

In the MTS method, we also start from Eqs. (8.57)-(8.59) (Yoshizawa et al. 1997). The essential point of the method is that almost all of the analyses are performed in physical space. This point leads to the simplicity of the mathematical manipulation and, at the same time, to the limit of its applicability. We expand the fluctuation f' in δ as

$$f' = \sum_{n=0}^{\infty} \delta^n f'_n. \tag{8.104}$$

We substitute Eq. (8.104) into Eqs. (8.57)-(8.59). Then the $O(1)$ parts obey

$$\frac{D\rho'_0}{D\tau} + \nabla_\xi \cdot (\rho'_0 \mathbf{u}'_0) + \bar{\rho}\nabla_\xi \cdot \mathbf{u}'_0 = 0, \tag{8.105}$$

$$\frac{Du'_{0i}}{D\tau} + (\mathbf{u}'_0 \cdot \nabla_\xi)\, u'_{0i} - \bar{\nu}\frac{\partial s'_{0ji}}{\partial \xi_j} + (\gamma - 1)\frac{\partial e'_0}{\partial \xi_i} + (\gamma - 1)\frac{E}{\bar{\rho}}\frac{\partial \rho'_0}{\partial \xi_i} = 0, \tag{8.106}$$

$$\frac{De'_0}{D\tau} + (\mathbf{u}'_0 \cdot \nabla_\xi)\, e'_0 - \bar{\lambda}\Delta_\xi e'_0 + (\gamma - 1)\, E\nabla_\xi \cdot \mathbf{u}'_0 = 0. \tag{8.107}$$

On the other hand, the $O(\delta)$ counterparts are

$$\frac{D\rho'_1}{D\tau} + \nabla_\xi \cdot (\mathbf{u}'_0 \rho'_1) = I_\rho, \tag{8.108}$$

$$\frac{Du'_{1i}}{D\tau} + (\mathbf{u}'_0 \cdot \nabla_\xi)\, u'_{1i} + (\mathbf{u}'_1 \cdot \nabla_\xi)\, u'_{0i} - \bar{\nu}\frac{\partial s'_{1ji}}{\partial \xi_j} = I_{ui}, \tag{8.109}$$

$$\frac{De'_1}{D\tau} + (\mathbf{u}'_0 \cdot \nabla_\xi)\, e'_1 - \bar{\lambda}\Delta_\xi e'_1 = I_e, \tag{8.110}$$

where

$$I_\rho = -(\mathbf{u}'_0 \cdot \nabla_X)\,\bar{\rho} - \rho'_0 \nabla_X \cdot \mathbf{U} - \frac{D\rho'_0}{DT} - \bar{\rho}\nabla_X \cdot \mathbf{u}'_0 \\ -\nabla_\xi \cdot (\rho_0 \mathbf{u}'_1) - \bar{\rho}\nabla_\xi \cdot \mathbf{u}'_1, \tag{8.111}$$

$$\begin{aligned}\mathbf{I}_u = &- (\mathbf{u'}_0 \cdot \nabla_X) \mathbf{U} - (\gamma - 1) \frac{\rho'_0}{\bar{\rho}} \nabla_X E - (\gamma - 1) \frac{e'_0}{\bar{\rho}} \nabla_X \bar{\rho} \\ &- \frac{\rho'_0}{\bar{\rho}} \frac{D\mathbf{U}}{DT} - \frac{D\mathbf{u'}_0}{DT} - (\gamma - 1) \nabla_X e'_0 - (\gamma - 1) \frac{E}{\bar{\rho}} \nabla_X \rho'_0, \\ &- (\gamma - 1) \nabla_\xi e'_1 - (\gamma - 1) \frac{E}{\bar{\rho}} \nabla_\xi \rho'_1, \end{aligned} \tag{8.112}$$

$$\begin{aligned} I_e = &- (\mathbf{u'}_0 \cdot \nabla_X) E - \frac{\rho'_0}{\bar{\rho}} \frac{DE}{DT} - (\gamma - 1) e'_0 \nabla_X \cdot \mathbf{U} \\ &- (\gamma - 1) \rho'_0 \frac{E}{\bar{\rho}} \nabla_X \cdot \mathbf{U} - \frac{De'_0}{DT} - (\gamma - 1) E \nabla_X \cdot \mathbf{u'}_0 \\ &- (\mathbf{u'}_1 \cdot \nabla_\xi) e'_0 - (\gamma - 1) E \nabla_\xi \cdot \mathbf{u'}_1. \end{aligned} \tag{8.113}$$

In the TSDIA formalism of Sec. 8.3, we needed the Green's functions for the $O(1)$ equations (8.105)-(8.107). These equations interact with one another due to effects of compressibility, resulting in a coupled system of equations for Green's functions. In order to reduce the complicatedness due to the coupling, we introduced the concept of the basic field $(\rho'_B, \mathbf{u'}_B, e'_B)$. We follow the same procedure and write

$$\frac{D\rho'_B}{D\tau} + \nabla_\xi \cdot (\rho'_B \mathbf{u'}_B) = 0, \tag{8.114}$$

$$\frac{Du'_{Bi}}{D\tau} + (\mathbf{u'}_B \cdot \nabla_\xi) u'_{Bi} - \bar{\nu} \frac{\partial s'_{Bji}}{\partial \xi_j} = 0, \tag{8.115}$$

$$\frac{De'_B}{D\tau} + (\mathbf{u'}_B \cdot \nabla_\xi) e'_B - \bar{\lambda} \Delta_\xi e'_B = 0. \tag{8.116}$$

Corresponding to this basic field, we define the differential operators

$$L_\rho f = \frac{Df}{D\tau} + \nabla_\xi \cdot (f \mathbf{u'}_B), \tag{8.117}$$

$$\begin{aligned} L_u f_i = &\frac{Df_i}{D\tau} + (\mathbf{u'}_B \cdot \nabla_\xi) f_i + (\mathbf{f} \cdot \nabla_\xi) u'_{Bi} \\ &- \bar{\nu} \frac{\partial}{\partial \xi_j} \left(\frac{\partial f_j}{\partial x_i} + \frac{\partial f_i}{\partial x_j} - \frac{2}{3} \nabla \cdot \mathbf{f} \delta_{ij} \right), \end{aligned} \tag{8.118}$$

$$L_e f = \frac{Df}{D\tau} + (\mathbf{u'}_B \cdot \nabla_\xi) f - \bar{\lambda} \Delta_\xi f. \tag{8.119}$$

We perform the expansion around the basic field $(\rho'_B, \mathbf{u}'_B, e'_B)$, as in Eq. (8.77). Then the $O(1)$ system consisting of Eqs. (8.105)-(8.107) is solved formally as

$$\rho'_0 = \rho'_B - L_\rho^{-1}\left(\bar{\rho}\nabla_\xi \cdot \mathbf{u}'_B\right) + \cdots, \tag{8.120}$$

$$\mathbf{u}'_0 = \mathbf{u}'_B - L_u^{-1}\left(\left(\gamma - 1\right)\nabla_\xi e'_B + \left(\gamma - 1\right)\frac{E}{\bar{\rho}}\nabla_\xi \rho'_B\right) + \cdots, \tag{8.121}$$

$$e'_0 = e'_B - L_e^{-1}\left(\left(\gamma - 1\right)E\nabla_\xi \cdot \mathbf{u}'_B\right) + \cdots. \tag{8.122}$$

In Eq. (8.122), the first part or e'_B survives in the solenoidal limit and is associated with the diffusion of the internal energy by turbulent motion. The second part expresses the internal-energy change due to dilatational fluctuating motion and will be neglected, compared with e'_B.

Similarly, the leading solution of the $O(\delta)$ system, Eqs. (8.108)-(8.110), is given by

$$\rho'_1 = L_\rho^{-1} I_\rho, \tag{8.123}$$

$$\mathbf{u}'_1 = L_u^{-1} \mathbf{I}_u, \tag{8.124}$$

$$e'_1 = L_e^{-1} I_e, \tag{8.125}$$

where $(\rho'_0, \mathbf{u}'_0, e'_0)$ in $(I_\rho, \mathbf{I}_u, I_e)$ are replaced with $(\rho'_B, \mathbf{u}'_B, e'_B)$.

B. Introduction of Characteristic Time Scales

Equations (8.120)-(8.125) are characterized by the integrals that may be written symbolically as

$$f = L_f^{-1} I_f = \int_{-\infty}^{\tau} G_f(\tau, s) I_f(s) ds, \tag{8.126}$$

where G_f is the Green's function expressing the inverse operator L_f^{-1}, and attention is focused on the time dependence for simplicity of discussion. One of the key steps of the MTS method is to approximate the time integral using a characteristic time scale τ_f, as is explained in Sec. 7.2.2; namely, we write

$$f = \tau_f I_f(\tau), \tag{8.127}$$

with

$$\tau_f = \int_{-\infty}^{\tau} G_f(\tau, s) ds. \tag{8.128}$$

In the solenoidal motion, we estimate the characteristic time scale τ_T as

$$\tau_T \propto K_0 / \varepsilon_0 , \tag{8.129}$$

using the Kolmogorov's inertial-range spectrum for $E_0(k)$, Eq. (7.19), that is related to the turbulent energy of the $O(1)$ field

$$K_0 \equiv \left\langle \mathbf{u}'^2_0/2 \right\rangle = \int E_0(k)dk \tag{8.130}$$

[ε_0 is the $O(1)$ part of ε]. Equation (8.129) leads to the solenoidal turbulent viscosity

$$\nu_T \propto K^2/\varepsilon, \tag{8.131}$$

after the renormalization procedure (see Sec. 7.2.3).

In the current compressible turbulence modeling, Eq. (8.129) is also used as the characteristic time scale. This is one of the primary reasons why it is difficult to include effects of compressibility explicitly in the modeling. In order to overcome this difficulty, we need to include compressibility effects explicitly in the course of the Markovianization of L_ρ^{-1}, L_u^{-1}, and L_e^{-1} that are defined by Eqs. (8.117)-(8.119), respectively.

In the study of compressible flow, the velocity field is often resolved into the solenoidal and compressible parts by means of the Helmholtz resolution. We denote each contribution to $E_0(k)$ by $E_S(k)$ and $E_C(k)$, respectively. We should note that such a division of the energy spectrum is not exact, except homogeneous turbulence. In the study of compressible isotropic turbulence, it was shown by Bertoglio et al. (1998) using the EDQNM approximation that the deviation from the Kolmogorov law becomes much more prominent in $E_C(k)$ with the increasing turbulent Mach number M_T defined as

$$M_T = \sqrt{\frac{\langle \mathbf{u}'^2_0 \rangle}{\bar{a}}} = \sqrt{\frac{2}{\gamma(\gamma-1)}\frac{K_0}{E}} \tag{8.132}$$

($\bar{a}$ is the mean sound speed).

In case that the solenoidal part of velocity fluctuations is a primary one, compared with the compressible part, the latter is supposed to obey the diffusion process due to the former that is similar to the scalar diffusion in turbulent motion (see Sec. 3.3). In the case, it is plausible to assume that $E_S(k)$ obeys almost the same power law as the Kolmogorov one, whereas the compressibility effects on the energy spectrum appear through $E_C(k)$. This picture of compressible turbulence leads to

$$E_S(k) = K_O \varepsilon_{0S}^{2/3} k^{-5/3}, \tag{8.133a}$$

$$E_C(k) = K'_O \varepsilon_{0D} \varepsilon_{0S}^{-1/3} k^{-5/3} \left(k\ell_C\right)^{-\beta}. \tag{8.133b}$$

Here K_O is the Kolmogorov constant, K'_O is the compressible counterpart of K_O, and ε_{0S} and ε_{0D} are the solenoidal and compressible parts of ε_0, respectively, which are defined as

$$\varepsilon_{0S} = \varepsilon_0 - \varepsilon_{0D}, \tag{8.134a}$$

$$\varepsilon_{0D} = (4/3)\,\nu \left\langle (\nabla \cdot \mathbf{u}'_0)^2 \right\rangle. \tag{8.134b}$$

In Eq. (8.133b), ℓ_C is the length scale characterizing energy-containing eddies in compressible turbulence, and the exponent β is dependent on M_T. In the comparison with the scalar diffusion, ε_{0D} corresponds to the destruction rate of a scalar variance. The energy spectral form (8.133) with ε_{0S} replaced with the total dissipation rate ε_0 was also adopted in the TSDIA study of compressible shear turbulence by Hamba and Blaisdell (1997).

We use the foregoing plausible assumption to examine effects of compressibility on the characteristic time scale. We apply Eq. (8.133) to a spectral expression

$$K_0 = \int_{\ell_C^{-1}}^{\infty} (E_S(k) + E_C(k))dk, \tag{8.135}$$

where the contribution of the interaction between solenoidal and compressible fluctuations to $E_0(k)$ is not treated explicitly. As a result, we have

$$K_0 = C_S \varepsilon_{0S}^{2/3} \ell_C^{2/3} \left(1 + \frac{C_C}{C_S}\frac{\varepsilon_{0D}}{\varepsilon_{0S}}\right) \tag{8.136a}$$

$$= C_S \varepsilon_0^{2/3} \ell_C^{2/3} \left(1 + \frac{C_C}{C_S}\frac{\varepsilon_{0D}}{\varepsilon_{0S}}\right) \left(1 + \frac{\varepsilon_{0D}}{\varepsilon_{0S}}\right)^{-2/3}, \tag{8.136b}$$

where

$$C_S = (3/2)\,K_O, \tag{8.137a}$$

$$C_C = \frac{3}{2+3\beta} K'_O. \tag{8.137b}$$

In obtaining Eq. (8.136), the essential point lies in not the detailed spectral form of $E_C(k)$ but its linear dependence on ε_{0D}.

In what follows, we shall neglect the dependence of β on M_T, and focus attention on the compressibility effects through ε_{0D}. As for $\varepsilon_{0D}/\varepsilon_{0S}$, we use a simple model

$$\varepsilon_{0D} \left/ \varepsilon_{0S} \right. = C_{DS} \left(K_{\rho 0 N} \left/ M_T^2 \right. \right), \tag{8.138}$$

where C_{DS} is a constant, and $K_{\rho 0 N}$ is the normalized density variance

$$K_{\rho 0 N} = K_{\rho 0} \left/ \bar{\rho}^2 \right., \tag{8.139}$$

with $K_{\rho 0} = \langle \rho_0^2 \rangle$. The relationship of Eq. (8.138) with the energy and density spectra will be referred to in *Note*. In the context of the dilatation effects on the enhancement of energy dissipation, the similar models were also proposed in the study of the compressible turbulence modeling of the three-equation type (Taulbee and VanOsdol 1991; Fujiwara and Arakawa 1993). On the other hand, a model

$$\varepsilon_{0D} / \varepsilon_{0S} = C'_{DS} M_T^2 \tag{8.140}$$

was proposed by Sarkar et al. (1991b), where C'_{DS} is a constant of $O(1)$.

We expand Eq. (8.136b) in $\varepsilon_{0D}/\varepsilon_{0S}$, and retain the terms up to its first order. We substitute Eq. (8.138) into $\varepsilon_{0D}/\varepsilon_{0S}$ in the resulting expression. Then we have

$$K_0 = C_S \varepsilon_0^{2/3} \ell_C^{2/3} \left(1 + A \frac{K_{\rho 0 N}}{M_T^2} \right), \tag{8.141}$$

where

$$A = \left(\frac{C_C}{C_S} - \frac{2}{3} \right) C_{DS}. \tag{8.142}$$

We solve Eq. (8.141) with respect to ℓ_C, and have

$$\ell_C = \frac{1}{C_S^{3/2}} \frac{K_0^{3/2}}{\varepsilon_0} \frac{1}{\left(1 + A \left(K_{\rho 0 N} / M_T^2\right)\right)^{3/2}} \tag{8.143a}$$

or

$$\frac{\ell_C}{\ell_{CS}} \propto \frac{1}{\left(1 + A \left(K_{\rho 0 N} / M_T^2\right)\right)^{3/2}}, \tag{8.143b}$$

where ℓ_{CS} is the solenoidal-type characteristic length scale of energy-containing eddies and is given by

$$\ell_{CS} \propto K_0^{3/2} / \varepsilon_0 . \tag{8.144}$$

Equation (8.143b) with positive A shows that effects of compressibility lead to the reduction of the integral scale. This finding is interesting in the context of Sarkar's (1995) indication that effects of compressibility are correlated well with the magnitude of the gradient Mach number M_G, which is defined by

$$M_G = \sqrt{S_{ij}^2} \ell_C / \bar{a} . \tag{8.145}$$

In the MTS method, we need to find the time scales characterizing L_ρ^{-1}, L_u^{-1}, and L_e^{-1} that occur in Eqs. (8.120)-(8.125). We note that $\ell_C/\sqrt{K}$ is a time scale in compressible turbulence, specifically, the one characterizing the energy-containing components of motion. From Eq. (8.143), we have

$$\left(L_\rho^{-1}, L_u^{-1}, L_e^{-1}\right) \equiv \left(\tau_{T\rho}, \tau_{Tu}, \tau_{Te}\right)$$
$$= \left(C_\rho, C_u, C_e\right) \frac{K_0 / \varepsilon_0}{\left(1 + A\left(K_{\rho 0 N} / M_T^2\right)\right)^{3/2}} \tag{8.146a}$$
$$\propto \left(C_\rho, C_u, C_e\right) \frac{\tau_T}{\left(1 + A\left(K_{\rho 0 N} / M_T^2\right)\right)^{3/2}}, \tag{8.146b}$$

where C_ρ, C_u, and C_e are numerical factors, and the solenoidal time scale τ_T is defined by Eq. (8.129). The numerical factors C_ρ etc. are left as unknown. The operators L_u^{-1} and L_e^{-1}, which are associated with C_u and C_e, respectively, survive in the solenoidal limit. As a result, these two constants may be estimated using the knowledge about turbulence models in incompressible turbulence, as will be shown later.

Note: Equation (8.138) may be regarded as an extension of Eq. (8.140). We consider the weak-compressibility case, in which the pressure fluctuation coming from compressible motion, p'_C, is of the same magnitude as the dynamic one p'_D. They are written as

$$p'_C \cong \bar{a}^2 \rho', \tag{8.147a}$$

$$p'_D \propto \bar{\rho} \mathbf{u}'^2. \tag{8.147b}$$

Here the former is obtained from the adiabatic relation and the definition of sound speed

$$p \propto \rho^\gamma, \; a = \sqrt{dp / d\rho}. \tag{8.148}$$

The balance of these two pressure effects results in

$$K_\rho \Big/ \bar{\rho}^2 \propto M_T^4, \tag{8.149}$$

where the turbulent Mach number M_T is defined by Eq. (8.132). We combine Eq. (8.138) with Eq. (8.149) to recover Eq. (8.140).

The modeling of the type (8.138) was proposed by Fujiwara and Arakawa (1993) using the DNS database of isotropic turbulence. In the DNS of homogeneous-shear turbulence, $\varepsilon_{0D}/\varepsilon_{0S}$ is highly dependent on the initial condition of density fluctuation, and $\varepsilon_{0D}/\varepsilon_{0S}$ cannot be related simply to M_T (Blaisdell et al. 1993). From the viewpoint of a spectral representation, Eq. (8.138) is equivalent to the density-variance spectrum

$$E_\rho(k) \propto M_T^2 \bar{\rho}^2 \varepsilon_{0D} \varepsilon_0^{-1} k^{-1} \left(\ell_C k\right)^{-\mu} \tag{8.150a}$$

combined with

$$K_{\rho 0} = \int_{\ell_C^{-1}}^{\infty} E_\rho(k) dk, \tag{8.150b}$$

where μ is a nondimensional coefficient (Hamba and Blaisdell 1997).

C. Evaluation of Correlation Functions

Our primary interest lies in the effect of compressibility on the Reynolds stress $\langle u'_i u'_j \rangle$. It is expanded as

$$\left\langle u'_i u'_j \right\rangle = \left\langle u'_{0i} u'_{0j} \right\rangle + \delta \left(\left\langle u'_{0i} u'_{1j} \right\rangle + \left\langle u'_{1i} u'_{0j} \right\rangle \right) + O\left(\delta^2\right). \tag{8.151}$$

We substitute Eqs. (8.121) and (8.124) into Eq. (8.151), and make the Markovianization of L_u^{-1} using τ_{Tu} in Eq. (8.146). Here we focus attention on the interaction with the mean field $(\bar{\rho}, \mathbf{U}, E)$. This approximation is consistent with the choice of τ_{Tu} related to the energy-containing components of motion. The resulting expression is

$$\begin{aligned}
&\left\langle u'_i u'_j \right\rangle - \frac{2}{3} K \delta_{ij} \\
&= \tau_{Tu} \left(- \left(\left\langle u'_{0i} u'_{0\ell} \right\rangle \frac{\partial U_j}{\partial x_\ell} + \left\langle u'_{0j} u'_{0\ell} \right\rangle \frac{\partial U_i}{\partial x_\ell} \right]_D \right. \\
&\quad - (\gamma - 1) \frac{1}{\bar{\rho}} \left(\left\langle \rho'_0 u'_{0j} \right\rangle \frac{\partial E}{\partial x_i} + \left\langle \rho'_0 u'_{0i} \right\rangle \frac{\partial E}{\partial x_j} - \frac{2}{3} \left\langle \rho'_0 u'_{0\ell} \right\rangle \frac{\partial E}{\partial x_\ell} \delta_{ij} \right) \\
&\quad - (\gamma - 1) \frac{1}{\bar{\rho}} \left(\left\langle e'_0 u'_{0j} \right\rangle \frac{\partial \bar{\rho}}{\partial x_i} + \left\langle e'_0 u'_{0i} \right\rangle \frac{\partial \bar{\rho}}{\partial x_j} - \frac{2}{3} \left\langle e'_0 u'_{0\ell} \right\rangle \frac{\partial \bar{\rho}}{\partial x_\ell} \delta_{ij} \right) \\
&\quad \left. - \frac{1}{\bar{\rho}} \left(\left\langle \rho'_0 u'_{0j} \right\rangle \frac{DU_i}{Dt} + \left\langle \rho'_0 u'_{0i} \right\rangle \frac{DU_j}{Dt} - \frac{2}{3} \left\langle \rho'_0 u'_{0\ell} \right\rangle \frac{DU_\ell}{Dt} \delta_{ij} \right) \right).
\end{aligned} \tag{8.152}$$

Here the main part of the second term on the left-hand side comes from the first $O(1)$ term of Eq. (8.151). We should note that $\mathbf{X}$ and T have been replaced with $\delta\mathbf{x}$ and δt, respectively, resulting in the disappearance of the scale parameter δ.

We assume that the basic field $(\rho'_B, \mathbf{u}'_B, e'_B)$ is statistically isotropic; namely, we write

$$\left\langle {\rho'_B}^2 \right\rangle = K_{\rho B}, \tag{8.153}$$

$$\left\langle u'_{Bi} u'_{Bj} \right\rangle = (2/3) K_B \delta_{ij}, \tag{8.154}$$

$$\left\langle {e'_B}^2 \right\rangle = K_{eB}, \tag{8.155}$$

$$\left\langle \rho'_B \mathbf{u}'_B \right\rangle = \left\langle e'_B \mathbf{u}'_B \right\rangle = 0. \tag{8.156}$$

We apply Eqs. (8.153)-(8.156) to Eq. (8.152) to express the latter using the mean field and the basic-filed quantities $K_{\rho B}$, K_B, and ε_B. Afterwards, we resort to the so-called renormalization procedure; namely, we perform the replacement

$$K_{\rho B} \to K_\rho,\ K_B \to K,\ \varepsilon_B \to \varepsilon,\ K_{eB} \to K_e. \tag{8.157}$$

As a result, we have the turbulent-viscosity representation

$$\left\langle u_i' u_j' \right\rangle - \frac{2}{3} K \delta_{ij} = -\nu_{TC} S_{ij}, \tag{8.158}$$

which comes from the first part of Eq. (8.152). The turbulent viscosity with the explicit dependence on the compressibility effect, ν_{TC}, is written as

$$\nu_{TC} = \frac{2}{3} \tau_{Tu} K = \frac{\nu_{TS}}{\left(\left[1 + A \left(K_{\rho N} / M_T^2 \right) \right) \right)^{3/2}}, \tag{8.159}$$

where use has been made of Eq. (8.146a), and the renormalized density variance $K_{\rho N}$ and the solenoidal turbulent viscosity ν_{TS} are defined as

$$K_{\rho N} = K_\rho \Big/ \bar{\rho}^2 , \tag{8.160}$$

$$\nu_{TS} = C_\nu \left(K^2 / \varepsilon \right),\ C_\nu = (2/3)\, C_u, \tag{8.161}$$

respectively.

In the limit of vanishing K_ρ, ν_{TC} tends to the incompressible form, Eq. (8.161). The numerical factor C_ν is usually chosen as 0.09, which leads to

$$C_u = 0.14. \tag{8.162}$$

This procedure of estimating the unknown constant is a key step of the MTS method not using Green's functions in the wavenumber space.

Similarly, we have

$$\langle \rho' \mathbf{u}' \rangle = -\frac{\nu_{TC}}{\sigma_\rho} \nabla \bar{\rho} - \frac{3}{2} (\gamma - 1)\, \nu_{TC} \frac{\bar{\rho}}{K} K_{\rho N} \nabla E - \frac{3}{2} \nu_{TC} \frac{\bar{\rho}}{K} K_{\rho N} \frac{D\mathbf{U}}{Dt}, \tag{8.163}$$

using Eqs. (8.120), (8.121), (8.123), and (8.124), where

$$\sigma_\rho = C_u / C_\rho\,. \tag{8.164}$$

In Eq. (8.163), the compressibility effects are explicitly taken into account through ν_{TC} and $K_{\rho N}$.

For $\langle e'\mathbf{u}'\rangle$, the counterpart of Eq. (8.163) is

$$\langle e'\mathbf{u}'\rangle = -\frac{\nu_{TC}}{\sigma_e}\nabla E - \frac{3}{2}(\gamma-1)\,\nu_{TC}\frac{K_e}{K}\frac{1}{\bar{\rho}}\nabla\bar{\rho}, \tag{8.165}$$

with

$$\sigma_e = C_u/C_e\,. \tag{8.166}$$

In the following, we shall aim at constructing a simple compressible turbulence model and limit the turbulent transport equations to those for (K_ρ, K, ε). Then we shall drop the second K_e-related effect. This approximation, however, does not imply that the effect is always unimportant. In combustion phenomena, large density gradients occur in the region adjacent to flame. One of their prominent features is the counter-gradient transport of heat; namely, heat is transported in the direction of positive temperature gradient. The second term is a promising candidate for its mechanism.

D. Nonlinear Corrections

In the foregoing analyses, a model for the turbulent viscosity ν_{TC} was derived, which explicitly includes the compressibility effects through $K_{\rho N}$ and M_T. There are, however, still some classes of flows whose important properties cannot be reproduced intrinsically within the framework of the turbulence-viscosity approximation to the Reynolds stress. For such flows, we need to introduce the corrections to the approximation. A straightforward method of obtaining the corrections to Eq. (8.158) is to perform the $O(\delta^2)$ calculation in Eq. (8.151). Such an analysis leads to a complicated mathematical one. One method that is much simpler but may capture some of the $O(\delta^2)$ results is the iterative use of Eq. (8.152). For the solenoidal motion, this method is explained in Sec. 7.2.3.

In the foregoing calculation, we started from the isotropic basic-field statistics, Eqs. (8.153)-(8.156), and obtained the simplest anisotropic effects generated by the mean velocity gradient, that is, Eqs. (8.158), (8.163), and (8.165). Now we proceed to the second iteration procedure; namely, we substitute these first iteration results into $\langle u'_{0i}u'_{0j}\rangle$, $\langle\rho'_0\mathbf{u}'_0\rangle$, and $\langle e'_0\mathbf{u}'_0\rangle$ on the right-hand side of Eq. (8.152). As a result, we are led to

$$\left\langle u'_i u'_j\right\rangle - \frac{2}{3}K\delta_{ij} = \sum_{n=1}^{4} M_{ij}^{(n)}. \tag{8.167}$$

Here $M_{ij}^{(n)}$'s are defined as

$$M_{ij}^{(1)} = -\nu_{TC}S_{ij}, \tag{8.168}$$

$$M_{ij}^{(2)} = \frac{6}{4}\frac{\nu_{TC}^2}{K}\left[S_{ik}S_{kj}\right]_D + \frac{3}{4}\frac{\nu_{TC}^2}{K}\left(S_{ik}\Omega_{kj} + S_{jk}\Omega_{ki}\right), \tag{8.169}$$

$$\begin{aligned} M_{ij}^{(3)} = &\frac{9}{2}(\gamma - 1)\frac{\nu_{TC}^2}{K^2}K_{\rho N}\left[\frac{DU_i}{Dt}\frac{\partial E}{\partial x_j} + \frac{DU_j}{Dt}\frac{\partial E}{\partial x_i}\right]_D \\ &+\frac{3}{2}\frac{1}{\sigma_\rho}\frac{1}{\bar{\rho}}\frac{\nu_{TC}^2}{K}\left[\frac{DU_i}{Dt}\frac{\partial \bar{\rho}}{\partial x_j} + \frac{DU_j}{Dt}\frac{\partial \bar{\rho}}{\partial x_i}\right]_D \\ &+\frac{9}{2}\frac{\nu_{TC}^2}{K^2}K_{\rho N}\left[\frac{DU_i}{Dt}\frac{DU_j}{Dt}\right]_D, \end{aligned} \tag{8.170}$$

$$\begin{aligned} M_{ij}^{(4)} = &\frac{2}{3}(\gamma - 1)\left(\frac{1}{\sigma_\rho} + \frac{1}{\sigma_e}\right)\frac{1}{\bar{\rho}}\frac{\nu_{TC}^2}{K}\left[\frac{\partial \bar{\rho}}{\partial x_i}\frac{\partial E}{\partial x_j} + \frac{\partial \bar{\rho}}{\partial x_j}\frac{\partial E}{\partial x_i}\right]_D \\ &+\frac{9}{2}(\gamma - 1)\frac{\nu_{TC}^2}{K^2}K_{\rho N}\left[\frac{\partial E}{\partial x_i}\frac{\partial E}{\partial x_j}\right]_D, \end{aligned} \tag{8.171}$$

where $[A_{ij}]_D$ is the deviatoric part of A_{ij} and is defined by Eq. (4.97), and the vorticity tensor Ω_{ij} is given by Eq. (4.101).

Equation (8.168) is the foregoing turbulent-viscosity representation. Equation (8.169) is the so-called nonlinear correction to the former. Here we should note that effects of compressibility enter through ν_{TC} automatically. The remaining two, $M_{ij}^{(3)}$ and $M_{ij}^{(4)}$, are associated with the corrections due to the mean density and internal-energy gradients. Near a shock wave, these gradients become large, and the importance of such corrections is supposed to increase. In such a case, $D\mathbf{U}/Dt$ may be replaced with $-(1/\bar{\rho})\nabla P$.

8.4.2. SIMPLIFIED COMPRESSIBLE MODEL

A. Three-Equation Model

In Sec. 8.4.1, we derived compressibility effects on some important correlations in the equations for the mean field $(\bar{\rho}, \mathbf{U}, E)$. They may be classified into two groups:

(a) Compressibility effects on the turbulent viscosity and the turbulent density and internal-energy diffusivity;

(b) Compressibility effects associated with the deviation of the Reynolds stress from a turbulent-viscosity approximation.

In the aerodynamical field where a high-speed compressible flow is one of the most important subjects, the mass-weighted averaging defined by Eq. (8.1b) is used in general, in place of the present ensemble averaging. Under this procedure, we do not have the terms corresponding to $\langle \rho' \mathbf{u}' \rangle$

and $\langle \rho' e' \rangle$ in Eqs. (8.10)-(8.12) for $\bar{\rho}$, $\mathbf{U}$, and E, resulting in the equations quite similar to the solenoidal counterparts. If some of the present findings combined with such equations may capture important properties of compressible turbulence, the fundamental concept developed here will be also useful in the study of mass-weighted-mean modeling.

From this reason, we adopt the following mean-field equations similar to the mass-weighted-mean counterparts

$$\frac{\partial \bar{\rho}}{\partial t} + \nabla \cdot (\bar{\rho} \mathbf{U}) = 0, \tag{8.172}$$

$$\frac{\partial}{\partial t} \bar{\rho} U_i + \frac{\partial}{\partial x_j} \bar{\rho} U_j U_i = -\frac{\partial P}{\partial x_i} + \frac{\partial}{\partial x_j} \left(-\bar{\rho} \left\langle u_j' u_i' \right\rangle \right) + \frac{\partial}{\partial x_j} \bar{\mu} S_{ji}, \tag{8.173}$$

$$\frac{\partial}{\partial t} \bar{\rho} E + \nabla \cdot (\bar{\rho} \mathbf{U} E) = \nabla \cdot (-\bar{\rho} \langle e' \mathbf{u}' \rangle) - P \nabla \cdot \mathbf{U} + \nabla \cdot \left(\bar{\kappa} \nabla \frac{E}{C_V} \right). \tag{8.174}$$

The compressibility effects on the Reynolds stress, the internal-energy transfer rate, etc. were derived in Sec. 8.4.1. In order to see their importance, we construct a simple model with attention focused on the compressibility effects through the turbulent viscosity and diffusivity (Yoshizawa et al. 1997). Namely, we write $\langle u_i' u_j' \rangle$ and $\langle e' \mathbf{u}' \rangle$ as

$$\left\langle u_i' u_j' \right\rangle = \frac{2}{3} K \delta_{ij} - \nu_{TC} S_{ij}, \tag{8.175a}$$

$$\nu_{TC} = C_\nu \frac{K^2 / \varepsilon}{\left(1 + A \left(K_{\rho N} / M_T^2\right)\right)^{3/2}}, \tag{8.175b}$$

$$\langle e' \mathbf{u}' \rangle = -\frac{\nu_{TC}}{\sigma_e} \nabla E, \tag{8.176}$$

from Eqs. (8.158), (8.159), and (8.165), where the renormalized density variance $K_{\rho N}$ and the turbulent Mach number M_T are given by Eqs. (8.160) and (8.132), respectively. The nonlinear corrections to Eq. (8.175a) were dropped since they were confirmed to be small, at least, in a fully-developed free-shear layer that is the primary subject in the following application of the present model. In Eq. (8.176), we should note that ν_{TC} / σ_e is the compressible version of the turbulent internal-energy diffusivity, and the $\nabla \bar{\rho}$-related part in Eq. (8.165) was dropped owing to the neglect of K_e. The importance of the part in relation to combustion was referred to in Sec. 8.4.1.C.

We need the transport equations for three turbulence quantities (K, ε, K_ρ). The equations for K and K_ρ are given by Eqs. (8.29) and (8.30), respectively. As a model equation for K, we adopt

$$\bar{\rho} \frac{DK}{Dt} = -\bar{\rho} \left\langle u_i' u_j' \right\rangle \frac{\partial U_j}{\partial x_i} - \bar{\rho} \varepsilon + \nabla \cdot \left(\left(\bar{\rho} \frac{\nu_{TS}}{\sigma_K} + \bar{\mu} \right) \nabla K \right), \tag{8.177}$$

where

$$\nu_{TS} = C_\nu \left(K^2/\varepsilon \right). \tag{8.178}$$

In the comparison with Eq. (8.29), we neglected the pressure-dilatation correlation $\langle p' \nabla \cdot \mathbf{u}' \rangle$, following the DNS indications (Vreman 1996). Moreover we dropped the $D\mathbf{U}/Dt$-related term since it is small in a free-shear flow where the streamwise variation of flow properties is not large. Considering that the production and dissipation terms are often of primary importance in turbulent shear flows, we neglected the compressibility effect on the diffusion term $\mathbf{T}_K$. Corresponding to Eq. (8.177), the equation for ε becomes

$$\bar{\rho}\frac{D\varepsilon}{Dt} = -C_{\varepsilon 1}\frac{\varepsilon}{K}\bar{\rho}\left\langle u_i' u_j' \right\rangle \frac{\partial U_j}{\partial x_i} - C_{\varepsilon 2}\bar{\rho}\frac{\varepsilon^2}{K} + \nabla \cdot \left(\left(\bar{\rho}\frac{\nu_{TS}}{\sigma_\varepsilon} + \bar{\mu} \right) \nabla \varepsilon \right). \tag{8.179}$$

Equations (8.177) and (8.179) for K and ε seem almost the same as the counterparts in the incompressible K-ε model, but the big difference comes from the compressibility effects on the Reynolds stress through ν_{TC}.

Finally, we model Eq. (8.30) for K_ρ as

$$\frac{DK_\rho}{Dt} = -2\left\langle \rho' \mathbf{u}' \right\rangle \cdot \nabla\bar{\rho} - 2K_\rho \nabla \cdot \mathbf{U}$$
$$-C_D \left(1 + A\frac{K_{\rho N}}{M_T^2} \right)^{3/2} \frac{\varepsilon}{K} K_\rho + \nabla \cdot \left(\frac{\nu_{TS}}{\sigma_{\rho\rho}} \nabla K_\rho \right), \tag{8.180}$$

where

$$\left\langle \rho' \mathbf{u}' \right\rangle = -\frac{\nu_{TC}}{\sigma_\rho}\nabla\bar{\rho} - \frac{3}{2}(\gamma - 1)\,\nu_{TC}\frac{\bar{\rho}}{K}K_{\rho N}\nabla E. \tag{8.181}$$

In Eq. (8.180), the first two terms are the exact expressions, whereas the last two terms, which correspond to the third and fifth terms in Eq. (8.30), are the models for the destruction and diffusion effects of K_ρ. The feature of Eq. (8.180) lies in the first and third terms. The first term contains effects of density fluctuation through ν_{TC} in $\langle \rho' \mathbf{u}' \rangle$, whereas the third term is proportional to $K_\rho/\tau_{T\rho}$ [$\tau_{T\rho}$ is given by the renormalized form of Eq. (8.146a)].

The above system of equations contains ten model constants as

$$C_\nu(=0.09),\ A,\ \sigma_e(=1.0),\ \sigma_K(=1.0),\ C_{\varepsilon 1}(=1.4),$$
$$C_{\varepsilon 2}(=1.9),\ \sigma_\varepsilon(=1.3),\ C_D,\ \sigma_{\rho\rho},\ \sigma_\rho. \tag{8.182}$$

Here the numerical values have been obtained from the solenoidal limit. The remaining constants A, C_D, $\sigma_{\rho\rho}$, and σ_ρ are to be determined through the application to some typical compressible flows such as a free-shear layer. From the analogy of σ_ρ, C_D, and $\sigma_{\rho\rho}$ with σ_e, $C_{\varepsilon 2}$, and $(\sigma_K, \sigma_\varepsilon)$, respectively, the former are inferred to be of $O(1)$.

B. Application to Free-Shear Layer

In order to examine the validity of the present model in a fully-developed free-shear layer, we choose the Cartesian coordinates (x, y, z), where the x direction is along the free streams, and y is taken as the transverse direction across the shear layer. The flow quantities of the faster free stream are denoted by attaching subscript 1, such as (U_1, ρ_1, E_1), whereas the slower counterparts are denoted using subscript 2 (see Fig. 8.1).

Two free streams are characterized typically by the ratio of U_2 to U_1, r_U, the density counterpart, r_ρ, and the convective Mach number M_C. They are written as

$$r_U = U_2 / U_1 , \tag{8.183a}$$

$$r_\rho = \rho_2 / \rho_1 , \tag{8.183b}$$

$$M_C = \frac{U_1 - U_2}{\bar{a}_1 + \bar{a}_2}, \tag{8.183c}$$

respectively, where $\bar{a}_1$ and $\bar{a}_2$ are the sound velocity of two streams. In order to describe the mean velocity of a free-shear layer, we introduce the nondimensional velocity U^* and the self-similar coordinate y^* as

$$U^* = \frac{U - U_2}{U_1 - U_2}, \tag{8.184a}$$

$$y^* = (y - y_{0.5}) / \delta. \tag{8.184b}$$

Here δ stands the distance between $y_{0.1}$ and $y_{0.9}$ (y_s denotes the y coordinate where the magnitude of U^* is s).

For the model constants, we adopt

$$C_D = 0.3, \ \sigma_{\rho\rho} = 1.0, \ \sigma_\rho = 2.0 \tag{8.185a}$$

from the foregoing analogy, but there is room for further tuning of the choice. For the remaining constant A, which has large influence on the computed results, we examine the following two cases:

$$A = 5, \ 10. \tag{8.185b}$$

The most typical phenomenon characterizing compressible mixing-layer flow is the drastic decrease in the growth rate with increasing M_C. As a quantity representing the thickness of the layer, we choose the vorticity thickness

$$\delta_\omega = \frac{U_1 - U_2}{(dU/dy)_{MAX}}. \tag{8.186}$$

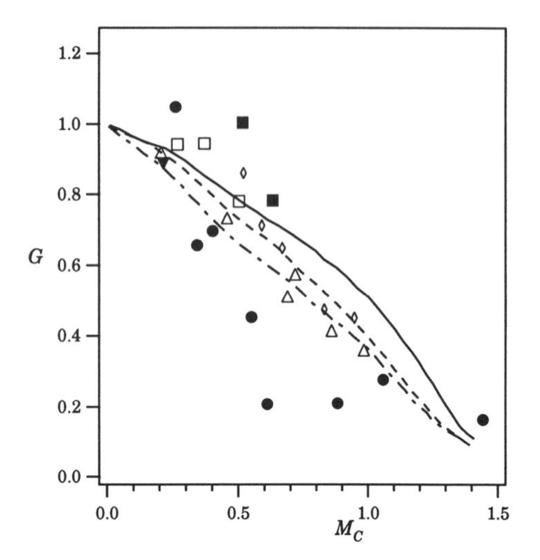

Figure 8.2. Relative growth rate G [Eq. (8.187)] versus the convective Mach number M_C. ——, Present ($A = 5, r_U = 0.1$); – – –, Present ($A = 10, r_U = 0.1$); - - - -, Present ($A = 5, r_U = 0.2$); Symbols, observations.

It is dependent on M_C, r_U, and M_ρ, in addition to x. Using Eq. (8.186), the growth rate normalized by the incompressible counterpart is given by

$$G = \frac{\delta'_\omega (M_C, r_U, r_\rho)}{\delta'_\omega (0, r_U, r_\rho)}. \tag{8.187a}$$

with

$$\delta'_\omega = \frac{d\delta_\omega}{dx}. \tag{8.187b}$$

The computed results are given in Fig. 8.2, and show that the steep decrease in the growth rate with increasing M_C is captured by the present model. The difference between the results for $A = 5$ and $A = 10$ is not so large, compared with the scattering of the experimental values.

References

Baldwin, B. S. and Lomax, H. (1978), AIAA Paper 78-257.

Blaisdell, G. A., Mansour, N. N., and Reynolds, W. C. (1993), J. Fluid Mech. **256**, 473.

Bertoglio, J.-P., Bataille, F., and Marion, J.-D. (1998), Phys Fluids **10** (to appear).

Coleman, G. N., Kim. J., and Moser, R. D. (1995), J. Fluid Mech. **305**, 159.

Fujiwara, H. (1996), Ph. D. Dissertation, University of Tokyo.
Fujiwara, H. and Arakawa, C. (1993), in *Ninth Symposium on Turbulent Shear Flows*, Kyoto, 22-2-1.
Gatski, T. B. (1997), in *New Tools in Turbulence Modeling*, edited by O. Métais and J. Ferziger (Springer, Berlin), p. 73.
Hamba, F. (1997), Private communication (Four-equation model for compressible turbulent flows, submitted to Phys. Fluids).
Hamba, F. and Blaisdell, G. A. (1997), Phys. Fluids A **9**, 2749.
Huang, P. G., Coleman, G. N., and Bradshaw, P. (1995), J. Fluid Mech. **305**, 185.
Kida, S. and Orszag, S. A. (1990). J. Sci. Comp. **5**, 85.
Lee, S., Lele, S. K., and Moin, P. (1993), J. Fluid Mech. **251**, 533.
Lee, S., Lele, S. K., and Moin, P. (1997), J. Fluid Mech. **340**, 225.
Lele, S. K. (1994), Annu. Rev. Fluid Mech. **26**, 211.
Liou, W. W., Shih, T. -H., and Duncan, B. S. (1995), Phys. Fluids **7**, 658.
Rubesin, M. W. (1989), AIAA Paper 89-0606.
Rubinstein, R. and Erlebacher, G. (1997), Phys. Fluids **9**, 3037.
Sandham, N. D. and Reynolds, W. C. (1991), J. Fluid Mech. **224**, 133.
Sarkar, S. (1992), Phys. Fluids A **4**, 2674.
Sarkar, S. (1995), J. Fluid Mech. **282**, 163.
Sarkar, S., Erlebacher, G., and Hussaini, M. Y. (1991a), Theor. Comput. Fluid Dyn. **2**, 291.
Sarkar, S., Erlebacher, G., Hussaini, M. Y., and Kreiss, H. O. (1991b), J. Fluid Mech. **227**, 473.
Taulbee, D. and VanOsdol, J. (1991), AIAA Pape 91-0524.
Vreman, A. W., Sandham, N. D., and Luo, K. H. (1996), J. Fluid Mech. **320**, 235.
Yoshizawa, A. (1986), Phys. Fluids **29**, 2152.
Yoshizawa, A. (1990), Phys. Fluids A **2**, 838.
Yoshizawa, A. (1992), Phys. Rev. A **46**, 3292.
Yoshizawa, A. (1995), Phys. Fluids **7**, 3105.
Yoshizawa, A., Liou, W. W., Yokoi, N., and Shih, T. -H. (1997), Phys. Fluids **9**, 3024.
Zeman, O. (1990), Phys. Fluids A **2**, 178.

CHAPTER 9

MAGNETOHYDRODYNAMIC TURBULENCE MODELING

9.1. Highly Electrically-Conducting Turbulent Flows

In turbulent fluid motion, a number of ordered or coherent structures can be detected, which may be classified according to their characteristic spatial scales. As is seen from Eq. (3.46), components of fluid motion with larger spatial scales possess longer time scales in general, and they have large influence on various flow aspects. In this sense, the component of turbulent motion that is abstracted through the ensemble averaging procedure often contain important global properties. In reality, the flow structure characterized by the mean velocity gradient governs the production mechanism of turbulent energy or the mechanism of draining energy from large- to small-scale components.

As the typical examples of the ordered flow structures controlling major flow properties, we may mention secondary vortical flows in a square-duct pipe (see Fig. 4.4) and the axial velocity profile of a swirling pipe flow with the local minimum at the central axis (see Fig. 4.5). In the former, the magnitude of those secondary flows is of at highest a few percent of the axial flow, but they have critical influence on the profile and magnitude of the axial velocity. In the latter, effects of swirling generate the specific axial velocity profile, resulting in the flow properties entirely different from a usual pipe flow. In order to clarify these properties, we need to understand the mechanism under which such mean or global flow structures are sustained against the destructive action due to turbulence that is one of the prominent features of highly fluctuating motions.

Ordered structures in turbulent motion are observed frequently in flow phenomena in natural sciences, in close relation to magnetic fields. In stellar objects represented by the Earth and the Sun, magnetic fields can be observed ubiquitously. It is not rare that these magnetic fields often govern some of the important events occurring in these objects. Such a familiar instance is sunspots. They are the cross sections of magnetic tubes that are generated inside the Sun, come up under effects of buoyancy, and break the photosphere (the layer adjacent to the solar outermost part). Sunspots of a few thousand Gauss have large influence on various activities in the region surrounding the Sun. The magnetic flux tubes that are the origin of sunspots are believed to be generated by the highly turbulent motion of electrically conducting fluids in the outer region called the solar convective zone. These magnetic phenomena are similar to the foregoing examples in the context of the sustainment of ordered structures in turbulence, but they contain various aspects that are still difficult to be explained.

The motion of an electrically conducting fluid is characterized by two nondimensional numbers, the kinetic Reynolds number R and the magnetic counterpart R_M. They are defined by Eq. (1.1) and

$$R_M = U_R L_R / \lambda_M , \tag{9.1}$$

where λ_M is the magnetic diffusivity. In the case of solar and other astrophysical phenomena, both R and R_M are huge. Earth's magnetic fields, which are also generated by the motion of electrically conducting fluids in the outer core beneath the mantle, are a rather exceptional example. In this case, R is very large, but R_M based on melted iron is inferred to be moderate or of $O(10^2)$ (this point will be stated later in more detail).

In contrast with astrophysical magnetic phenomena, the flow situation with large R_M is rarely encountered in laboratory experiments. A familiar example of an electrically conducting fluid in laboratory is mercury. There λ_M is about $0.8 \times 10^6 \mathrm{cm}^2\mathrm{s}^{-1}$, resulting in

$$R_M = 1.3 \times 10^{-6} U_R L_R. \tag{9.2}$$

From the magnitude of U_R and L_R realized in a laboratory system, R_M usually becomes very small. Its sole exception is the experiments related to the confinement of plasmas in nuclear fusion. In this case, λ_M is very small because of large electrical conductivity of high-temperature plasmas consisting of mainly electrons and ions.

In summary, the turbulent motion of electrically conducting fluids is divided into two categories; high- and low-R_M phenomena. In this book, we focus attention on high-R_M phenomena that are closely connected with magnetic fields in astrophysical and fusion phenomena. The sustainment mechanism of magnetic fields in turbulent motion is generally called turbulent dynamo, whose theoretical investigation is the primary theme of this and next chapters.

9.2. Magnetohydrodynamic Approximation

9.2.1. TWO-FLUID SYSTEM OF EQUATIONS

We consider that an electrically conducting fluid consists of electron and ion gases. We assume one species of ion, and write the mass and charge of ion and electron as

$$\mathrm{Mass}: \; m_E, \; m_I \; ; \tag{9.3a}$$

$$\text{Charge}: \ -e,\ Ze, \tag{9.3b}$$

where it should be noted that

$$m_E/m_I \cong O\left(10^{-3}\right) \tag{9.4}$$

(subscripts E and I denote electron and ion, respectively). Under the magnetohydrodynamic (MHD) approximation of electron and ion gases, the fundamental properties of each gas are characterized by

$$\text{Number density}: \ n_E,\ n_I; \tag{9.5a}$$

$$\text{Mass density}: \ \rho_{ME}\left(=n_E m_E\right),\ \rho_{MI}\left(=n_I m_I\right); \tag{9.5b}$$

$$\text{Charge density}: \ \rho_{CE}\left(=-n_E e\right),\ \rho_{CI}\left(=n_I Ze\right); \tag{9.5c}$$

$$\text{Velocity}: \ v_E,\ v_I; \tag{9.5d}$$

$$\text{Pressure}: \ p_E,\ p_I. \tag{9.5e}$$

The mass conservation law for each of electron and ion gases has the same mathematical form as the hydrodynamic one, Eq. (2.10), and is written as

$$\frac{\partial \rho_{ME}}{\partial t} + \nabla \cdot \left(\rho_{ME}\mathbf{v}_E\right) = 0, \tag{9.6}$$

$$\frac{\partial \rho_{MI}}{\partial t} + \nabla \cdot \left(\rho_{MI}\mathbf{v}_I\right) = 0. \tag{9.7}$$

Similarly, the counterparts of the hydrodynamic momentum equation (2.29) are

$$\frac{\partial}{\partial t}\rho_{ME}v_{Ei} + \frac{\partial}{\partial x_j}\rho_{ME}v_{Ej}v_{Ei} = -\frac{\partial p_E}{\partial x_i} + \rho_{CE}\left(\mathbf{e} + \mathbf{v}_E \times \mathbf{b}\right)_i + C_{EIi}, \tag{9.8}$$

$$\frac{\partial}{\partial t}\rho_{MI}v_{Ii} + \frac{\partial}{\partial x_j}\rho_{MI}v_{Ij}v_{Ii} = -\frac{\partial p_I}{\partial x_i} + \rho_{CI}\left(\mathbf{e} + \mathbf{v}_I \times \mathbf{b}\right)_i - C_{EIi}. \tag{9.9}$$

In each of Eqs. (9.8) and (9.9), the second term on the right-hand side is the Lorentz force, and $\mathbf{e}$ and $\mathbf{b}$ are the electric field and the magnetic field, respectively. In their last terms, C_{EIi} expresses the force occurring from the collision between electron and ion, and is written as

$$\mathbf{C}_{EI} = -\rho_{ME}\left(\mathbf{v}_E - \mathbf{v}_I\right)\nu_{EI}, \tag{9.10}$$

using the collision frequency ν_{EI} (the number of collision per unit time). Since $-\rho_{ME}(\mathbf{v}_E - \mathbf{v}_I)$ represents the momentum gain by the electron gas per unit volume in one collision, the gain per unit time is given by Eq. (9.10), resulting in the loss of the ion momentum under the law of action and reaction. In close relation to $\mathbf{C}_{EI}$, the resistive force generated by the collision among electrons or ions themselves was dropped here. This force corresponds to the viscous term in the hydrodynamic equation (2.29) and will be supplemented later.

For the pressure of each gas, p_E and p_I, we simply assume that they are perfect gases. Then we have

$$p_E = R\rho_{ME}\theta_E, \tag{9.11}$$

$$p_I = R\rho_{MI}\theta_I, \tag{9.12}$$

where R is the gas constant, and θ_E and θ_I are the temperature of each gas.

For describing $\mathbf{e}$ and $\mathbf{b}$, we add the Maxwell system of equations, which is given by

$$\frac{\partial \mathbf{b}}{\partial t} = -\nabla \times \mathbf{e}, \tag{9.13}$$

$$\nabla \times \mathbf{b} - \varepsilon_0\mu_0\frac{\partial \mathbf{e}}{\partial t} = \mu_0\mathbf{j}, \tag{9.14}$$

$$\nabla \cdot \mathbf{b} = 0, \tag{9.15}$$

$$\varepsilon_0\nabla \cdot \mathbf{e} = \rho_{CE} + \rho_{CI}, \tag{9.16}$$

where ε_0 is the dielectric constant, μ_0 is the magnetic permeability, and $\mathbf{j}$ is the electric current density, which is written as

$$\mathbf{j} = \rho_{CE}\mathbf{v}_E + \rho_{CI}\mathbf{v}_I. \tag{9.17}$$

In order to close the foregoing system consisting of Eqs. (9.6)-(9.9) and (9.13)-(9.16), we need to add the energy conservation laws concerning θ_E and θ_I or each internal energy, $e_E(= C_{VE}\theta_E)$ and $e_I(= C_{VI}\theta_E)$ (C_{VE} and C_{VI} are the specific heats of electron and ion gases at constant volume). The hydrodynamic counterpart is given by Eq. (2.44). In what follows, we shall proceed to the one-fluid description. Therefore we omit the details of the energy conservation law of each of electron and ion gases. A detailed account of plasma and two-fluid system is given by Miyamoto (1989).

9.2.2. ONE-FLUID SYSTEM OF EQUATIONS

In a variety of MHD phenomena in astrophysics and nuclear fusion, the behaviors of the mixture of electron and ion gases or plasma are often dependent highly on the characteristics of each gas. In such a case, it is indispensable to deal with the foregoing two fluid system. Concerning the dynamo, however, the theoretical study still remains at the stage far from sufficiently exploited. Therefore its development in a simpler system is necessary and is expected to provide a useful theoretical basis for the study in a more complicated system such as the two-fluid one.

In this and next chapters, we shall adopt the one-fluid MHD system, and pay attention to the total contribution of electron and ion gases. For this purpose, we define the total mass density ρ, the effective fluid velocity $\mathbf{u}$, the total pressure p, and the total charge density ρ_C as

$$\rho = \rho_{ME} + \rho_{MI}, \tag{9.18}$$

$$\mathbf{u} = \frac{\rho_{ME}\mathbf{v}_E + \rho_{MI}\mathbf{v}_I}{\rho}, \tag{9.19}$$

$$p = p_E + p_I, \tag{9.20}$$

$$\rho_C = \rho_{CE} + \rho_{CI} = -e\left(n_E - Zn_I\right). \tag{9.21}$$

A key assumption in the one-fluid approximation is the quasi-neutrality

$$\rho_C \cong 0; \tag{9.22a}$$

namely, we have

$$n_E \cong Zn_I. \tag{9.22b}$$

The time scale relevant to the relaxation of a nonneutral state due to charge separation is measured using the speed of light, $c_0(= 1/\sqrt{\varepsilon_0\mu_0})$, and its effects occur through the $\partial\mathbf{e}/\partial t$-related term in Eq. (9.14). The term is neglected under the approximation (9.22).

A. Mass Conservation Law

We combine Eq. (9.6) with Eq. (9.7) and make use of Eqs. (9.18) and (9.19). Then we have the same equation as the hydrodynamic one

$$\frac{\partial\rho}{\partial t} + \nabla\cdot(\rho\mathbf{u}) = 0. \tag{9.23}$$

Equation (9.18) is rewritten as

$$\rho = \rho_{MI}\left(1 + Z\frac{m_E}{m_I}\right). \tag{9.24}$$

From the mass-ratio relation (9.4), Eq. (9.24) is reduced to

$$\rho \cong \rho_{MI}, \tag{9.25}$$

under the approximation of dropping the terms of $O(m_E/m_I)$ since $Z = O(1)$. In the following derivation of the one-fluid system, this approximation will play a key role.

B. Momentum Conservation Law
We rewrite the fluid velocity $\mathbf{u}$ as

$$\mathbf{u} = \mathbf{v}_I \left(1 + \frac{Z\,(m_E/m_I)}{1 + Z\,(m_E/m_I)} \left(\frac{\mathbf{v}_E}{\mathbf{v}_I} - 1 \right) \right) \cong \mathbf{v}_I. \tag{9.26}$$

Equation (9.26) signifies that $\mathbf{u}$ is essentially the velocity of ion gas.

From Eqs. (9.8) and (9.9), we have

$$\frac{\partial}{\partial t}\rho u_i + \frac{\partial}{\partial x_j}\left(\rho_{MI} v_{Ij} v_{Ii} + \rho_{ME} v_{Ej} v_{Ei}\right) = -\frac{\partial p}{\partial x_i} + \rho_C e_i + (\mathbf{j} \times \mathbf{b})_i, \tag{9.27}$$

where use has been made of Eqs. (9.17) and (9.21), and we should note the disappearance of the effects of collision, $\mathbf{C}_{EI}$. Under the mass-ratio relation (9.4), we use Eqs. (9.25) and (9.26) to approximate

$$\rho_{MI} v_{Ij} v_{Ii} + \rho_{ME} v_{Ej} v_{Ei} \cong \rho u_j u_i. \tag{9.28}$$

From the quasi-neutrality condition (9.22) and Eq. (9.28), we finally reach the familiar one-fluid MHD equation

$$\frac{\partial}{\partial t}\rho u_i + \frac{\partial}{\partial x_j}\rho u_j u_i = -\frac{\partial p}{\partial x_i} + (\mathbf{j} \times \mathbf{b})_i. \tag{9.29}$$

In Eqs. (9.8) and (9.9), the effects of collision among ions or electrons themselves are neglected. In neutral molecules, the collision effects are usually modeled in terms of the isotropic viscosity μ, as in the second part of Eq. (2.28). From the comparison with observations, the model gives rise to no critical shortfall under the condition that the typical flow scale concerned is much larger than the mean-free path of molecular motion.

On the other hand, electrons and ions show entirely different behaviors in the directions along and normal to magnetic field lines. In this situation, the simple isotropic-viscosity representation for the effects of collision among electrons or ions themselves is not so good, specifically, at the scales where the electric field due to charge separation becomes important. In the present one-fluid system, these effects have already been neglected, and the fluid motion with large spatial scales is the primary interest in turbulent dynamo. Therefore the modeling of collision effects based on an isotropic viscosity model may be considered a plausible approximation. After supplementing Eq. (9.29) with such collision effects, we have

$$\begin{aligned}\frac{\partial}{\partial t}\rho u_i + \frac{\partial}{\partial x_j}\rho u_j u_i &= -\frac{\partial p}{\partial x_i} + (\mathbf{j} \times \mathbf{b})_i \\ &+ \frac{\partial}{\partial x_j}\mu\left(\frac{\partial u_j}{\partial x_i} + \frac{\partial u_i}{\partial x_j} - \frac{2}{3}\nabla\cdot\mathbf{u}\delta_{ij}\right).\end{aligned} \tag{9.30}$$

C. Ohm's Law

We have derived the momentum conservation law for plasma motion, but it is essentially the same as the one for ion gas, Eq. (9.9), since ion is much heavier than electron. Next we pay attention to Eq. (9.8) for electron gas, and simplify it to abstract an expression for electric currents. We cast Eq. (9.8) into the form

$$\mathbf{e} + v_E \times \mathbf{b} + \frac{m_E \nu_{EI}}{e} (\mathbf{v}_E - \mathbf{v}_I) + \frac{1}{n_E e} \nabla p_E = \mathbf{R}_{TO}, \tag{9.31}$$

where $\mathbf{R}_{TO}$ denotes the residual terms related to the inertia or advection (subscript O denotes Ohm). The inertia effect of electron is considered small since their mass is small.

From Eq. (9.17) and the quasi-neutrality condition (9.22), we have

$$\mathbf{j} = -n_E e (\mathbf{v}_E - \mathbf{v}_I), \tag{9.32}$$

resulting in

$$\mathbf{v}_E = \mathbf{u} - \frac{1}{n_E e} \mathbf{j} \tag{9.33}$$

[use has been made of Eq. (9.26)]. We combine Eq. (9.33) with Eq. (9.31), and have

$$\mathbf{e} + \left(\mathbf{u} - \frac{1}{n_E e} \mathbf{j} \right) \times \mathbf{b} - \frac{m_E \nu_{EI}}{n_E e^2} \mathbf{j} + \frac{1}{n_E e} \nabla p_E = 0. \tag{9.34}$$

We regard Eq. (9.34) as an equation for $\mathbf{j}$, and rewrite it as

$$\mathbf{j} = \sigma (\mathbf{e} + \mathbf{u} \times \mathbf{b}) + \frac{\sigma}{\rho_{CE}} \mathbf{j} \times \mathbf{b} - \frac{\sigma}{\rho_{CE}} \nabla p_E, \tag{9.35}$$

where σ is the so-called electric conductivity, which is defined as

$$\sigma = \frac{n_E e^2}{m_E \nu_{EI}}. \tag{9.36}$$

On retaining the first part on the right-hand side of Eq. (9.35), we are led to the familiar Ohm's law

$$\mathbf{j} = \sigma (\mathbf{e} + \mathbf{u} \times \mathbf{b}). \tag{9.37}$$

A prominent feature of this relation is that the motion of plasma generates the electric current normal to the magnetic field. The second and third terms in Eq. (9.35) contain the charge density of electron gas, ρ_{CE}. Its description is beyond the scope of the one-fluid MHD system, and those terms will be dropped. From the viewpoint of the turbulent dynamo detailed later, the second term related to $\mathbf{j} \times \mathbf{b}$ does not always bring the aspect entirely different from $\mathbf{u} \times \mathbf{b}$. Both of these effects lead to the generation of $\mathbf{j}$ normal to $\mathbf{b}$. The most important task in turbulent dynamo is to seek a mechanism different from this process, such as the generation of the electric current parallel to the magnetic field, as will be shown later.

D. Maxwell System of Equations

Under the quasi-neutrality condition, the $\mathbf{e}$ effects coming from charge separation are totally neglected. As this consequence, the Maxwell system of equations is much simplified as

$$\frac{\partial \mathbf{b}}{\partial t} = -\nabla \times \mathbf{e}, \tag{9.38}$$

$$\nabla \times \mathbf{b} = \mu_0 \mathbf{j}. \tag{9.39}$$

The electric field $\mathbf{e}$ in Eq. (9.38) may be eliminated using the Ohm's law (9.37).

E. Energy Conservation Law

The energy conservation law may be written using the effective one-fluid internal energy e_T, as is entirely similar to the hydrodynamic counterpart. In the MHD case, the law may be written similarly as

$$\frac{\partial}{\partial t} \rho e_T + \nabla \cdot (\rho e_T \mathbf{u}) = \nabla \cdot (\kappa \nabla \theta) - p \nabla \cdot \mathbf{u} + \phi, \tag{9.40}$$

where θ is the one-fluid temperature, and the dissipation function ϕ is defined as

$$\phi = \frac{1}{2} \mu \left(\left(\frac{\partial u_j}{\partial x_i} + \frac{\partial u_i}{\partial x_j} \right)^2 - \frac{4}{3} (\nabla \cdot \mathbf{u})^2 \right) + \frac{1}{\sigma} \mathbf{j}^2. \tag{9.41}$$

Here the second part represents the conversion of magnetic to thermal energy due to the Joule heating.

Finally, the total pressure p, the one-fluid internal energy e_T, and the corresponding temperature θ are connected with one another using the relation of a perfect gas as

$$e_T = C_V \theta, \tag{9.42}$$

$$p = R \rho \theta = (\gamma - 1) \rho e_T. \tag{9.43}$$

9.2.3. CONSTANT-DENSITY ONE-FLUID SYSTEM

The degree of importance of density variation or fluid compressibility is highly dependent on phenomena. In thermal nuclear fusion represented by tokamaks, the profiles of electron and ion density have large influence on the duration time of plasma confinement. On the other hand, fluid compressibility is not always considered to play a critical role in the generation and sustainment of astrophysical magnetic fields, although it contributes much to the generation of fluid motion through effects of buoyancy. This point is closely related to fact that the Maxwell system of equations is not dependent explicitly on the fluid density. In this and next chapters with attention focused on the generation mechanism of magnetic fields in turbulent motion, the constant-density MHD system is adopted as the theoretical starting point in the study of turbulent dynamo.

A. Constant-Density System in Alfvén-Velocity Units

In the case of constant density, the mass and momentum conservation laws, Eqs. (9.23) and (9.30), are reduced to

$$\nabla \cdot \mathbf{u} = 0, \tag{9.44}$$

$$\frac{\partial \mathbf{u}}{\partial t} + (\mathbf{u} \cdot \nabla) \mathbf{u} = -\frac{1}{\rho} \nabla p + \frac{1}{\rho} \mathbf{j} \times \mathbf{b} + \nu \Delta \mathbf{u}, \tag{9.45}$$

where $\nu = \mu/\rho$. On the other hand, the Maxwell system is unchanged and is given by Eqs. (9.37)-(9.39).

We perform the transformation

$$\mathbf{u} \to \mathbf{u}, \ \ p/\rho \to p, \tag{9.46a}$$

$$\mathbf{b}/\sqrt{\rho\mu_0} \to \mathbf{b}, \ \ \mathbf{j}/\sqrt{\rho/\mu_0} \to \mathbf{j}, \ \ \mathbf{e}/\sqrt{\rho\mu_0} \to \mathbf{e}. \tag{9.46b}$$

Under this transformation, we have

$$\frac{\partial \mathbf{u}}{\partial t} + (\mathbf{u} \cdot \nabla) \mathbf{u} = -\nabla p + \mathbf{j} \times \mathbf{b} + \nu \Delta \mathbf{u}, \tag{9.47}$$

$$\frac{\partial \mathbf{b}}{\partial t} = -\nabla \times \mathbf{e}, \tag{9.48}$$

$$\mathbf{j} = \nabla \times \mathbf{b} \tag{9.49a}$$

$$= \frac{1}{\lambda_M} (\mathbf{e} + \mathbf{u} \times \mathbf{b}). \tag{9.49b}$$

In Eq. (9.49b), λ_M is the magnetic diffusivity defined as

$$\lambda_M = 1/(\sigma\mu_0), \tag{9.50}$$

and is simply called the resistivity. The physical meaning of λ_M as the magnetic diffusivity will be clear below. A general account of the magnetohydrodynamic approximation and its aspects at low magnetic Reynolds number are detailed by Moreau (1990).

Equation (9.47) indicates that new $\mathbf{b}$ has the same dimension as $\mathbf{u}$, that is, the dimension of velocity. This system of units is usually called Alfvén-velocity units and is relevant to the study of turbulent dynamo, specifically, in the case of constant density. We should note that λ_M is the sole physical parameter occurring in the Maxwell system, Eqs. (9.48) and (9.49).

Equation (9.47) is cast into the form

$$\frac{\partial \mathbf{u}}{\partial t} + (\mathbf{u}\cdot\nabla)\,\mathbf{u} - (\mathbf{b}\cdot\nabla)\,\mathbf{b} = -p_M + \nu\Delta\mathbf{u}, \tag{9.51}$$

where p_M is defined as

$$p_M = p + \frac{\mathbf{u}^2}{2}, \tag{9.52}$$

which may be called the MHD pressure. In obtaining Eq. (9.51), we used the relation

$$(\nabla\times\mathbf{b})\times\mathbf{b} = (\mathbf{b}\cdot\nabla)\,\mathbf{b} - \nabla\left(\frac{\mathbf{b}^2}{2}\right) \tag{9.53}$$

We eliminate $\mathbf{e}$ from Eqs. (9.48) and (9.49) to have

$$\frac{\partial \mathbf{b}}{\partial t} = \nabla\times(\mathbf{u}\times\mathbf{b}) + \lambda_M\Delta\mathbf{b} \tag{9.54}$$

or

$$\frac{\partial \mathbf{b}}{\partial t} + (\mathbf{u}\cdot\nabla)\,\mathbf{b} - (\mathbf{b}\cdot\nabla)\,\mathbf{u} = \lambda_M\Delta\mathbf{b}, \tag{9.55}$$

where use has been made of

$$\nabla\times(\nabla\times\mathbf{b}) = -\Delta\mathbf{b} + \nabla\,(\nabla\cdot\mathbf{b})\,, \tag{9.56a}$$

$$\nabla\times(\mathbf{u}\times\mathbf{b}) = (\mathbf{b}\cdot\nabla)\,\mathbf{u} - (\mathbf{u}\cdot\nabla)\,\mathbf{b} + \mathbf{u}\,(\nabla\cdot\mathbf{b}) - \mathbf{b}\,(\nabla\cdot\mathbf{u})\,, \tag{9.56b}$$

in the combination with the solenoidal condition on $\mathbf{u}$ and $\mathbf{b}$.

In turbulent dynamo one of whose primary subjects is magnetic fields in stellar objects, effects of frame rotation become important. In the frame rotating with the angular velocity $\mathbf{\Omega}_F$, Eqs. (9.47) and (9.51) are changed into

$$\frac{\partial \mathbf{u}}{\partial t} + (\mathbf{u}\cdot\nabla)\,\mathbf{u} + 2\mathbf{\Omega}_F\times\mathbf{u} = -\nabla p + \mathbf{j}\times\mathbf{b} + \nu\Delta\mathbf{u}, \tag{9.57a}$$

$$\frac{\partial \mathbf{u}}{\partial t} + (\mathbf{u}\cdot\nabla)\,\mathbf{u} - (\mathbf{b}\cdot\nabla)\,\mathbf{b} + 2\mathbf{\Omega}_F\times\mathbf{u} = -\nabla p_M + \nu\Delta\mathbf{u}, \tag{9.57b}$$

respectively. In correspondence to this transformation, the vorticity $\boldsymbol{\omega}(= \nabla \times \mathbf{u})$ obeys the transformation

$$\boldsymbol{\omega} \rightarrow \boldsymbol{\omega} + 2\boldsymbol{\Omega}_F, \tag{9.58}$$

as is explained Eq. (2.79). From the importance of effects of frame rotation in astrophysical phenomena, this transformation will play a key role in the following analysis. In the context of the two-scale direct-interaction approximation (TSDIA) analysis of helicity effects in Sec. 6.7.2, the evaluation of $\boldsymbol{\Omega}_F$ effects is generally simpler than that of $\boldsymbol{\Omega}$ ones, and the latter effects may be obtained through the transformation (9.58). In what follows, this fact will be often used.

B. Elsasser's Variables

Equations (9.51) and (9.55) have a very similar mathematical structure. This point becomes much clearer under the use of Elsasser's variables. These variables are defined as

$$\boldsymbol{\phi} = \mathbf{u} + \mathbf{b}, \ \boldsymbol{\psi} = \mathbf{u} - \mathbf{b}, \tag{9.59}$$

resulting in

$$\mathbf{u} = \left(\boldsymbol{\phi} + \boldsymbol{\psi}\right)/2, \ \mathbf{b} = \left(\boldsymbol{\phi} - \boldsymbol{\psi}\right)/2. \tag{9.60}$$

Using Eq. (9.59), we may cast Eqs. (9.51) and (9.55) into the form nearly symmetric to each other as

$$\frac{\partial \boldsymbol{\phi}}{\partial t} + \left(\boldsymbol{\psi} \cdot \nabla\right) \boldsymbol{\phi} = -\nabla p_M + \frac{\nu + \lambda_M}{2} \Delta \boldsymbol{\phi} + \frac{\nu - \lambda_M}{2} \Delta \boldsymbol{\psi}, \tag{9.61}$$

$$\frac{\partial \boldsymbol{\psi}}{\partial t} + \left(\boldsymbol{\phi} \cdot \nabla\right) \boldsymbol{\psi} = -\nabla p_M + \frac{\nu + \lambda_M}{2} \Delta \boldsymbol{\psi} + \frac{\nu - \lambda_M}{2} \Delta \boldsymbol{\phi}. \tag{9.62}$$

In astrophysical magnetic phenomena, both the Reynolds number R [Eq. (1.1)] and the Magnetic Reynolds number R_M [Eq. (9.1)] are usually huge. Specifically, in case that the difference between them is not so critical, Eqs. (9.61) and (9.62) are simplified into the completely symmetric form

$$\frac{\partial \boldsymbol{\phi}}{\partial t} + \left(\boldsymbol{\psi} \cdot \nabla\right) \boldsymbol{\phi} = -\nabla p_M + \nu \Delta \boldsymbol{\phi}, \tag{9.63}$$

$$\frac{\partial \boldsymbol{\psi}}{\partial t} + \left(\boldsymbol{\phi} \cdot \nabla\right) \boldsymbol{\psi} = -\nabla p_M + \nu \Delta \boldsymbol{\phi}. \tag{9.64}$$

The analysis of MHD turbulence using statistical theories such as the TSDIA leads to a formidably complicated mathematical manipulation. In this case, the use of Eqs. (9.63) and (9.64) in place of the original counterparts, Eqs. (9.51) and (9.55), greatly simplifies the analysis.

Note: In the hydrodynamic case, the vorticity $\boldsymbol{\omega}$ obeys

$$\frac{\partial \boldsymbol{\omega}}{\partial t} = \nabla \times (\mathbf{u} \times \boldsymbol{\omega}) + \nu \Delta \boldsymbol{\omega}. \tag{9.65}$$

Equation (9.65) is very similar to the magnetic induction equation (9.54) at a first glance. From this resemblance, the properties of magnetic fields are often inferred by the analogy with the behaviors of $\boldsymbol{\omega}$. The critical difference between them, however, lies in the fact that $\boldsymbol{\omega}$ is linked with $\mathbf{u}$ through $\boldsymbol{\omega} = \nabla \times \mathbf{u}$, unlike $\mathbf{b}$. On the contrary, the close relationship between $\mathbf{u}$ and $\mathbf{b}$ is clear from Elsasser's variables $\boldsymbol{\phi}$ and $\boldsymbol{\psi}$. In reality, the importance of the correspondence

$$\mathbf{u} \leftrightarrow \mathbf{b}, \; \boldsymbol{\omega} \leftrightarrow \mathbf{j} \tag{9.66}$$

will become clear in the following study about turbulent dynamo.

C. Conservation Properties

In the hydrodynamic system of constant density, it is shown that the total amounts of kinetic energy and helicity, that is,

$$\int_V \left(\mathbf{u}^2/2\right) dV, \tag{9.67a}$$

$$\int_V \mathbf{u} \cdot \boldsymbol{\omega} dV, \tag{9.67b}$$

are conserved in the absence of the molecular viscosity. In the MHD system of constant density, we should note that expression (9.67) is not conserved. The counterparts of expression (9.67) are the total amounts of MHD energy and cross helicity, which are given by

$$\int_V \frac{\mathbf{u}^2 + \mathbf{b}^2}{2} dV, \tag{9.68a}$$

$$\int_V \mathbf{u} \cdot \mathbf{b} dV. \tag{9.68b}$$

In order to see this point, we construct the equations governing two quantities (9.68a) and (9.68b) from Eqs. (9.51) and (9.55). They are written as

$$\frac{\partial}{\partial t} \int_V \frac{\mathbf{u}^2 + \mathbf{b}^2}{2} dV = - \int_V \nabla \cdot \left(\left(\frac{\mathbf{u}^2 + \mathbf{b}^2}{2} + p_M \right) \mathbf{u} - (\mathbf{u} \cdot \mathbf{b}) \, \mathbf{b} \right) dV$$
$$+ \int_V \left(-\nu \left(\frac{\partial u_j}{\partial x_i} \right)^2 - \lambda_M \left(\frac{\partial b_j}{\partial x_i} \right)^2 + \nu \Delta \frac{\mathbf{u}^2}{2} + \lambda_M \Delta \frac{\mathbf{b}^2}{2} \right) dV, \tag{9.69a}$$

$$\frac{\partial}{\partial t}\int_V \mathbf{u}\cdot\mathbf{b}dV = -\int_V \nabla\cdot\left(\left(-\frac{\mathbf{u}^2+\mathbf{b}^2}{2}+p_M\right)\mathbf{b}+(\mathbf{u}\cdot\mathbf{b})\,\mathbf{u}\right)dV$$
$$+\int_V\left(-(\nu+\lambda_M)\frac{\partial u_j}{\partial x_i}\frac{\partial b_j}{\partial x_i}+\nu\frac{\partial}{\partial x_i}\left(b_j\frac{\partial u_j}{\partial x_i}\right)\right.$$
$$\left.+\lambda_M\frac{\partial}{\partial x_i}\left(u_j\frac{\partial b_j}{\partial x_i}\right)\right). \tag{9.69b}$$

The first term on each of the right-hand sides is reduced to the integral at the surface surrounding the volume V by using the Gauss' integral theorem (2.6). Therefore these two quantities are conserved in the absence of ν and λ_M so long as there are no inflows of MHD energy and cross helicity through a boundary. In the context of Elsasser's variables,

$$\int \phi^2 dV \text{ and } \int \psi^2 dV \tag{9.70}$$

are conserved.

In the MHD case, it will be shown that the residual helicity defined by

$$-\mathbf{u}\cdot\boldsymbol{\omega}+\mathbf{b}\cdot\mathbf{j} \tag{9.71}$$

is an important quantity, in place of kinetic helicity $\mathbf{u}\cdot\boldsymbol{\omega}$. Its total amount, however, is not conserved. As the other quantity conserved in the absence of molecular effects, we may mention the total amount of magnetic helicity, which is given by

$$\int_V \mathbf{a}\cdot\mathbf{b}dV, \tag{9.72}$$

where $\mathbf{a}$ is the so-called vector potential and is related to $\mathbf{b}$ as

$$\mathbf{b} = \nabla\times\mathbf{a}. \tag{9.73}$$

From the viewpoint of turbulent dynamo, the role of the magnetic helicity $\mathbf{a}\cdot\mathbf{b}$ is smaller than the residual one (9.71).

Finally, we should note that expressions (9.68b), (9.71), and (9.72) are pseudo-scalars and change their signs under the reflection of the coordinate system, that is, $\mathbf{x}\to -\mathbf{x}$. Such quantities are useful for distinguishing between the turbulence properties in the northern and southern hemispheres of a rotating stellar object, for they can take opposite signs in the two hemispheres.

9.3. Cowling's Anti-Dynamo Theorem

As the most familiar magnetic fields of a stellar object, we consider geomagnetic fields. The interior of the Earth is roughly divided into three parts:

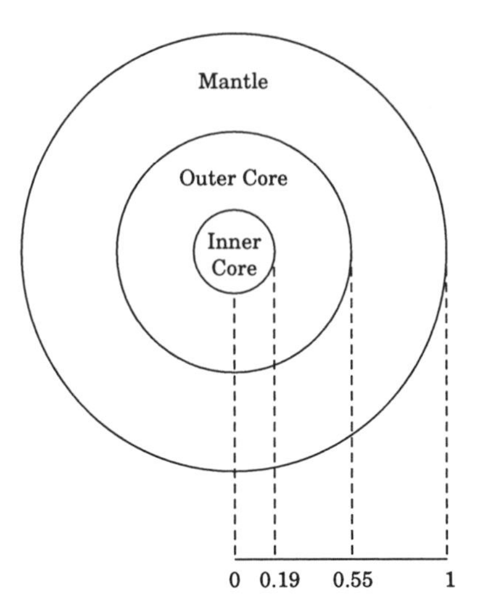

Figure 9.1. Interior of the Earth.

the inner core of solid iron, the outer core of melted iron, and the mantle with silicon as the main ingredient (see Fig. 9.1). The ratios of the inner- and outer-core radii, r_I and r_O, to Earth's radius $r_E(\cong 6400\text{km})$ are

$$r_I/r_E \cong 0.19, \;\; r_O/r_E \cong 0.55. \tag{9.74}$$

Geomagnetic fields are generated and sustained by the motion of melted iron. We divide $\mathbf{b}$ into the toroidal and poloidal components, $\mathbf{b}_T$ and $\mathbf{b}_P$. In the spherical coordinate system (r, θ, ϕ), these components are written as

$$\mathbf{b}_T = b_\phi \mathbf{e}_\phi, \tag{9.75a}$$

$$\mathbf{b}_P = b_r \mathbf{e}_r + b_\theta \mathbf{e}_\theta, \tag{9.75b}$$

where $\mathbf{e}_r$, $\mathbf{e}_\theta$, and $\mathbf{e}_\phi$ are the unit vector in each of three directions (see Fig. 6.3). In the geomagnetic case, $\mathbf{b}_T$ is confined inside the outer core owing to the electrically non-conducting mantle and is not directly measurable. On the other hand, we can measure $\mathbf{b}_P$ at the surface. Specifically, the dipole component axisymmetric around the axis of rotation is strongest. As the driving forces for the motion of melted iron, we may mention two kinds of effects. One is the buoyancy force due to the temperature difference between the inner core and the mantle. Another is the force caused by the motion of silicon rising through much heavier melted iron.

9.3.1. INCOMPATIBILITY OF AXISYMMETRIC MAGNETIC FIELDS WITH AXISYMMETRIC FLUID MOTION

In correspondence to the axisymmetric magnetic field, we consider the axisymmetric fluid motion $\mathbf{u}$, which is written as

$$\mathbf{u} = \mathbf{u}_P + \mathbf{u}_T. \tag{9.76}$$

In this situation, we can state that

Axisymmetric magnetic fields cannot be sustained stationarily by axisymmetric fluid motion.

This is the famous Cowling's (1934) anti-dynamo theorem.

In order to prove this theorem simply, we consider the situation depicted in Fig. 9.2 (Hoyng 1992), where the toroidal electric current $\mathbf{j}_T$ is along the path C with vanishing $\mathbf{b}_P$. Along C, we have

$$\mathbf{j}_T = (1/\lambda_M)(\mathbf{e} + \mathbf{u} \times \mathbf{b})_T = (1/\lambda_M)\,\mathbf{e}_T. \tag{9.77}$$

From the Stokes' integral theorem (2.62) and Eq. (9.48), we have

$$\oint_C \mathbf{j}_T \cdot d\mathbf{s} = \frac{1}{\lambda_M}\int_S (\nabla \times \mathbf{e}) \cdot \mathbf{n} dS = -\frac{1}{\lambda_M}\int_S \frac{\partial \mathbf{b}}{\partial t} \cdot \mathbf{n} dS, \tag{9.78}$$

which gives

$$\oint_C \mathbf{j}_T \cdot d\mathbf{s} = 0 \tag{9.79}$$

in the stationary state. From the condition of axisymmetry, Eq. (9.79) leads to vanishing of $\mathbf{j}_T$ itself, resulting in vanishing of $\mathbf{b}_P$. Namely, axisymmetric $\mathbf{b}$ may not be sustained by axisymmetric $\mathbf{u}$.

9.3.2. PLAUSIBILITY OF TURBULENT DYNAMO

From the Cowling's anti-dynamo theorem, nonaxisymmetric fluid motion is indispensable for the stationary sustainment of highly symmetric global magnetic fields. The occurrence of asymmetric fluid motion is closely related to the magnitude of the Reynolds number R [Eq. (1.1)]. How magnetic fields respond to such asymmetric fluid motion depends on the magnitude of the magnetic Reynolds number R_M [Eq. (9.1)].

In the case of geomagnetic fields, we adopt the radius of the outer core as L_R, and the kinematic viscosity of melted iron at the surface as ν. Then we have

$$L_R = O\left(10^6 \mathrm{m}\right), \ \nu = O\left(10^{-6} \mathrm{m}^2 \mathrm{s}^{-1}\right) \tag{9.80}$$

(Melchior 1986). The latter is of the same order as for water. This fact suggests that some of the dynamical properties of melted iron can be estimated from those of water. On the other hand, it is difficult to estimate

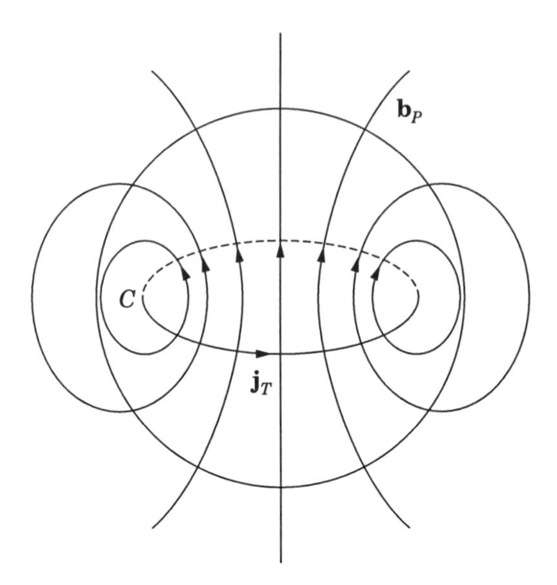

Figure 9.2. Cowling's anti-dynamo.

the velocity in the outer core that is not directly measurable. It is usually inferred from the observation of the westward-migration speed of magnetic fields as

$$U_R = O\left(10^{-4}\mathrm{ms}^{-1}\right). \tag{9.81}$$

Using these values, we have

$$R = O\left(10^{6}\right). \tag{9.82}$$

This R ensures that the fluid motion in the outer core is highly turbulent.

In Eq. (9.50) for λ_M, we use the value of melted iron at the surface as σ, and the value of μ_0 in a vacuum, resulting in

$$\lambda_M = O\left(1\mathrm{m}^2\mathrm{s}^{-1}\right). \tag{9.83}$$

From Eqs. (9.80), (9.81), and (9.83), we have

$$R_M = O\left(10^{2}\right), \tag{9.84}$$

which is much smaller than Eq. (9.82). Therefore the magnetic fields are not always in a highly fluctuating state, but it is doubtless that they as well as the fluid motion are far from an axisymmetric state.

As the next familiar magnetic fields in a stellar object, we consider solar magnetic fields (Priest 1982). The Sun is composed mainly of hydrogen (about 90 %) and helium (about 10 %), which are in a highly ionized state

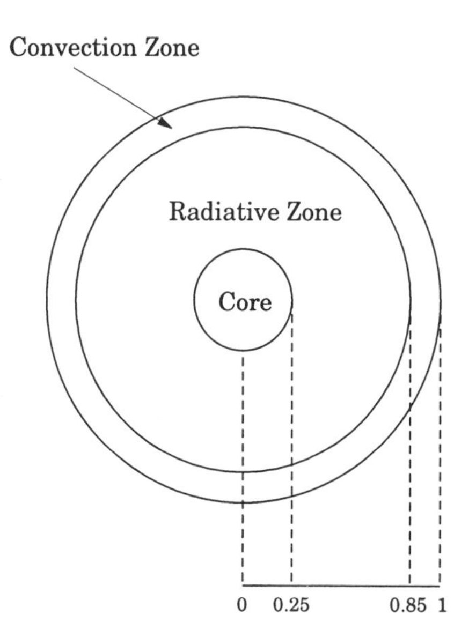

Figure 9.3. Interior of the Sun.

owing to the high temperature. Its interior is divided into three parts: the core, the radiative zone, and the convection zone (see Fig. 9.3). The ratios of the radii of the core and the radiative zone, r_C and r_R, to the solar one $r_S(\cong 700\text{Mm})$ are

$$r_C/r_S \cong 0.25, \quad r_R/r_S \cong 0.86, \tag{9.85}$$

where 1Mm= 1000km. In the core, nuclear energy is converted to thermal one, and the latter is transferred to the convection zone by radiation across the intermediate radiative zone.

From the viewpoint of macroscopic dynamical motion, the radiative zone is at rest, whereas the outermost convection zone is in a highly turbulent fluid motion. For instance, the local motions called granules and supergranules are observed at the solar surface, and their speed often reaches $O(10^3 \text{ms}^{-1})$. The resistivity of the convection zone is very small because of its ionization. As a result, both R and R_M are huge, and the convection zone may be regarded as being in a highly turbulent state in both the dynamical and magnetic sense.

From the foregoing discussions, the fluid motions generating the magnetic fields in the Earth and the Sun are in the state far from the axisymmetric one. As the method for understanding these generation mechanisms, we have two concepts. One is laminar dynamo. In it, the deviation of the state from the axisymmetric one is important, but the modes close to the axisymmetric one dominates the dynamo process (Braginskii 1964). Another is turbulent dynamo. The state is regarded as highly turbulent and is characterized by a broad range of kinetic and magnetic energy spectra (Moffatt 1978; Krause and Rädler 1980; Roberts 1993).

In this book, we shall adopt the concept of turbulent dynamo. In the geomagnetic case, the magnitude of R_M is moderate in its current estimate, as is seen from Eq. (9.84). This situation, however, is rather exceptional in a variety of astrophysical and fusion-related magnetic phenomena, and the concept of turbulent dynamo is expected to possess a wide range of applicability.

9.4. Ensemble-Mean System of Equations

We focus attention on the constant-density one-fluid MHD system given in Sec. 9.2.3. The shortfalls originating from the constraint of constant density are not so small, but the use of this simple system is helpful to the development of turbulent dynamo on a firm theoretical basis.

We adopt the ensemble-averaging procedure to abstract global properties of turbulent state, as is the same as for the hydrodynamic system. Then we write

$$f = F + f', \ F = \langle f \rangle , \tag{9.86}$$

where

$$f = (\mathbf{u}, \ \boldsymbol{\omega}, \ p, \ \mathbf{b}, \ \mathbf{j}, \ \mathbf{e}) , \tag{9.87a}$$

$$F = (\mathbf{U}, \ \boldsymbol{\Omega}, \ P, \ \mathbf{B}, \ \mathbf{J}, \ \mathbf{E}) , \tag{9.87b}$$

$$f' = (\mathbf{u}', \ \boldsymbol{\omega}', \ p', \ \mathbf{b}', \ \mathbf{j}', \ \mathbf{e}') . \tag{9.87c}$$

9.4.1. MEAN-FIELD EQUATIONS

We apply the averaging procedure to Eqs. (9.47)-(9.49), and have

$$\frac{DU_i}{Dt} = -\frac{\partial}{\partial x_i}\left(P + \frac{\langle \mathbf{b}'^2 \rangle}{2}\right) + (\mathbf{J} \times \mathbf{B})_i - \frac{\partial R_{ji}}{\partial x_j} + \nu \Delta U_i, \tag{9.88}$$

$$\frac{\partial \mathbf{B}}{\partial t} = -\nabla \times \mathbf{E}, \tag{9.89}$$

$$\mathbf{J} = \nabla \times \mathbf{B} \tag{9.90a}$$

$$= (1/\lambda_M) (\mathbf{E} + \mathbf{U} \times \mathbf{B} + \mathbf{E}_M) , \tag{9.90b}$$

with the solenoidal condition

$$\nabla \cdot \mathbf{U} = \nabla \cdot \mathbf{B} = 0, \tag{9.91}$$

where

$$\frac{D}{Dt} = \frac{\partial}{\partial t} + \mathbf{U} \cdot \nabla. \tag{9.92}$$

Equations (9.89) and (9.90) are reduced to

$$\frac{\partial \mathbf{B}}{\partial t} = \nabla \times (\mathbf{U} \times \mathbf{B} + \mathbf{E}_M) + \lambda_M \Delta \mathbf{B}. \tag{9.93}$$

In Eq. (9.88), R_{ij} is the MHD counterpart of the Reynolds stress (2.101), and is given by

$$R_{ij} = \left\langle u'_i u'_j - b'_i b'_j \right\rangle. \tag{9.94}$$

On the other hand, $\mathbf{E}_M$ in Eqs. (9.90b) and (9.93) is defined as

$$\mathbf{E}_M = \langle \mathbf{u}' \times \mathbf{b}' \rangle, \tag{9.95}$$

which is usually named the turbulent electromotive force. It may be regarded as the electric field generated by turbulence effects.

In the context of the Cowling's anti-dynamo theorem, Eq. (9.78) is replaced with

$$\oint_C \mathbf{J}_T \cdot d\mathbf{s} = -\frac{1}{\lambda_M} \int_S \frac{\partial \mathbf{B}}{\partial t} \cdot \mathbf{n} dS + \frac{1}{\lambda_M} \oint_C \mathbf{E}_M \cdot d\mathbf{s}; \tag{9.96}$$

namely, stationary $\mathbf{B}$ does not always lead to vanishing $\mathbf{J}_T$ because of newly occurring $\mathbf{E}_M$. As this result, axisymmetric $\mathbf{U}$ does not contradict the axisymmetric global component of magnetic fields, $\mathbf{B}$, that is often clearly measurable (a typical example of the latter is the dipole component of geomagnetic fields).

9.4.2. TURBULENCE EQUATIONS

From Eqs. (9.51), (9.55), (9.88), and (9.93), the velocity and magnetic-field fluctuations, $\mathbf{u}'$ and $\mathbf{b}'$, obey

$$\begin{aligned} &\frac{Du'_i}{Dt} + \frac{\partial}{\partial x_j}\left(u'_j u'_i - b'_j b'_i - R_{ji}\right) + \frac{\partial p'_M}{\partial x_i} - \nu \Delta u'_i \\ &\quad = B_j \frac{\partial b'_i}{\partial x_j} - u'_j \frac{\partial U_i}{\partial x_j} + b'_j \frac{\partial B_i}{\partial x_j}, \end{aligned} \tag{9.97}$$

$$\begin{aligned} &\frac{Db'_i}{Dt} + \frac{\partial}{\partial x_j}\left(u'_j b'_i - u'_i b'_j - \epsilon_{ij\ell} E_{M\ell}\right) - \lambda_M \Delta b'_i \\ &\quad = B_j \frac{\partial u'_i}{\partial x_j} - u'_j \frac{\partial B_i}{\partial x_j} + b'_j \frac{\partial U_i}{\partial x_j}, \end{aligned} \tag{9.98}$$

with the solenoidal condition

$$\nabla \cdot \mathbf{u}' = \nabla \cdot \mathbf{b}' = 0. \tag{9.99}$$

In Sec. 9.2.3, we stated that the total amounts of MHD energy and cross helicity are conserved in the absence of molecular effects. This fact signifies that $\left(\mathbf{u}^2 + \mathbf{b}^2\right)/2$ and $\mathbf{u} \cdot \mathbf{b}$ cascade in the presence of these effects, as is entirely similar to the kinetic energy $\mathbf{u}^2/2$ in constant-density hydrodynamic flow. The turbulence parts of $\left(\mathbf{u}^2 + \mathbf{b}^2\right)/2$ and $\mathbf{u} \cdot \mathbf{b}$, which are given by

$$K = \left\langle \mathbf{u}'^2 + \mathbf{b}'^2 \right\rangle /2, \tag{9.100}$$

$$W = \left\langle \mathbf{u}' \cdot \mathbf{b}' \right\rangle, \tag{9.101}$$

obey the equations whose mathematical structures are very clear. They are given by

$$\frac{DZ}{Dt} = P_Z - \varepsilon_Z + \nabla \cdot \mathbf{T}_Z \;\; (Z = K \text{ or } W). \tag{9.102}$$

Here P_Z, ε_Z, and $\mathbf{T}_Z$ represent the production, destruction, and transport rates of Z, respectively. In the K equation, they are defined as

$$P_K = -R_{ij} \frac{\partial U_j}{\partial x_i} - \mathbf{E}_M \cdot \mathbf{J}, \tag{9.103}$$

$$\varepsilon_K \equiv \varepsilon = \nu \left\langle \left(\frac{\partial u'_j}{\partial x_i} \right)^2 \right\rangle + \lambda_M \left\langle \left(\frac{\partial b'_j}{\partial x_i} \right)^2 \right\rangle, \tag{9.104}$$

$$\mathbf{T}_K = W\mathbf{B} + \mathbf{T}'_K, \tag{9.105a}$$

$$\mathbf{T}'_K = -\left\langle \left(\frac{\mathbf{u}'^2 + \mathbf{b}'^2}{2} + p'_M \right) \mathbf{u}' + \left(\mathbf{u}' \cdot \mathbf{b}' \right) \mathbf{b}' \right\rangle. \tag{9.105b}$$

On the other hand, the counterparts in the W equation are

$$P_W = -R_{ij} \frac{\partial B_j}{\partial x_i} - \mathbf{E}_M \cdot \mathbf{\Omega}, \tag{9.106}$$

$$\varepsilon_W = \left(\nu + \lambda_M \right) \left\langle \frac{\partial u'_j}{\partial x_i} \frac{\partial b'_j}{\partial x_i} \right\rangle, \tag{9.107}$$

$$\mathbf{T}_W = K\mathbf{B} + \mathbf{T}'_W, \tag{9.108a}$$

$$\mathbf{T}'_W = -\left\langle \left(\mathbf{u}' \cdot \mathbf{b}' \right) \mathbf{u}' - \left(\frac{\mathbf{u}'^2 + \mathbf{b}'^2}{2} - p'_M \right) \mathbf{b}' \right\rangle. \tag{9.108b}$$

The mathematical structures of the K and W equations are very similar to Eqs. (2.111) and (2.132) in the hydrodynamic case. This resemblance is shared by all the equations for the turbulence parts of the quantities that possess the conservation property in the absence of molecular effects. Specifically, the close correspondence between each two of Eqs. (9.103)-(9.105) and (9.106)-(9.108), respectively, is prominent. For instance, Eqs. (9.103) and (9.106) may be rewritten as

$$P_K = -(1/2)R_{ij}S_{ij} - \mathbf{E}_M \cdot \mathbf{J}, \tag{9.109}$$

$$P_W = -(1/2)R_{ij}M_{ij} - \mathbf{E}_M \cdot \mathbf{\Omega}, \tag{9.110}$$

where S_{ij} and M_{ij} are the velocity- and magnetic-strain tensors, respectively, and are given by

$$S_{ij} = \frac{\partial U_j}{\partial x_i} + \frac{\partial U_i}{\partial x_j}, \tag{9.111}$$

$$M_{ij} = \frac{\partial B_j}{\partial x_i} + \frac{\partial B_i}{\partial x_j}. \tag{9.112}$$

The mean velocity and magnetic-field gradients are divided into the symmetric and anti-symmetric parts as

$$\frac{\partial U_j}{\partial x_i} = \frac{1}{2}S_{ij} + \frac{1}{2}\epsilon_{ij\ell}\Omega_\ell, \tag{9.113}$$

$$\frac{\partial B_j}{\partial x_i} = \frac{1}{2}M_{ij} + \frac{1}{2}\epsilon_{ij\ell}J_\ell \tag{9.114}$$

[see Eq. (2.61)]. Namely, Eq. (9.110) is obtained from Eq. (9.109) by replacing the velocity strain and the electric current with their counterparts, that is, the magnetic strain and the vorticity, respectively.

In the context of the residual helicity (9.71), the turbulent residual helicity H, which is defined as

$$H = \left\langle -\mathbf{u}' \cdot \boldsymbol{\omega}' + \mathbf{b}' \cdot \mathbf{j}' \right\rangle, \tag{9.115}$$

obeys a very complicated equation

$$\frac{DH}{Dt} = \sum_{n=1}^{9} \Lambda_n + R_H, \tag{9.116}$$

where

$$\Lambda_1 = \left\langle u'_j u'_i + b'_j b'_i \right\rangle \frac{\partial \Omega_i}{\partial x_j}, \tag{9.117}$$

$$\Lambda_2 = \langle \mathbf{u}' \times \boldsymbol{\omega}' \rangle \cdot \boldsymbol{\Omega}, \tag{9.118}$$

$$\Lambda_3 = -\nabla \cdot \left(\left\langle \mathbf{u}'^2/2 \right\rangle \boldsymbol{\Omega} \right), \tag{9.119}$$

$$\Lambda_4 = \left\langle \frac{\partial u_i'}{\partial x_j} j_i' + \frac{\partial \omega_i'}{\partial x_j} b_i' - \omega_i' \frac{\partial b_i'}{\partial x_j} - u_i' \frac{\partial j_i'}{\partial x_j} \right\rangle B_j, \tag{9.120}$$

$$\Lambda_5 = \left\langle b_j' j_i' \right\rangle \frac{\partial U_i}{\partial x_j}, \tag{9.121}$$

$$\Lambda_6 = -\left\langle \omega_i' b_j' \right\rangle \frac{\partial B_i}{\partial x_j}, \tag{9.122}$$

$$\Lambda_7 = -\langle \mathbf{u}' \times \mathbf{j}' \rangle \cdot \mathbf{J}, \tag{9.123}$$

$$\Lambda_8 = \left\langle u_j' b_i' + u_i' b_j' \right\rangle \frac{\partial J_i}{\partial x_j}, \tag{9.124}$$

$$\Lambda_9 = -2 \left(-\nu \left\langle \frac{\partial u_j'}{\partial x_i} \frac{\partial \omega_j'}{\partial x_i} \right\rangle + \lambda_M \left\langle \frac{\partial b_j'}{\partial x_i} \frac{\partial j_j'}{\partial x_i} \right\rangle \right). \tag{9.125}$$

The residual part R_H consists of the terms of the third order in $\mathbf{u}'$, $\boldsymbol{\omega}'$, $\mathbf{b}'$, and $\mathbf{j}'$, and is not dependent directly on the mean quantities $(\mathbf{U}, \boldsymbol{\Omega}, \mathbf{B}, \mathbf{J})$. Moreover, R_H cannot be expressed in a divergence form, unlike the last terms in Eq. (9.102). This residual-helicity equation should be compared with Eq. (2.132) for $\langle \mathbf{u}' \cdot \boldsymbol{\omega}' \rangle$ in the hydrodynamic case. The complicatedness of the former arises from the fact that the total amount of $-\mathbf{u} \cdot \boldsymbol{\omega} + \mathbf{b} \cdot \mathbf{j}$ is not conserved in the absence of molecular effects. This situation will become a stumbling block for the helicity- or alpha-dynamo modeling.

Note: Between K and W, we have the relationship

$$|W| / K = 2 \left| \langle \mathbf{u}' \cdot \mathbf{b}' \rangle \right| \Big/ \langle \mathbf{u}'^2 + \mathbf{b}'^2 \rangle \leq 1. \tag{9.126}$$

It is important in two aspects. One is that the proper model equations for K and W should obey this inequality. Another is associated with the generation mechanism of magnetic fields under the cross-helicity effects, which will be detailed in Chapter 10.

9.5. Quasi-Kinematic and Counter-Kinematic Approaches

In turbulent dynamo, the primary interest lies in the global structure of magnetic fields generated by global fluid motion and the former feedback effect on the latter. This mutual interaction is described through the mean field $(\mathbf{U}, \mathbf{B})$, which obeys the MHD system, Eqs. (9.88)-(9.90). In order to close the system, we need to relate the Reynolds stress and the turbulent electromotive force, R_{ij} and $\mathbf{E}_M$, to $\mathbf{U}$, $\mathbf{B}$, and the statistical quantities characterizing the MHD turbulent state. This situation is entirely similar to the case of hydrodynamic turbulent shear flow, except the further complexity arising from the nonlinear coupling between velocity and magnetic fluctuations.

In a long history of the study about turbulent dynamo, the mathematical form of $\mathbf{E}_M$ has attracted much more attention, compared with R_{ij}. As a result, $\mathbf{E}_M$ has been examined mainly from the kinematic standpoint. The kinematic derivation of $\mathbf{E}_M$ using the smoothing or quasi-linear approximation is detailed by Moffatt (1978). This method is suited for the analysis of dynamo effects at low R_M. The MHD dynamo dealing with both $\mathbf{E}_M$ and R_{ij} is detailed by Krause and Rädler (1980) with emphasis put on $\mathbf{E}_M$. This method is founded on the concept of two-scales variables and is similar to the two-scale direct-interaction approximation (TSDIA) in Chapter 6, although a clear scale parameter for distinguishing between these two scales is not introduced. There a general mathematical form for $\mathbf{E}_M$ is also sought by tensor analysis, and the relationship with gravity and frame rotation is given.

In this section, we shall examine $\mathbf{E}_M$ and R_{ij} from the viewpoint of both quasi-kinematic and counter-kinematic dynamos. These approaches are an introduction to the MHD methods such as the TSDIA in Sec. 9.6.

9.5.1. QUASI-KINEMATIC APPROACH

In the kinematic approach, the statistical properties of fluid motion are assumed to be known, and their effects on $\mathbf{E}_M$ are the primary interest. This approach is insufficient in the analysis of the interaction between velocity and magnetic fields, but it is instrumental to abstracting some essential ingredients of dynamo effects by a rather simple mathematical manipulation.

A. Two Spatial and Temporal Variables

In what follows, we shall investigate $\mathbf{E}_M$ from the standpoint of the two-scale variables adopted in the TSDIA method, unlike the familiar kinematic approach. First, we introduce a small scale parameter δ, and define two spatial and temporal variables

$$\boldsymbol{\xi}(=\mathbf{x}),\ \mathbf{X}(=\delta\mathbf{x});\ \tau(=t),\ T(=\delta t), \tag{9.127}$$

resulting in

$$\nabla = \nabla_\xi + \delta\nabla_X,\ \frac{\partial}{\partial t} = \frac{\partial}{\partial \tau} + \delta\frac{\partial}{\partial T} \tag{9.128}$$

$[\nabla_\xi = (\partial/\partial\xi_i)$ and $\nabla_X = (\partial/\partial X_i)]$. With the aid of Eq. (9.127), we distinguish the fast variation of fluctuations from the slow one of the mean field (see Secs. 6.2 and 6.3.1), and write

$$f = F(\mathbf{X};T) + f'(\boldsymbol{\xi},\mathbf{X};\tau,T). \tag{9.129}$$

We apply Eqs. (9.127)-(9.129) to Eq. (9.98), and have

$$\begin{aligned}
&\frac{\partial b'_i}{\partial \tau} + U_j \frac{\partial b'_i}{\partial \xi_j} + \frac{\partial}{\partial \xi_j}\left(u'_j b'_i - u'_i b'_j\right) - \lambda_M \Delta_\xi b'_i \\
&= B_j \frac{\partial u'_i}{\partial \xi_j} + \delta\left(-u'_j \frac{\partial B_i}{\partial X_j} + b'_j \frac{\partial U_i}{\partial X_j} - \frac{Db'_i}{DT} + B_j \frac{\partial u'_i}{\partial X_j}\right. \\
&\left. - \frac{\partial}{\partial X_j}\left(u'_j b'_i - u'_i b'_j - \epsilon_{ij\ell} E_{M\ell}\right)\right),
\end{aligned} \tag{9.130}$$

where $D/DT = \partial/\partial T + \mathbf{U}\cdot\nabla_X$, and the resistivity effect of $O(\delta)$ was dropped.

B. Fourier Representation

We introduce the Fourier representation concerning the fast spatial variable $\boldsymbol{\xi}$ as

$$f'(\boldsymbol{\xi},\mathbf{X};\tau,T) = \int f'(\mathbf{k},\mathbf{X};\tau,T)\exp\left(-i\mathbf{k}\cdot(\boldsymbol{\xi} - \mathbf{U}\tau)\right)d\mathbf{k}, \tag{9.131}$$

and deal with the fast variation of fluctuations as waves or eddies. Here the factor $\exp(-i\mathbf{k}\cdot(\boldsymbol{\xi} - \mathbf{U}\tau))$ signifies that the fast-varying motion of turbulent eddies is observed in the frame moving with $\mathbf{U}$.

We apply Eq. (9.131) to Eq. (9.130), and have

$$\begin{aligned}
&\frac{\partial b'_i(\mathbf{k};\tau)}{\partial \tau} + \lambda_M k^2 b'_i(\mathbf{k};\tau) - iN_{ij\ell}(\mathbf{k}) \iint \delta(\mathbf{k}-\mathbf{p}-\mathbf{q}) d\mathbf{p} d\mathbf{q} \\
&\times u'_j(\mathbf{p};\tau) b'_\ell(\mathbf{q};\tau) = -i(\mathbf{k}\cdot\mathbf{B}) u'_i(\mathbf{k};\tau) + \delta\left(-u'_j(\mathbf{k};\tau)\frac{\partial B_i}{\partial X_j}\right. \\
&\left. + b'_j(\mathbf{k};\tau)\frac{\partial U_i}{\partial X_j} - \frac{Db'_i(\mathbf{k};\tau)}{DT_I} + B_j \frac{\partial u'_i(\mathbf{k};\tau)}{\partial X_{Ij}} + N_{Bi}\right),
\end{aligned} \tag{9.132}$$

where D/DT_I and ∇_{XI} are defined by Eq. (6.36), and

$$N_{ij\ell}(\mathbf{k}) = k_j \delta_{i\ell} - k_\ell \delta_{ij}, \tag{9.133}$$

and $\mathbf{N}_B$ denotes the contribution that is nonlinear in $\mathbf{u}'$ and $\mathbf{b}'$, and is not dependent directly on the mean field $(\mathbf{U},\mathbf{B})$. We focus attention on the interaction between the mean field and fluctuations, and neglect $\mathbf{N}_B$ in what follows. Moreover, the dependence on the slow variables $\mathbf{X}$ and T is not written explicitly, except when necessary.

C. Scale-Parameter Expansion

Considering that effects of the mean-field gradients occur in $O(\delta)$, we expand

$$\mathbf{b}'(\mathbf{k};\tau)=\sum_{n=0}^{\infty}\delta^n \mathbf{b}'_n(\mathbf{k};\tau). \tag{9.134}$$

The $O(1)$ field $\mathbf{b}'_0(\mathbf{k};\tau)$ obeys

$$\frac{\partial b'_{0i}(\mathbf{k};\tau)}{\partial\tau}+\lambda_M k^2 b'_{0i}(\mathbf{k};\tau)-iN_{ij\ell}(\mathbf{k})\iint\delta(\mathbf{k}-\mathbf{p}-\mathbf{q})d\mathbf{p}d\mathbf{q} \\ \times u'_j(\mathbf{p};\tau)b'_{0\ell}(\mathbf{q};\tau)=-i(\mathbf{k}\cdot\mathbf{B})u'_i(\mathbf{k};\tau). \tag{9.135}$$

On the other hand, the equation governing the $O(\delta)$ field $\mathbf{b}'_1(\mathbf{k};\tau)$ is

$$\frac{\partial b'_{1i}(\mathbf{k};\tau)}{\partial\tau}+\lambda_M k^2 b'_{1i}(\mathbf{k};\tau) \\ -iN_{ij\ell}(\mathbf{k})\iint\delta(\mathbf{k}-\mathbf{p}-\mathbf{q})d\mathbf{p}d\mathbf{q}u'_j(\mathbf{p};\tau)b'_{1\ell}(\mathbf{q};\tau) \\ =-u'_j(\mathbf{k};\tau)\frac{\partial B_i}{\partial X_j}+b'_{0j}(\mathbf{k};\tau)\frac{\partial U_i}{\partial X_j}-\frac{Db'_{0i}(\mathbf{k};\tau)}{DT_I}+B_j\frac{\partial u'_i(\mathbf{k};\tau)}{\partial X_{Ij}}. \tag{9.136}$$

In order to solve Eqs. (9.135) and (9.136), we introduce the Green's function $G'_{Mij}(\mathbf{k};\tau,\tau')$, which is similar to Eqs. (3.100) and (6.50) and obeys

$$\frac{\partial G'_{Mij}(\mathbf{k};\tau,\tau')}{\partial\tau}+\lambda_M k^2 G'_{Mij}(\mathbf{k};\tau,\tau')-iN_{i\ell m}(\mathbf{k})\iint\delta(\mathbf{k}-\mathbf{p}-\mathbf{q})d\mathbf{p}d\mathbf{q} \\ \times u'_\ell(\mathbf{p};\tau)G'_{Mmj}(\mathbf{q};\tau,\tau')=\delta_{ij}\delta(\tau-\tau') \tag{9.137}$$

[note the remarks made on Eq. (3.100)]. Using Eq. (9.137), Eq. (9.135) may be integrated as

$$b'_{0i}(\mathbf{k};\tau)=b'_{Bi}(\mathbf{k};\tau)-i(\mathbf{k}\cdot\mathbf{B})\int_{-\infty}^{\tau}d\tau_1 G'_{Mij}(\mathbf{k};\tau,\tau_1)u'_j(\mathbf{k};\tau_1), \tag{9.138}$$

where $\mathbf{b}'_B(\mathbf{k};\tau)$, which may be called the basic field, is governed by Eq. (9.135) with the right-hand side dropped. Entirely similarly, the $O(\delta)$ solution is given by

$$b'_{1i}(\mathbf{k};\tau)=-\frac{\partial B_j}{\partial X_\ell}\int_{-\infty}^{\tau}d\tau_1 G'_{Mij}(\mathbf{k};\tau,\tau_1)u'_\ell(\mathbf{k};\tau_1) \\ +\frac{\partial U_j}{\partial X_\ell}\int_{-\infty}^{\tau}d\tau_1 G'_{Mij}(\mathbf{k};\tau,\tau_1)b'_{0\ell}(\mathbf{k};\tau_1). \tag{9.139}$$

In obtaining Eq. (9.139), attention has been focused on the first two terms on the right-hand side of Eq. (9.136), which are dependent directly on the mean field.

D. Statistics of Fluid Motion

In the kinematic dynamo, we need to postulate statistical properties of fluid motion. We assume that the velocity field is statistically isotropic, but that it lacks reflectional symmetry. Namely, we write

$$\frac{\left\langle u_i'(\mathbf{k};\tau)u_j'(\mathbf{k}';\tau')\right\rangle}{\delta\left(\mathbf{k}+\mathbf{k}'\right)} = D_{ij}(\mathbf{k})Q_K(k;\tau,\tau') + \frac{i}{2}\frac{k_\ell}{k^2}\epsilon_{ij\ell}H_K(k;\tau,\tau'), \quad (9.140)$$

where $D_{ij}(\mathbf{k})$ is the so-called solenoidal projection operator defined as

$$D_{ij}(\mathbf{k}) = \delta_{ij} - \left(k_i k_j/k^2\right). \quad (9.141)$$

Equation (9.140) gives

$$\left\langle \mathbf{u}'^2/2\right\rangle = \int Q_K(k;\tau,\tau)d\mathbf{k}, \quad (9.142)$$

$$\langle \mathbf{u}'\cdot\boldsymbol{\omega}'\rangle = \int H_K(k;\tau,\tau)d\mathbf{k}. \quad (9.143)$$

Equation (9.143) comes from the second part of Eq. (9.140) and represents the breakage of statistical reflectional symmetry of velocity fluctuations. The importance of such a pseudo-scalar quantity is referred to in Sec. 2.5.3, and the relationship with effects of frame rotation is detailed in Sec. 6.7.2 in the context of turbulent shear flow. In correspondence to Eq. (9.140), the Green's function $G'_{Mij}(\mathbf{k};\tau,\tau')$, which obeys Eq. (9.137), may be assumed to be statistically isotropic. Then we have

$$\left\langle G'_{Mij}(\mathbf{k};\tau,\tau')\right\rangle = \delta_{ij}G_M(k;\tau,\tau'). \quad (9.144)$$

In the usual kinematic dynamo, we pay no attention to the correlation between velocity and magnetic fields. In real-world dynamo, such a correlation always exists. In order to retain this feature, we write

$$\frac{\left\langle u_i'(\mathbf{k};\tau)b'_{Bj}(\mathbf{k}';\tau')\right\rangle}{\delta\left(\mathbf{k}+\mathbf{k}'\right)} = D_{ij}(\mathbf{k})W(k;\tau,\tau'), \quad (9.145)$$

resulting in

$$\langle \mathbf{u}'\cdot\mathbf{b}'_B\rangle = 2\int W(k;\tau,\tau)d\mathbf{k}. \quad (9.146)$$

Equation (9.146) is usually called the turbulent cross helicity, and such an effect will be shown to bring an interesting dynamo effect, which has been missing in the long history of turbulent-dynamo research. In this respect, it is more proper to call the present method retaining Eq. (9.145) a quasi-kinematic dynamo.

E. Turbulent Electromotive Force

Up to $O(\delta)$, $\mathbf{E}_M$ is expanded as

$$E_{Mi} = \epsilon_{ij\ell}\left(\left\langle u'_j b'_{0\ell}\right\rangle + \delta\left(\left\langle u'_j b'_{1\ell}\right\rangle\right)\right). \qquad (9.147)$$

We substitute Eqs. (9.138) and (9.139) into Eq. (9.147), and make use of the statistical properties of the fluid motion, Eqs. (9.140), (9.144), and (9.145). This calculation is simple, and gives

$$\mathbf{E}_M = \alpha_K \mathbf{B} - \beta_K \mathbf{J} + \gamma_K \mathbf{\Omega}. \qquad (9.148)$$

Here use has been made of Eq. (2.60), and the dimensional coefficients α_K, β_K, and γ_K are written as

$$\alpha_K = -(1/3)\int d\mathbf{k}\int_{-\infty}^{\tau} d\tau_1 G_M(k;\tau,\tau_1) H_K(k;\tau,\tau_1), \qquad (9.149)$$

$$\beta_K = (2/3)\int d\mathbf{k}\int_{-\infty}^{\tau} d\tau_1 G_M(k;\tau,\tau_1) Q_K(k;\tau,\tau_1), \qquad (9.150)$$

$$\gamma_K = (2/3)\int d\mathbf{k}\int_{-\infty}^{\tau} d\tau_1 G_M(k;\tau,\tau_1) W(k;\tau,\tau_1). \qquad (9.151)$$

In Eq. (9.148), the artificial scale parameter δ has disappeared automatically through the replacement of $\mathbf{X} \to \delta\mathbf{x}$ and $T \to \delta t$.

In order to understand the physical meaning of each term in Eq. (9.148), we substitute it into the mean Ohm's law (9.90b), and have

$$\mathbf{J} = \frac{1}{\lambda_M + \beta_K}\left(\mathbf{E} + \mathbf{U}\times\mathbf{B} + \alpha_K\mathbf{B} + \gamma_K\mathbf{\Omega}\right). \qquad (9.152)$$

The physical meaning of β_K is clear; namely, β_K represents the enhancement of resitivity due to turbulent motion, and is usually called the turbulent or anomalous resistivity. Equation (9.150) indicates that the isotropic and reflectionally symmetric property of turbulence, which is closely related to the turbulent kinetic energy, contributes to the effective increase in the resistivity.

In Eq. (9.152), the α_K-related part generates the mean electric current parallel or anti-parallel to the mean magnetic field. This situation makes sharp contrast with the second part $\mathbf{U}\times\mathbf{B}$ that comes from the $\mathbf{b}$-related one in the original Ohm's law (9.49b). In the latter, the induced electric current is always normal to $\mathbf{b}$. The electric-current generation mechanism due to the first term in Eq. (9.148) is called the alpha or helicity dynamo. Its mechanism is usually explained using Fig. 9.4. The minus sign in Eq. (9.149) indicates that positive $\langle\mathbf{u}'\cdot\boldsymbol{\omega}'\rangle$ tends to induce $\mathbf{J}$ anti-parallel to original $\mathbf{B}$.

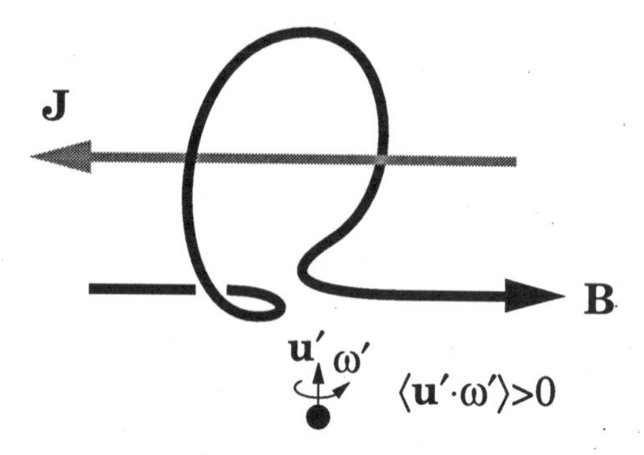

Figure 9.4. Alpha or helicity dynamo.

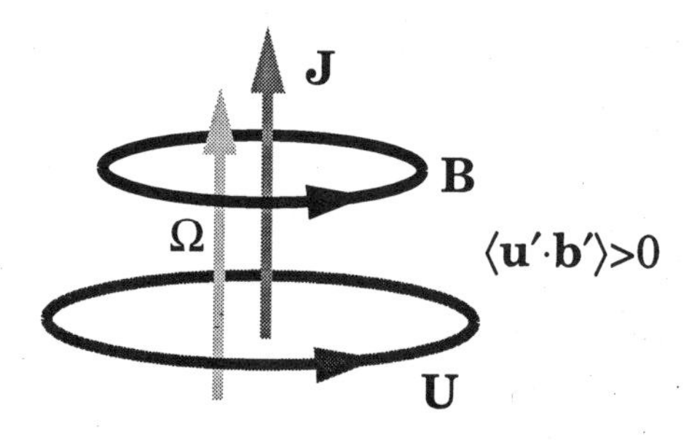

Figure 9.5. Cross-helicity dynamo.

The explanation in Fig. 9.4, however, is not so clear as is widely believed, for nonvanishing $\langle \mathbf{u}' \cdot \boldsymbol{\omega}' \rangle$ does not simply lead to such a looped magnetic-field line normal to original $\mathbf{B}$ as is depicted there.

The last term in Eq. (9.148), γ_K, indicates the alignment between $\mathbf{J}$ and $\boldsymbol{\Omega}$, which is equivalent to that between $\mathbf{B}$ and $\mathbf{U}$ because of the relation $\mathbf{J} = \nabla \times \mathbf{B}$ and $\boldsymbol{\Omega} = \nabla \times \mathbf{U}$. The latter alignment is characterized by their inner product or the mean cross helicity, $\mathbf{U} \cdot \mathbf{B}$. We consider the relationship between $\mathbf{U} \cdot \mathbf{B}$ and its turbulence counterpart or the turbulent cross helicity, $W (= \langle \mathbf{u}' \cdot \mathbf{b}' \rangle)$. As is referred to in Sec. 9.2.3, the total amount of cross helicity is conserved in the absence of molecular effects, and the cross helicity cascades towards the small-scale components of motion in the presence of those effects. This fact indicates that nonvanishing $\mathbf{U} \cdot \mathbf{B}$ is accompanied by nonvanishing W, and that the sign and magnitude of W are closely associated with the degree of the alignment between $\mathbf{U}$ and $\mathbf{B}$. These circumstances are shown schematically in Fig. 9.5. Positive W tends to generate $\mathbf{J}$ parallel to $\boldsymbol{\Omega}$.

In the context of the cross-helicity dynamo, we should recall the transformation (9.58). In a rotating frame, Eqs. (9.148) and (9.152) are replaced with

$$\mathbf{E}_M = \alpha_K \mathbf{B} - \beta_K \mathbf{J} + \gamma_K \left(\boldsymbol{\Omega} + 2\boldsymbol{\Omega}_F \right), \tag{9.153}$$

$$\mathbf{J} = \frac{1}{\lambda_M + \beta_K} \left(\mathbf{E} + \mathbf{U} \times \mathbf{B} + \alpha_K \mathbf{B} + \gamma_K \left(\mathbf{\Omega} + 2\mathbf{\Omega}_F \right) \right). \tag{9.154}$$

From these results, the cross-helicity effect is sensitive to frame rotation.

F. Reynolds Stress

In Eq. (9.88) for $\mathbf{U}$, the counterpart of $\mathbf{E}_M$ is R_{ij} [Eq. (9.94)]. In the quasi-kinematic approach, the kinetic part $\langle u'_i u'_j \rangle$ is prescribed by Eq. (9.140), and its magnetic counterpart $\langle b'_i b'_j \rangle$ is the sole quantity to be examined. Its calculation based on Eqs. (9.138) and (9.139) is simple, resulting in

$$-\left\langle b'_i b'_j \right\rangle = -(1/3) \left\langle \mathbf{b}'^2 \right\rangle \delta_{ij} - \nu_{TM} S_{ij} + \nu_{MM} M_{ij}, \tag{9.155}$$

where

$$\nu_{TM} = (2/3) \int d\mathbf{k} \int_{-\infty}^{\tau} d\tau_1 G_M(k;\tau,\tau_1) Q_M(k;\tau,\tau_1), \tag{9.156}$$

$$\nu_{MM} = (2/3) \int d\mathbf{k} \int_{-\infty}^{\tau} d\tau_1 G_M(k;\tau,\tau_1) W(k;\tau,\tau_1), \tag{9.157}$$

and the velocity- and magnetic-strain tensors, S_{ij} and M_{ij}, are defined as Eqs. (9.111) and (9.112), respectively. Moreover, $Q_M(k;\tau,\tau')$ in ν_{MM} is the magnetic counterpart of $Q_K(k;\tau,\tau')$, and is given by

$$\frac{\left\langle b'_{Bi}(\mathbf{k};\tau) b'_{Bj}(\mathbf{k}';\tau') \right\rangle}{\delta\left(\mathbf{k} + \mathbf{k}'\right)} = D_{ij}(\mathbf{k}) Q_M(k;\tau,\tau'). \tag{9.158}$$

Here the helical property or nonmirrorsymmetric part of $\mathbf{b}'$ was dropped, and this point will be discussed later.

From Eqs. (9.94) and (9.155), we may see that ν_{TM} is a kind of turbulent viscosity generated by isotropic magnetic-field fluctuations. In the hydrodynamic case, the turbulent viscosity is related to isotropic velocity fluctuations, as is given by Eq. (6.104). The ν_{MM}-related term in Eq. (9.155) expresses the feedback effect of the generated mean magnetic field on the mean velocity. This effect has also been missing in the current turbulent dynamo. Its physical meaning will be discussed later.

9.5.2. COUNTER-KINEMATIC APPROACH

In the quasi-kinematic dynamo, we examined the dynamo effects under a prescribed fluid motion. There the feedback effects of $\mathbf{b}'$ on $\mathbf{u}'$ were neglected. In order to see these effects, we consider the counter dynamo process under prescribed magnetic fields.

A. Scale-Parameter Expansion

We apply the two-scale variables (9.127) to Eq. (9.97) with the solenoidal condition (9.99), and have

$$\frac{\partial u'_i}{\partial \xi_i} = -\delta \frac{\partial u'_i}{\partial X_i}, \tag{9.159}$$

$$\begin{aligned}
&\frac{\partial u'_i}{\partial \tau} + U_j \frac{\partial u'_i}{\partial \xi_j} + \frac{\partial}{\partial \xi_j}\left(u'_j u'_i - b'_j b'_i\right) + \frac{\partial p'_M}{\partial \xi_i} - \nu \Delta_\xi u'_i \\
&= B_j \frac{\partial b'_i}{\partial \xi_j} + \delta \left(-u'_j \frac{\partial U_i}{\partial X_j} + b'_j \frac{\partial B_i}{\partial X_j} - \frac{D u'_i}{DT} + B_j \frac{\partial b'_i}{\partial X_j} \right. \\
&\left. - \frac{\partial p'_M}{\partial X_i} - \frac{\partial}{\partial X_j}\left(u'_j u'_i - b'_j b'_i - R_{ji}\right) \right).
\end{aligned} \tag{9.160}$$

We further apply the Fourier representation (9.131) to these equations. As a result, we have

$$\mathbf{k} \cdot \mathbf{u}'(\mathbf{k};\tau) = \delta \left(-i \frac{\partial u'_i(\mathbf{k};\tau)}{\partial X_{Ii}} \right). \tag{9.161}$$

$$\begin{aligned}
&\frac{\partial u'_i(\mathbf{k};\tau)}{\partial \tau} + \nu k^2 u'_i(\mathbf{k};\tau) - i k_i p'_M(\mathbf{k};\tau) \\
&- i k_j \iint \delta(\mathbf{k}-\mathbf{p}-\mathbf{q}) d\mathbf{p} d\mathbf{q} u'_i(\mathbf{p};\tau) u'_j(\mathbf{q};\tau) \\
&= -i(\mathbf{k}\cdot\mathbf{B}) b'_i(\mathbf{k};\tau) - i k_j \iint \delta(\mathbf{k}-\mathbf{p}-\mathbf{q}) d\mathbf{p} d\mathbf{q} b'_i(\mathbf{p};\tau) b'_j(\mathbf{q};\tau) \\
&+ \delta \left(-u'_j(\mathbf{k};\tau) \frac{\partial U_i}{\partial X_j} + b'_j(\mathbf{k};\tau) \frac{\partial B_i}{\partial X_j} - \frac{D u'_i(\mathbf{k};\tau)}{D T_I} \right. \\
&+ B_j \frac{\partial b'_i(\mathbf{k};\tau)}{\partial X_{Ij}} - \frac{\partial p'_M(\mathbf{k};\tau)}{\partial X_{Ii}} - \iint \delta(\mathbf{k}-\mathbf{p}-\mathbf{q}) d\mathbf{p} d\mathbf{q} \\
&\left. \times \frac{\partial}{\partial X_{Ij}} \left(u'_i(\mathbf{p};\tau) u'_j(\mathbf{q};\tau) - b'_i(\mathbf{p};\tau) b'_j(\mathbf{q};\tau) \right) \right).
\end{aligned} \tag{9.162}$$

In the absence of magnetic fields, Eq. (9.162) is reduced to the hydrodynamic counterpart (6.34).

From the scale-parameter expansion

$$\mathbf{u}'(\mathbf{k};\tau) = \sum_{n=0}^{\infty} \delta^n \mathbf{u}'_n(\mathbf{k};\tau), \tag{9.163}$$

the $O(1)$ system is given by

$$\mathbf{k}\cdot\mathbf{u}'_0(\mathbf{k};\tau)=0, \tag{9.164}$$

$$\frac{\partial u'_{0i}(\mathbf{k};\tau)}{\partial\tau}+\nu k^2u'_{0i}(\mathbf{k};\tau)-ik_ip'_{M0}(\mathbf{k};\tau)$$
$$-ik_j\iint\delta(\mathbf{k}-\mathbf{p}-\mathbf{q})d\mathbf{p}d\mathbf{q}u'_{0i}(\mathbf{p};\tau)u'_{0j}(\mathbf{q};\tau)$$
$$=-i(\mathbf{k}\cdot\mathbf{B})b'_i(\mathbf{k};\tau)-ik_j\iint\delta(\mathbf{k}-\mathbf{p}-\mathbf{q})d\mathbf{p}d\mathbf{q}b'_i(\mathbf{p};\tau)b'_j(\mathbf{q};\tau). \tag{9.165}$$

We use Eq. (9.164) to eliminate $p'_{M0}(\mathbf{k};\tau)$ from Eq. (9.165) (this procedure in the hydrodynamic case is detailed in Sec. 6.3.3 in the context of the TSDIA). The resulting equation is

$$\frac{\partial u'_{0i}(\mathbf{k};\tau)}{\partial\tau}+\nu k^2u'_{0i}(\mathbf{k};\tau)-iM_{ij\ell}(\mathbf{k})\iint\delta(\mathbf{k}-\mathbf{p}-\mathbf{q})d\mathbf{p}d\mathbf{q}$$
$$\times u'_{0j}(\mathbf{p};\tau)u'_{0\ell}(\mathbf{q};\tau)=-i(\mathbf{k}\cdot\mathbf{B})b'_i(\mathbf{k};\tau)$$
$$-iM_{ij\ell}(\mathbf{k})\iint\delta(\mathbf{k}-\mathbf{p}-\mathbf{q})d\mathbf{p}d\mathbf{q}b'_j(\mathbf{p};\tau)b'_\ell(\mathbf{q};\tau), \tag{9.166}$$

where

$$M_{ij\ell}(\mathbf{k})=(1/2)\left(k_jD_{i\ell}(\mathbf{k})+k_\ell D_{ij}(\mathbf{k})\right). \tag{9.167}$$

In the $O(\delta)$ system, Eq. (9.161) is given by

$$\mathbf{k}\cdot\mathbf{u}'_1(\mathbf{k};\tau)=-i\frac{\partial u'_{0i}(\mathbf{k};\tau)}{\partial X_{Ii}}. \tag{9.168}$$

This equation can be reduced to a more manageable form with the aid of the transformation (Hamba 1987)

$$\mathbf{u}'_1(\mathbf{k};\tau)=-i\frac{\mathbf{k}}{k^2}\frac{\partial u'_{0i}(\mathbf{k};\tau)}{\partial X_i}+\mathbf{u}'_{S1}(\mathbf{k};\tau), \tag{9.169}$$

where $\mathbf{u}'_{S1}(\mathbf{k};\tau)$ obeys the familiar solenoidal condition concerning $\mathbf{k}$ as

$$\mathbf{k}\cdot\mathbf{u}'_{S1}(\mathbf{k};\tau)=0. \tag{9.170}$$

We substitute Eq. (9.169) into the equation for $\mathbf{u}'_1(\mathbf{k};\tau)$, and eliminate $p'_{M1}(\mathbf{k};\tau)$ using Eq. (9.170) (these procedures are also detailed in Sec. 6.3.3). As a result, we have

$$\frac{\partial u'_{S1i}(\mathbf{k};\tau)}{\partial\tau}+\nu k^2u'_{S1i}(\mathbf{k};\tau)-2iM_{i\ell m}(\mathbf{k})\iint\delta(\mathbf{k}-\mathbf{p}-\mathbf{q})d\mathbf{p}d\mathbf{q}$$
$$\times u'_{0\ell}(\mathbf{p};\tau)u'_{S1m}(\mathbf{q};\tau)=-D_{i\ell}(\mathbf{k})u'_{0j}(\mathbf{k};\tau)\frac{\partial U_\ell}{\partial X_j}+D_{i\ell}(\mathbf{k})\,b'_j(\mathbf{k};\tau)\frac{\partial B_\ell}{\partial X_j}$$
$$-\frac{Du'_{0i}(\mathbf{k};\tau)}{DT_I}+B_j\frac{\partial b'_i(\mathbf{k};\tau)}{\partial X_{Ij}}+N_{Ui}, \tag{9.171}$$

where $\mathbf{N}_U$ is given by the last two terms in Eq. (6.61) that are nonlinear in $\mathbf{u'}_0$ (its details are not necessary in the later discussion).

B. Perturbational Solution

The $O(1)$ equation (9.166) is nonlinear in $\mathbf{u'}_0$, and it is difficult to solve it directly. Therefore we first drop the $\mathbf{b'}_0$-related terms on the right-hand side, and denote the part of $\mathbf{u'}_0$ obeying the equation, by $\mathbf{u'}_B$ (the basic field). We expand $\mathbf{u'}_0$ around $\mathbf{u'}_B$, and solve Eq. (9.166). At this time, we introduce the Green's function $G'_{Kij}(\mathbf{k};\tau,\tau')$, whose governing equation is obtained by linearizing the $\mathbf{u'}_B$ equation and is given by

$$\frac{\partial G'_{Kij}(\mathbf{k};\tau,\tau')}{\partial \tau} + \nu k^2 G'_{Kij}(\mathbf{k};\tau,\tau') - 2iM_{i\ell m}(\mathbf{k}) \iint \delta(\mathbf{k}-\mathbf{p}-\mathbf{q}) d\mathbf{p} d\mathbf{q} \times u'_{B\ell}(\mathbf{p};\tau) G'_{Kmj}(\mathbf{q};\tau,\tau') = \delta_{ij}\delta(\tau-\tau') \tag{9.172}$$

(subscript K denotes kinetic).

Using Eq. (9.172), the perturbational solution of Eq. (9.166) may be written as

$$u'_{0i}(\mathbf{k};\tau) = u'_{Bi}(\mathbf{k};\tau) - i\,(\mathbf{k}\cdot\mathbf{B}) \int_{-\infty}^{\tau} d\tau_1 G'_{Kij}(\mathbf{k};\tau,\tau_1) b'_j(\mathbf{k};\tau_1) + N_{U0i}, \tag{9.173}$$

where $\mathbf{N}_{U0}$ represents the terms nonlinear in $\mathbf{u'}_B$ and $\mathbf{b'}$.

We substitute Eq. (9.173) into Eq. (9.171) for $\mathbf{u'}_1(\mathbf{k};\tau)$, and may solve the resulting equation in entirely the same manner. Specifically, we focus attention on the contributions from the first two terms on the right-hand side that are directly linked with the mean field $(\mathbf{U},\mathbf{B})$. Then we have

$$\begin{aligned} u'_{1i}(\mathbf{k};\tau) = & -i\frac{k_i}{k^2}\frac{\partial u'_{Bj}(\mathbf{k};\tau)}{\partial X_{Ij}} \\ & - D_{\ell m}(\mathbf{k})\frac{\partial U_\ell}{\partial X_j}\int_{-\infty}^{\tau} d\tau_1 G'_{Kim}(\mathbf{k};\tau,\tau_1) u'_{Bj}(\mathbf{k};\tau_1) \\ & + D_{\ell m}(\mathbf{k})\frac{\partial B_\ell}{\partial X_j}\int_{-\infty}^{\tau} d\tau_1 G'_{Kim}(\mathbf{k};\tau,\tau_1) b'_j(\mathbf{k};\tau_1). \end{aligned} \tag{9.174}$$

C. Statistics of Magnetic Fields

In the counter-kinematic approach, we need to postulate statistical properties of magnetic fields. As is similar to the kinematic dynamo, we write

$$\frac{\left\langle b'_i(\mathbf{k};\tau) b'_j(\mathbf{k}';\tau')\right\rangle}{\delta(\mathbf{k}+\mathbf{k}')} = D_{ij}(\mathbf{k}) Q_M(k;\tau,\tau') + \frac{i}{2}\frac{k_\ell}{k^2}\varepsilon_{ij\ell} H_M(k;\tau,\tau'), \tag{9.175}$$

$$\left\langle G'_{Kij}(\mathbf{k};\tau,\tau')\right\rangle = \delta_{ij} G_K(k;\tau,\tau'), \tag{9.176}$$

$$\frac{\left\langle u'_{Bi}(\mathbf{k};\tau)b'_j(\mathbf{k}';\tau')\right\rangle}{\delta(\mathbf{k}+\mathbf{k}')} = D_{ij}(\mathbf{k})W(k;\tau,\tau'). \tag{9.177}$$

Equation (9.177) leads to an expression similar to Eq. (9.146), and Eq. (9.175) gives

$$\left\langle \mathbf{b}'^2/2\right\rangle = \int Q_M(k;\tau,\tau)d\mathbf{k}, \tag{9.178}$$

$$\langle \mathbf{b}'\cdot\mathbf{j}'\rangle = \int H_M(k;\tau,\tau)d\mathbf{k}. \tag{9.179}$$

Equations (9.178) and (9.179) are the magnetic counterparts of Eqs. (9.142) and (9.143).

In Sec. 2.5.3, the physical meaning of the kinetic helicity $\mathbf{u}\cdot\boldsymbol{\omega}$ is discussed, and it is shown to be an indicator of helical or spiral structures of fluid motion. In the correspondence (9.66) between $(\mathbf{u},\boldsymbol{\omega})$ and $(\mathbf{b},\mathbf{j})$, $\mathbf{b}\cdot\mathbf{j}$ may be regarded as an indicator of such structures of magnetic-field lines. Larger $|\mathbf{b}\cdot\mathbf{j}|$ leads to shorter pitches of magnetic spirals. In this sense, $\mathbf{b}\cdot\mathbf{j}$ may be called the magnetic helicity.

D. Turbulent Electromotive Force

In this case, $\mathbf{E}_M$ is written as

$$E_{Mi} = \varepsilon_{ij\ell}\left(\left\langle u'_{0j}b'_\ell\right\rangle + \delta\left(\left\langle u'_{1j}b'_\ell\right\rangle\right)\right), \tag{9.180}$$

up to $O(\delta)$. The calculation of Eq. (9.180) based on Eqs. (9.173)-(9.177) is entirely similar to that of Eq. (9.147) and leads to the same mathematical form

$$\mathbf{E}_M = \alpha_M\mathbf{B} - \beta_M\mathbf{J} + \gamma_M\boldsymbol{\Omega}. \tag{9.181}$$

The coefficients α_M, β_M, and γ_M are given by

$$\alpha_M = (1/3)\int d\mathbf{k}\int_{-\infty}^{\tau} d\tau_1 G_K(k;\tau,\tau_1)H_M(k;\tau,\tau_1), \tag{9.182}$$

$$\beta_M = (1/3)\int d\mathbf{k}\int_{-\infty}^{\tau} d\tau_1 G_K(k;\tau,\tau_1)Q_M(k;\tau,\tau_1), \tag{9.183}$$

$$\gamma_M = (1/3)\int d\mathbf{k}\int_{-\infty}^{\tau} d\tau_1 G_K(k;\tau,\tau_1)W(k;\tau,\tau_1), \tag{9.184}$$

respectively.

In the comparison between Eqs. (9.182)-(9.184) and Eqs. (9.149)-(9.151), β_M and γ_M are specifically similar to their counterparts, β_K and γ_K, including their signs. A small difference related to β_M and β_K is that β_M is generated by magnetic-field fluctuations, whereas β_K is by velocity ones. In both the cases, mirrorsymmetric, isotropic fluctuations give rise to the increase in the resistivity of plasma.

On the other hand, a greater difference exists between α_K [Eq. (9.149)] and α_M [Eq. (9.182)]. The former arises from the turbulent kinetic helicity $\langle \mathbf{u}' \cdot \boldsymbol{\omega}' \rangle$, and its physical meaning is explained using Fig. 9.4. The latter is associated with the turbulent magnetic helicity $\langle \mathbf{b}' \cdot \mathbf{j}' \rangle$. It is not so easy to understand its effect using an illustration such as Fig. 9.4. The occurrence of $\langle \mathbf{b}' \cdot \mathbf{j}' \rangle$, however, is natural in the context of the intrinsic property of the alpha effect that $\mathbf{J}$ is aligned with $\mathbf{B}$. The mean field $\mathbf{B}$ and $\mathbf{J}$ represent the large-scale components of $\mathbf{b}$ and $\mathbf{j}$, respectively. The former alignment signifies that some degree of correlation exists between $\mathbf{b}$ and $\mathbf{j}$, resulting in nonvanishing of $\langle \mathbf{b}' \cdot \mathbf{j}' \rangle$. This viewpoint is consistent with the finding that positive $\langle \mathbf{b}' \cdot \mathbf{j}' \rangle$ contributes to $\mathbf{J}$ parallel $\mathbf{B}$. In contrast with this situation, positive $\langle \mathbf{u}' \cdot \boldsymbol{\omega}' \rangle$ generates $\mathbf{J}$ anti-parallel to $\mathbf{B}$.

As a quantity characterizing helical properties of magnetic fields, we mentioned $\mathbf{a} \cdot \mathbf{b}$ in Sec. 9.2.3.C, and its total amount is conserved in the absence of λ_M. From the viewpoint of the conservation law, $\mathbf{a} \cdot \mathbf{b}$ is expected to be more important than $\mathbf{b} \cdot \mathbf{j}$. In the context of turbulent dynamo, however, $\langle \mathbf{b}' \cdot \mathbf{j}' \rangle$ is linked with the alpha effect and plays a greater role in the combination with $\langle \mathbf{u}' \cdot \boldsymbol{\omega}' \rangle$.

E. Reynolds Stress

In the counter-kinematic approach, the primary part of the Reynolds stress R_{ij} comes from $\langle u_i' u_j' \rangle$. It is similar to Eq. (9.155) and is given by

$$\left\langle u_i' u_j' \right\rangle = (1/3) \left\langle \mathbf{u}'^2 \right\rangle \delta_{ij} - \nu_{TK} S_{ij} + \nu_{MK} M_{ij}, \tag{9.185}$$

which is similar to Eq. (9.155). Here ν_{TK} and ν_{MK} are

$$\nu_{TK} = (7/15) \int d\mathbf{k} \int_{-\infty}^{\tau} d\tau_1 G_K(k; \tau, \tau_1) Q_K(k; \tau, \tau_1), \tag{9.186}$$

$$\nu_{MK} = (7/15) \int d\mathbf{k} \int_{-\infty}^{\tau} d\tau_1 G_K(k; \tau, \tau_1) W(k; \tau, \tau_1), \tag{9.187}$$

and $Q_K(k; \tau, \tau')$ is the kinematic counterpart of $Q_M(k; \tau, \tau')$ [Eq. (9.158)] and is defined as

$$\frac{\left\langle u'_{Bi}(\mathbf{k}; \tau) u'_{Bj}(\mathbf{k}'; \tau') \right\rangle}{\delta \left(\mathbf{k} + \mathbf{k}' \right)} = D_{ij}(\mathbf{k}) Q_K(k; \tau, \tau'). \tag{9.188}$$

This derivation is essentially the same as in Sec. 6.3.4.A.

The turbulent viscosity ν_{TK} is generated by mirrorsymmetric, isotropic velocity fluctuations. This situation is similar to ν_{TM} in Eq. (9.155), apart from the difference between velocity and magnetic fluctuations. The last term in Eq. (9.185), which corresponds to that in Eq. (9.155), is the feedback effect due to the magnetic field.

9.5.3. SUMMARY OF DYNAMO EFFECTS

In Secs. 9.5.1 and 9.5.2, we examined the turbulent electromotive force $\mathbf{E}_M$ and the Reynolds stress R_{ij} from two viewpoints: the quasi-kinematic approach and the counter-kinematic one. In the former, effects of fluid motion on these quantities are examined. In the latter, effects of fluid motion generated by magnetic fields on them are discussed. In MHD turbulent dynamo, such two types of effects occur in their combination. The simple addition of these two findings is not right, but it is expected to shed light on the primary properties of turbulent dynamo.

With this reservation in mind, we combine the results by the two approaches. From Eqs. (9.148) and (9.181), the alpha effect, which produces $\mathbf{J}$ aligned with $\mathbf{B}$, is closely related to the difference between the effects of kinetic and magnetic helicity, apart from the details of the Green's functions $G_K(k;\tau,\tau')$ and $G_M(k;\tau,\tau')$. The quantity

$$H(k;\tau,\tau') = -H_K(k;\tau,\tau') + H_M(k;\tau,\tau') \tag{9.189}$$

is called the turbulent residual helicity spectrum in correspondence to Eq. (9.115). Its importance was pointed out by Pouquet et al. (1976) in the study of homogeneous MHD turbulence using the eddy-damped quasi-normal Markovianized (EDQNM) approximation. The relationship of the turbulent residual helicity with $\mathbf{E}_M$ was also derived by Chen and Montgomery (1987) using the approach based on the clear-cut separation between large and small-scale fluctuations.

In the context of helicity effects, we should note that they do not occur in R_{ij} in the analysis up to $O(\delta)$. In the hydrodynamic case, it is shown that the effect of kinetic helicity appears in the combination of the $O(\delta)$ and frame-rotation terms or through the $O(\delta^2)$ ones, as is shown by Eq. (6.262). Specifically, the inhomogeneity of the kinetic helicity plays a central role, unlike the alpha effect in $\mathbf{E}_M$.

The turbulent resistivity β_K [Eq. (9.150)] and β_M [Eq. (9.183)] are similar to the turbulent viscosity ν_{TM} [Eq. (9.156)] and ν_{TK} [Eq. (9.186)] in their physical structures. The combination of each pair implies that the turbulent resistivity and viscosity are expressed in terms of the turbulent MHD energy spectrum

$$Q(k;\tau,\tau') = Q_K(k;\tau,\tau') + Q_M(k;\tau,\tau'). \tag{9.190}$$

Effects of the turbulent cross helicity in γ_K [Eq. (9.151)], γ_M [Eq. (9.184)], ν_{MM} [Eq. (9.157)], and ν_{MK} [Eq. (9.187)] are more similar to one another, compared with the foregoing effects. This property comes from the fact that the cross-helicity effects are associated originally with the interaction between fluid motion and magnetic fields, resulting in the less difference between $\mathbf{E}_M$ and R_{ij}.

9.6. TSDIA Approach

Some important properties of $\mathbf{E}_M$ and R_{ij} were clarified through the quasi-kinematic and counter-kinematic approaches. Such methods, however, are not sufficient for drawing definite conclusions about these quantities, for instance, about the importance of the residual helicity H [Eq. (9.189)] rather than each of the kinetic and magnetic parts. In order to eradicate such ambiguity, we need to treat Eqs. (9.132), (9.161), and (9.162) in a simultaneous manner, but the mathematical complexity occurring in the application of the TSDIA to these equations is formidable. A method of alleviating this difficulty is the use of Elsasser's variables $\boldsymbol{\phi}$ and $\boldsymbol{\psi}$ that are defined by Eq. (9.59). In what follows, we shall adopt this approach to perform a more accurate analysis of dynamo effects (Yoshizawa 1985, 1990).

9.6.1. FORMALISM BASED ON ELSASSER'S VARIABLES

A. Fundamental Equations

Elsasser's variables $\boldsymbol{\phi}$ and $\boldsymbol{\psi}$ obey Eqs. (9.61) and (9.62). In what follows, we shall pay less attention to molecular effects, and adopt a simpler system consisting of Eqs. (9.63) and (9.64). These variables are divided into mean and fluctuations, as in Eq. (9.86):

$$\boldsymbol{\phi} = \boldsymbol{\Phi} + \boldsymbol{\phi}', \ \boldsymbol{\psi} = \boldsymbol{\Psi} + \boldsymbol{\psi}'. \tag{9.191}$$

Under Eq. (9.191), $\mathbf{E}_M$ and R_{ij} may be expressed in the form

$$E_{Mi} = -(1/2)\epsilon_{ij\ell} R_{Ej\ell}, \tag{9.192}$$

$$R_{ij} = (1/2)\left(R_{Eij} + R_{Eji}\right), \tag{9.193}$$

where R_{Eij} is defined as

$$R_{Eij} = \left\langle \phi'_i \psi'_j \right\rangle. \tag{9.194}$$

Equation (9.194) may be called Elsasser's Reynolds stress since it enters the equations for the mean field $(\boldsymbol{\Phi}, \boldsymbol{\Psi})$ that are derived from Eqs. (9.63) and (9.64).

In order to examine R_{Eij}, we consider the equations for $\boldsymbol{\phi}'$ and $\boldsymbol{\psi}'$. In the kinematic and counter-kinematic approaches, we had no effects of helicity on the Reynolds stress R_{ij} up to $O(\delta)$. In the hydrodynamic case, such effects occur through the $O(\delta)$ effects combined with the frame-rotation one or the $O(\delta^2)$ effects, as was noted in Sec. 6.7.2 [recall the equivalence between the mean vorticity and frame rotation through Eq. (9.58)]. As this result, helicity effects can be found more simply from the analysis based on a rotating frame. In a frame rotating with the angular velocity $\mathbf{\Omega}_F$, $\boldsymbol{\phi}'$ obeys

$$\frac{\partial \phi'_i}{\partial t} + \Psi_j \frac{\partial \phi'_i}{\partial x_j} + \frac{\partial}{\partial x_j}\left(\psi'_j \phi'_i - R_{Eij}\right) + \frac{\partial p'_M}{\partial x_i} - \nu \Delta \phi'_i$$
$$= -\epsilon_{ij\ell}\Omega_{Fj}\left(\phi'_\ell + \psi'_\ell\right) - \psi'_j \frac{\partial \Phi_i}{\partial x_j}, \tag{9.195}$$

with the solenoidal condition

$$\nabla \cdot \boldsymbol{\phi}' = 0. \tag{9.196}$$

The $\boldsymbol{\psi}'$ counterpart may be found through the exchange of variables

$$\boldsymbol{\phi}' \to \boldsymbol{\psi}', \ \boldsymbol{\psi}' \to \boldsymbol{\phi}', \ \mathbf{\Phi} \to \mathbf{\Psi}, \ \mathbf{\Psi} \to \mathbf{\Phi}, \ R_{Eij} \to R_{Eji}. \tag{9.197}$$

This symmetry of equations contributes much to the reduction of the mathematical complexity in the TSDIA analysis of MHD turbulent shear flow.

We apply the two-scale description given by Eqs. (9.127)-(9.129) to Eqs. (9.195) and (9.196). Then we have

$$\frac{\partial \phi'_i}{\partial \tau} + \Psi_j \frac{\partial \phi'_i}{\partial \xi_j} + \frac{\partial}{\partial \xi_j}\psi'_j \phi'_i + \frac{\partial p'_M}{\partial \xi_i} - \nu \Delta_\xi \phi'_i$$
$$= -\epsilon_{ij\ell}\Omega_{Fj}\left(\phi'_\ell + \psi'_\ell\right) + \delta \left(-\psi'_j \frac{\partial \Phi_i}{\partial X_j} - \frac{\partial \phi'_i}{\partial T} - \Psi_j \frac{\partial \phi'_i}{\partial X_j} \right.$$
$$\left. - \frac{\partial}{\partial X_j}\left(\psi'_j \phi'_i - R_{Eij}\right) - \frac{\partial p'_M}{\partial X_i} \right), \tag{9.198}$$

$$\frac{\partial \phi'_i}{\partial \xi_i} = -\delta \frac{\partial \phi'_i}{\partial X_i}. \tag{9.199}$$

As is entirely similar to the quasi-kinematic and counter-kinematic approaches, we apply the Fourier transformation (9.131) to Eqs. (9.198) and (9.199). Then we have

$$\frac{\partial \phi'_i(\mathbf{k};\tau)}{\partial \tau} + \nu k^2 \phi'_i(\mathbf{k};\tau) - i k_i p'_M(\mathbf{k};\tau)$$
$$- i k_j \iint \delta(\mathbf{k} - \mathbf{p} - \mathbf{q}) d\mathbf{p} d\mathbf{q} \psi'_j(\mathbf{p};\tau)\phi'_i(\mathbf{q};\tau)$$

$$= -i(\mathbf{k}\cdot\mathbf{B})\phi'_i(\mathbf{k};\tau) - \epsilon_{ij\ell}\Omega_{Fj}\left(\phi'_\ell(\mathbf{k};\tau) + \psi'_\ell(\mathbf{k};\tau)\right)$$

$$+\delta\left(-\psi'_j(\mathbf{k};\tau)\frac{\partial\Phi_i}{\partial X_j} - \frac{D\phi'_i(\mathbf{k};\tau)}{DT_I} + B_j\frac{\partial\phi'_i(\mathbf{k};\tau)}{\partial X_{Ij}} - \frac{\partial p'_M(\mathbf{k};\tau)}{\partial X_{Ii}}\right.$$

$$\left. - \frac{\partial}{\partial X_{Ii}}\iint \delta(\mathbf{k}-\mathbf{p}-\mathbf{q})d\mathbf{p}d\mathbf{q}\psi'_j(\mathbf{p};\tau)\phi'_i(\mathbf{q};\tau)\right), \tag{9.200}$$

$$\mathbf{k}\cdot\boldsymbol{\phi}'(\mathbf{k};\tau) = \delta\left(-i\frac{\partial\phi'_i(\mathbf{k};\tau)}{\partial X_{Ii}}\right). \tag{9.201}$$

In Eq. (9.200), $\mathbf{B}$ itself occurs since $\mathbf{U}$ is partly eliminated in the course of the Fourier transformation. The $\boldsymbol{\psi}'$ counterpart of Eqs. (9.200) and (9.201) may be obtained through the replacement (9.197). In the procedure, we should note that $\mathbf{B}$ in Eq. (9.200) is replaced with $-\mathbf{B}$ because the replacement of $\boldsymbol{\Phi}$ with $\boldsymbol{\Psi}$ corresponds to that of $\mathbf{B}$ with $-\mathbf{B}$.

The solenoidal condition (9.201) may be changed into a more manageable form following the transformation

$$\boldsymbol{\phi}'(\mathbf{k};\tau) = \boldsymbol{\phi}'_S(\mathbf{k};\tau) + \delta\left(-i\frac{\mathbf{k}}{k^2}\frac{\partial\phi'_i(\mathbf{k};\tau)}{\partial X_{Ii}}\right). \tag{9.202}$$

Then $\boldsymbol{\phi}'_S(\mathbf{k};\tau)$ obeys the usual solenoidal condition concerning $\mathbf{k}$ as

$$\mathbf{k}\cdot\boldsymbol{\phi}'_S(\mathbf{k};\tau) = 0. \tag{9.203}$$

B. Scale-Parameter Expansion

We expand $\boldsymbol{\phi}'$ and $\boldsymbol{\psi}'$ in the form

$$\boldsymbol{\phi}'(\mathbf{k};\tau) = \sum_{n=0}^{\infty}\delta^n\boldsymbol{\phi}'_n(\mathbf{k};\tau),\ \boldsymbol{\phi}'_S(\mathbf{k};\tau) = \sum_{n=0}^{\infty}\delta^n\boldsymbol{\phi}'_{Sn}(\mathbf{k};\tau), \tag{9.204a}$$

$$p'_M(\mathbf{k};\tau) = \sum_{n=0}^{\infty}\delta^n p'_{Mn}(\mathbf{k};\tau), \tag{9.204b}$$

and the similar expressions for $\boldsymbol{\psi}'$. The $O(1)$ field obeys

$$\frac{\partial\phi'_{0i}(\mathbf{k};\tau)}{\partial\tau} + \nu k^2\phi'_{0i}(\mathbf{k};\tau) - ik_i p'_{M0}(\mathbf{k};\tau)$$

$$-ik_j\iint\delta(\mathbf{k}-\mathbf{p}-\mathbf{q})d\mathbf{p}d\mathbf{q}\psi'_{0j}(\mathbf{p};\tau)\phi'_{0i}(\mathbf{q};\tau)$$

$$= -i(\mathbf{k}\cdot\mathbf{B})\phi'_{0i}(\mathbf{k};\tau) - \epsilon_{ij\ell}\Omega_{Fj}\left(\phi'_{0\ell}(\mathbf{k};\tau) + \psi'_{0\ell}(\mathbf{k};\tau)\right), \tag{9.205}$$

$$\mathbf{k}\cdot\boldsymbol{\phi}'_0(\mathbf{k};\tau) = 0. \tag{9.206}$$

We apply Eq. (9.206) to Eq. (9.205), and have

$$p'_{M0}(\mathbf{k};\tau) = -\frac{k_i k_j}{k^2} \iint \delta(\mathbf{k}-\mathbf{p}-\mathbf{q}) d\mathbf{p} d\mathbf{q} \psi'_{0j}(\mathbf{p};\tau)\phi'_{0i}(\mathbf{q};\tau) - i\epsilon_{ij\ell}\Omega_{Fj}\frac{k_i}{k^2}\left(\phi'_{0\ell}(\mathbf{k};\tau)+\psi'_{0\ell}(\mathbf{k};\tau)\right). \qquad (9.207)$$

From Eq. (9.207), Eq. (9.205) is reduced to

$$\frac{\partial \phi'_{0i}(\mathbf{k};\tau)}{\partial \tau} + \nu k^2 \phi'_{0i}(\mathbf{k};\tau) - iZ_{ij\ell}(\mathbf{k}) \iint \delta(\mathbf{k}-\mathbf{p}-\mathbf{q}) d\mathbf{p} d\mathbf{q} \psi'_{0j}(\mathbf{p};\tau)\phi'_{0\ell}(\mathbf{q};\tau) = -i(\mathbf{k}\cdot\mathbf{B})\phi'_{0i}(\mathbf{k};\tau) - \epsilon_{mj\ell}\Omega_{Fj}D_{im}(\mathbf{k})\left(\phi'_{0\ell}(\mathbf{k};\tau)+\psi'_{0\ell}(\mathbf{k};\tau)\right), \qquad (9.208)$$

where

$$Z_{ij\ell}(\mathbf{k}) = k_j D_{i\ell}(\mathbf{k}). \qquad (9.209)$$

In Eq. (9.208), we should note that the third term on the left-hand side is not symmetric with respect to j and ℓ, unlike the hydrodynamic counterpart in Eq. (6.46).

Next, we proceed to the equations for the $O(\delta)$ field. They have a number of terms and their final expressions are complicated. In order to reduce this complexity, we focus attention mainly on the terms linear in fluctuations since such terms occur in the combination with the mean field $(\mathbf{\Phi}, \mathbf{\Psi})$ and the angular velocity of a frame rotation, $\mathbf{\Omega}_F$. Under this approximation, the $O(\delta)$ field obeys

$$\frac{\partial \phi'_{1i}(\mathbf{k};\tau)}{\partial \tau} + \nu k^2 \phi'_{1i}(\mathbf{k};\tau) - ik_i p'_{M1}(\mathbf{k};\tau) - ik_j \iint \delta(\mathbf{k}-\mathbf{p}-\mathbf{q}) d\mathbf{p} d\mathbf{q} \psi'_{0j}(\mathbf{p};\tau)\phi'_{1i}(\mathbf{q};\tau) = -i(\mathbf{k}\cdot\mathbf{B})\phi'_{1i}(\mathbf{k};\tau) - \epsilon_{ij\ell}\Omega_{Fj}\left(\phi'_{1\ell}(\mathbf{k};\tau)+\psi'_{1\ell}(\mathbf{k};\tau)\right) - \psi'_{0j}(\mathbf{k};\tau)\frac{\partial \Phi_i}{\partial X_j} - \frac{D\phi'_{0i}(\mathbf{k};\tau)}{DT_I} + B_j \frac{\partial \phi'_{0i}(\mathbf{k};\tau)}{\partial X_{Ij}} - \frac{\partial p'_{M0}(\mathbf{k};\tau)}{\partial X_{Ii}}, \qquad (9.210)$$

$$\boldsymbol{\phi}'_1(\mathbf{k};\tau) = \boldsymbol{\phi}'_{S1}(\mathbf{k};\tau) - i\frac{\mathbf{k}}{k^2}\frac{\partial \phi'_{0i}(\mathbf{k};\tau)}{\partial X_{Ii}}, \qquad (9.211)$$

with

$$\mathbf{k}\cdot\boldsymbol{\phi}'_{S1}(\mathbf{k};\tau) = 0. \qquad (9.212)$$

In Eq. (9.210), we dropped the terms

$$-ik_j \iint \delta(\mathbf{k}-\mathbf{p}-\mathbf{q}) d\mathbf{p} d\mathbf{q} \psi'_{1j}(\mathbf{p};\tau)\phi'_{0i}(\mathbf{q};\tau), \qquad (9.213a)$$

$$-\frac{\partial}{\partial X_{Ij}} \iint \delta(\mathbf{k}-\mathbf{p}-\mathbf{q}) d\mathbf{p} d\mathbf{q} \psi'_{0j}(\mathbf{p};\tau) \phi'_{0i}(\mathbf{q};\tau). \tag{9.213b}$$

We solve the ψ'_1 counterpart of Eq. (9.211) and substitute it into Eq. (9.213a), resulting in an expression nonlinear in ϕ'_0 and ψ'_0. On the other hand, we retained p'_{M0} since it includes the term linear in them, as may be seen from Eq. (9.207).

We substitute Eq. (9.211) into Eq. (9.210), and eliminate p'_{M1} with the aid of Eq. (9.212). As a result, ϕ'_1 obeys

$$\begin{aligned}
&\frac{\partial \phi'_{S1i}(\mathbf{k};\tau)}{\partial \tau} + \nu k^2 \phi'_{S1i}(\mathbf{k};\tau) \\
&\quad -iZ_{ij\ell}(\mathbf{k}) \iint \delta(\mathbf{k}-\mathbf{p}-\mathbf{q}) d\mathbf{p} d\mathbf{q} \psi'_{0j}(\mathbf{p};\tau) \phi'_{S1\ell}(\mathbf{q};\tau) \\
&\quad = -D_{i\ell}(\mathbf{k}) \psi'_{0j}(\mathbf{k};\tau) \frac{\partial \Phi_\ell}{\partial X_j} - \frac{D\phi'_{0i}(\mathbf{k};\tau)}{DT_I} + B_j \frac{\partial \phi'_{0i}(\mathbf{k};\tau)}{\partial X_{Ij}} \\
&\quad +i\epsilon_{mj\ell}\Omega_{Fj} \frac{k_\ell}{k^2} D_{im}(\mathbf{k}) \frac{\partial}{\partial X_{In}} \left(\phi'_{0n}(\mathbf{k};\tau) + \psi'_{0n}(\mathbf{k};\tau)\right) \\
&\quad +i\epsilon_{mj\ell}\Omega_{Fj} \frac{k_m}{k^2} D_{in}(\mathbf{k}) \frac{\partial}{\partial X_{In}} \left(\phi'_{0\ell}(\mathbf{k};\tau) + \psi'_{0\ell}(\mathbf{k};\tau)\right) \\
&\quad -i(\mathbf{k}\cdot\mathbf{B}) \phi'_{S1i}(\mathbf{k};\tau) - \epsilon_{mj\ell}\Omega_{Fj} D_{im}(\mathbf{k}) \\
&\qquad \times \left(\phi'_{S1\ell}(\mathbf{k};\tau) + \psi'_{S1\ell}(\mathbf{k};\tau)\right).
\end{aligned} \tag{9.214}$$

C. Introduction of Green's Functions

In the TSDIA formalism detailed in Chapter 6, we assume the statistical isotropy of fluctuations that are not directly dependent on effects of mean shear, buoyancy, and frame rotation. Equation (9.208) for the $O(1)$ field still depends on the effects of mean magnetic field and frame rotation. These effects are primary factors generating anisotropy of fluctuations. Then we introduce the basic field (ϕ'_B, ψ'_B) free from their direct influence, as is similar to Secs. 9.5.1.C and 9.5.2.B; namely, the field obeys

$$\begin{aligned}
&\frac{\partial \phi'_{Bi}(\mathbf{k};\tau)}{\partial \tau} + \nu k^2 \phi'_{Bi}(\mathbf{k};\tau) \\
&\quad -iZ_{ij\ell}(\mathbf{k}) \iint \delta(\mathbf{k}-\mathbf{p}-\mathbf{q}) d\mathbf{p} d\mathbf{q} \psi'_{Bj}(\mathbf{p};\tau) \phi'_{B\ell}(\mathbf{q};\tau) = 0,
\end{aligned} \tag{9.215}$$

and the ψ'_B counterpart that is obtained by the replacement (9.197). In correspondence to these equations, we introduce the Green's functions $G'_{\phi ij}$ and $G'_{\psi ij}$ that satisfy

$$\frac{\partial G'_{\phi ij}(\mathbf{k};\tau,\tau')}{\partial \tau} + \nu k^2 G'_{\phi ij}(\mathbf{k};\tau,\tau') - iZ_{i\ell m}(\mathbf{k}) \iint \delta(\mathbf{k}-\mathbf{p}-\mathbf{q}) d\mathbf{p} d\mathbf{q}$$

$$\times \psi'_{B\ell}(\mathbf{p};\tau) G'_{\phi mj}(\mathbf{q};\tau,\tau') = \delta_{ij}\delta\left(\tau-\tau'\right), \tag{9.216}$$

and the $G'_{\psi ij}$ counterpart.

We expand ϕ'_0 and ψ'_0 around ϕ'_B and ψ'_B as

$$\phi'_0(\mathbf{k};\tau) = \phi'_B(\mathbf{k};\tau) + \sum_{m=1}^{\infty} \phi'_{0m}(\mathbf{k};\tau), \tag{9.217a}$$

$$\psi'_0(\mathbf{k};\tau) = \psi'_B(\mathbf{k};\tau) + \sum_{m=1}^{\infty} \psi'_{0m}(\mathbf{k};\tau), \tag{9.217b}$$

and solve Eq. (9.208) in an iterative manner with ϕ'_B and ψ'_B as the leading terms (ϕ'_{0m} is the solution arising from the mth iteration). The first-iteration part ϕ'_{01} is given by

$$\begin{aligned}\phi'_{01i}(\mathbf{k};\tau) = &-i(\mathbf{k}\cdot\mathbf{B}) \int_{-\infty}^{\tau} d\tau_1 G'_{\phi ij}\left(\mathbf{k};\tau,\tau_1\right) \phi'_{Bj}\left(\mathbf{k};\tau_1\right) \\ &-\epsilon_{mj\ell}\Omega_{Fj} D_{mn}(\mathbf{k}) \int_{-\infty}^{\tau} d\tau_1 G'_{\phi in}\left(\mathbf{k};\tau,\tau_1\right) \\ &\times \left(\phi'_{B\ell}\left(\mathbf{k};\tau_1\right) + \psi'_{B\ell}\left(\mathbf{k};\tau_1\right)\right). \end{aligned} \tag{9.218}$$

We substitute Eq. (9.218) into Eq. (9.214) for ϕ'_{S1}. Here we pay special attention to the leading contributions to ϕ'_{S1}, which arise from Eq. (9.214) with (ϕ'_0, ψ'_0) replaced with (ϕ'_B, ψ'_B). As this consequence, we have

$$\phi'_1(\mathbf{k};\tau) = \phi'_{S1}(\mathbf{k};\tau) - i\frac{\mathbf{k}}{k^2}\frac{\partial \phi'_{Bi}(\mathbf{k};\tau)}{\partial X_{Ii}}, \tag{9.219}$$

with

$$\begin{aligned}\phi'_{S1i}(\mathbf{k};\tau) = &-\frac{\partial \Phi_\ell}{\partial X_j} D_{\ell m}(\mathbf{k}) \int_{-\infty}^{\tau} d\tau_1 G'_{\phi im}\left(\mathbf{k};\tau,\tau_1\right) \psi'_{Bj}\left(\mathbf{k};\tau_1\right) \\ &- \int_{-\infty}^{\tau} d\tau_1 G'_{\phi ij}\left(\mathbf{k};\tau,\tau_1\right) \frac{D\phi'_{Bj}\left(\mathbf{k};\tau_1\right)}{DT_I} \\ &+ B_j \int_{-\infty}^{\tau} d\tau_1 G'_{\phi i\ell}\left(\mathbf{k};\tau,\tau_1\right) \frac{\partial \phi'_{B\ell}\left(\mathbf{k};\tau_1\right)}{\partial X_{Ij}} \\ &+ i\epsilon_{mj\ell}\Omega_{Fj}\frac{k_\ell}{k^2} D_{mh}(\mathbf{k}) \int_{-\infty}^{\tau} d\tau_1 G'_{\phi ih}\left(\mathbf{k};\tau,\tau_1\right) \\ &\times \frac{\partial}{\partial X_{In}} \left(\phi'_{Bn}\left(\mathbf{k};\tau_1\right) + \psi'_{Bn}\left(\mathbf{k};\tau_1\right)\right) \\ &+ i\epsilon_{mj\ell}\Omega_{Fj}\frac{k_m}{k^2} D_{hn}(\mathbf{k}) \int_{-\infty}^{\tau} d\tau_1 G'_{\phi ih}\left(\mathbf{k};\tau,\tau_1\right) \end{aligned}$$

$$\times \frac{\partial}{\partial X_{In}} \left(\phi'_{B\ell} \left(\mathbf{k}; \tau_1 \right) + \psi'_{B\ell} \left(\mathbf{k}; \tau_1 \right) \right)$$

$$- i(\mathbf{k} \cdot \mathbf{B}) \int_{-\infty}^{\tau} d\tau_1 G'_{\phi ij} \left(\mathbf{k}; \tau, \tau_1 \right) \phi'_{S1j} \left(\mathbf{k}; \tau_1 \right)$$

$$- \epsilon_{mj\ell} \Omega_{Fj} D_{mn}(\mathbf{k}) \int_{-\infty}^{\tau} d\tau_1 G'_{\phi in} \left(\mathbf{k}; \tau, \tau_1 \right)$$

$$\times \left(\phi'_{S1\ell} \left(\mathbf{k}; \tau_1 \right) + \psi'_{S1\ell} \left(\mathbf{k}; \tau_1 \right) \right). \tag{9.220}$$

The right-hand side of Eq. (9.220) still contains $\boldsymbol{\phi}'_{S1}$ and $\boldsymbol{\psi}'_{S1}$. We solve it by the iteration method entirely similar to Eq. (9.217). The leading part of $\boldsymbol{\phi}'_{S1}$ may be obtained by dropping the last two terms in Eq. (9.220).

D. Statistics of Basic Field

We assume that the basic field $(\boldsymbol{\phi}'_B, \boldsymbol{\psi}'_B)$ is statistically isotropic but antimirrorsymmetric. The assumption of isotropy is plausible since it is not explicitly dependent on effects such as the mean field and frame rotation. Then we write

$$\frac{\langle \vartheta_{Bi}(\mathbf{k};\tau) \chi_{Bj}(\mathbf{k}';\tau') \rangle}{\delta(\mathbf{k}+\mathbf{k}')} = D_{ij}(\mathbf{k}) Q_{\vartheta\chi}(k;\tau,\tau') + \frac{i}{2} \frac{k_\ell}{k^2} \epsilon_{ij\ell} H_{\vartheta\chi}(k;\tau,\tau'), \tag{9.221}$$

$$\left\langle G'_{\vartheta ij}(\mathbf{k};\tau,\tau') \right\rangle = \delta_{ij} G_{\vartheta}(k;\tau,\tau'), \tag{9.222}$$

where subscripts ϑ and χ represent either one of ϕ and ψ or one of u and b ($\mathbf{u}'_B$ and $\mathbf{b}'_B$ are the basic-field counterparts of $\boldsymbol{\phi}'_B$ and $\boldsymbol{\psi}'_B$). The correlation functions based on $\boldsymbol{\phi}'_B$ and $\boldsymbol{\psi}'_B$ are related to those based on $\mathbf{u}'_B$ and $\mathbf{b}'_B$ as

$$Q_{\phi\phi}(k;\tau,\tau') + Q_{\psi\psi}(k;\tau,\tau') = 2\left(Q_{uu}(k;\tau,\tau') + Q_{bb}(k;\tau,\tau')\right), \tag{9.223a}$$

$$Q_{\phi\phi}(k;\tau,\tau') - Q_{\psi\psi}(k;\tau,\tau') = 2\left(Q_{ub}(k;\tau,\tau') + Q_{bu}(k;\tau,\tau')\right), \tag{9.223b}$$

$$Q_{\phi\psi}(k;\tau,\tau') + Q_{\psi\phi}(k;\tau,\tau') = 2\left(Q_{uu}(k;\tau,\tau') - Q_{bb}(k;\tau,\tau')\right), \tag{9.223c}$$

$$Q_{\phi\psi}(k;\tau,\tau') - Q_{\psi\phi}(k;\tau,\tau') = 2\left(Q_{bu}(k;\tau,\tau') - Q_{ub}(k;\tau,\tau')\right), \tag{9.223d}$$

and

$$H_{\phi\phi}(k;\tau,\tau') + H_{\psi\psi}(k;\tau,\tau') = 2\left(H_{uu}(k;\tau,\tau') + H_{bb}(k;\tau,\tau')\right), \tag{9.224a}$$

$$H_{\phi\phi}(k;\tau,\tau') - H_{\psi\psi}(k;\tau,\tau') = 2\left(H_{ub}(k;\tau,\tau') + H_{bu}(k;\tau',\tau)\right), \tag{9.224b}$$

$$H_{\phi\psi}(k;\tau,\tau') + H_{\psi\phi}(k;\tau,\tau') = 2\left(H_{uu}(k;\tau,\tau') - H_{bb}(k;\tau,\tau')\right), \tag{9.224c}$$

$$H_{\phi\psi}(k;\tau,\tau') - H_{\psi\phi}(k;\tau,\tau') = 2\left(H_{bu}(k;\tau,\tau') - H_{ub}(k;\tau,\tau')\right). \tag{9.224d}$$

From Eqs. (9.223) and (9.224), we have

$$\left\langle \left(\mathbf{u}'^2_B + \mathbf{b}'^2_B \right) /2 \right\rangle = \int (Q_{uu}(k;\tau,\tau) + Q_{bb}(k;\tau,\tau))d\mathbf{k},$$
$$= (1/2) \int (Q_{\phi\phi}(k;\tau,\tau) + Q_{\psi\psi}(k;\tau,\tau))d\mathbf{k}, \quad (9.225)$$

$$\langle \mathbf{u}'_B \cdot \mathbf{b}'_B \rangle = 2 \int Q_{ub}(k;\tau,\tau)d\mathbf{k}$$
$$= (1/2) \int (Q_{\phi\phi}(k;\tau,\tau) - Q_{\psi\psi}(k;\tau,\tau))d\mathbf{k}, \quad (9.226)$$

$$\langle -\mathbf{u}'_B \cdot \boldsymbol{\omega}'_B + \mathbf{b}'_B \cdot \mathbf{j}'_B \rangle = \int (-H_{uu}(k;\tau,\tau) + H_{bb}(k;\tau,\tau))d\mathbf{k}$$
$$= -(1/2) \int (H_{\phi\psi}(k;\tau,\tau) + H_{\psi\phi}(k;\tau,\tau))d\mathbf{k}. \quad (9.227)$$

An example of the quantities less important from the viewpoint of dynamo is

$$\langle \mathbf{u}'_B \cdot \mathbf{j}'_B \rangle = \int H_{ub}(k;\tau,\tau)d\mathbf{k}$$
$$= (1/4) \int (H_{\phi\phi}(k;\tau,\tau) - H_{\psi\psi}(k;\tau,\tau))d\mathbf{k}. \quad (9.228)$$

Finally, we define

$$G_S(k;\tau,\tau') = (1/2) \left(G_\phi(k;\tau,\tau') + G_\psi(k;\tau,\tau') \right), \quad (9.229a)$$

$$G_A(k;\tau,\tau') = (1/2) \left(G_\phi(k;\tau,\tau') - G_\psi(k;\tau,\tau') \right). \quad (9.229b)$$

They denote the mirrorsymmetric and anti-mirrorsymmetric parts of G_ϕ and G_ψ, respectively.

9.6.2. CALCULATION OF CORRELATION FUNCTIONS

We use the $O(1)$ solution (9.218) and the $O(\delta)$ solution (9.219) with Eq. (9.220), and calculate the correlation functions that play an important role in turbulent dynamo. Our primary interest lies in effects of the mean magnetic field itself, the mean velocity and magnetic-field gradients, the frame rotation, etc. In the hydrodynamic TSDIA, the method of calculating correlation functions is detailed in Chapter 6, specifically, in Sec. 6.3.4. In this MHD case, the mathematical manipulation is similar but more complicated (Yoshizawa 1985, 1990). We shall give final results about important correlation functions.

In order to show the results compactly, we introduce the following abbreviation of time and wavenumber integration:

$$I_n\left\{A\right\} = \int k^{2n} A(k, \mathbf{x}; \tau, \tau, t) d\mathbf{k}; \tag{9.230a}$$

$$I_n\left\{A, B\right\} = \int k^{2n} d\mathbf{k} \int_{-\infty}^{\tau} d\tau_1 A\left(k, \mathbf{x}; \tau, \tau_1, t\right) B\left(k, \mathbf{x}; \tau, \tau_1, t\right). \tag{9.230b}$$

A. Turbulent Electromotive Force
First, we calculate Eq. (9.194). The turbulent electromotive force $\mathbf{E}_M$ may be obtained from Eq. (9.192). The resulting expression is

$$\mathbf{E}_M = \alpha \mathbf{B} - \beta \mathbf{J} + \gamma \mathbf{\Omega} + 2\gamma_F \mathbf{\Omega}_F. \tag{9.231}$$

The coefficients α, β, γ, and γ_F are written as

$$\alpha = (1/3) \left(I_0 \left\{G_S, -H_{uu} + H_{bb}\right\} - I_0 \left\{G_A, H_{ub} - H_{bu}\right\}\right), \tag{9.232}$$

$$\beta = (1/3) \left(I_0 \left\{G_S, Q_{uu} + Q_{bb}\right\} - I_0 \left\{G_A, Q_{ub} + Q_{bu}\right\}\right), \tag{9.233}$$

$$\gamma = (1/3) \left(I_0 \left\{G_S, Q_{ub} + Q_{bu}\right\} - I_0 \left\{G_A, Q_{uu} + Q_{bb}\right\}\right), \tag{9.234}$$

$$\gamma_F = (2/3) \left(I_0 \left\{G_S, Q_{bu}\right\} - I_0 \left\{G_A, Q_{uu}\right\}\right). \tag{9.235}$$

Equation (9.231) is of the same form as Eqs. (9.148) and (9.181), except the last frame-rotation effect. We are in a position to examine the similarity and difference among their corresponding coefficients. The counterparts of Eq. (9.232) are given by Eqs. (9.149) and (9.182). In the latter, α_K and α_M are related to the turbulent kinetic and magnetic helicity spectra, H_K and H_M, respectively. In the context of helicity spectra, Eq. (9.232) indicates that those spectra occur in the form of the residual-helicity spectrum, as is inferred in Sec. 9.5.3. The second part of Eq. (9.232), however, shows that there is room for the contribution of the velocity-current correlation that is given by Eq. (9.228). In the stationary state of MHD turbulence, this contribution vanishes since

$$H_{bu}(k; \tau, \tau') = H_{bu}(k; \tau', \tau) = H_{ub}(k; \tau, \tau') \tag{9.236}$$

(we should recall that H_{bu} is dependent on the time difference $\tau - \tau'$ in the stationary state). Therefore it may be concluded that α is related mainly to the turbulent residual helicity.

In the comparison between Eqs. (9.233) and (9.150) or (9.183), the first part of the former shows the importance of turbulent MHD energy (the sum of turbulent kinetic and magnetic energy) in the generation of the turbulent resistivity. The second part expresses the contribution from the cross-helicity effect in the combination with G_A (the Green's function lacking mirrorsymmetry). In general, Green's functions are closely associated with characteristic time scales of MHD turbulence. The mirrorsymmetric part G_S is supposed to play a bigger role than the anti-mirrorsymmetric one G_A since time scales are originally pure scalars. In what follows, we shall drop the latter effects.

Under the approximation of dropping G_A effects, Eq. (9.234) is essentially the same as Eqs. (9.151) and (9.184) as long as velocity and magnetic fluctuations are statistically stationary. In this case, we should emphasize that

$$\gamma = \gamma_F \tag{9.237}$$

from Eqs. (9.234) and (9.235) since $Q_{ub} = Q_{bu}$, as is similar to Eq. (9.236). Then Eq. (9.231) is reduced to

$$\mathbf{E}_M = \alpha \mathbf{B} - \beta \mathbf{J} + \gamma \left(\mathbf{\Omega} + 2\mathbf{\Omega}_F \right), \tag{9.238}$$

which is consistent with the vorticity transformation (9.58).

B. Reynolds Stress

From Eq. (9.193), the Reynolds stress R_{ij} may be derived as

$$R_{ij} = \frac{2}{3} K_R \delta_{ij} - \nu_T S_{ij} + \nu_M M_{ij} + 2 \left[\Omega_{Fi} \Gamma_j + \Omega_{Fj} \Gamma_i \right]_D . \tag{9.239}$$

Here $[A_{ij}]_D$ is the deviatoric part of A_{ij} defined by Eq. (4.97), and K_R is the turbulent residual energy given by

$$K_R = \left\langle \left(\mathbf{u}'^2 - \mathbf{b}'^2 \right) /2 \right\rangle . \tag{9.240}$$

The other three coefficients, ν_T (turbulent viscosity), ν_M, and $\mathbf{\Gamma}$, are expressed in the form

$$\nu_T = (7/5) \beta, \tag{9.241}$$

$$\nu_M = (7/5) \gamma, \tag{9.242}$$

$$\mathbf{\Gamma} = (1/15) \left(I_{-1} \left\{ G_S, \nabla H_{uu} \right\} - I_{-1} \left\{ G_A, \nabla H_{bu} \right\} \right) . \tag{9.243}$$

A remarkable point of R_{ij} lies in Eqs. (9.241) and (9.242). From Eqs. (9.150) and (9.156), we can see that the turbulent resistivity and viscosity are related to kinetic and magnetic energy spectra, respectively. The similar situation also holds in Eqs. (9.183) and (9.186). We cannot find, however, a definite relationship between these two pairs, such as Eqs. (9.241) and (9.242). The simple relationship in the latter is not accidental, but it is tightly linked with the use of Elsasser's variables. Under their use, the origin of $\beta \mathbf{J}$ is the same as that of $\nu_T S_{ij}$, and these two terms arise from the symmetric and anti-symmetric parts of Elsasser's Reynolds stress R_{Eij} [Eq. (9.194)].

The last part in Eq. (9.239) is the effect of helicity on R_{ij}, specifically, that of kinetic helicity. Here ∇H_{uu} expresses the slow spatial variation of statistical properties of fluctuations or the inhomogeneity of turbulence state. Such an effect in the hydrodynamic case is discussed in detail in Sec. 6.7.2. From the equivalence between vorticity and frame rotation, Eq. (9.58), R_{ij} should be written as

$$R_{ij} = \frac{2}{3} K_R \delta_{ij} - \nu_T S_{ij} + \nu_M M_{ij} + \left[(\Omega_i + 2\Omega_{Fi}) \Gamma_j + (\Omega_j + 2\Omega_{Fj}) \Gamma_i \right]_D , \tag{9.244}$$

which is the MHD counterpart of Eq. (6.262). Here the newly-added $\boldsymbol{\Omega}$-related terms are of $O(\delta^2)$ in the TSDIA formalism since $\boldsymbol{\Gamma}$ is dependent on ∇H_{uu}, and their derivation needs the tedious mathematical manipulation. In this sense, the analysis in a rotating frame, which is combined with the transformation (9.58), is instrumental to the reduction of such complicatedness.

C. Supplementary Results

As the quantities characterizing the strength of velocity and magnetic fluctuations, we may mention the turbulent MHD energy K and its residual counterpart K_R. They are written in terms of Elsasser's variables $\boldsymbol{\phi}'$ and $\boldsymbol{\psi}'$ as

$$K = \left\langle \left(\boldsymbol{\phi}'^2 + \boldsymbol{\psi}'^2 \right) / 2 \right\rangle , \tag{9.245}$$

$$K_R = \left\langle \left(\boldsymbol{\phi}' \cdot \boldsymbol{\psi}' \right) / 2 \right\rangle . \tag{9.246}$$

We have

$$\begin{aligned} K = {} & I_0 \{Q_{uu} + Q_{bb}\} \\ & - \left(I_0 \left\{ G_S, \frac{D}{Dt} (Q_{uu} + Q_{bb}) \right\} + I_0 \left\{ G_A, \frac{D}{Dt} (Q_{ub} + Q_{bu}) \right\} \right) \\ & + \frac{1}{3} (\boldsymbol{\Omega} + 2\boldsymbol{\Omega}_F) \cdot \left(I_{-1} \{G_S, \nabla H_{uu}\} + I_{-1} \{G_A, \nabla H_{bu}\} \right) \\ & + \mathbf{B} \cdot \left(I_0 \{G_S, (\nabla Q_{ub} + Q_{bu})\} + I_0 \{G_A, (\nabla Q_{uu} + Q_{bb})\} \right) , \end{aligned} \tag{9.247}$$

$$\begin{aligned} K_R = {} & I_0 \{Q_{uu} - Q_{bb}\} \\ & - \left(I_0 \left\{ G_S, \frac{D}{Dt} (Q_{uu} - Q_{bb}) \right\} + I_0 \left\{ G_A, \frac{D}{Dt} (Q_{bu} - Q_{ub}) \right\} \right) \end{aligned}$$

$$+\frac{1}{3}(\boldsymbol{\Omega}+2\boldsymbol{\Omega}_F)\cdot\left(I_{-1}\left\{G_S,\nabla H_{uu}\right\}-I_{-1}\left\{G_A,\nabla H_{bu}\right\}\right)$$
$$-\mathbf{B}\cdot\left(I_0\left\{G_S,\nabla\left(Q_{bu}-Q_{ub}\right)\right\}-I_0\left\{G_A,\nabla\left(Q_{uu}-Q_{bb}\right)\right\}\right). \quad (9.248)$$

The former should be compared with the hydrodynamic counterpart, Eq. (6.263), and their main parts resemble each other, except the **B**-related term.

The turbulent cross helicity W, which is written as

$$W=\langle\mathbf{u}'\cdot\mathbf{b}'\rangle=\left\langle\left(\phi'^2-\psi'^2\right)/4\right\rangle, \quad (9.249)$$

is given by

$$W=2I_0\left\{Q_{ub}\right\}$$
$$-\left(I_0\left\{G_S,\frac{D}{Dt}\left(Q_{ub}+Q_{bu}\right)\right\}+I_0\left\{G_A,\frac{D}{Dt}\left(Q_{uu}+Q_{bb}\right)\right\}\right)$$
$$+\frac{1}{3}(\boldsymbol{\Omega}+2\boldsymbol{\Omega}_F)\cdot\left(I_{-1}\left\{G_S,\nabla H_{bu}\right\}+I_{-1}\left\{G_A,\nabla H_{uu}\right\}\right)$$
$$+\mathbf{B}\cdot\left(I_0\left\{G_S,\nabla\left(Q_{uu}+Q_{bb}\right)\right\}+I_0\left\{G_A,\nabla\left(Q_{ub}+Q_{bu}\right)\right\}\right). \quad (9.250)$$

Finally, we refer to the turbulent residual helicity H. We may express H in terms of Elsasser's variables ϕ' and ϕ' as

$$H=-(1/2)\left\langle\phi'\cdot\omega'_\psi+\psi'\cdot\omega'_\phi\right\rangle, \quad (9.251)$$

where

$$\omega'_\phi=\nabla\times\phi',\ \omega'_\psi=\nabla\times\psi'. \quad (9.252)$$

Then H is written as

$$H=I_0\left\{-H_{uu}+H_{bb}\right\}$$
$$-\frac{1}{3}\left(I_0\left\{G_S,\frac{D}{Dt}\left(-H_{uu}+H_{bb}\right)\right\}+I_0\left\{G_A,\frac{D}{Dt}\left(H_{bu}-H_{ub}\right)\right\}\right)$$
$$-\frac{8}{27}(\boldsymbol{\Omega}+2\boldsymbol{\Omega}_F)\cdot\left(I_0\left\{G_S,\nabla Q_{uu}\right\}-I_0\left\{G_A,\nabla Q_{bu}\right\}\right)$$
$$+\frac{1}{3}\mathbf{B}\cdot\left(I_0\left\{G_S,\nabla\left(H_{bu}-H_{ub}\right)\right\}+I_0\left\{G_A,\nabla\left(-H_{uu}+H_{bb}\right)\right\}\right). \quad (9.253)$$

Various MHD effects that have been obtained by the foregoing statistical approaches shed light on the generation mechanisms of global magnetic fields in turbulent motion. These results, however, are expressed in terms of statistical properties in wavenumber space and are not applicable to real-world dynamo phenomena. For the study of such phenomena, we need to construct a closed MHD system of equations dependent on one-point quantities only. In Sec. 9.7, we shall discuss the one-point modeling of turbulent dynamo. There we shall make full use of the Markovianized two-scale (MTS) method that is detailed in Chapter 7. By this method, we may easily find the relationship of various correlation functions with the mean field $(\mathbf{U}, \mathbf{B})$. A typical example of the quantities to which the MTS method is difficult to apply is Λ_4 [Eq. (9.120)] in the H equation (9.116). Its reason will become clear later.

Equation (9.120) is related to Elsasser's variables $\boldsymbol{\phi}'$ and $\boldsymbol{\psi}'$ as

$$\Lambda_4 = \frac{1}{2} B_j \left\langle \phi'_i \frac{\partial \omega'_{\psi i}}{\partial x_j} - \psi'_i \frac{\partial \omega'_{\phi i}}{\partial x_j} + \omega'_{\phi i} \frac{\partial \psi'_i}{\partial x_j} - \omega'_{\psi i} \frac{\partial \phi'_i}{\partial x_j} \right\rangle . \tag{9.254}$$

The TSDIA analysis gives

$$\Lambda_4 = -\mathbf{B} \cdot (h_1 \mathbf{B} - h_2 \mathbf{J} + h_3 (\boldsymbol{\Omega} + 2\boldsymbol{\Omega}_F)) , \tag{9.255}$$

where

$$h_1 = (4/3) \left(I_1 \{G_S, -H_{uu} + H_{bb}\} - I_1 \{G_A, H_{ub} - H_{bu}\} \right) , \tag{9.256}$$

$$h_2 = (32/27) \left(I_1 \{G_S, Q_{uu} + Q_{bb}\} - I_1 \{G_A, Q_{ub} + Q_{bu}\} \right) , \tag{9.257}$$

$$h_3 = (32/27) \left(I_1 \{G_S, Q_{ub} + Q_{bu}\} - I_1 \{G_A, Q_{uu} + Q_{bb}\} \right) . \tag{9.258}$$

The calculation of Eq. (9.255) is rather complicated, and the simplification of geometrical factor such as

$$\epsilon_{ij\ell} k_j k_m D_{\ell h}(\mathbf{k}) D_{in}(\mathbf{k}) \cong (8/27) \epsilon_{ij\ell} \delta_{jm} \delta_{\ell h} \delta_{in} \tag{9.259}$$

based on $D_{ij}(\mathbf{k}) \cong (2/3)\delta_{ij}$ was made. Therefore the numerical factors obtained are less accurate, compared with $\mathbf{E}_M$ and R_{ij}. In the last term in Eq. (9.255), we first obtained the $\boldsymbol{\Omega}_F$-related part and then applied the transformation (9.58).

From the comparison between Eqs. (9.232)-(9.234) and Eqs. (9.256)-(9.258), we may easily see the correspondence

$$\alpha \leftrightarrow h_1, \ \beta \leftrightarrow h_2, \ \gamma \leftrightarrow h_3, \tag{9.260}$$

which also suggests

$$\mathbf{E}_M \leftrightarrow h_1 \mathbf{B} - h_2 \mathbf{J} + h_3 (\boldsymbol{\Omega} + 2\boldsymbol{\Omega}_F) . \tag{9.261}$$

This relationship will be used in Sec. 9.7 in the context of the one-point modeling.

9.7. MTS Method and One-Point Dynamo Modeling

In Secs. 9.5 and 9.6, we elucidated the mathematical structures of some important turbulence quantities such as $\mathbf{E}_M$ and R_{ij}. They are expressed in terms of statistical properties of MHD turbulence in wavenumber space. In order to reduce these results to the form applicable to real-world phenomena, it is indispensable to rewrite them using the one-point quantities in physical space. We take Eq. (9.150) as a simple example and illustrate the reduction of an expression in wavenumber space to a one-point counterpart. We write

$$\frac{Q_K(k;\tau,\tau')}{\left\langle \mathbf{u}'^2/2 \right\rangle} = g_Q(k;\tau,\tau'), \tag{9.262}$$

with the constraint

$$\int g_Q(k;\tau,\tau)d\mathbf{k} = 1. \tag{9.263}$$

Then Eq. (9.150) is reduced to

$$\beta_K = \tau_\beta \left\langle \mathbf{u}'^2/2 \right\rangle. \tag{9.264}$$

There τ_β is defined by

$$\tau_\beta = \frac{2}{3}\int dk \int_{-\infty}^{\tau} d\tau_1 G_M\left(k;\tau,\tau_1\right) g_Q\left(k;\tau,\tau_1\right), \tag{9.265}$$

and represents the time scale characterizing MHD fluctuations, specifically, their energy-containing parts since the contributions from low-wavenumber components are dominant in general. Equation (9.264) indicates that the one-point modeling of β_K may be done by finding a proper time scale related to such components. In what follows, we shall resort to the MTS method and perform the one point modeling of turbulent dynamo from a more systematic viewpoint.

9.7.1. APPLICATION OF MTS METHOD

The MTS method was detailed in Chapter 7 and was applied to the turbulence modeling of compressible flows in Chapter 8. In what follows, we shall use this method to make the one-point modeling of turbulent dynamo (Yoshizawa 1996).

A. Fundamental Equations

We start from Eqs. (9.97) and (9.98), and focus attention on the effects of the mean field $(\mathbf{U}, \mathbf{B})$ on their right-hand sides. We formally integrate these equations using integral operators, and then replaced the operators with proper characteristic time scales, as will be shown later. In this procedure, we can estimate the correlation functions such as $\mathbf{E}_M$ and R_{ij} that are dependent directly on $\mathbf{u}'$ and $\mathbf{b}'$. The correlation functions in the H equation (9.116), Λ_n's, depend on $\boldsymbol{\omega}'$ and $\mathbf{j}'$, besides $\mathbf{u}'$ and $\mathbf{b}'$. Therefore we need to derive these equations from Eqs. (9.97) and (9.98).

The equation for $\boldsymbol{\omega}'$ is similar to the equations for $\mathbf{u}'$ and $\mathbf{b}'$, and is given by

$$\frac{D\omega_i'}{Dt} + \frac{\partial}{\partial x_j}\left(u_j'\omega_i' - \omega_j' u_i' - b_j' j_i' + j_j' b_i' + \epsilon_{i\ell m}\frac{\partial R_{jm}}{\partial x_\ell}\right) - \nu\Delta\omega_i'$$
$$= -u_j'\frac{\partial \Omega_i}{\partial x_j} + \Omega_j\frac{\partial u_i'}{\partial x_j} + \omega_j'\frac{\partial U_i}{\partial x_j} + B_j\frac{\partial j_i'}{\partial x_j}.$$
$$+ b_j'\frac{\partial J_i}{\partial x_j} - J_j\frac{\partial b_i'}{\partial x_j} - j_j'\frac{\partial B_i}{\partial x_j}. \tag{9.266}$$

The counterpart for $\mathbf{j}'$, however, is much more complicated and is written as

$$\frac{Dj_i'}{Dt} + \frac{\partial}{\partial x_j}\left(u_j' j_i' - b_j'\omega_i'\right) - \lambda\Delta j_i'$$
$$= -u_j'\frac{\partial J_i}{\partial x_j} + B_j\frac{\partial \omega_i'}{\partial x_j} + b_j'\frac{\partial \Omega_i}{\partial x_j} + R_{Ci}. \tag{9.267}$$

Here $\mathbf{R}_C$ consists of the terms such as

$$\epsilon_{i\ell m}\frac{\partial^2 u_j'}{\partial x_j \partial x_\ell} b_m', \quad \epsilon_{i\ell m}\frac{\partial^2 b_j'}{\partial x_j \partial x_\ell} u_m', \tag{9.268}$$

which do not have any counterparts in Eqs. (9.97), (9.98), and (9.266) for $\mathbf{u}'$, $\mathbf{b}'$, and $\boldsymbol{\omega}'$, respectively (subscript C denotes current). The appearance of $\mathbf{R}_C$ is linked with the situation that we have no $\mathbf{j}$-related conservation property, unlike $\mathbf{u}'$, $\mathbf{b}'$, and $\boldsymbol{\omega}'$. Concerning the latter three quantities, the conservation properties in the MHD or hydrodynamic case are referred to in Sec. 9.2.3.C. In what follows, we shall concentrate attention on the interaction between the mean field and fluctuations and neglect the contribution of $\mathbf{R}_C$.

B. Scale-Parameter Expansion

As is similar to the previous approaches, we apply two spatial and temporal variables (9.127) to Eqs. (9.97), (9.98), (9.266), and (9.267). Then we have Eq. (9.160) for $\mathbf{u}'$, Eq. (9.130) for $\mathbf{b}'$, and the counterparts for $\boldsymbol{\omega}'$ and $\mathbf{j}'$. In what follows, we shall give the equations related to $\mathbf{u}'$ and $\mathbf{b}'$ explicitly.

We expand $\mathbf{u}'$, $\mathbf{b}'$, $\boldsymbol{\omega}'$, and $\mathbf{j}'$ as

$$f' = \sum_{n=0}^{\infty} \delta^n f'_n. \tag{9.269}$$

We substitute Eq. (9.269) to Eqs. (9.160), (9.130), etc. Then the $O(1)$ field obeys

$$\frac{\partial u'_{0i}}{\partial \tau} + U_j \frac{\partial u'_{0i}}{\partial \xi_j} + \frac{\partial}{\partial \xi_j} u'_{0j} u'_{0i} + \frac{\partial p'_{M0}}{\partial \xi_i} - \nu \Delta_\xi u'_{0i} = B_j \frac{\partial b'_{0i}}{\partial \xi_j} + \frac{\partial}{\partial \xi_j} b'_{0j} b'_{0i}, \tag{9.270}$$

$$\frac{\partial b'_{0i}}{\partial \tau} + U_j \frac{\partial b'_{0i}}{\partial \xi_j} + \frac{\partial}{\partial \xi_j} \left(u'_{0j} b'_{0i} - u'_{0i} b'_{0j} \right) - \lambda_M \Delta_\xi b'_{0i} = B_j \frac{\partial u'_{0i}}{\partial \xi_j}. \tag{9.271}$$

On the other hand, the equations for the $O(\delta)$ field are written as

$$\begin{aligned}
&\frac{\partial u'_{1i}}{\partial \tau} + U_j \frac{\partial u'_{1i}}{\partial \xi_j} + \frac{\partial}{\partial \xi_j} \left(u'_{0j} u'_{1i} + u'_{1j} u'_{0i} \right) + \frac{\partial p'_{M1}}{\partial \xi_i} - \nu \Delta_\xi u'_{1i} \\
&= -u'_{0j} \frac{\partial U_i}{\partial X_j} + b'_{0j} \frac{\partial B_i}{\partial X_j} - \frac{D u'_{0i}}{DT} + B_j \frac{\partial b'_{0i}}{\partial X_j} + B_j \frac{\partial b'_{1i}}{\partial \xi_j} \\
&+ \frac{\partial}{\partial \xi_j} \left(b'_{0j} b'_{1i} + b'_{1j} b'_{0i} \right) - \frac{\partial}{\partial X_j} \left(u'_{0j} u'_{0i} - b'_{0j} b'_{0i} - R_{ji} \right) \\
&- \frac{\partial p'_{M0}}{\partial X_i},
\end{aligned} \tag{9.272}$$

$$\begin{aligned}
&\frac{\partial b'_{1i}}{\partial \tau} + U_j \frac{\partial b'_{1i}}{\partial \xi_j} + \frac{\partial}{\partial \xi_j} \left(u'_{0j} b'_{1i} - u'_{0i} b'_{1j} \right) - \lambda_M \Delta_\xi b'_{1i} \\
&= -u'_{0j} \frac{\partial B_i}{\partial X_j} + b'_{0j} \frac{\partial U_i}{\partial X_j} - \frac{D b'_{0i}}{DT} + B_j \frac{\partial u'_{0i}}{\partial X_j} + B_j \frac{\partial u'_{1i}}{\partial \xi_j}. \\
&- \frac{\partial}{\partial \xi_j} \left(u'_{1j} b'_{0i} - u'_{1i} b'_{0j} \right) - \frac{\partial}{\partial X_j} \left(u'_{0j} b'_{0i} - u'_{0i} b'_{0j} - \epsilon_{ij\ell} E_{M\ell} \right).
\end{aligned} \tag{9.273}$$

The quantities $\mathbf{u}'_0$, $\mathbf{b}'_0$, $\boldsymbol{\omega}'_0$, and $\mathbf{j}'_0$ are coupled with one another in a complicated manner. This situation becomes more prominent in the quantities $\mathbf{u}'_1$, $\mathbf{b}'_1$, $\boldsymbol{\omega}'_1$, and $\mathbf{j}'_1$. In order to reduce the analytical complicatedness and abstract important properties of dynamo effects in a clear form, we introduce the basic field $(\mathbf{u}'_B, \mathbf{b}'_B, \boldsymbol{\omega}'_B, \mathbf{j}'_B)$. It obeys Eqs. (9.270) and (9.271) with their right-hand sides dropped or

$$\frac{\partial u'_{Bi}}{\partial \tau} + U_j \frac{\partial u'_{Bi}}{\partial \xi_j} + \frac{\partial}{\partial \xi_j} u'_{Bj} u'_{Bi} + \frac{\partial p'_{MB}}{\partial \xi_i} - \nu \Delta_\xi u'_{Bi} = 0, \tag{9.274}$$

$$\frac{\partial b'_{Bi}}{\partial \tau} + U_j \frac{\partial b'_{Bi}}{\partial \xi_j} + \frac{\partial}{\partial \xi_j}\left(u'_{Bj} b'_{Bi} - u'_{Bi} b'_{Bj}\right) - \lambda_M \Delta_\xi b'_{Bi} = 0, \tag{9.275}$$

and the similar expressions for $(\boldsymbol{\omega}'_B, \mathbf{j}'_B)$. We expand $(\mathbf{u}'_0, \mathbf{b}'_0, \boldsymbol{\omega}'_0, \mathbf{j}'_0)$ around $(\mathbf{u}'_B, \mathbf{b}'_B, \boldsymbol{\omega}'_B, \mathbf{j}'_B)$ as

$$f'_0 = f'_B + \sum_{m=1}^{\infty} f'_{0m}, \tag{9.276}$$

and solve Eqs. (9.270), (9.271), etc. by an iteration method, where f'_{0m} is the solution arising from the mth iteration. Then $\mathbf{u}'_{01}$ and $\mathbf{b}'_{01}$ are governed by

$$\frac{\partial u'_{01i}}{\partial \tau} + U_j \frac{\partial u'_{01i}}{\partial \xi_j} + \frac{\partial}{\partial \xi_j}\left(u'_{Bj} u'_{01i} + u'_{01j} u'_{Bi}\right) + \frac{\partial p'_{M01}}{\partial \xi_i} - \nu \Delta_\xi u'_{01i} \equiv L_u u'_{01i} = I_{u0i}, \tag{9.277}$$

$$\frac{\partial b'_{01i}}{\partial \tau} + U_j \frac{\partial b'_{01i}}{\partial \xi_j} + \frac{\partial}{\partial \xi_j}\left(u'_{Bj} b'_{01i} - u'_{Bi} b'_{01j}\right) - \lambda_M \Delta_\xi b'_{01i} \equiv L_b b'_{01i} = I_{b0i}, \tag{9.278}$$

where

$$I_{u0i} = B_j \frac{\partial b'_{Bi}}{\partial \xi_j} + \frac{\partial}{\partial \xi_j} b'_{Bj} b'_{Bi}, \tag{9.279a}$$

$$I_{b0i} = B_j \frac{\partial u'_{Bi}}{\partial \xi_j} - \frac{\partial}{\partial \xi_j}\left(u'_{01j} b'_{Bi} - u'_{01i} b'_{Bj}\right). \tag{9.279b}$$

We introduce the inverse operators of L_u and L_b, L_u^{-1} and L_b^{-1}, and integrate Eqs. (9.277) and (2.278) formally. Then the $O(1)$ field is given by

$$\mathbf{u}'_0 = \mathbf{u}'_B + L_u^{-1} \mathbf{I}_{u0} + \cdots, \tag{9.280}$$

$$\mathbf{b}'_0 = \mathbf{b}'_B + L_b^{-1} \mathbf{I}_{b0} + \cdots. \tag{9.281}$$

The counterparts of $\boldsymbol{\omega}'_0$ and $\mathbf{j}'_0$ are written as

$$\boldsymbol{\omega}'_0 = \boldsymbol{\omega}'_B + L_\omega^{-1} \mathbf{I}_{\omega 0} + \cdots, \tag{9.282}$$

$$\mathbf{j}'_0 = \mathbf{j}'_B + L_j^{-1} \mathbf{I}_{j0} + \cdots. \tag{9.283}$$

Here the operators L_ω^{-1} and L_j^{-1} are defined as

$$L_\omega A_i = \frac{\partial A_i}{\partial \tau} + U_j \frac{\partial A_i}{\partial \xi_j} + \frac{\partial}{\partial \xi_j}\left(u'_{Bj} A_i - A_j u'_{Bi}\right) - \nu \Delta_\xi A_i, \tag{9.284}$$

$$L_j A_i = \frac{\partial A_i}{\partial \tau} + U_\ell \frac{\partial A_i}{\partial \xi_\ell} + \frac{\partial}{\partial \xi_\ell} u'_{B\ell} A_i - \lambda_M \Delta_\xi A_i, \tag{9.285}$$

and $\mathbf{I}_{\omega 0}$ and $\mathbf{I}_{j0}$ are

$$I_{\omega 0 i} = B_j \frac{\partial j'_{Bi}}{\partial \xi_j} + \Omega_j \frac{\partial u'_{Bi}}{\partial \xi_j} - J_j \frac{\partial b'_{Bi}}{\partial \xi_j} + \frac{\partial}{\partial \xi_j} \left(b'_{Bj} j'_{Bi} - j'_{Bj} b'_{Bi} \right), \tag{9.286a}$$

$$I_{j0i} = B_\ell \frac{\partial \omega'_{Bi}}{\partial \xi_\ell} + \frac{\partial}{\partial \xi_j} b'_{Bj} \omega'_{Bi}. \tag{9.286b}$$

We substitute Eqs. (9.280) and (9.281) into Eqs. (9.272) and (9.273). The leading contributions to the $O(\delta)$ equations arise from these equations with $\mathbf{u}'_0$ and $\mathbf{b}'_0$ replaced with $\mathbf{u}'_B$ and $\mathbf{b}'_B$, respectively. In this approximation, we should note that the same operators as for Eqs. (9.277) and (9.278), L_u and L_b, occur for $\mathbf{u}'_1$ and $\mathbf{b}'_1$. As this result, we have

$$\mathbf{u}'_1 = L_u^{-1} \mathbf{I}_{u1}, \tag{9.287}$$

$$\mathbf{b}'_1 = L_b^{-1} \mathbf{I}_{b1}, \tag{9.288}$$

where

$$I_{u1i} = -u'_{Bj} \frac{\partial U_i}{\partial X_j} + b'_{Bj} \frac{\partial B_i}{\partial X_j}, \tag{9.289a}$$

$$I_{b1i} = -u'_{Bj} \frac{\partial B_i}{\partial X_j} + b'_{Bj} \frac{\partial U_i}{\partial X_j}. \tag{9.289b}$$

In Eqs. (9.289), we focused attention on the interaction between the mean field and fluctuations, and retained a small number of terms. The inclusion of more terms may be done similarly.

Under the same degree of approximation as for Eqs. (9.287) and (9.288), we have

$$\boldsymbol{\omega}'_1 = L_\omega^{-1} \mathbf{I}_{\omega 1}, \tag{9.290}$$

$$\mathbf{j}'_1 = L_j^{-1} \mathbf{I}_{j1}, \tag{9.291}$$

where

$$I_{\omega 1 i} = \omega'_{Bj} \frac{\partial U_i}{\partial X_j} - j'_{Bj} \frac{\partial B_i}{\partial X_j} - u'_{Bj} \frac{\partial \Omega_i}{\partial X_j} + b'_{Bj} \frac{\partial J_i}{\partial X_j}, \tag{9.292a}$$

$$I_{j1i} = -u'_{B\ell} \frac{\partial J_i}{\partial X_\ell} + b'_{B\ell} \frac{\partial \Omega_i}{\partial X_\ell}. \tag{9.292b}$$

C. Introduction of Characteristic Time Scales

The inverse operators L_u^{-1}, L_b^{-1}, etc. correspond to the Green's functions G_K, G_ϕ, etc. in the methods that are explained in Secs. 9.5 and 9.6. From the viewpoint of the applicability to real-world dynamo phenomena, it is necessary to derive a one-point system of dynamo equations in which all quantities are written in terms of local variables in physical space. For this purpose, we perform the Markovianization of L_u^{-1}, L_b^{-1}, etc. In the MTS method, we pay attention to the fact that L_u^{-1}, L_b^{-1}, etc. are related to characteristic turbulence time scales, and approximate these inverse operators simply as

$$L_u^{-1},\ L_b^{-1},\ L_\omega^{-1},\ L_j^{-1} \cong \tau_{TB}, \tag{9.293}$$

using a proper time scale of the basic field $(\mathbf{u}'_B, \mathbf{b}'_B, \boldsymbol{\omega}'_B, \mathbf{j}'_B)$, τ_{TB}. Here it is more appropriate to distinguish among the time scales of L_u^{-1}, L_b^{-1}, etc., but we adopt this choice for constructing a dynamo model as simple as possible.

In evaluating correlation functions with the aid of Eqs. (9.280)-(9.283), (9.287), (9.288), (9.290), and (9.291), we need to prescribe the statistical property of the basic field $(\mathbf{u}'_B, \mathbf{b}'_B, \boldsymbol{\omega}'_B, \mathbf{j}'_B)$. The equations for the field are not dependent directly on the mean filed $(\mathbf{U}, \mathbf{B})$ generating anisotropy. It is reasonable to postulate that the basic field is statistically isotropic, although not mirrorsymmetric. As the simplest choice, we write

$$\left\langle u'_{Bi} u'_{Bj} \right\rangle = (1/3) \left\langle \mathbf{u}'^2_B \right\rangle \delta_{ij}, \tag{9.294a}$$

$$\left\langle b'_{Bi} b'_{Bj} \right\rangle = (1/3) \left\langle \mathbf{b}'^2_B \right\rangle \delta_{ij}, \tag{9.294b}$$

$$\left\langle u'_{Bi} b'_{Bj} \right\rangle = (1/3) \left\langle \mathbf{u}'_B \cdot \mathbf{b}'_B \right\rangle \delta_{ij}, \tag{9.294c}$$

$$\left\langle u'_{Bi} \omega'_{Bj} \right\rangle = (1/3) \left\langle \mathbf{b}'_B \cdot \boldsymbol{\omega}'_B \right\rangle \delta_{ij}, \tag{9.294d}$$

$$\left\langle b'_{Bi} j'_{Bj} \right\rangle = (1/3) \left\langle \mathbf{b}'_B \cdot \mathbf{j}'_B \right\rangle \delta_{ij}. \tag{9.294e}$$

Equation (9.294) corresponds to the physical-space representations of Eqs. (9.140) and (9.175), but it is more limited, compared with Eq. (9.221) based on Elsasser's variables. In the latter, there is room for the occurrence of the contributions from $\langle \mathbf{u}'_B \cdot \mathbf{j}'_B \rangle$, as is seen from Eq. (9.228). Within the framework of statistical isotropy, there is no legitimate reason for excluding such contributions. The simplicity of the resulting model is a very important requirement from the standpoint of a dynamo model applicable to real-world dynamo phenomena. What quantities should be preferentially selected depends highly on the degree of importance of the effects arising from them.

For later convenience, we introduce the turbulent MHD energy K_B, the corresponding residual energy K_{RB}, and the turbulent cross helicity W_B of the basic field as

$$K_B = \left\langle \left(\mathbf{u}'^2_B + \mathbf{b}'^2_B \right) /2 \right\rangle , \tag{9.295a}$$

$$K_{RB} = \left\langle \left(\mathbf{u}'^2_B - \mathbf{b}'^2_B \right) /2 \right\rangle , \tag{9.295b}$$

$$W_B = \left\langle \mathbf{u}'_B \cdot \mathbf{b}'_B \right\rangle . \tag{9.295c}$$

Their helicity counterparts, the turbulent residual helicity H_B and the turbulent total helicity H_{TB}, are given by

$$H_B = \left\langle -\mathbf{u}'_B \cdot \boldsymbol{\omega}'_B + \mathbf{b}'_B \cdot \mathbf{j}'_B \right\rangle , \tag{9.296a}$$

$$H_{TB} = \left\langle \mathbf{u}'_B \cdot \boldsymbol{\omega}'_B + \mathbf{b}'_B \cdot \mathbf{j}'_B \right\rangle . \tag{9.296b}$$

The full counterparts of K_B, W_B, H_B, K_{RB}, and H_{TB} are given by Eqs. (9.100), (9.101), (9.115), (9.240), and

$$H_T = \left\langle \mathbf{u}' \cdot \boldsymbol{\omega}' + \mathbf{b}' \cdot \mathbf{j}' \right\rangle . \tag{9.297}$$

D. Calculation of Correlation Functions

D.1. Renormalization. We use the solution obtained in Sec. 9.7.1.B, and calculate some important correlation functions under the condition on the basic field, (9.294). As a result, those correlation functions are expressed in terms of K_B, W_B, etc. and the characteristic turbulence time scale τ_{TB} in addition to the mean field $(\mathbf{U}, \mathbf{B})$. The quantities given by Eq. (9.294) are not sufficient for modeling τ_{TB} since the pure scalars are K_B and K_{RB} only. As the other quantity characterizing MHD turbulence, we introduce the dissipation rate of K_B, ε_B, which is defined as

$$\varepsilon_B = \nu \left\langle \left(\frac{\partial u'_{Bj}}{\partial x_i} \right)^2 \right\rangle + \lambda_M \left\langle \left(\frac{\partial b'_{Bj}}{\partial x_i} \right)^2 \right\rangle . \tag{9.298}$$

Considering the importance of ε in the K equation (9.102), the choice of ε_B is proper [we should recall that Eq. (9.102) possesses a firm mathematical basis in relation to the conservation law]. Then we model

$$\tau_{TB} = C_\tau \left(K_B / \varepsilon_B \right) , \tag{9.299}$$

where C_τ is a positive constant.

The foregoing solution is still a lower-order one from the viewpoint of a perturbational expansion, as is clear from Eqs. (9.280) and (9.287). In order to retain the higher-order effects in such a perturbational solution, we apply the renormalization procedure to the results obtained using the solution; namely, we perform the replacement

$$K_B \to K, \; K_{RB} \to K_R, \; W_B \to W, \; H_B \to H, \; \text{etc.} \tag{9.300}$$

As this consequence, we have

$$\tau_{TB} \to \tau_T = C_\tau \left(K/\varepsilon \right). \tag{9.301}$$

D.2. Turbulent Electromotive Force and Reynolds Stress. The correlation functions of the second order in $\mathbf{u}'$ and $\mathbf{b}'$ may be expressed as

$$\left\langle u_i' u_j' \right\rangle = \frac{1}{3} \left(K + K_R \right) \delta_{ij} - \frac{1}{3} \tau_T \left(K + K_R \right) S_{ij} + \frac{1}{3} \tau_T W M_{ij}, \tag{9.302}$$

$$\left\langle b_i' b_j' \right\rangle = \frac{1}{3} \left(K - K_R \right) \delta_{ij} + \frac{1}{3} \tau_T \left(K - K_R \right) S_{ij} - \frac{1}{3} \tau_T W M_{ij}, \tag{9.303}$$

$$\left\langle u_i' b_j' \right\rangle = \frac{1}{3} W \delta_{ij} + \frac{1}{6} \tau_T H \epsilon_{ij\ell} B_\ell + \frac{1}{3} \tau_T W \Omega_{ij} - \frac{1}{6} \tau_T \left(K + K_R \right) \frac{\partial B_j}{\partial x_i} + \frac{1}{6} \tau_T \left(K - K_R \right) \frac{\partial B_i}{\partial x_j}. \tag{9.304}$$

Using Eqs. (9.302)-(9.304), we have $\mathbf{E}_M$ and R_{ij} as

$$\mathbf{E}_M = \alpha \mathbf{B} - \beta \mathbf{J} + \gamma \boldsymbol{\Omega}, \tag{9.305}$$

$$R_{ij} = \frac{2}{3} K_R \delta_{ij} - \nu_T S_{ij} + \nu_M M_{ij}, \tag{9.306}$$

where

$$\alpha = (1/3) \tau_T H = C_\alpha \left(K/\varepsilon \right) H, \tag{9.307}$$

$$\beta = \nu_T = (2/3) \tau_T K = C_\beta \left(K^2/\varepsilon \right), \tag{9.308}$$

$$\gamma = \nu_M = (2/3) \tau_T W = C_\gamma \left(K/\varepsilon \right) W, \tag{9.309}$$

where

$$C_\alpha = C_\tau / 3, \tag{9.310a}$$

$$C_\beta = C_\gamma = (2/3) C_\tau. \tag{9.310b}$$

The first relations of Eqs. (9.308) and (9.309) are similar to those by the TSDIA based on Elsasser's variables, Eqs. (9.241) and (9.242). In this respect, the TSDIA and its much simpler physical-space version, the MTS, give rise to no large difference.

In order to clarify the degree of importance of the cross helicity (W) on $\mathbf{E}_M$ and R_{ij} and estimate the magnitude of model constants, the computer simulation of an artificial MHD turbulence in simple geometry was made by Hamba (1992). In the simulation, the MHD turbulence, which is homogeneous in two directions and inhomogeneous in the remaining one, is generated in a cubic box by imposing a specific driving force. Its prominent finding is that the cross-helicity effect is much more dominant than the helicity one. Using the numerical database, C_β and C_γ were given as

$$C_\beta = 0.055, \ C_\gamma = 0.039. \tag{9.311}$$

In this estimate, Eqs. (9.241) and (9.242), which are more accurate than the first relations of Eqs. (9.308) and (9.309), were adopted as the relationship between (β, γ) and (ν_T, ν_M). In the present MTS method, C_β is equal to C_γ, as is seen from Eq. (9.310b).

In the simulation by Hamba (1992), the helicity effect is much smaller than the cross-helicity one, and the estimate of C_α was not done so reliably as that of C_β and C_γ. Equation (9.310) suggests

$$C_\alpha \cong C_\beta/2 \text{ or } C_\gamma/2, \tag{9.312}$$

which leads to

$$C_\alpha \cong 0.02 \tag{9.313}$$

from Eq. (9.311).

In the MTS method, the constants occurring in the characteristic time scales are left as unknown. The merit of the method, however, is that a number of model constants appearing in various correlation functions are related to one another through a few constants. In the present case, we have three constants in $\mathbf{E}_M$, C_α, C_β, and C_γ. By estimating one of them by other methods, we can estimate C_τ, resulting in the estimate of the remaining constants. This merit is fully utilized in the compressible turbulence modeling that is detailed in Chapter 8. In Eq. (9.293), we introduced one time scale, but we can adopt more time scales and refine the foregoing model expressions.

D.3. Equation for Turbulent-Residual Helicity. In order to construct a closed dynamo system of equations, we need to derive the model transport equations for K, W, H, and ε since all of the coefficients in $\mathbf{E}_M$ and R_{ij} are related to these quantities. The exact equations for the first two are given by Eq. (9.102) with Eqs. (9.103)-(9.108). Their mathematical structures are clear and the number of the terms to be modeled is small; namely, they are $\mathbf{T}'_K$, ε_W, and $\mathbf{T}'_W$. As has already been stressed, this situation is associated with the conservation properties concerning the total amounts of the MHD energy and the cross helicity. On the other hand, H and ε are not linked with such properties, and their equations are very complicated, as is seen from Eq. (9.116) with Eqs. (9.117)-(9.125), and the hydrodynamic counterpart (4.26).

We are in a position to estimate $\Lambda_n (n = 1\text{-}9)$. Here it is difficult to analyze Λ_4 and Λ_9 by using the present MTS method. In this method, the time scales characterizing the energy-containing or low-wavenumber components of fluctuations play a key role in place of Green's functions. On the other hand, Λ_9 are related to molecular effects, and Λ_4 is dependent on the derivatives of $\boldsymbol{\omega}'$ and $\mathbf{j}'$. The components of fluctuations that are much smaller than the energy-containing ones contribute to these two quantities. The choice of the time scales relevant to energy-containing motions is not appropriate for their analysis. Concerning Λ_4, it was analyzed using the TSDIA and is given by Eq. (9.255).

With the aid of the MTS method, we have

$$\Lambda_1 = -\frac{2}{3}\tau_T K_R S_{ij}\frac{\partial \Omega_i}{\partial x_j}, \tag{9.314}$$

$$\Lambda_2 = \tau_T \left(-\frac{1}{3}(K + K_R)\nabla \times \boldsymbol{\Omega} + \frac{1}{3}W\nabla \times \mathbf{J} \right) \cdot \boldsymbol{\Omega}, \tag{9.315}$$

$$\Lambda_3 = -(1/2)\nabla \cdot \left((K + K_R)\,\boldsymbol{\Omega} \right), \tag{9.316}$$

$$\Lambda_5 = \tau_T \left(-\frac{1}{3}W\frac{\partial J_i}{\partial x_j} + \frac{1}{3}(K - K_R)\frac{\partial \Omega_i}{\partial x_j} + \frac{1}{6}(H + H_T)\frac{\partial U_j}{\partial x_i} \right) \frac{\partial U_i}{\partial x_j}, \tag{9.317}$$

$$\Lambda_6 = -\tau_T \left(\frac{1}{12}(H + H_T)\,\epsilon_{ij\ell}J_\ell - \frac{1}{6}H_T M_{ij} - \frac{1}{3}W\frac{\partial \Omega_i}{\partial x_j} + \frac{1}{3}(K - K_R)\frac{\partial J_i}{\partial x_j} \right) \frac{\partial B_i}{\partial x_j}, \tag{9.318}$$

$$\Lambda_7 = \tau_T \left(-\frac{1}{3}(K + K_R)\nabla \times \mathbf{J} + \frac{1}{3}W\nabla \times \boldsymbol{\Omega} - \frac{1}{6}(H + H_T)\,\mathbf{J} \right) \cdot \mathbf{J}, \tag{9.319}$$

$$\Lambda_8 = \frac{2}{3}\tau_T K_R M_{ij}\frac{\partial J_i}{\partial x_j}. \tag{9.320}$$

Under the approximation of neglecting the spatial derivatives of K, W, etc., Λ_2 may be rewritten as

$$\Lambda_2 \cong -\frac{1}{2}\frac{\partial R_{ji}}{\partial x_j}\Omega_i. \tag{9.321}$$

This approximation is reasonable since attention is focused on effects of the mean field in obtaining $\mathbf{E}_M$, R_{ij}, etc. In the context of the foregoing one-point expressions for Λ_n's, it is reasonable to model Λ_4 [Eq. (9.255)] as

$$\Lambda_4 = -C_{HB}\left(\varepsilon^2 \big/ K^3\right)\mathbf{B}\cdot\mathbf{E}_M, \tag{9.322}$$

from dimensional analysis and the correspondence shown by Eq. (9.261), where C_{HB} is a positive constant.

D.4. Supplementary Remarks on Cross-Helicity Effects. We examine effects of the cross helicity in relation to the turbulent diffusion effects. The second and third terms in R_{ij} [Eq. (9.306)] and $\mathbf{E}_M$ [(9.305)] give the contributions

$$\frac{D\mathbf{U}}{Dt}: \ \nu_T\Delta\mathbf{U} - \nu_M\Delta\mathbf{B}; \tag{9.323a}$$

$$\frac{D\mathbf{B}}{Dt}: \ \beta\Delta\mathbf{B} - \gamma\Delta\mathbf{U}, \tag{9.323b}$$

where the spatial variation of coefficients ν_T etc. has been neglected for simplicity of discussion. The first terms in Eq. (9.323) play the role of destroying global structures in fluid motion and magnetic fields.

Positive W (the positive correlation between $\mathbf{u}'$ and $\mathbf{b}'$) indicates that $\mathbf{U}$ and $\mathbf{B}$ tend to be aligned with each other in the same direction, as is noted in Sec. 9.5.1.E. The second term in Eq. (9.323a), which expresses the feedback effect of $\mathbf{B}$ on $\mathbf{U}$, indicates that induced locally large $\mathbf{B}$ tends to reduce the diffusion effect of the first term and keep the global fluid motion from being destroyed further. As a result, the energy comes back from magnetic to kinetic parts, and $\mathbf{B}$ cannot continue to grow indefinitely. In other words, the second term in Eq. (9.323a) represents a kind of saturation effect on $\mathbf{B}$. The entirely similar discussion can be made for negative W and Eq. (9.323b) [see Yoshizawa and Yokoi (1996) for more detailed discussions on the related effects].

9.7.2. ONE-POINT DYNAMO MODELING

We are in a position to present a closed system of dynamo equations by summarizing all the foregoing results. First, the mean field $(\mathbf{U}, \mathbf{B})$ obeys

$$\frac{DU_i}{Dt} = -\frac{\partial}{\partial x_i}\left(P + \frac{\langle \mathbf{b}'^2\rangle}{2}\right) + (\mathbf{J}\times\mathbf{B})_i - \frac{\partial R_{ji}}{\partial x_j} + \nu\Delta U_i, \tag{9.324}$$

$$\frac{\partial \mathbf{B}}{\partial t} = -\nabla \times \mathbf{E}, \tag{9.325}$$

$$\mathbf{J} = \nabla \times \mathbf{B} = \frac{1}{\lambda_M} \left(\mathbf{E} + \mathbf{U} \times \mathbf{B} + \mathbf{E}_M \right), \tag{9.326}$$

with the solenoidal condition

$$\nabla \cdot \mathbf{U} = \nabla \cdot \mathbf{B} = 0. \tag{9.327}$$

In order to model the Reynolds stress R_{ij} and the turbulent electromotive force $\mathbf{E}_M$, we postulate the quantities characterizing MHD turbulence state. In the present model, we adopt

$$K[\text{Eq. (9.100)}], \ W[\text{Eq. (9.101)}], \ H[\text{Eq. (9.115)}], \ \varepsilon[\text{Eq. (9.104)}]. \tag{9.328}$$

The difference between the turbulent kinetic and magnetic energy, K_R [Eq. (9.240)], which is called the residual energy, is also an important quantity indicating the relative strength of kinetic and magnetic fluctuations. Here we have excluded K_R from the simplicity of the resulting model, but it is one of the quantities to be included in refining the present model. The total turbulent helicity H_T has also been dropped from the similar reason.

Under the choice (9.328), R_{ij} and $\mathbf{E}_M$ are given by

$$R_{ij} = \frac{2}{3} K_R \delta_{ij} - \nu_T S_{ij} + \nu_M M_{ij}, \tag{9.329}$$

$$\mathbf{E}_M = \alpha \mathbf{B} - \beta \mathbf{J} + \gamma \mathbf{\Omega}, \tag{9.330}$$

where

$$\alpha = C_\alpha \left(K/\varepsilon \right) H, \tag{9.331}$$

$$\beta = (5/7)\nu_T \left(\cong \nu_T\right) = C_\beta \left(K^2/\varepsilon \right), \tag{9.332}$$

$$\gamma = (5/7)\nu_M \left(\cong \nu_M\right) = C_\gamma \left(K/\varepsilon \right) W. \tag{9.333}$$

The turbulence equations consist of those for K, W, H, and ε. The first two equations are

$$\frac{DZ}{Dt} = P_Z - \varepsilon_Z + \nabla \cdot \mathbf{T}_Z \ \ (Z = K \text{ or } W), \tag{9.334}$$

with $\varepsilon_K \equiv \varepsilon$, where

$$P_K = -R_{ij} \frac{\partial U_j}{\partial x_i} - \mathbf{E}_M \cdot \mathbf{J}, \tag{9.335a}$$

$$\mathbf{T}_K = W\mathbf{B} + \left(\nu_T/\sigma_K \right) \nabla K, \tag{9.335b}$$

and

$$P_W = -R_{ij}\frac{\partial B_j}{\partial x_i} - \mathbf{E}_M \cdot \mathbf{\Omega}, \tag{9.336a}$$

$$\varepsilon_W = C_W\,(\varepsilon/K)\,W, \tag{9.336b}$$

$$\mathbf{T}_W = K\mathbf{B} + (\nu_T/\sigma_W)\,\nabla W \tag{9.336c}$$

(σ_K, C_W, and σ_W are model constants).

In Eq. (9.335b), we adopted the simplest gradient-diffusion model for $\mathbf{T}'_K$ [Eq. (9.105b)]. Namely, $(\mathbf{u}'^2+\mathbf{b}'^2)\mathbf{u}'/2$ may be regarded as the transport rate of $(\mathbf{u}'^2+\mathbf{b}'^2)/2$ by the velocity fluctuation $\mathbf{u}'$, and such transport is supposed to occur from the region with large K to the one with small K. This type of modeling is frequently used in the hydrodynamic case, as is detailed in Secs. 4.2.2 and 6.6.1. In the latter case, the contribution from the pressure fluctuation is shown to be much smaller. In $\mathbf{T}'_K$, we have one more term, $\langle(\mathbf{u}'\cdot\mathbf{b}')\mathbf{b}'\rangle$, but its modeling is left as unknown. We have applied the similar modeling to $\mathbf{T}'_W$ [Eq. (9.108b)]. The original equation for the W destruction rate ε_W [Eq. (9.107)] is very complicated because of the lack in the connection with the conservation law, as is entirely similar to ε. Of ε_W and ε, the latter is indispensable for the construction of the characteristic time scale τ_T [Eq. (9.301)]. In order to reduce the complexity of the resulting dynamo system, we adopt the simplest model for ε_W, Eq. (9.336b).

As has already been noted, the equations governing H and ε inevitably become phenomenological. We write

$$\frac{DH}{Dt} = -\frac{1}{2}\frac{\partial R_{ji}}{\partial x_j}\Omega_i - C_{HB}\frac{\varepsilon^2}{K^3}\mathbf{B}\cdot\mathbf{E}_M - C_H\frac{\varepsilon}{K}H + \nabla\cdot\left(-\frac{1}{2}K\mathbf{\Omega} + \frac{\nu_T}{\sigma_H}\nabla H\right), \tag{9.337}$$

$$\frac{D\varepsilon}{Dt} = C_{\varepsilon 1}\frac{\varepsilon}{K}P_K - C_{\varepsilon 2}\frac{\varepsilon^2}{K} + \nabla\cdot\left(\frac{\nu_T}{\sigma_\varepsilon}\nabla\varepsilon\right), \tag{9.338}$$

where C_{HB}, C_H, σ_H, $C_{\varepsilon 1}$, $C_{\varepsilon 2}$, and σ_ε are model constants.

The terms in Eq. (9.337) have been selected from the following reasons. First, we retain the terms that survive in the hydrodynamic limit or in the case of the dominance of kinetic effects. They are Λ_2 [Eq. (9.321)], and Λ_3 [Eq. (9.316)]. Under the choice (9.328), K_R is dropped, resulting in the neglect of Λ_1 [Eq. (9.314)]. Next, we retain the terms important in the opposite case, that is, in the absence of global fluid motion $\mathbf{U}$. Such a typical term is Λ_4 [Eq. (9.322)], and its significance will be referred to in the context of the force-free state of Chapter 10. The third term is the modeling of Λ_9 [Eq. (9.125)] representing the H destruction due to molecular effects.

In the last term of Eq. (9.337), the ∇H-related part expresses the modeling of the terms that are of the third order in $\mathbf{u}'$ and $\mathbf{b}'$ and are denoted by R_H in Eq. (9.116). In this modeling, we should note that all of these third-order terms cannot be written in the divergence form, unlike the counterparts in the K and W equations that are linked with the conservation laws.

The model equation (9.338) for ε is a straightforward extension of the hydrodynamic counterpart (4.37), whose theoretical derivation is discussed in Sec. 6.6.2. In the present MHD case, effects of magnetic fields are taken into account through R_{ij} and the magnetic part of $\mathbf{E}_M$ in the turbulent energy production rate P_K [Eq. (9.335a)].

Finally, the foregoing turbulent-dynamo system contains twelve model constants; namely they are

$$R_{ij},\ E_M:\ C_\alpha,\ C_\beta,\ C_\gamma; \tag{9.339a}$$

$$K\ \text{equation}:\ \sigma_K; \tag{9.339b}$$

$$W\ \text{equation}:\ C_W,\ \sigma_W; \tag{9.339c}$$

$$H\ \text{equation}:\ C_{HB},\ C_H,\ \sigma_W; \tag{9.339d}$$

$$\varepsilon\ \text{equation}:\ C_{\varepsilon 1},\ C_{\varepsilon 2},\ \sigma_\varepsilon. \tag{9.339e}$$

Some of these constants were estimated as

$$C_\alpha \cong 0.02,\ C_\beta \cong 0,05,\ C_\gamma \cong 0.04,\ C_W > 1\,(\cong 1)\,, \tag{9.340a}$$

using the computer-experiment database (Hamba 1992). Here the constraint on C_W is due to the discussion in the following *Note.* In the hydrodynamic limit, we have

$$\sigma_K \cong 1,\ C_{\varepsilon 1} \cong 1.4,\ C_{\varepsilon 2} \cong 1.9,\ \sigma_\varepsilon \cong 1.3. \tag{9.340b}$$

Under Elsasser's variables, the K and W equations correspond to those for $\langle \phi'^2 + \psi'^2 \rangle/4$ and $\langle \phi'^2 - \psi'^2 \rangle/4$, respectively. This close relationship indicates $\sigma_W \cong \sigma_K$, that is,

$$\sigma_W \cong 1. \tag{9.340c}$$

Concerning the remaining constants attached to the H equation, C_{HB}, C_H, and σ_H, their estimate comparable to Eq. (9.340) is yet to be made.

Note: In the homogeneous MHD turbulence with vanishing $\mathbf{U}$ and $\mathbf{B}$, Eq. (9.334) is reduced to

$$\frac{\partial K}{\partial t} = -\varepsilon, \tag{9.341}$$

$$\frac{\partial W}{\partial t} = -C_W \frac{\varepsilon}{K} W. \tag{9.342}$$

From these equations, we have

$$\frac{\partial}{\partial t} \frac{W}{K} = -\left(C_W - 1 \right) \frac{\varepsilon W}{K^2}. \tag{9.343}$$

In the case of $C_W < 1$, initially nonvanishing $|W/K|$ continues to grow and eventually exceeds one. This result contradicts the important constraint (9.126). Therefore we need the condition $C_W > 1$.

9.8. Supplementary Remarks

We should refer to the lowest-order parts of fluctuations whose statistics are postulated as Eqs. (9.140), (9.175), (9.221), and (9.294). In the present chapter, we made the analysis of their effects on the generation mechanism of the mean or global magnetic field, without delving into their investigation itself. Such an approach seems curious for the readers who are much interested in the characteristics of small-scale fluctuations. This approach, however, is due to the fact that the details of these fluctuations tend to be shielded in the turbulent dynamo with attention focused on the relationship with the mean field, specifically, in the one-point modeling.

In order to understand the foregoing situation, we consider the MHD counterpart of the Kolmogorov's $-5/3$ power-law spectrum in hydrodynamic turbulence. In the MHD case, we may mention two kinds of energy spectra. One is the familiar Kolmogorov's type, which is

$$E(k) \propto \varepsilon^{2/3} k^{-5/3}. \tag{9.344}$$

Another is the $-3/2$ power spectrum such as Eq. (3.153). In Sec. 3.5.3.D, the spectrum arises from the overestimate of sweeping effects of larger-scale fluctuations on small-scale ones. It is eventually rejected in the context of the Galilean invariance of the hydrodynamic system of equations under the transformation

$$\mathbf{x} \to \mathbf{x} + \mathbf{U}_\infty t, \ t \to t, \ \mathbf{u} \to \mathbf{u} + \mathbf{U}_\infty \tag{9.345}$$

($\mathbf{U}_\infty$ is a constant velocity).

The MHD system is also invariant under the transformation (9.345), but not concerning the transformation

$$\mathbf{b} \to \mathbf{b} + \mathbf{B}_\infty \tag{9.346}$$

($\mathbf{B}_\infty$ is a constant magnetic field). This fact indicates the possibility that the MHD inertial-range spectrum is subject to the direct influence of larger-scale components of fluctuations. We denote their contribution to the turbulent energy by K_C. Then the MHD counterpart of Eq. (3.153) is given by

$$E(k) \propto K_C^{1/4} \varepsilon^{1/2} k^{-3/2}. \tag{9.347}$$

This type of spectrum was first pointed by Kraichnan (1965).

These two spectra surely give rise to a significant difference concerning the behaviors for large k. We see what difference may occur in the context of one-point turbulent dynamo. Two representative turbulence quantities, $\mathbf{E}_M$ and R_{ij}, arise from the energy-containing components of fluctuations. We denote their characteristic spatial scale by ℓ_C, and may define it as

$$K = \int_{2\pi/\ell_C}^{\infty} E(k) dk. \tag{9.348}$$

From Eqs. (9.344) and (9.347), we have

$$\ell_C \propto K^{3/2}/\varepsilon, \tag{9.349a}$$

$$\ell_C \propto \frac{K^2}{K_C^{1/2}\varepsilon}, \tag{9.349b}$$

respectively.

The time scale characterizing the energy-containing fluctuations, τ_C, may be given by

$$\tau_C = \ell_C \Big/ \sqrt{K}\ . \tag{9.350}$$

From Eqs. (9.349a) and (9.349b), we have

$$\tau_C \propto K/\varepsilon, \tag{9.351a}$$

$$\tau_C \propto \left(\frac{K}{K_C}\right)^{1/2} \frac{K}{\varepsilon}, \tag{9.351b}$$

respectively. Considering that K_C is the primary portion of K, the difference between K and K_C is small, and it does not matter so much which of these two τ_C's should be adopted as the characteristic time scale τ_{TB} in Eq. (9.293). What is more important from the viewpoint of turbulent dynamo is the difference between the kinetic and magnetic spectra in close relation to the residual turbulent energy K_R. In the preset dynamo system, this effect is neglected. In the absence of the mean magnetic field, it was discussed by Kato and Yoshizawa (1993) in relation to accretion disks, and this work was extended by Nakao (1997) to the case of a given mean magnetic field.

The recent developments in the study on small-scale MHD turbulence are detailed in the book by Biskamp (1993).

References

Biskamp, D. (1993), *Nonlinear Magnetohydrodynamics* (Cambridge U. P., Cambridge).
Braginskii, S. (1964), JETP **20**, 1462.
Chen, H. and Montgomery, D. (1987) Plasma Phys. and Controlled Fusion **29**, 205.
Cowling, T. G. (1934), Mon. Not. Astron. Soc. **94**, 39.
Hamba, F. (1987), J. Phys. Soc. Jpn. **56**, 79.
Hamba, F. (1992), Phys. Fluids A **4**, 441.
Hoyng, P. (1993), in *The Sun: A Laboratory for Astrophysics*, edited by J. T. Schmelz and J. C. Brown (Kluwer, Dordrecht), p. 99.
Kato, S. and Yoshizawa, A. (1993), Publ. Astron. Soc. Jpn. **45**, 103.
Kraichnan, R. H. (1965), Phys. Fluids **8**, 1385.
Krause, F. and Rädler, K. -H. (1980), *Mean-Field Magnetohydrodynamics and Dynamo Theory* (Pergamon, Oxford).
Melchior, P. (1986), *The Physics of the Earth's Core* (Pergamon, Oxford).
Miyamoto, K. (1989), *Plasma Physics for Controlled Fusion* (The MIT, Cambridge).
Moffatt, H. K. (1978), *Magnetic Field Generation in Electrically Conducting Fluids* (Cambridge U. P., Cambridge).
Moreau, R. (1990), *Magnetohydrodynamics* (Kluwer, Dordrecht).
Nakao, Y. (1997), Publ. Astron. Soc. Jpn. **49**, 659.
Pouquet, A., Frisch, U., and Léorat, J. (1976), J. Fluid Mech. **77**, 321.
Priest, E. (1982), *Solar Magnetohydrodynamics* (D. Reidel, Dordrecht).
Roberts, P. H. (1993), in *Astrophysical Fluid Dynamics*, edited by J. -P. Zahn and J. Zinn-Justin (Elsevier, Amsterdam), p. 229.
Yoshizawa, A. (1985), Phys. Fluids **28**, 3313.
Yoshizawa, A. (1990), Phys. Fluids B **2**, 1589.
Yoshizawa, A. (1996), J. Phys. Soc. Jpn. **65**, 124.
Yoshizawa, A. and Yokoi, N. (1996), Phys. Plasmas **3**, 3604.

CHAPTER 10

GLOBAL MAGNETIC FIELDS IN TURBULENT MOTION

10.1. Introductory Remarks

In Chapter 9, we examined turbulence effects on the global or mean components of magnetic fields and fluid motion. They are characterized by the turbulent electromotive force $\mathbf{E}_M$ and the Reynolds stress R_{ij}. These two quantities are related to the mean components and the turbulence characteristics, and the construction of the transport equations for the latter leads to a closed system of turbulent dynamo. In the approach, attention is focused on the analytical insight into the global magnetic-field generation processes that are common to various types of turbulent motion of electrically conducting fluids. Therefore the highly time-dependent properties of turbulence associated with small-scale components are beyond the reach of turbulent dynamo.

In this point, the computer experiment based on the primitive MHD equations makes sharp contrast with the turbulent dynamo. In the computer experiment of dynamo, the MHD system of equations given in Sec. 9.2.2 or 9.2.3 is usually supplemented with some driving forces such as the thermal buoyancy force. With the advance in the computer capability, the computer experiment is making remarkable progress (its recent progress is given in the works by Glatzmaier and Roberts (1995), Kageyama and Sato (1995), and the works cited therein). A great merit of computer experiments is that the small-scale properties of MHD turbulence can be captured with the global structures of MHD flows. In real-world phenomena, however, the nondimensional physical parameters represented by the kinetic and magnetic Reynolds numbers are large. The computer experiment of turbulence whose physical parameters are close to real-world ones is beyond the scope of present and foreseeable-future computers. The situation is entirely similar to that in hydrodynamic turbulent flow, and the merit and limitation of such an approach have been fully recognized in the latter.

In the computer experiment of hydrodynamic flow, which is often called the direct numerical simulation (DNS), one of its primary roles is to provide the numerical databases for the construction of one-point turbulence models including subgrid-scale models, although the Reynolds number of these databases is at best moderate. The computer experiment of MHD flow may play a similar role, but it is not so simple to abstract the global structures of magnetic fields in a mathematical form from a large amount of numerical data. In this situation, the suggestions by the turbulent dynamo may provide a significant guide for seeking those structures from the numerical data. Therefore these two entirely different approaches, the computer experiment and the turbulent dynamo, are regarded as supplementary to each other.

In the present chapter, we shall make full use of the results about $\mathbf{E}_M$ and R_{ij} in Chapter 9, and seek the global structures of magnetic fields under some typical circumstances such as the dominance of fluid motion or magnetic effects. Through these discussions, we shall clarify some typical features of dynamo processes that are common to a variety of turbulence phenomena.

10.2. Helicity Dynamo

10.2.1. FEATURES OF HELICITY EFFECTS

A. Force-Free Field

In the absence of the global fluid motion represented by the mean velocity $\mathbf{U}$, it is sufficient to consider the equation for $\mathbf{B}$, which consists of Eqs. (9.325), (9.326), and (9.330). From the latter two equations, we have

$$\beta\mathbf{J} - \alpha\mathbf{B} = \mathbf{E}, \tag{10.1}$$

where the molecular resistivity λ_M has been dropped, compared with the turbulent counterpart β. In Eq. (10.1), we consider the case

$$|\beta\mathbf{J}| \cong |\alpha\mathbf{B}| \gg |\mathbf{E}|. \tag{10.2}$$

Then we have

$$\mathbf{J} = (\alpha/\beta)\,\mathbf{B}. \tag{10.3}$$

This is the most typical manifestation of the helicity or alpha effect that generates the global electric current aligned with the global magnetic field. Equation (10.3) leads to

$$\mathbf{J} \times \mathbf{B} = 0, \tag{10.4}$$

and the Lorentz force in the mean velocity equation (9.324) vanishes. Then this magnetic field exerts no force to the fluid motion, and it is consistent with the starting assumption of no global fluid motion. The magnetic field described by Eq. (10.3) is called the force-free field.

We consider the relationship of the alignment between $\mathbf{J}$ and $\mathbf{B}$ with the mathematical property of the coefficient α. As may be seen from Eq. (9.331) or the more accurate expression (9.232), α is related to the turbulent residual helicity H or its spectrum. In the absence of $\mathbf{U}$, the main source of energy resides in the mean magnetic field $\mathbf{B}$. In this case, the magnetic part of the turbulent energy K, $\langle \mathbf{b}'^2/2 \rangle$, is supposed to be much larger than the kinetic part $\langle \mathbf{u}'^2/2 \rangle$ since the magnetic energy directly cascades from $\mathbf{B}$ to $\mathbf{b}'$ and $\mathbf{u}'$ gets the energy from the interaction with $\mathbf{b}'$. Then it is highly probable that the magnetic part of H, $\langle \mathbf{b}' \cdot \mathbf{j}' \rangle$, is dominant, compared with the kinetic one $\langle \mathbf{u}' \cdot \boldsymbol{\omega}' \rangle$. This dominance may be considered to be linked with the high alignment between $\mathbf{J}$ and $\mathbf{B}$.

In real-world phenomena, the mean electric field $\mathbf{E}$ cannot always be dropped even under the condition (10.2). In this case, we have

$$\mathbf{J} \times \mathbf{B} = (1/\ \beta)\, \mathbf{E} \times \mathbf{B}, \tag{10.5}$$

from Eq. (10.1). The condition of vanishing $\mathbf{U}$ is fulfilled as long as Eq. (10.5) balances with the mean pressure gradient or the first term on the right-hand side of the mean equation (9.324).

B. Reversed-Field Pinches of Plasmas

As a typical example of magnetic fields showing the force-free state, we consider reversed-field pinches (RFP's) of plasmas in controlled fusion (Bodin and Newton 1980). The representative approach to plasma confinement by magnetic fields is tokamaks in torus geometry (Fig. 10.1). In the approach, the toroidal magnetic field $\mathbf{B}_T$ generated by external coils wrapped with a torus is much stronger than the poloidal one $\mathbf{B}_P$ coming from a toroidal plasma current. Plasmas in a torus can move freely along this strong toroidal field, but the motion normal to it is highly limited owing to the tension of the magnetic field. The relationship between $\mathbf{B}_T$ and $\mathbf{B}_P$ in tokamaks is characterized by

$$q \equiv \frac{a}{R} \frac{|\mathbf{B}_T|}{|\mathbf{B}_P|} > 1, \tag{10.6}$$

where R and a are the major and minor radii, respectively. The quantity q is usually called the safety factor (Miyamoto 1989). From this constraint, the minor radius a cannot be made much smaller than R. In the context of q, the confinement approach that is in a directly opposite position to tokamaks is RFP's. In the approach, the poloidal magnetic field is relatively much larger and is nearly comparable to the toroidal one. The RFP state is characterized by

$$q \ll 1, \tag{10.7}$$

and the minor radius a can be chosen to be much smaller than the major one R. As this result, the torus is often approximated by a circular cylinder in the theoretical investigation of RFP's.

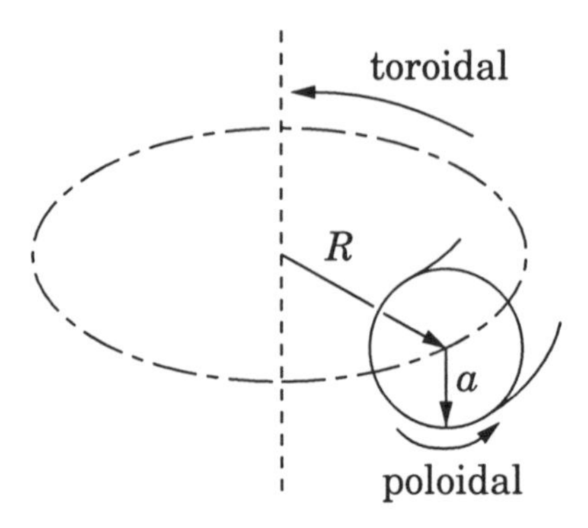

Figure 10.1. Toroidal geometry in controlled fusion.

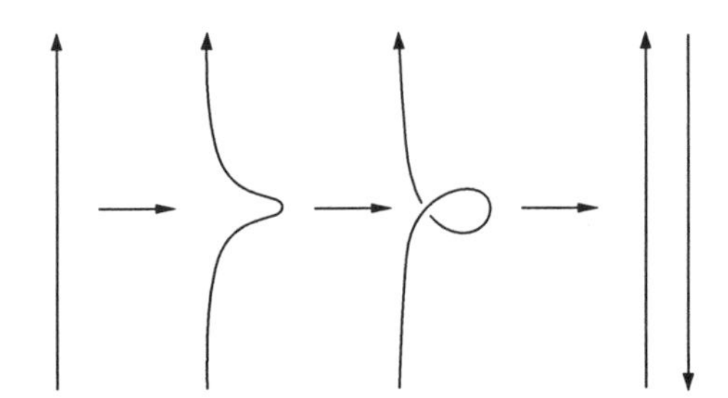

Figure 10.2. Reversal of toroidal magnetic field due to kink instability and reconnection.

The most prominent feature of RFP's is the reversal of the toroidal magnetic field at the outer edge of plasma. This phenomenon may be explained intuitively as follows. In the initial setting-up phase, the poloidal field generated by a strong plasma current interacts with the toroidal magnetic field, and a large deformation of the latter due to kink and sausage instability is induced. Such deformation leads to the formation of the loop of a magnetic-field line because of its tension, as in Fig. 10.2, and the reconnection of the field line results in the reversal of the toroidal magnetic field near the wall of a torus. In the situation, the continuation of plasma currents is equivalent to that of the reversed toroidal magnetic field and is indispensable for plasma confinement. The deeper reversal of the toroidal field leads to the longer sustainment of plasma confinement. For the progress in the observational and computer-experiment RFP studies, the reader may consult the works by Ji et al (1996) and Nagata et al. (1995) (and the works cited therein).

In RFP's, the toroidal plasma current eventually generates the reversed toroidal magnetic field. This alignment between the current and magnetic field is a primary incentive of the investigation of RFP's using the helicity dynamo, as was first pointed out by Gimblett and Watkins (1975). We seek a solution of Eq. (10.3) in cylindrical geometry and examine its relationship with RFP's. To this end, we employ the cylindrical coordinates (r, θ, z), where the θ and z directions represent the poloidal and toroidal ones, respectively. We assume the constancy of α and take the curl of Eq. (10.3). Then we have

$$\Delta\mathbf{B} + \kappa^2\mathbf{B} = 0, \tag{10.8}$$

where

$$\kappa = \alpha/\beta \tag{10.9}$$

(we should recall that κ is a pseudo-scalar). Under the condition of the axisymmetry and z independence of $\mathbf{B}$, we are led to

$$\frac{d^2B_z}{dr^2} + \frac{1}{r}\frac{dB_z}{dr} + \left(\kappa^2 - \frac{1}{r^2}\right)B_z = 0, \tag{10.10a}$$

$$B_\theta = -\frac{1}{\kappa}\frac{dB_z}{dr}, \tag{10.10b}$$

from Eqs. (10.3) and (10.8). The solution of Eq. (10.10) is

$$B_z/B_z(0) = J_0(\kappa r), \tag{10.11a}$$

$$B_\theta/B_z(0) = J_1(\kappa r), \tag{10.11b}$$

where $B_z(0)$ is the value of B_z at the central axis, and J_n is the first Bessel function of the nth order.

The Bessel function J_n possesses a infinite number of points, s_{nm}'s, at which $J_n(s_{nm}) = 0$. The first zero point s_{01} is about 2.4. This fact indicates that the toroidal magnetic field reverses its sign at the edge of plasma under the condition

$$|\kappa a| > 2.4. \tag{10.12}$$

This situation is depicted schematically in Fig. 10.3. The solution of the Bessel-function type, Eq. (10.11), captures well the feature of the global magnetic fields in RFP's and has been a guiding principle in understanding the essence of RFP's experimentally and theoretically.

Equation (10.11) with constant κ suffers from a shortfall near the edge; $\mathbf{J}$ coming from it does not always vanish there, and this finding contradicts the observations of RFP's. Within the framework of turbulent dynamo, α and β are determined by K, H, and ε through Eqs. (9.331) and (9.332), and κ is dependent on location. The turbulent dynamo model presented in Sec. 9.7.2 was applied by Hamba (1990) to RFP's, and their primary features were confirmed to be reproduced under spatially varying κ. The

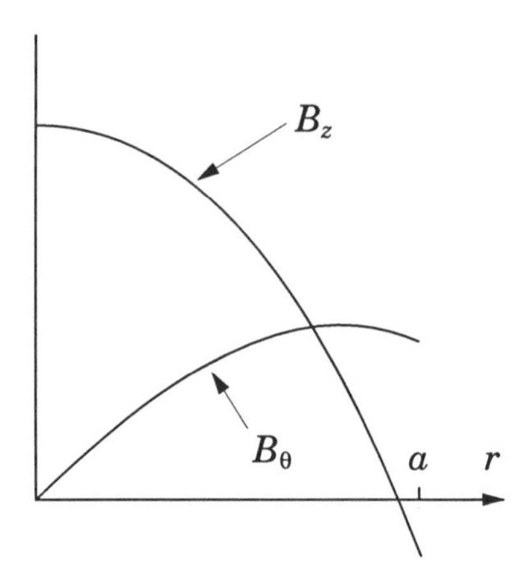

Figure 10.3. Global magnetic fields in RFP's.

importance of the helicity dynamo has been recognized well in relation to astrophysical magnetic fields. Their observational information, specifically, the quantitative one, is greatly limited, and detailed and accurate data on magnetic fields are much more available in RFP's. In this sense, RFP's are an interesting subject from the standpoint of the helicity-dynamo modeling, apart from the degree of utility as an approach to controlled fusion.

In the relationship of the turbulent dynamo with RFP's, we should emphasize that Eq. (10.3) for the force-free field was originally derived by Taylor (1974) on an entirely different theoretical basis. This point is explained in the following *Note.* In the method, the constancy of κ is one of the primary theoretical ingredients.

Note: The derivation of Eq. (10.3) by Taylor is deeply rooted on the recognition that the force-free field is a kind of equilibrium state in the magnetic-field relaxation process. It is considered to be the state of minimum energy under some constraints. From the fact that the total amount of magnetic helicity, $\int \mathbf{A} \cdot \mathbf{B} dV$, is conserved in the absence of molecular effects, Taylor considered that the force-free field corresponds to the state described by the condition

$$\text{minimum} \int \left(\mathbf{B}^2/2\right) dV \text{ under constant} \int \mathbf{A} \cdot \mathbf{B} dV, \tag{10.13}$$

where $\mathbf{A}$ is the vector potential and is related to $\mathbf{B}$ as $\mathbf{B} = \nabla \times \mathbf{A}$. In the comparison between these two integrals, the former contains higher-wavenumber components than the latter, owing to the relation between $\mathbf{B}$ and $\mathbf{A}$. As a result, the former is affected more strongly by small-scale destruction effects.

We write the condition (10.13) in the variational form

$$\delta\left(\int\left(\frac{\mathbf{B}^2}{2}-\frac{\kappa}{2}\mathbf{A}\cdot\mathbf{B}\right)dV\right)=0, \tag{10.14}$$

with $\mathbf{A}$ fixed at the volume surface (κ is a constant multiplier). With the aid of partial integration, Eq. (10.14) is reduced to

$$\int(\nabla\times\mathbf{B}-\kappa\mathbf{B})\cdot\delta\mathbf{A}dV=0, \tag{10.15}$$

resulting in Eq. (10.3) [note the relation $\nabla\cdot(\mathbf{a}\times\mathbf{b})=\mathbf{b}\cdot\nabla\times\mathbf{a}-\mathbf{a}\cdot\nabla\times\mathbf{b}$].

10.2.2. TYPICAL MAGNETIC-FIELD GENERATION PROCESSES

In the presence of global fluid motion, Eqs. (9.326) and (9.330) give

$$\mathbf{J}=(1/\beta)\left(\mathbf{E}+\mathbf{U}\times\mathbf{B}+\alpha\mathbf{B}\right), \tag{10.16}$$

where the cross-helicity and molecular-resistivity terms have been dropped. Equation (10.16) is combined with Eq. (9.325) to give

$$\frac{\partial\mathbf{B}}{\partial t}=\nabla\times\left(\mathbf{U}\times\mathbf{B}+\alpha\mathbf{B}-\beta\nabla\times\mathbf{B}\right). \tag{10.17}$$

The magnetic field generated by Eq. (10.17) has influence on Eq. (9.324) for the fluid motion through the Lorentz force $\mathbf{J}\times\mathbf{B}$. As was seen in Sec. 10.2.1, however, the direct contribution of the helicity or alpha dynamo to $\mathbf{J}\times\mathbf{B}$ is small, and its essential feature lies in the generation process of $\mathbf{B}$. In this sense, it is significant to examine Eq. (10.17) from the kinematic viewpoint.

A. Differential Rotation in Helicity Dynamo

In dynamo phenomena, the generation process of magnetic fields in a stellar object is a central subject. We consider Eq. (10.17) in spherical geometry, and adopt the spherical coordinates (r,θ,ϕ), as in Fig. 10.4 (the axis of rotation is given by $\theta=0$). In this context, we also use the cylindrical coordinates (σ,ϕ,z) with the z axis coincident with the axis of rotation. In the case of a rotating stellar object, the primary motion is the toroidal one in the ϕ direction. We write $\mathbf{U}$ in the combination of the toroidal and poloidal components as

$$\mathbf{U}=U_\phi\mathbf{e}_\phi+\mathbf{U}_P, \tag{10.18}$$

with

$$\mathbf{U}_P=U_r\mathbf{e}_r+U_\theta\mathbf{e}_\theta, \tag{10.19}$$

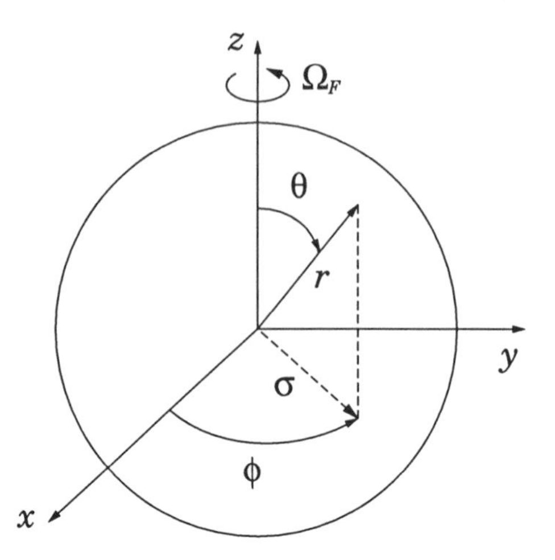

Figure 10.4. Spherical and cylindrical coordinate systems.

where $\mathbf{e}_r$, $\mathbf{e}_\theta$, and $\mathbf{e}_\phi$ are the unit vectors in three directions. The mean vorticity $\boldsymbol{\Omega}$ is expressed as

$$\boldsymbol{\Omega} = 2\Omega_{Fz}\mathbf{e}_z + \boldsymbol{\Omega}_D. \tag{10.20}$$

Here $\boldsymbol{\Omega}_F(= \Omega_{Fz}\mathbf{e}_z)$ is the angular velocity vector of the solid-rotation part of the stellar object, whereas $\boldsymbol{\Omega}_D$ represents the deviation of $\boldsymbol{\Omega}$ from it and depends on r and θ or σ and z under the condition of the axisymmetry of $\mathbf{U}$.

The deviation part $\boldsymbol{\Omega}_D$ represents the so-called differential rotation, and its physical importance lies in the following two points:

(i) Distortion of global magnetic-field lines through the term $\nabla \times (\mathbf{U} \times \mathbf{B})$;
(ii) Sustainment of turbulent state due to the turbulent-energy production term P_K [Eq. (9.335a)], which guarantees nonvanishing α and β.

In the conventional turbulent dynamo, much emphasis has been put on the former aspect. In what follows, we shall see helicity effects from this viewpoint. The latter aspect will be discussed later in relation to the cross-helicity dynamo.

We divide $\mathbf{B}$ into the toroidal and the poloidal parts as

$$\mathbf{B} = B_\phi\mathbf{e}_\phi + \mathbf{B}_P, \tag{10.21}$$

as is similar to Eq. (10.18). Under the condition of axisymmetry, $\mathbf{B}_P$ is written in the form

$$\mathbf{B}_P = \nabla \times (A_\phi\mathbf{e}_\phi), \tag{10.22}$$

using the toroidal component of the vector potential $\mathbf{A}$, A_ϕ (see *Note*). From Eq. (10.17), we have

$$\frac{\partial B_\phi}{\partial t} + \sigma\left(\mathbf{U}_P \cdot \nabla\right)\frac{B_\phi}{\sigma} = \alpha\left(\nabla \times \mathbf{B}_P\right)_\phi + \sigma\left(\mathbf{B}_P \cdot \nabla\right)\frac{U_\phi}{\sigma} + \beta\left(\Delta - \frac{1}{\sigma^2}\right)B_\phi, \tag{10.23}$$

$$\frac{\partial A_\phi}{\partial t} + \sigma\left(\mathbf{U}_P \cdot \nabla\right)\left(\sigma A_\phi\right) = \alpha B_\phi + \beta\left(\Delta - \frac{1}{\sigma^2}\right)A_\phi, \tag{10.24}$$

where the spatial variation of α and β has been dropped for simplicity of discussion, and we recall the relation $\sigma = r\sin\theta$.

We briefly see the role of each term on the right-hand sides of Eqs. (10.23) and (10.24). In Eq. (10.23), the first term represents

α process: Poloidal field
→ Toroidal current due to the Ampère law
→ Toroidal field due to the helicity effect. (10.25)

This is the typical magnetic-field generation process of the helicity dynamo.

The second term is

ω process: Distortion of poloidal field
→ Toroidal field. (10.26)

The physical meaning of this process may be explained as follows. As a typical case, we consider a magnetic-field tube parallel to the rotation or z axis. The quantity U_ϕ/σ expresses the local angular velocity of a fluid blob rotating around the z axis. The quantity $(\mathbf{B}_P \cdot \nabla)U_\phi/\sigma$ signifies the change of the angular velocity along the tube. At high Reynolds numbers, magnetic fields are nearly frozen in the blob and move with it. In case that $(\mathbf{B}_P \cdot \nabla)U_\phi/\sigma$ is positive, the upper part of the magnetic tube moves faster in the toroidal direction than the lower counterpart. This deformation of the tube leads to the generation of the toroidal component of magnetic fields. In this process, the tube is stretched, and its cross section decreases, resulting in the increase in the strength of magnetic tension. The over-stretched tube overcomes the stretching by the fluid motion and, in turn, shrinks (recall the motion of a spring). This situation indicates that the ω process has oscillatory behaviors in time. The third term represents the familiar destruction effect of global magnetic fields by turbulent motion.

In Eq. (10.24), the first term corresponds to the counterpart in Eq. (10.23), which represents

α process: Toroidal field
→ Poloidal field through the generation of A_ϕ (10.27)

[recall Eq. (10.22)]. The second term is the same destruction effect as the third one in Eq. (10.23).

Note: Under the condition of axisymmetry, the curl of the poloidal part of an arbitrary vector $\mathbf{F}$, $\mathbf{F}_P$, may be written as

$$\nabla \times \mathbf{F}_P = \mathbf{e}_\phi \frac{1}{r} \left(\frac{\partial}{\partial r} (r F_r) - \frac{\partial F_\phi}{\partial \theta} \right), \tag{10.28}$$

which gives a toroidal field. Moreover, the curl of a toroidal field generates a poloidal one, as is seen from

$$\nabla \times (F_\phi \mathbf{e}_\phi) = \mathbf{e}_r \frac{1}{r \sin\theta} \left(\frac{\partial}{\partial \theta} (\sin\theta F_\phi) \right) + \mathbf{e}_\theta \frac{1}{r} \left(-\frac{\partial}{\partial r} (r F_\phi) \right). \tag{10.29}$$

B. α^2 Dynamo
In Eq. (10.23), we have two $\mathbf{B}$-generating terms, as was seen above. The relative importance of the first to second terms is characterized by the nondimensional quantity

$$D_{\alpha/\omega} = \left| \frac{\alpha_R}{L_R \Omega_R} \right|. \tag{10.30}$$

Here α_R, L_R, and Ω_R are the reference values of α, the length, and the angular velocity of the mean fluid motion, respectively.

In the case of large $D_{\alpha/\omega}$, the generation of B_ϕ from $\mathbf{B}_P$ occurs through the α process (10.25). The counter process (10.27), that is, the generation of $\mathbf{B}_P$ from B_ϕ is also done by the α process in Eq. (10.24). Under these circumstances, magnetic fields are sustained through the alpha effect only. The combination of such two α effects is usually called the α^2 dynamo. The α effect itself is not related to the stretching of magnetic-field lines, unlike the ω process. Therefore the oscillatory behaviors of generated magnetic fields are weak, and the α^2 dynamo is relevant to the description of the stationary aspect of dynamo processes.

C. α-ω Dynamo
In the case of small $D_{\alpha/\omega}$, the generation of B_ϕ from $\mathbf{B}_P$ is dominated by the differential rotation of a fluid [the ω process (10.26)] The generation of $\mathbf{B}_P$ is done through the α process (10.27), as is similar to the α^2 dynamo. The combination of these two processes leads to the so-called α-ω dynamo. In solar magnetic phenomena, the differential rotation in the convective zone is prominent, and much more attention has been paid to the α-ω dynamo in their study.

In Secs. 9.1 and 9.3.2, we simply referred to solar magnetic fields related to sunspots (the internal structure of the Sun is depicted schematically in Fig. 9.3). The representative features of sunspot-related phenomena may be summarized as follows (Priest 1982):

(i) Occurrence of sunspots of a few thousand Gauss;
(ii) Restriction of sunspots to the low-latitude region;
(iii) Law of polarity;
(iv) 11-year oscillation in the sunspot number and the reversal of polarity near the sunspot maximum;
(v) Drift of sunspot occurrence towards the equator.

Of these phenomena, the 11-year oscillation is specifically prominent, and this behavior is entirely different from Earth's global magnetic fields whose reversal time scale is much longer. The 11-year oscillation is an important subject of the α-ω dynamo with oscillatory properties, and the dynamo was confirmed to reproduce such an oscillation under the carefully chosen fluid motion and α coefficient. This point is reviewed in detail in the works by Parker (1979) and Priest (1982). The other features such as (iii) will be referred to in the context of the cross-helicity dynamo.

10.3. Cross-Helicity Dynamo

10.3.1. FEATURES OF CROSS-HELICITY EFFECTS

In Sec. 10.2, we focused attention on the magnetic-field generation process due to the helicity or alpha effect, and dropped the cross-helicity one. The prominent difference between these two effects is that the former is related to the fluctuation of vorticity, whereas the latter is tightly linked with the mean fluid motion through the mean vorticity or frame rotation [we should note Eq. (9.58)]. In this sense, the cross-helicity effect may become important in the MHD dynamo in which the interaction between $\mathbf{U}$ and $\mathbf{B}$ is not negligible. In order to abstract the role of $\mathbf{U}$ in $\mathbf{E}_M$, we pay special attention to the cross-helicity effect, and neglect the helicity one. Then we have

$$\mathbf{E} = -\mathbf{U} \times \mathbf{B} + \beta \mathbf{J} - \gamma \mathbf{\Omega}, \tag{10.31}$$

from Eqs. (9.326) and (9.330) (the molecular resistivity λ_M was dropped).

In Eq. (10.31), we examine the state in which

$$|\beta \mathbf{J}| \cong |\gamma \mathbf{\Omega}| \gg |\mathbf{E}| . \tag{10.32}$$

This situation is not so unusual. For instance, we consider the magnetic-field generation process in a spherical stellar object. The global fluid motion and magnetic field, $\mathbf{U}$ and $\mathbf{B}$, may be regarded as approximately axisymmetric around the rotation axis. In this case, there is no specific mechanism for sustaining the toroidal loop voltage arising from E_ϕ, unlike fusion devises such as tokamaks and RFP's, and we may put

$$E_\phi = 0. \tag{10.33}$$

The primary fluid motion is a toroidal one coming from the rotation of the object. Therefore it is reasonable to assume the relation (10.32), resulting in

$$\mathbf{U} \times \mathbf{B} - \beta \mathbf{J} + \gamma \boldsymbol{\Omega} = 0. \tag{10.34}$$

For simplicity of discussion, we assume the local homogeneity, that is, the local constancy of β and γ. This assumption is not so critical except the specific region such as the vicinity of the equatorial plane in a spherical object. Across the plane, the pseudo-scalar γ changes its sign, whereas β does not. The assumption of the local homogeneity, however, does not rule out the slow spatial variation of β and γ, but it corresponds to the approximation of neglecting direct effects of $\nabla\beta$ and $\nabla\gamma$. This assumption will be referred to in the following *Note*. Under the local homogeneity, Eq. (10.34) has a very simple and useful solution, which is given by

$$\mathbf{B} = \frac{\gamma}{\beta}\mathbf{U} = \frac{C_\gamma}{C_\beta}\frac{W}{K}\mathbf{U}, \tag{10.35}$$

$$\mathbf{J} = \frac{\gamma}{\beta}\boldsymbol{\Omega} = \frac{C_\gamma}{C_\beta}\frac{W}{K}\boldsymbol{\Omega}. \tag{10.36}$$

Here we should note that $\mathbf{U} \times \mathbf{B}$ in Eq. (10.34) vanishes identically because of Eq. (10.35). The second relation in each of Eqs. (10.35) and (10.36) was obtained using Eqs. (9.332) and (9.333).

In astrophysical phenomena, rotational motion is observed ubiquitously. This fact indicates the importance of the toroidal component of the global or mean velocity. Equation (10.35) gives a clear mechanism for generating a toroidal magnetic field under the rotational motion. A toroidal field is usually a primary part of astrophysical magnetic fields, as is represented by sunspots. In the context of the constraint (9.126) on W, we have

$$\left|\frac{\mathbf{B}}{\mathbf{U}}\right| = \frac{C_\gamma}{C_\beta}\left|\frac{W}{K}\right| \le \frac{C_\gamma}{C_\beta} \; (\cong 0.7) \tag{10.37}$$

[recall Eq. (9.340a)]. Equation (10.37) gives the upper limit of the strength of the magnetic fields generated by the fluid motion under the cross-helicity dynamo. In Earth's magnetic fields, the toroidal component is confined within the outer core covered with the electrically nonconducting mantle, as is noted in Sec. 9.3. On the other hand, solar magnetic fields are generated by the fluid motion in the outermost part or the convection zone, and their toroidal components are directly observable as sunspots etc. Equation (10.37) will be examined later in the context of solar magnetic fields.

From Eqs. (10.35) and (10.36), the Lorentz force in the mean velocity equation (9.324), $\mathbf{J} \times \mathbf{B}$, is given by

$$\mathbf{J} \times \mathbf{B} = \left(\frac{\gamma}{\beta}\right)^2 \mathbf{\Omega} \times \mathbf{U} = \left(\frac{C_\gamma}{C_\beta}\right)^2 \left(\frac{W}{K}\right)^2 \mathbf{\Omega} \times \mathbf{U}. \tag{10.38}$$

Except some specific cases, Eq. (10.38) gives nonvanishing contributions to the mean fluid motion.

The effect of magnetic fields on fluid motion also occurs through the Reynolds stress R_{ij} [Eq. (9.329)]. From Eq. (10.35), we have

$$R_{ij} = \frac{2}{3} K_R \delta_{ij} - \nu_{eff} S_{ij}, \tag{10.39}$$

where ν_{eff} is the effective turbulent viscosity defined as

$$\nu_{eff} = \left(1 - \left(\frac{C_\gamma}{C_\beta}\right)^2 \left(\frac{W}{K}\right)^2\right) \nu_T. \tag{10.40}$$

From the second inequality in Eq. (10.37), ν_{eff} obeys

$$0 < \nu_{eff} / \nu_T < 1. \tag{10.41}$$

Positive ν_{eff} guarantees its physical meaning as an effective turbulent viscosity, resulting in the decrease in the turbulent-viscosity effect.

Note: In obtaining Eq. (10.35), we neglected the spatial variation of β and γ. In a spherical object, γ changes its sign across its equatorial plane since it is a pseudo-scalar. In order to see whether or not this approximation contradicts the properties of $\mathbf{B}$, we consider the solenoidal condition that is its most important property. From Eq. (10.35), we have

$$\nabla \cdot \mathbf{B} = \frac{\gamma}{\beta} \nabla \cdot \mathbf{U} + \mathbf{U} \cdot \nabla \left(\frac{\gamma}{\beta}\right). \tag{10.42}$$

Here the first term vanishes from the solenoidal condition on $\mathbf{U}$. Near the equatorial plane, the spatial change of (γ/β) is largest in the direction normal to it, whereas $\mathbf{U}$ is in the toroidal direction. Therefore Eq. (10.42) vanishes, and the local-constancy approximation gives rise to no critical shortfall.

10.3.2. DIFFERENTIAL ROTATION AND TURBULENT STATE

At a first glance, Eq. (10.35) seems to indicate that the solid rotation given by the first term of Eq. (10.20) can generate the toroidal magnetic field, but this is not the case. Such a motion cannot sustain the turbulent state leading to nonvanishing of β and γ. In order to see this point, we consider the production rate of the turbulent energy K, P_K, that is defined by (9.335a). Using Eq. (10.35), P_K may be rewritten as

$$P_K = (1/2)\nu_{eff} S_{ij}^2. \tag{10.43}$$

In case that the dominant part of the fluid motion is the toroidal one U_ϕ, S_{ij}^2 may be written as

$$S_{ij}^2 = 2\left(\sigma \frac{d}{d\sigma}\frac{U_\phi}{\sigma}\right)^2 \tag{10.44}$$

in the cylindrical coordinates (σ, ϕ, z) depicted in Fig. 10.4. Equation (10.43) vanishes under the solid rotation that is given by $U_\phi = \Omega_{Fz}\sigma$.

Next, we consider the production rate of the turbulent cross helicity W, P_W, that is given by Eq. (9.336a). In the second part, $\mathbf{E}_M$ vanishes under Eq. (10.35) with the helicity term $\alpha\mathbf{B}$ dropped. Then we have

$$P_W = \frac{\gamma}{\beta} P_K, \tag{10.45}$$

and nonvanishing P_K leads to nonvanishing P_W. In the present turbulent-viscosity and -resistivity approximation, P_K is always positive, and the sign of P_W is coincident with that of W; namely, we have

$$P_W > 0 \text{ for } W > 0, \tag{10.46a}$$

$$P_W < 0 \text{ for } W < 0. \tag{10.46b}$$

From this relation, it may be concluded that positive (or negative) W continues to be generated in the region with positive (or negative) W at the initial stage of MHD turbulence as long as the turbulence is sustained by fluid motions such as differential rotation.

10.3.3. RELATIONSHIP WITH TYPICAL ASTROPHYSICAL MAGNETIC FIELDS

As the geometry characterizing astrophysical objects, we may mention the spherical geometry represented by the Sun and the Earth, and the circular-disk one, that is, accretion disks surrounding high-mass objects such as active galactic nuclei and neutron stars. In what follows, we shall examine the magnetic fields generated in each geometry.

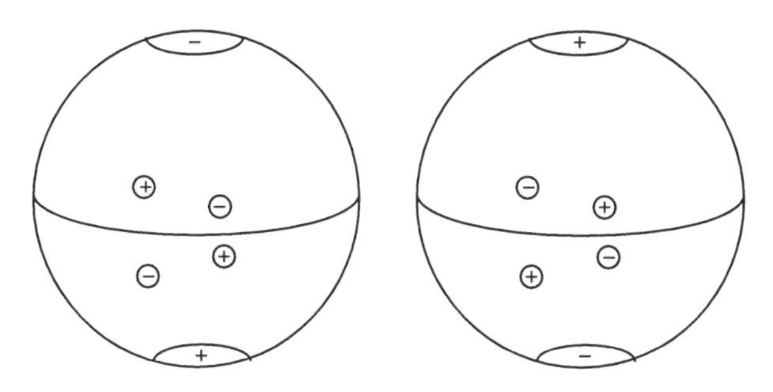

Figure 10.5. Polarity rule of sunspots.

A. Origin of Sunspots

One of the prominent phenomena in solar magnetic fields is sunspots. They originate in the intense toroidal magnetic-flux tube that is generated in the convection zone (see Fig. 9.3). Such a tube rises up owing to effects of buoyancy forces and breaks through the photosphere adjacent to the outer boundary of the convection zone. The cross section of the tube becomes visible as sunspots, which are usually observed in pair. The polarity in the northern hemisphere is opposite to the southern counterpart, as in Fig. 10.5. This fact indicates that the direction of the toroidal magnetic field is opposite in the two hemispheres. The magnetic-field strength of large sunspots reaches a few thousand Gauss. The polarity of the poloidal magnetic field near the northern pole is coincident with that of the leading sunspot. These properties of sunspots are called the polarity rule. In what follows, we shall use Eq. (10.35) to examine some of these properties (Yoshizawa 1993). The relationship with the polarity near the poles will be discussed in Sec. 10.4 from the viewpoint of the combination of the cross-helicity and helicity effects.

A.1. Estimate of Hydrogen Number Density. The primary part of the global solar fluid motion in the convection zone is the toroidal rotational one. Its velocity U_ϕ is related to the toroidal component of magnetic fields, B_ϕ, as

$$B_\phi = \frac{C_\gamma}{C_\beta}\frac{W}{K}U_\phi, \tag{10.47}$$

from Eq. (10.35). We retain the solid-rotation part in Eq. (10.20) as

$$\boldsymbol{\Omega} \cong 2\boldsymbol{\Omega}_F (= 2\Omega_{Fz}\mathbf{e}_z), \tag{10.48}$$

resulting in

$$U_\phi \cong \Omega_{Fz} r \sin\theta. \tag{10.49}$$

This approximation does not deny the importance of the differential-rotation part. In the Sun, the differential rotation is dominant, specifically, in the low-latitude region. This point is considered to be connected with the fact that the occurrence of sunspots is limited to the low-latitude region. In the context of the present turbulent dynamo, the differential rotation is linked with the sustainment of the MHD turbulence state, as is discussed in Sec. 10.3.2. The magnitude of K and W is highly dependent on the strength of differential rotation.

In Eq. (10.47), the sign of W at location (r, θ, ϕ) in the northern hemisphere is opposite to that at location $(r, \pi - \theta, \phi)$ in the southern one. This fact indicates that the direction of the toroidal field in the former is opposite to that of the latter, resulting in the polarity of sunspots in Fig. 10.5.

We are in a position to see Eq. (10.47) in a more quantitative manner. A typical velocity related to the solar rotational motion is the equatorial speed, which is about $2000\mathrm{ms}^{-1}$. On adopting

$$U_\phi \cong 1000\mathrm{ms}^{-1}, \quad C_\gamma / C_\beta \cong 1 \tag{10.50}$$

in Eq. (10.35), we have

$$B_\phi \cong 10^3 \left(W/K \right) \left(\mathrm{ms}^{-1} \right). \tag{10.51}$$

The magnetic field in Gauss units, B_ϕ^* , is related to B_ϕ in Alfvén velocity units as

$$B_\phi^* \cong 0.4 \times 10^{-12} n^{1/2} B_\phi (\mathrm{Gauss}), \tag{10.52}$$

from the relation (9.46b), where n is the number density of hydrogen (Priest 1982). We use Eqs. (10.51) and (10.52), which lead to

$$B_\phi^* \cong 0.4 \times 10^{-9} (W/K) n^{1/2} \ (\mathrm{Gauss}). \tag{10.53}$$

As an example, we choose $B_\phi^* \cong 2000$ Gauss and have

$$n \cong 3 \times 10^{27} \mathrm{m}^{-3} \text{ for } W/K \cong 0.1, \tag{10.54a}$$

$$n \cong 10^{26} \mathrm{m}^{-3} \text{ for } W/K \cong 0.5, \tag{10.54b}$$

$$n \cong 0.8 \times 10^{26} \mathrm{m}^{-3} \text{ for } W/K \cong 0.8. \tag{10.54c}$$

From observations, n is estimated as

$$n = O\left(10^{32}\right) \mathrm{m}^{-3} \text{ at the center,} \tag{10.55a}$$

$$n = O\left(10^{23}\right) \mathrm{m}^{-3} \text{ in the photosphere.} \tag{10.55b}$$

Therefore the estimate from the present model is consistent with the occurrence of the toroidal magnetic field of $O(10^3)$ Gauss.

A.2. Restriction to Lower-Latitude Region. The toroidal magnetic field B_ϕ, which is the origin of sunspots, is proportional to $\sin\theta$, as may be seen from Eqs. (10.47) and (10.49). Therefore it tends to strengthen and weaken in the low- and high-latitude regions, respectively. In Eq. (10.44), $(d/d\sigma)(U_\phi/\sigma)$ represents the strength of differential rotation, which is biggest near the equatorial plane, and $\sigma(= r\sin\theta)$ is also biggest there. From Eqs. (10.43) and (10.45), P_K and P_W are proportional to S_{ij}^2. These facts indicate that the MHD turbulent state tends to be strong at low latitude and weak at high one, resulting in the restriction of sunspots to the low- and middle-latitude regions.

A.3. Meridional Flows. We look at the effects of the magnetic fields induced in the convection zone on its fluid motion. The Lorentz force due to the cross-helicity effect is given by Eq. (10.38). We combine it with Eqs. (10.48) and (10.49), and have

$$\mathbf{J}\times\mathbf{B} = (\gamma/\beta)^2 r\Omega_{Fz}^2\left(-2\sin^2\theta, -\sin 2\theta, 0\right), \tag{10.56}$$

in the spherical coordinates (r,θ,ϕ). The θ component of the Lorentz force is negative and positive in the northern and southern hemispheres, respectively, and gives a poleward driving force.

A poleward meridional flow is observed at the surface of the convective zone, and its observed velocity is about $10\mathrm{ms}^{-1}$ (Priest 1982). The flow is inferred to contribute to the poleward drift of the magnetic flux, but the origin of the poleward flow is not clear. Equation (10.56) contributes to the induction of the poleward meridional flow and is an interesting candidate for the force driving the flow. The velocity is estimated from the balance with the resistive force due to the turbulent viscosity ν_T, as

$$U_\theta \cong -\tau(\gamma/\beta)^2 r\Omega_{Fz}^2 \sin 2\theta. \tag{10.57}$$

Here τ is given by

$$\tau \cong H_S^2/\nu_T\,, \tag{10.58}$$

which represents the diffusion time scale based on the turbulent viscosity and the scale height H_S (a characteristic vertical length scale in the convection zone).

B. Galactic Magnetic Fields

The progress in optical and radio polarization measuring techniques has made possible a reliable observation of interstellar magnetic fields of our Galaxy (Milky Way) and some external galaxies. The strength of these magnetic fields is about $1 \sim 10\mu$ Gauss, and they are very weak, compared

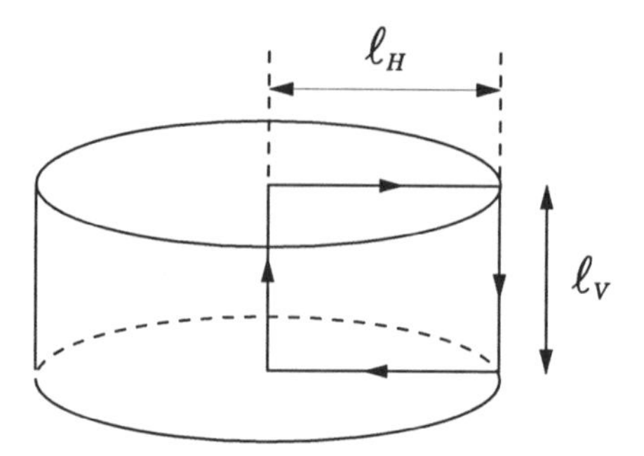

Figure 10.6. Path of integral.

with the solar magnetic fields represented by sunspots of a few thousand Gauss, and Earth's dipole one of a few Gauss. For sustaining these weak fields, however, a formidably large electromotive force is necessary because of huge spatial scales.

We consider the disk geometry as a model of a galaxy (see Fig. 10.6). We apply the induction equation (9.325) to the rectangle with the side length ℓ_H and ℓ_V, and use the Stokes' integral theorem (2.62) (subscripts H and V denote horizontal and vertical, respectively). The voltage between the central and outer parts of a galaxy, V_G, may be estimated as

$$V_G \cong (\ell_H \ell_V / T_G) B_\phi^*, \tag{10.59}$$

where T_G is a characteristic galactic time scale. We measure V_G and B_ϕ^* in volt and Gauss, respectively, and employ

$$\ell_H \cong 10\text{kpc}, \ \ell_V \cong 1\text{kpc}, \ T_G \cong 10^{10}\text{year}. \tag{10.60}$$

Here 1kpc = 10^3pc (1pc is about 3.3 light years). From Eqs. (10.59) and (10.60), we have

$$V_G \cong 10^{13} \text{ (volt)} \tag{10.61}$$

(Yokoi 1996a).

As the cause of such a huge electromotive force, we may mention the following possibility:

(i) Primordial magnetic fields closely related to the generation process of each Galaxy;
(ii) Global magnetic fields generated by turbulent motion.

In what follows, we shall see galactic magnetic fields from the latter viewpoint.

The geometrical structures of galaxies can be divided into several classes according their structures. The most typical structures are ellipticals, ordinary spirals, and barred ones. Magnetic fields in spiral galaxies are detailed

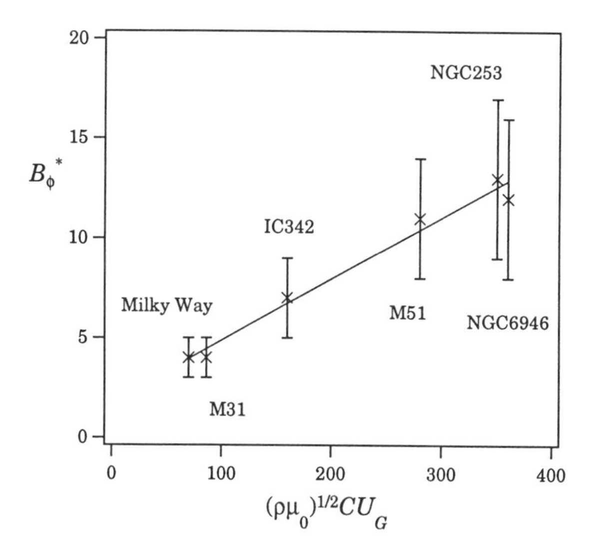

Figure 10.7. Magnetic field (μG) and rotational speed in galaxies.

by Ruzumaikin et al. (1987) and Wielebinski and Krause (see also the works cited therein). Galaxies are rotating with constant velocity in their great portion, except the central part. The rotational speed of each galaxy, U_G, and the magnitude of its magnetic field, B_ϕ^*, are plotted in Fig. 10.7 (Yokoi 1996b). These results indicate the monotonic increase in B_ϕ^* with U_G, and give an interesting support to Eq. (10.47).

C. Accretion Disks

It is widely accepted that a high-mass object is located at the center of an active galaxy. The object is surrounded by an interstellar gas of circular-disk form. The latter accretes onto the former, while rotating around it. Such disk structures are called accretion disks. Through this process, the kinetic and gravitational energy are supplied to the central high-mass object. At the same time, the angular momentum possessed by the accreting gas is transported towards the center. Therefore the object cannot attain the stationary or quasi-stationary state without some proper mechanism for disposing of the angular momentum. The traditional concept on the angular-momentum disposal is based on the Reynolds stress by turbulent motion. In this concept, the Reynolds stress is related linearly to the mean pressure [Shakura and Sunaev (1973); see Horiuchi and Kato (1990) for the theoretical investigation of this model], and the angular momentum is transported by turbulence towards the outer part of a disk, oppositely to the mass transport.

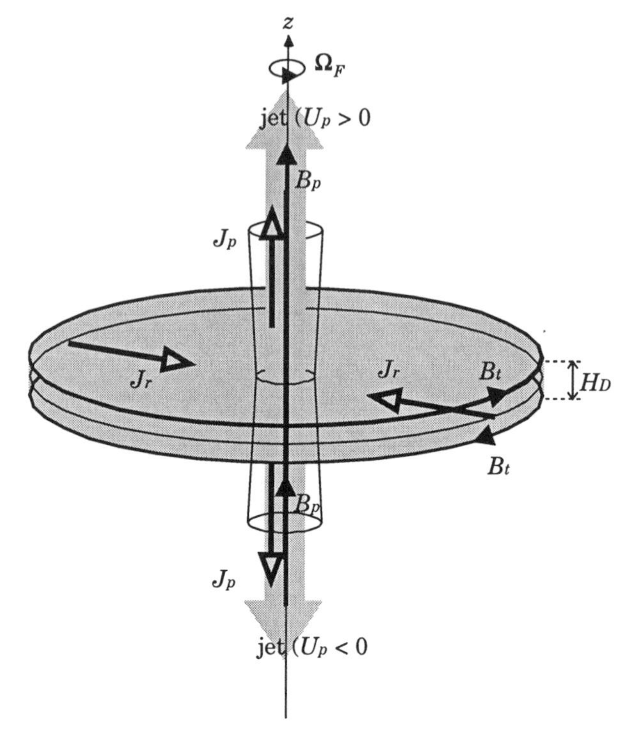

Figure 10.8. Accretion disk and bipolar jets.

The prominent phenomenon observed ubiquitously around accretion disks is the occurrence of bipolar jets. They are ejected in the directions normal to the disks, as is depicted schematically in Fig. 10.8. Their speed usually ranges from a few ten kms^{-1} to a few thousand kms^{-1}, and the one of fastest jets reaches a few percent of light speed. These jets may be considered a promising candidate for the mechanism of disposing of the angular momentum by carrying it away. A driving mechanism of bipolar jets is yet to be established, but it is believed that magnetic fields generated around the disks play a key role in the occurrence of jets. The developments in the study of accretion disks and jets are reviewed by Begelman and Blandford (1984) and Kato et al. (1998).

C.1. Generation of Toroidal Magnetic Field and Condition on Jet Occurrence. The most typical motion around a high-mass body is the so-called Keplerian motion. In the cylindrical coordinates (σ, ϕ, z) with the z axis coincident with the rotation axis, the Keplerian angular velocity Ω_K obeys

$$m\sigma\Omega_K^2 = G\frac{Mm}{\sigma^2}, \tag{10.62}$$

where M is the mass of the central object, m is the mass of a portion of rotating gases around it, and G is the gravitational constant. Equation (10.62) leads to the differential rotation

$$\Omega_K \propto \sigma^{-3/2}, \tag{10.63}$$

which gives the velocity of the toroidal motion

$$U_\phi \propto \sigma^{-1/2}. \tag{10.64}$$

In Sec. 10.3.2, it is shown that the MHD turbulence state with non-vanishing K and W is sustained by differential rotation. In the geometrical structure in Fig. 10.8, it is reasonable to assume that W is anti-symmetric with respect to the midplane since it is a pseudo-scalar. Moreover, Eq. (10.46a) indicates that positive (or negative) W continues to be generated in the region with positive (or negative) W at the initial stage. We assume the situation in which

$$W > 0 \text{ for } 0 < z < H_D/2, \tag{10.65a}$$

$$W < 0 \text{ for } -H_D/2 < z < 0, \tag{10.65b}$$

with $W(-z) = -W(z)$, where H_D is the width of the disk. From Eqs. (10.35) and (10.65), the positive and negative toroidal magnetic fields are generated in the upper and lower regions, respectively, as in Fig. 10.8. This magnetic-field configuration implies that the electric currents flow towards the center of the disk. From the conservation of electric currents, those currents result in the poloidal one nearly normal to the disk, $\mathbf{J}_P$. Moreover, Eq. (10.36) with $\boldsymbol{\Omega}$ approximated by $2\Omega_K \mathbf{e}_z$ also contributes to the generation of the poloidal current of the quadruple type.

From the above arguments, the magnetic fields near the center of the disk consist of the contribution from Eq. (10.35), $B_{\phi,U}$, and the field generated by $\mathbf{J}_P$, $B_{\phi,J}$; namely, we have

$$B_\phi = B_{\phi,U} + B_{\phi,J} \tag{10.66}$$

(Yoshizawa and Yokoi 1993). The condition under which bipolar jets are driven by magnetic fields is that the magnetic energy is comparable to the gravitational one; namely, we put

$$B_\phi^2/2 \cong \phi_G, \tag{10.67}$$

where ϕ_G is the gravitational potential. This condition was examined by Nishino and Yokoi (1998) solving the cross-helicity dynamo model given in Sec. 9.7.2, with effects of the turbulent helicity dropped. In this study, the fluid motion is assumed to consist of the outer Keplerian and inner solid rotation, and the homogeneity of turbulent state is assumed in the upper and lower regions, respectively. As this result, the condition (10.67) leads to the estimate of jet speeds that are consistent with the observational ones in active galactic nuclei and protostars.

C.2. Poloidal Magnetic Fields. We consider bipolar jets. Their velocity is characterized by the poloidal part $\mathbf{U}_P$, which is

$$U_{Pz} > 0 \text{ for } z > 0, \tag{10.68a}$$

$$U_{Pz} < 0 \text{ for } z < 0. \tag{10.68b}$$

We combine Eq. (10.35) with this velocity under the condition (10.65) on W. Then we have

$$B_{Pz} > 0 \text{ for all } z; \tag{10.69}$$

namely, the induced poloidal magnetic field is of the dipole type. In this and foregoing discussions, Eq. (10.35) plays a key role. Its greatest feature is the alignment between the fluid velocity and the resulting magnetic field. As a work related to the validity of Eq. (10.35), we may mention the computer experiment by Shibata and Uchida (1990). In the experiment, the MHD system of equations was solved under the condition of axisymmetry around the rotation axis of a disk. An initial uniform poloidal magnetic field threading the disk vertically is pulled towards the center with the contraction of the disk and is concurrently twisted. In this course, the reconnection of magnetic fields occurs, and their tension launches gas vertically to the disk. Each component of the velocity and the magnetic field is given in Fig. 10.9, which shows the striking similarity between the spatial profiles of the velocity and magnetic field.

10.4. Combined Effects of Cross-Helicity and Helicity Dynamos

In Sec. 10.3, we focused attention on the mechanism of generating a toroidal magnetic field by a rotational or toroidal fluid motion. In general, fluid motion contains a poloidal velocity component, which leads to a poloidal magnetic field under Eq. (10.35). In the combination with the helicity dynamo, however, the poloidal component of magnetic fields occurs without considering any special poloidal fluid motion. In what follows, we shall pursue two typical mechanisms leading to the generation of a poloidal field (Yoshizawa and Yokoi 1996). Before proceeding to such discussions, we consider the influence of the extent of a flow region on the magnetic-field generation processes.

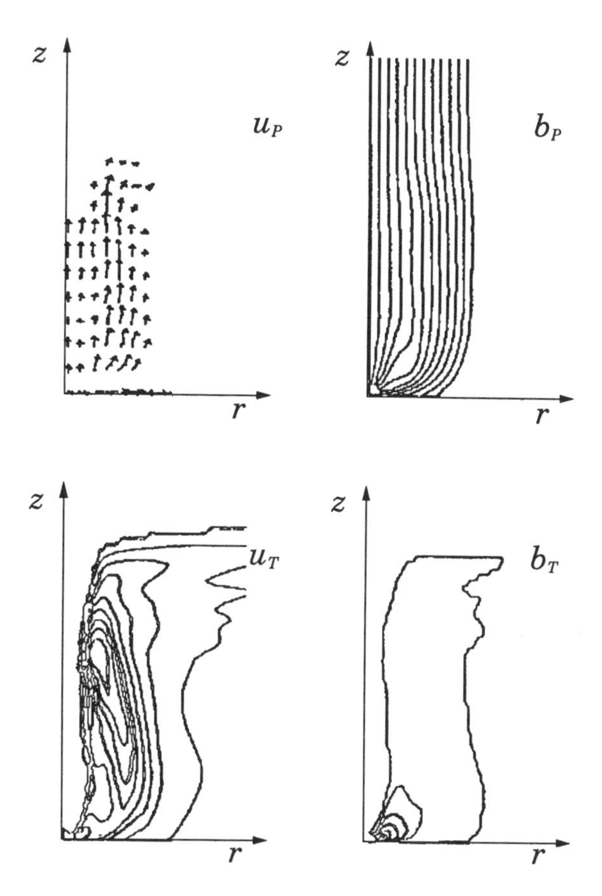

Figure 10.9. Toroidal and poloidal components of velocity and magnetic fields in an accretion disk.

10.4.1. INFLUENCE OF EXTENT OF FLOW REGION

We consider a rotating spherical region, as is depicted schematically in Fig. 10.10. The outer flow region corresponds to the convection zone in the Sun or to the outer core in the Earth (the mantle is not illustrated therein). Fluid motion subject to buoyancy effects in a rotating spherical object has been studied extensively since the pioneering work by Busse (1970). The most prominent feature of the motion is the occurrence of convection columns along the rotation axis. The number of the columns increases with strengthening of rotational effects.

The computer experiment of the MHD system of equations has revealed a number of interesting properties related to the column formation (see Glatzmaier 1985; Glatzmaier and Roberts 1995; Kageyama and Sato 1995; the works cited therein). The column appears in pairs. One column rotates in the same direction as the rotation axis, and the fluid inside it sinks from

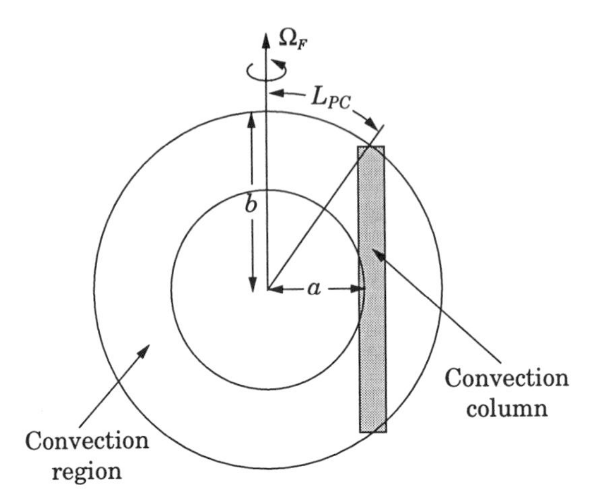

Figure 10.10. Convection column nearest to the rotation axis.

the column ends to the equatorial plane. In another of the pair rotating in the opposite direction, the fluid goes from the equatorial plane to the ends. As a result, the kinetic helicity $\mathbf{u} \cdot \boldsymbol{\omega}$ is negative in the northern hemisphere, whereas it is positive in the southern hemisphere. The results of these computer experiments show that the generation of poloidal magnetic fields is deeply associated with the formation of columns. This fact indicates the relevance of helicity effects to the poloidal-field generation process.

A quantity characterizing the extent of a flow region is the distance between the pole and the end of the column, L_{PC}. The quantity nondimensionalized by the radius b is given by

$$\Lambda \equiv \frac{L_{PC}}{b} = \sin^{-1}\frac{a}{b}, \tag{10.70}$$

with

$$\Lambda_{MAX} = \pi/2. \tag{10.71}$$

In the case of $\Lambda/\Lambda_{MAX} \ll 1$, the flow region occupies a large potion of the whole region, and the fluid motion near the equatorial plane can exert influence to the pole region through the motion in the columns. On the other hand, in the case of $\Lambda/\Lambda_{MAX} \cong 1$, the flow region becomes a thin shell one, and the pole region is separated from the low-latitude one.

From the inner structures of the Earth and the Sun illustrated in Figs. 9.1 and 9.3, we have

$$\Lambda_E/\Lambda_{MAX} = 0.22, \ \Lambda_S/\Lambda_{MAX} = 0.63 \tag{10.72}$$

(E and S denote Earth and Sun, respectively). Equation (10.72) indicates that the solar pole region is subject to the magnetic-field generation mechanism entirely different from Earth's counterpart.

10.4.2. GENERATION OF POLOIDAL MAGNETIC FIELD

From Eqs. (9.325) and (9.326), the stationary induction equation is given by

$$\nabla \times (\mathbf{U} \times \mathbf{B} + \alpha \mathbf{U} - \beta \mathbf{J} + \gamma \mathbf{\Omega}) = 0. \tag{10.73}$$

In what follows, we shall separate some of the terms in Eq. (10.73), and examine the relationship of each term with the poloidal-field generation mechanism.

A. Poloidal Magnetic Field under Dominance of Toroidal One

We bear in mind the situation with $\Lambda/\Lambda_{MAX} \ll 1$, which corresponds to Earth's inner structure. In this case, the toroidal fluid motion coming from the rotational motion is strong specifically in the low- to middle-latitude region. This motion generates a toroidal magnetic field under the cross-helicity dynamo, as is given by Eq. (10.35). Assuming that such a toroidal field is dominant, we write

$$\mathbf{B} = \mathbf{B}_0 + \mathbf{B}_1, \tag{10.74a}$$

$$\mathbf{J} = \mathbf{J}_0 + \mathbf{J}_1, \tag{10.74b}$$

where $\mathbf{B}_0$ and $\mathbf{J}_0$ are given by Eqs. (10.35) and (10.36); namely, they are

$$\mathbf{B}_0 = (\gamma/\beta)\mathbf{U}, \tag{10.75}$$

$$\mathbf{J}_0 = (\gamma/\beta)\mathbf{\Omega}, \tag{10.76}$$

with the fluid motion represented by

$$\mathbf{U} = (0, 0, U_\phi(r, \theta)) \tag{10.77}$$

and Eq. (10.48) for $\mathbf{\Omega}$ (recall that the differential-rotation part is indispensable for the sustainment of nonvanishing β and γ). The remaining parts $\mathbf{B}_1$ and $\mathbf{J}_1$ obey

$$\nabla \times (\mathbf{U} \times \mathbf{B}_1 - \beta \mathbf{J}_1) = -\nabla \times (\alpha \mathbf{B}_0), \tag{10.78}$$

from Eq. (10.73).

We apply the Stokes' integral theorem (2.62) to Eq. (10.78) that represents the interaction with the helicity effect (the closed line C is chosen as an azimuthal circle with fixed r and θ). Then we have

$$(\mathbf{U} \times \mathbf{B}_1)_\phi - \beta J_{1\phi} = -\alpha B_{0\phi} = -\frac{\alpha\gamma}{\beta} U_\phi, \tag{10.79}$$

under the condition of axisymmetry. Noting that $(\mathbf{U} \times \mathbf{B}_1)_\phi = 0$ under Eq. (10.77), we have

$$J_{1\phi} = \frac{\alpha}{\beta} B_{0\phi} = \frac{\alpha\gamma}{\beta^2} U_\phi. \tag{10.80}$$

Namely, the toroidal current $J_{1\phi}$ is generated from the rotational motion through the intermediary of the toroidal magnetic field. This current results in the poloidal magnetic field under the Ampère law.

Here we should note that the poloidal field arising from Eq. (10.80) is of the dipole type. For, both α and γ are pseudo-scalars, and their product $\alpha\gamma$ is a pure scalar, and $J_{1\phi}$ is in the same direction in the northern and southern hemispheres. This gives a theoretical picture of the generation mechanism of toroidal and magnetic fields with the toroidal one being dominant. In the Earth corresponding to $\Lambda/\Lambda_{MAX} \ll 1$, the toroidal field is estimated to be much larger from the indirect observational results, compared with the poloidal one of a few Gauss.

B. Local Poloidal Magnetic Fields in Pole Regions

In Eqs. (10.75) and (10.76), the mean rotational motion plays a key role, although the term $\mathbf{U}\times\mathbf{B}_0$ vanishes eventually in the magnetic induction equation. This treatment may be regarded as plausible in the low- to middle-latitude region with high rotational velocity. In the case of $\Lambda/\Lambda_{MAX} \ll 1$, it is highly probable that the resulting toroidal magnetic field has influence on the pole region through the combination with the helicity effect such as Eq. (10.80). On the other hand, in the solar case of $\Lambda/\Lambda_{MAX} \cong 1$, such an influence is weak, as is discussed in Sec. 10.4.1. Then we approximate Eq. (10.73) as

$$\nabla \times (\alpha\mathbf{B} - \beta\mathbf{J} + \gamma\boldsymbol{\Omega}) = 0, \tag{10.81}$$

in the vicinity of the pole region with small $\mathbf{U}$. Here we should note that small $\mathbf{U}$ does not always mean small $\boldsymbol{\Omega}$.

In Eq. (10.20), the differential rotation is usually strongest near the equator, whereas it is very weak in the high-latitude region. Then we approximate $\boldsymbol{\Omega}$ by the solid-rotation part, as in Eq. (10.48), resulting in

$$\nabla \times (\alpha\mathbf{B} - \beta\mathbf{J} + 2\gamma\boldsymbol{\Omega}_F) = 0. \tag{10.82}$$

The special solution of Eq. (10.82) is given by

$$\mathbf{B} = -\,(2\gamma/\alpha)\,\boldsymbol{\Omega}_F, \tag{10.83}$$

which leads to vanishing of $\mathbf{J}$ under the condition of the local constancy of α and γ.

The field given by Eq. (10.83) is parallel or anti-parallel to the axis of rotation since γ/α is a pure scalar, and leads to a poloidal field of the dipole type. The alignment of magnetic fields with the rotation axis was also pointed out by Gilman (1983) performing the computer experiment of the MHD system in a spherical region. Equation (10.83) will be discussed in the context of the sunspot polarity law.

C. Sunspot Polarity Law

A number of important properties of solar magnetic fields have been studied theoretically since the pioneering work by Parker (1979) [see also Priest (1982) and the works cited therein]. One of the interesting properties of sunspots is the so-called polarity rule, which is shown in Fig. 10.5. The polarity of all the leading sunspots in one hemisphere is the same and is coincident with the polarity near the pole. Specifically, the poloidal magnetic field near the pole is of a few Gauss, which is very weak, compared with sunspots whose magnetic field reaches a thousand Gauss.

We are in a position to consider the foregoing polarity rule with the aid of the analytical solution obtained in Secs. 10.4.2.A and B. In this context, we should recall that solar magnetic fields show highly time-dependent properties such as the polarity reversal with the period of 11 years. The aim of the following discussion is to see that the present dynamo solution is consistent with the polarity rule in a stationary sense.

We examine the left-hand side of Fig. 10.5. We assume

$$\text{Northern hemisphere}: \ \gamma > 0, \ \alpha > 0, \tag{10.84a}$$

$$\text{Southern hemisphere}: \ \gamma < 0, \ \alpha < 0. \tag{10.84b}$$

Under the condition (10.84), Eq. (10.75) generates the positive and negative toroidal magnetic fields in the northern and southern hemispheres, respectively. The rising-up of such fields as a result of buoyancy effects leads to the sunspots' polarity depicted at the left-hand side of Fig. 10.5. Equation (10.80), which arises from the combination with helicity dynamo, results in the positive poloidal field, and its polarity contradicts that in Fig. 10.5.

The solar convection zone is a thin shell, as is noted in Sec. 10.4.1, and it is more reasonable to consider that the magnetic field near the pole is determined by the dynamo process intrinsic to the pole region. Under the condition (10.84), Eq. (10.83) leads to the negative poloidal field near the north pole, and obeys the polarity rule at the left-hand side of Fig. 10.5. Entirely similarly, the polarity at the right-hand side can be reproduced under the condition

$$\text{Northern hemispere}: \ \gamma < 0, \ \alpha > 0, \tag{10.85a}$$

$$\text{Southern hemispere}: \ \gamma > 0, \ \alpha < 0. \tag{10.85b}$$

Here we should note that the polarity reversal is made through that of γ, but not of α. This conclusion is consistent with the finding of the computer experiments of the MHD system of equations. In these experiments, negative $\mathbf{u} \cdot \boldsymbol{\omega}$ and positive $\mathbf{b} \cdot \mathbf{j}$ are always dominant in the northern hemisphere, irrespective of the polarity, resulting in the positive residual helicity

$-\mathbf{u}\cdot\boldsymbol{\omega}+\mathbf{b}\cdot\mathbf{j}$ there (Gilman 1983; Glatzmaier 1985; Kageyama and Sato 1995). Therefore the reversal through the intermediary of helicity effects is not probable.

10.5. Rotation Effects on Turbulence Transport

10.5.1. ROLES OF FRAME AND DIFFERENTIAL ROTATION

Global rotational motion plays a central role in turbulent dynamo, specifically, in both the α-ω and cross-helicity dynamos. It contributes to the generation of the toroidal component of magnetic fields. In order to recall and clarify the role of differential rotational motion, we write the mean velocity $\mathbf{U}$ in the form

$$\mathbf{U}=\boldsymbol{\Omega}_F\times\mathbf{r}+\mathbf{U}_D \tag{10.86}$$

(subscript D denotes differential). The corresponding mean vorticity $\boldsymbol{\Omega}$ is

$$\boldsymbol{\Omega}=2\boldsymbol{\Omega}_F+\boldsymbol{\Omega}_D, \tag{10.87}$$

with

$$\boldsymbol{\Omega}_D=\nabla\times\mathbf{U}_D. \tag{10.88}$$

From Eqs. (10.86) and (10.87), the mean velocity gradient $\partial U_j/\partial x_i$ is separated into the symmetric and anti-symmetric parts as

$$\frac{\partial U_j}{\partial x_i}=\frac{1}{2}S_{Dij}+\frac{1}{2}\epsilon_{ij\ell}\left(2\Omega_{F\ell}+\Omega_{D\ell}\right), \tag{10.89}$$

where

$$S_{Dij}=\frac{\partial U_{Dj}}{\partial x_i}+\frac{\partial U_{Di}}{\partial x_j}. \tag{10.90}$$

From the constraint (9.126) on $|W/K|$, the magnitude of induced $\mathbf{B}$ is heavily dependent on that of $\mathbf{U}$. In a rotating stellar object, the first part in Eq. (10.86) is usually much larger than the second and gives a primary contribution, at least, in the context of the magnitude of $\mathbf{U}$. On the other hand, the second or differential-rotation part is the sole source of the mean velocity strain tensor, as is given by the first term in Eq. (10.89). The MHD turbulent state, which leads to nonvanishing α and γ, is sustained by the strain effects through P_K and P_W, Eqs. (10.43) and (10.45). This fact indicates that what plays a central role in turbulent dynamo is the strain part of the differential rotational motion, but not the vorticity counterpart. In the latter context, the solid-rotation part has greater influence on the magnetic-field generation process through the mean velocity $\mathbf{U}$.

From the foregoing discussions, differential rotational motion bears the two effects that seem to contradict each other at a first glance. One is the enhancement of turbulent mixing or transport effects expressed by the concepts of turbulent viscosity (ν_T) and resistivity (β), and it tends to destroy global ordered structures of fluid motion and magnetic fields. Another is the generation of such structures that results from the sustainment of the MHD turbulent state characterized by nonvanishing α and γ. The degree of the balance of these two effects strongly affects the details of the resulting structures.

10.5.2. SUPPRESSION OF TURBULENCE DUE TO PLASMA ROTATION

A. High-Mode Confinement in Tokamaks

In Sec. 10.2.1.B, we referred to RFP's in controlled fusion. A similar but more developed plasma-confinement approach is tokamaks. The difference between these two approaches is represented typically by the safety factor q, as is shown by Eqs. (10.6) and (10.7). A breakthrough in tokamaks was brought by the so-called high-mode confinement, which is usually abbreviated as H modes, whereas the usual modes are called L modes. The transition from L to H modes is characterized by the drastic suppression of fluctuations of magnetic fields and density of electron and ion. This leads to the decrease in the energy transport to the outside of the plasma, and the plasma confinement is greatly improved. The occurrence of H modes is accompanied by that of the poloidal plasma rotation and the radial electric field. The critical importance of the latter was first pointed out theoretically by Itoh and Itoh (1988) and Shaing and Crume (1989), and the L to H transition was related to a bifurcation of plasma state.

In the cylindrical approximation to a torus with the coordinates (r, θ, z), the poloidal direction corresponds to the θ one. In the comparison with observations, however, this approximation is not sufficient, and the toroidicity is still an important ingredient in tokamak confinement. With this reservation in mind, the cylindrical approximation is often employed in the theoretical investigation of H modes. The modes originate in a very narrow region near the plasma edge, and their influence spreads towards the central region. The latter stage is often called VH (very high) modes, which are also accompanied by the toroidal plasma rotation (the motion in the z direction in the cylindrical approximation). The poloidal plasma rotation and radial electric field are tightly linked each other, and it is not always so clear which of these two is the more essential cause of H modes. In order to see the close connection between the poloidal plasma rotation and radial electric field, we consider the situation

$$\mathbf{u} = (0, u_\theta, 0)\,, \quad \mathbf{b} = (0, b_\theta, b_z)\,, \tag{10.91}$$

resulting in

$$\mathbf{u} \times \mathbf{b} = (u_\theta b_z, 0, 0)\,. \tag{10.92}$$

From the Ohm's law (9.37), Eq. (10.92) is equivalent to the occurrence of the radial component of the electric field $\mathbf{e}$ in the context of the current generation.

The developments in the experimental and theoretical studies on H modes are reviewed in detail in Wooton et al. (1990), Stambaugh et al. (1990), and Itoh and Itoh (1996).

B. H Modes and Turbulent Dynamo

Rotational motion has great influence on global ordered structures in hydrodynamic turbulent flow. Such a typical example is a swirling flow in a pipe that is referred to in Sec. 4.2.5.B. The swirling motion imposed at the entrance of a pipe generates the axial-velocity profile with the local minimum at the central axis. The profile is entirely different from the counterpart with the maximum velocity at the central axis in the absence of swirling. It continues to be maintained as long as the swirling motion is kept. This observational finding indicates that the transport effect enhanced by turbulent motion, which usually contributes to the destruction of order structures, is suppressed by the effect of rotational motion. In Sec. 6.7.2.A, the suppression mechanism is examined on the basis of helicity effects.

In the context of the suppression of turbulent transport, the effect of poloidal plasma rotation seems to be similar to the foregoing rotation effect in a swirling pipe flow. In H modes, however, the region with plasma rotation is very narrow in the radial direction, at least, at the early stage. This steep radial velocity gradient leads to large velocity strain, which contributes to the enhancement of turbulent fluctuations through Eq. (10.43) for the turbulent-energy production rate P_K. Nevertheless, an entirely different phenomenon, that is, the drastic suppression of turbulent transport occurs in H modes. In what follows, we shall seek a mechanism of turbulence suppression due to poloidal plasma rotation from the standpoint of the turbulent dynamo based on cross-helicity effects.

In advance of the discussion based on the turbulent dynamo, we should remark the following points. The present dynamo model is founded on the one-fluid MHD system of equations. Various plasma behaviors in tokamaks are sensitive to each spatial profile of the temperature and density of electron and ion. Such properties are originally beyond the scope of the one-fluid description. We may separate the characteristics of H modes into following two aspects:

(i) Effects of plasma poloidal rotation on turbulent transports;
(ii) Effects of electric field on plasma behaviors.

In the one-fluid MHD system, the electric field is not treated explicitly, as is seen from the induction equation (9.54). Therefore the aspect (ii) is not within the scope of the present MHD dynamo, and attention will be focused on the aspect (i). For the explicit treatment of the electric field, we need a turbulent dynamo including effects of the breakage of charge neutrality. The extension of the MHD approximation based on the replacement of the right-hand side of Eq. (9.29) with the counterpart of Eq. (9.27) and the study about the transport suppression due to effects of electric field and charge nonneutrality were discussed by Yoshizawa and Yokoi (1998).

B.1. Suppression of Turbulence Due to Cross-Helicity Effects. We consider the situation in which the L-H transition has been established and the plasma rotation has started. To begin with, we examine the generation of the turbulent cross helicity W on the basis of Eqs. (9.334) and (9.336a). In tokamaks, the magnetic shear is small, compared with RFP's, and the second part in Eq. (9.336a) becomes a primary one. We use Eq. (9.330) for $\mathbf{E}_M$ with the helicity term $\alpha \mathbf{B}$ dropped, and have

$$P_W \cong \beta \mathbf{J} \cdot \mathbf{\Omega} - \gamma \mathbf{\Omega}^2. \tag{10.93}$$

The turbulent resistivity β is always positive. Therefore Eq. (9.334) indicates that W with the same sign as $\mathbf{J} \cdot \mathbf{\Omega}$ is generated under the constraint

$$1 - \frac{\gamma \mathbf{\Omega}^2}{\beta \mathbf{J} \cdot \mathbf{\Omega}} > 0; \tag{10.94}$$

namely, we have

$$W > 0 \text{ for } \mathbf{J} \cdot \mathbf{\Omega} > 0, \tag{10.95a}$$

$$W < 0 \text{ for } \mathbf{J} \cdot \mathbf{\Omega} < 0. \tag{10.95b}$$

Expression (10.95) signifies that, in the presence of the toroidal plasma current J_z, W may be generated by the poloidal plasma motion leading to nonvanishing Ω_z. The cross helicity thus generated nearly balances the destruction rate ε_W [Eq. (9.336b)], resulting in a steady or quasi-steady state. The constraint (10.94) will be referred to below.

Next, we use Eqs. (9.334) and (9.335a), and examine how generated W affects the magnitude of the MHD turbulent energy K. Considering that the magnetic strain is small, as is similar to Eq. (10.93), we have

$$P_K \cong \beta \mathbf{J}^2 + P'_K, \tag{10.96}$$

with

$$P'_K = \frac{1}{2} \nu_T S_{ij}^2 - \gamma \mathbf{J} \cdot \mathbf{\Omega}. \tag{10.97}$$

The first term in Eq. (10.96) is the production rate of K due to the turbulent-resistivity (β) effect. The turbulent energy generated by it is supplied by the mean magnetic field $\mathbf{B}$ and is dissipated through ε, eventually leading to the anomalous heat transport.

Equation (10.97) is the K production rate newly occurring from the poloidal plasma motion. At the start of the H modes, such a motion is localized at the plasma edge and possesses a steep velocity strain. The first term corresponds to the increase in the production rate due to this strain effect. The second term expresses the effect of plasma rotation on the K production rate. The coefficient γ is proportional to W [Eq. (9.333)]. Expression (10.95) combined with this fact indicates that $\gamma \mathbf{J} \cdot \mathbf{\Omega}$ is always positive, contributing to the reduction of the K production rate. Namely, the latter cross-helicity effect tends to eliminate the contribution from the velocity strain (Yoshizawa 1991). This point was confirmed by Yokoi (1996b) through the numerical computation of the present cross-helicity dynamo model.

B.2. Feedback Effects on Plasma Motion. In the foregoing discussions, we assumed the occurrence of poloidal plasma motion. We now consider its feedback effects on the plasma motion. In the situation that the primary part of plasma currents is the toroidal component, we may write

$$\mathbf{J} = (0, 0, J_z), \ \mathbf{B} = (0, B_\theta, B_z), \tag{10.98}$$

where B_z is generated by external coils, and B_θ arises from J_z. In the cylindrical approximation to a torus, all the quantities are assumed to be axisymmetric or independent of θ. In this situation, we have

$$(\mathbf{J} \times \mathbf{B})_\theta = (\mathbf{J} \times \mathbf{B})_z = 0; \tag{10.99}$$

namely, the feedback effects on the poloidal motion through the Lorentz force do not occur.

Next, we consider the feedback effects through fluctuations. In Eq. (9.324) for $\mathbf{U}$, the effect of turbulence is given by the third term on the right-hand side, that is,

$$(\nabla \cdot \mathbf{R})_i \equiv \frac{\partial R_{ji}}{\partial x_j}. \tag{10.100}$$

From Eq. (9.329), $(\nabla \cdot \mathbf{R})_\theta$ may be written as

$$(\nabla \cdot \mathbf{R})_\theta \cong -\nu_M \frac{dJ_z}{dr}, \tag{10.101}$$

where we paid attention to the M_{ij}-related term corresponding to the feedback effect of the magnetic field, and used the relation $\Delta \mathbf{B} = -\nabla \times \mathbf{J}$ (for simplicity of discussion, the spatial change of ν_M was discarded).

We consider the situation in which both J_z and Ω_z are positive. Towards the plasma edge, the plasma current usually decreases, resulting in $dJ_z/dr < 0$. From expression (10.95), positive W is generated since $\mathbf{J} \cdot \mathbf{\Omega} (\cong J_z \Omega_z)$ is positive in this case. Then we have positive ν_M from Eq. (9.333), and Eq. (10.101) becomes positive and contributes to the driving of the poloidal rotation. When $\mathbf{\Omega}$ strengthens so that the constraint (10.94) is violated, negative W is generated in the case of positive $\mathbf{J} \cdot \mathbf{\Omega}$, resulting in the retardation of the plasma rotation. The similar discussion can be made for positive J_z and negative Ω_z.

The feedback effect on the z or toroidal component of $\nabla \cdot \mathbf{R}$ is given by

$$(\nabla \cdot \mathbf{R})_z \cong -\nu_M \frac{1}{r} \frac{d}{dr} (r J_\theta) . \tag{10.102}$$

The sign of the resulting force is dependent on the spatial profile of the poloidal current.

References

Begelman, M. C. and Blandford, R. D. (1984), Rev. Mod. Phys. **56**, 265.

Bodin, H. A. B. and Newton, A. A. (1980), Nucl. Fusion **20**, 1255.

Busse, F. (1970), J. Fluid Mech. **44**, 441.

Gilman, G. A. (1983), Astrophys. J. Suppl. Ser. **53**, 243.

Gimblett, C. G. and Watkins, M. L. (1975), in *Proceedings of the Seventh European Conference on Controlled Fusion and Plasma Physics* (Ecole de Polytechnique Fédéral de Lausanne, Lausanne), Vol. I, p. 103.

Glatzmaier, G. A. (1985), Astrophys. J. **291**, 300.

Glatzmaier, G. A. and Roberts, P. H. (1995), Nature **377**, 203.

Hamba, F. (1990), Phys. Fluids B **2**, 3064.

Horiuchi, T. and Kato, S. (1990), Publ. Astron. Soc. Jpn. **42**, 661.

Itoh, S-I. and Itoh, K. (1988), Phys. Rev. Lett. **60**, 2276.

Itoh, K. and Itoh, S-I. (1996), Plasma Phys. and Controlled Fusion **38**, 1.

Ji, H., Prager, S. C., Almagiri, A. F., Sariff, J. S., Yagi, Y., Hirano, Y., Hattori, K., and Toyama, H. (1996), Phys. Plasmas **3**, 1935.

Kageyama, A. and Sato, T. (1995), Phys. Plasmas **2**, 1421.

Kato, S., Fukue, J., and Miuneshige, S. (1998), *Black-Hole Accretion Disks* (Kyoto U. P., Kyoto).

Miyamoto, K. (1989), *Plasma Physics for Nuclear Fusion* (The MIT, Cambridge).

Nagata, A., Sakamoto, H., Sato. K. I., Ashida, H., and Amano, T. (1995), Phys. Plasmas **2**, 1182.

Nishino, S. and Yokoi, N. (1998), private communication (Analysis of toroidal magnetic fields in accretion disks using the cross-helicity dynamo and estimate of jet velocity, submitted to Publ. Astron. Soc. Jpn.).

Parker, E. N. (1979), *Cosmical Magnetic Fields* (Clarendon, Oxford).
Priest, E. (1982), *Solar Magnetohydrodynamics* (D. Reidel, Dordrecht).
Ruzumaikin, A. A., Shukurov, A. M., and Sokoloff, D. D. (1988), *Magnetic Fields of Galaxies* (Kluwer, Dordrecht).
Shaing, K. C. and Crume Jr, E. C. (1989), Phys. Rev. Lett. **63**, 2369.
Shakura, N. I. and Sunyaev, R. A. (1973), Astron. Astrophys. **24**, 337.
Shibata, K. and Uchida, Y. (1990), Publ. Astron. Soc. Jpn. **42**, 39.
Stambaugh, R. D., Wolfe, S. M., Hawryluk, R. J., Harris, J. H., Biglari, H., Prager, S. C., Goldston, R. J., Fonck, R. J., Ohkawa, T., Logan, B. G., and Oktay, E. (1990), Phys. Fluids B **2**, 2941.
Taylor, J. B. (1974), Phys. Rev. Lett. **33**, 1139.
Wielebinski, R. and Krause, F. (1993), Astron. Astrophys. Rev. **4**, 449.
Wooton, A. J., Carreras, B. A., Matsumoto, H., McGuire, K., Peebles, W. A., Ritz, Ch. P., Terry, P. W., and Zweben, J. (1990), Phys. Fluids B **2**, 2879.
Yokoi, N. (1996a), Astron. Astrophys. **311**, 731.
Yokoi, N. (1996b), J. Phys. Soc. Jpn. **65**, 2353.
Yoshizawa, A. (1991), Phys. Fluids B **3**, 2723.
Yoshizawa, A. (1993), Publ. Astron. Soc. Jpn. **45**, 129.
Yoshizawa, A. and Yokoi, N. (1993), Astrophys. J. **407**, 540.
Yoshizawa, A. and Yokoi, N. (1996), Phys. Plasmas **3**, 3604.
Yoshizawa, A. and Yokoi, N. (1998), Phys. Plasmas **5**, No.8 (to appear; Statistical analysis of turbulent-transport suppression based on the magnetohydrodynamic approximation with electric-field effects incorporated).

INDEX

Mechanics

FLUID MECHANICS AND ITS APPLICATIONS

Series Editor: R. Moreau

Aims and Scope of the Series

The purpose of this series is to focus on subjects in which fluid mechanics plays a fundamental role. As well as the more traditional applications of aeronautics, hydraulics, heat and mass transfer etc., books will be published dealing with topics which are currently in a state of rapid development, such as turbulence, suspensions and multiphase fluids, super and hypersonic flows and numerical modelling techniques. It is a widely held view that it is the interdisciplinary subjects that will receive intense scientific attention, bringing them to the forefront of technological advancement. Fluids have the ability to transport matter and its properties as well as transmit force, therefore fluid mechanics is a subject that is particularly open to cross fertilisation with other sciences and disciplines of engineering. The subject of fluid mechanics will be highly relevant in domains such as chemical, metallurgical, biological and ecological engineering. This series is particularly open to such new multidisciplinary domains.

1. M. Lesieur: *Turbulence in Fluids.* 2nd rev. ed., 1990 ISBN 0-7923-0645-7
2. O. Métais and M. Lesieur (eds.): *Turbulence and Coherent Structures.* 1991 ISBN 0-7923-0646-5
3. R. Moreau: *Magnetohydrodynamics.* 1990 ISBN 0-7923-0937-5
4. E. Coustols (ed.): *Turbulence Control by Passive Means.* 1990 ISBN 0-7923-1020-9
5. A.A. Borissov (ed.): *Dynamic Structure of Detonation in Gaseous and Dispersed Media.* 1991 ISBN 0-7923-1340-2
6. K.-S. Choi (ed.): *Recent Developments in Turbulence Management.* 1991 ISBN 0-7923-1477-8
7. E.P. Evans and B. Coulbeck (eds.): *Pipeline Systems.* 1992 ISBN 0-7923-1668-1
8. B. Nau (ed.): *Fluid Sealing.* 1992 ISBN 0-7923-1669-X
9. T.K.S. Murthy (ed.): *Computational Methods in Hypersonic Aerodynamics.* 1992 ISBN 0-7923-1673-8
10. R. King (ed.): *Fluid Mechanics of Mixing.* Modelling, Operations and Experimental Techniques. 1992 ISBN 0-7923-1720-3
11. Z. Han and X. Yin: *Shock Dynamics.* 1993 ISBN 0-7923-1746-7
12. L. Svarovsky and M.T. Thew (eds.): *Hydroclones.* Analysis and Applications. 1992 ISBN 0-7923-1876-5
13. A. Lichtarowicz (ed.): *Jet Cutting Technology.* 1992 ISBN 0-7923-1979-6
14. F.T.M. Nieuwstadt (ed.): *Flow Visualization and Image Analysis.* 1993 ISBN 0-7923-1994-X
15. A.J. Saul (ed.): *Floods and Flood Management.* 1992 ISBN 0-7923-2078-6
16. D.E. Ashpis, T.B. Gatski and R. Hirsh (eds.): *Instabilities and Turbulence in Engineering Flows.* 1993 ISBN 0-7923-2161-8
17. R.S. Azad: *The Atmospheric Boundary Layer for Engineers.* 1993 ISBN 0-7923-2187-1
18. F.T.M. Nieuwstadt (ed.): *Advances in Turbulence IV.* 1993 ISBN 0-7923-2282-7
19. K.K. Prasad (ed.): *Further Developments in Turbulence Management.* 1993 ISBN 0-7923-2291-6
20. Y.A. Tatarchenko: *Shaped Crystal Growth.* 1993 ISBN 0-7923-2419-6

Kluwer Academic Publishers – Dordrecht / Boston / London

Mechanics

FLUID MECHANICS AND ITS APPLICATIONS

Series Editor: R. Moreau

21. J.P. Bonnet and M.N. Glauser (eds.): *Eddy Structure Identification in Free Turbulent Shear Flows.* 1993 ISBN 0-7923-2449-8
22. R.S. Srivastava: *Interaction of Shock Waves.* 1994 ISBN 0-7923-2920-1
23. J.R. Blake, J.M. Boulton-Stone and N.H. Thomas (eds.): *Bubble Dynamics and Interface Phenomena.* 1994 ISBN 0-7923-3008-0
24. R. Benzi (ed.): *Advances in Turbulence V.* 1995 ISBN 0-7923-3032-3
25. B.I. Rabinovich, V.G. Lebedev and A.I. Mytarev: *Vortex Processes and Solid Body Dynamics.* The Dynamic Problems of Spacecrafts and Magnetic Levitation Systems. 1994 ISBN 0-7923-3092-7
26. P.R. Voke, L. Kleiser and J.-P. Chollet (eds.): *Direct and Large-Eddy Simulation I.* Selected papers from the First ERCOFTAC Workshop on Direct and Large-Eddy Simulation. 1994 ISBN 0-7923-3106-0
27. J.A. Sparenberg: *Hydrodynamic Propulsion and its Optimization.* Analytic Theory. 1995 ISBN 0-7923-3201-6
28. J.F. Dijksman and G.D.C. Kuiken (eds.): *IUTAM Symposium on Numerical Simulation of Non-Isothermal Flow of Viscoelastic Liquids.* Proceedings of an IUTAM Symposium held in Kerkrade, The Netherlands. 1995 ISBN 0-7923-3262-8
29. B.M. Boubnov and G.S. Golitsyn: *Convection in Rotating Fluids.* 1995 ISBN 0-7923-3371-3
30. S.I. Green (ed.): *Fluid Vortices.* 1995 ISBN 0-7923-3376-4
31. S. Morioka and L. van Wijngaarden (eds.): *IUTAM Symposium on Waves in Liquid/Gas and Liquid/Vapour Two-Phase Systems.* 1995 ISBN 0-7923-3424-8
32. A. Gyr and H.-W. Bewersdorff: *Drag Reduction of Turbulent Flows by Additives.* 1995 ISBN 0-7923-3485-X
33. Y.P. Golovachov: *Numerical Simulation of Viscous Shock Layer Flows.* 1995 ISBN 0-7923-3626-7
34. J. Grue, B. Gjevik and J.E. Weber (eds.): *Waves and Nonlinear Processes in Hydrodynamics.* 1996 ISBN 0-7923-4031-0
35. P.W. Duck and P. Hall (eds.): *IUTAM Symposium on Nonlinear Instability and Transition in Three-Dimensional Boundary Layers.* 1996 ISBN 0-7923-4079-5
36. S. Gavrilakis, L. Machiels and P.A. Monkewitz (eds.): *Advances in Turbulence VI.* Proceedings of the 6th European Turbulence Conference. 1996 ISBN 0-7923-4132-5
37. K. Gersten (ed.): *IUTAM Symposium on Asymptotic Methods for Turbulent Shear Flows at High Reynolds Numbers.* Proceedings of the IUTAM Symposium held in Bochum, Germany. 1996 ISBN 0-7923-4138-4
38. J. Verhás: *Thermodynamics and Rheology.* 1997 ISBN 0-7923-4251-8
39. M. Champion and B. Deshaies (eds.): *IUTAM Symposium on Combustion in Supersonic Flows.* Proceedings of the IUTAM Symposium held in Poitiers, France. 1997 ISBN 0-7923-4313-1
40. M. Lesieur: *Turbulence in Fluids.* Third Revised and Enlarged Edition. 1997 ISBN 0-7923-4415-4; Pb: 0-7923-4416-2

Kluwer Academic Publishers – Dordrecht / Boston / London

Mechanics

***FLUID* MECHANICS AND ITS APPLICATIONS**

Series Editor: R. Moreau

41. L. Fulachier, J.L. Lumley and F. Anselmet (eds.): *IUTAM Symposium on Variable Density Low-Speed Turbulent Flows.* Proceedings of the IUTAM Symposium held in Marseille, France. 1997 ISBN 0-7923-4602-5
42. B.K. Shivamoggi: *Nonlinear Dynamics and Chaotic Phenomena.* An Introduction. 1997 ISBN 0-7923-4772-2
43. H. Ramkissoon, *IUTAM Symposium on Lubricated Transport of Viscous Materials.* Proceedings of the IUTAM Symposium held in Tobago, West Indies. 1998 ISBN 0-7923-4897-4
44. E. Krause and K. Gersten, *IUTAM Symposium on Dynamics of Slender Vortices.* Proceedings of the IUTAM Symposium held in Aachen, Germany. 1998 ISBN 0-7923-5041-3
45. A. Biesheuvel and G.J.F. van Heyst (eds.): *In Fascination of Fluid Dynamics.* A Symposium in honour of Leen van Wijngaarden. 1998 ISBN 0-7923-5078-2
46. U. Frisch (ed.): *Advances in Turbulence VII.* Proceedings of the Seventh European Turbulence Conference, held in Saint-Jean Cap Ferrat, 30 June–3 July 1998. 1998 ISBN 0-7923-5115-0
47. E.F. Toro and J.F. Clarke, *Numerical Methods for Wave Propagation.* Selected Contributions from the Workshop held in Manchester, UK. 1998 ISBN 0-7923-5125-8
48. A. Yoshizawa: *Hydrodynamic and Magnetohydrodynamic Turbulent Flows.* Modelling and Statistical Theory. 1998 ISBN 0-7923-5225-4

Kluwer Academic Publishers – Dordrecht / Boston / London

Mechanics

SOLID MECHANICS AND ITS APPLICATIONS

Series Editor: G.M.L. Gladwell

Aims and Scope of the Series

The fundamental questions arising in mechanics are: *Why?, How?,* and *How much?* The aim of this series is to provide lucid accounts written by authoritative researchers giving vision and insight in answering these questions on the subject of mechanics as it relates to solids. The scope of the series covers the entire spectrum of solid mechanics. Thus it includes the foundation of mechanics; variational formulations; computational mechanics; statics, kinematics and dynamics of rigid and elastic bodies; vibrations of solids and structures; dynamical systems and chaos; the theories of elasticity, plasticity and viscoelasticity; composite materials; rods, beams, shells and membranes; structural control and stability; soils, rocks and geomechanics; fracture; tribology; experimental mechanics; biomechanics and machine design.

1. R.T. Haftka, Z. Gürdal and M.P. Kamat: *Elements of Structural Optimization.* 2nd rev.ed., 1990 ISBN 0-7923-0608-2
2. J.J. Kalker: *Three-Dimensional Elastic Bodies in Rolling Contact.* 1990 ISBN 0-7923-0712-7
3. P. Karasudhi: *Foundations of Solid Mechanics.* 1991 ISBN 0-7923-0772-0
4. *Not published*
5. *Not published.*
6. J.F. Doyle: *Static and Dynamic Analysis of Structures.* With an Emphasis on Mechanics and Computer Matrix Methods. 1991 ISBN 0-7923-1124-8; Pb 0-7923-1208-2
7. O.O. Ochoa and J.N. Reddy: *Finite Element Analysis of Composite Laminates.* ISBN 0-7923-1125-6
8. M.H. Aliabadi and D.P. Rooke: *Numerical Fracture Mechanics.* ISBN 0-7923-1175-2
9. J. Angeles and C.S. López-Cajún: *Optimization of Cam Mechanisms.* 1991 ISBN 0-7923-1355-0
10. D.E. Grierson, A. Franchi and P. Riva (eds.): *Progress in Structural Engineering.* 1991 ISBN 0-7923-1396-8
11. R.T. Haftka and Z. Gürdal: *Elements of Structural Optimization.* 3rd rev. and exp. ed. 1992 ISBN 0-7923-1504-9; Pb 0-7923-1505-7
12. J.R. Barber: *Elasticity.* 1992 ISBN 0-7923-1609-6; Pb 0-7923-1610-X
13. H.S. Tzou and G.L. Anderson (eds.): *Intelligent Structural Systems.* 1992 ISBN 0-7923-1920-6
14. E.E. Gdoutos: *Fracture Mechanics.* An Introduction. 1993 ISBN 0-7923-1932-X
15. J.P. Ward: *Solid Mechanics.* An Introduction. 1992 ISBN 0-7923-1949-4
16. M. Farshad: *Design and Analysis of Shell Structures.* 1992 ISBN 0-7923-1950-8
17. H.S. Tzou and T. Fukuda (eds.): *Precision Sensors, Actuators and Systems.* 1992 ISBN 0-7923-2015-8
18. J.R. Vinson: *The Behavior of Shells Composed of Isotropic and Composite Materials.* 1993 ISBN 0-7923-2113-8
19. H.S. Tzou: *Piezoelectric Shells.* Distributed Sensing and Control of Continua. 1993 ISBN 0-7923-2186-3

Kluwer Academic Publishers – Dordrecht / Boston / London

Mechanics

SOLID MECHANICS AND ITS APPLICATIONS

Series Editor: G.M.L. Gladwell

20. W. Schiehlen (ed.): *Advanced Multibody System Dynamics.* Simulation and Software Tools. 1993 ISBN 0-7923-2192-8
21. C.-W. Lee: *Vibration Analysis of Rotors.* 1993 ISBN 0-7923-2300-9
22. D.R. Smith: *An Introduction to Continuum Mechanics.* 1993 ISBN 0-7923-2454-4
23. G.M.L. Gladwell: *Inverse Problems in Scattering.* An Introduction. 1993 ISBN 0-7923-2478-1
24. G. Prathap: *The Finite Element Method in Structural Mechanics.* 1993 ISBN 0-7923-2492-7
25. J. Herskovits (ed.): *Advances in Structural Optimization.* 1995 ISBN 0-7923-2510-9
26. M.A. González-Palacios and J. Angeles: *Cam Synthesis.* 1993 ISBN 0-7923-2536-2
27. W.S. Hall: *The Boundary Element Method.* 1993 ISBN 0-7923-2580-X
28. J. Angeles, G. Hommel and P. Kovács (eds.): *Computational Kinematics.* 1993 ISBN 0-7923-2585-0
29. A. Curnier: *Computational Methods in Solid Mechanics.* 1994 ISBN 0-7923-2761-6
30. D.A. Hills and D. Nowell: *Mechanics of Fretting Fatigue.* 1994 ISBN 0-7923-2866-3
31. B. Tabarrok and F.P.J. Rimrott: *Variational Methods and Complementary Formulations in Dynamics.* 1994 ISBN 0-7923-2923-6
32. E.H. Dowell (ed.), E.F. Crawley, H.C. Curtiss Jr., D.A. Peters, R. H. Scanlan and F. Sisto: *A Modern Course in Aeroelasticity.* Third Revised and Enlarged Edition. 1995 ISBN 0-7923-2788-8; Pb: 0-7923-2789-6
33. A. Preumont: *Random Vibration and Spectral Analysis.* 1994 ISBN 0-7923-3036-6
34. J.N. Reddy (ed.): *Mechanics of Composite Materials.* Selected works of Nicholas J. Pagano. 1994 ISBN 0-7923-3041-2
35. A.P.S. Selvadurai (ed.): *Mechanics of Poroelastic Media.* 1996 ISBN 0-7923-3329-2
36. Z. Mróz, D. Weichert, S. Dorosz (eds.): *Inelastic Behaviour of Structures under Variable Loads.* 1995 ISBN 0-7923-3397-7
37. R. Pyrz (ed.): *IUTAM Symposium on Microstructure-Property Interactions in Composite Materials.* Proceedings of the IUTAM Symposium held in Aalborg, Denmark. 1995 ISBN 0-7923-3427-2
38. M.I. Friswell and J.E. Mottershead: *Finite Element Model Updating in Structural Dynamics.* 1995 ISBN 0-7923-3431-0
39. D.F. Parker and A.H. England (eds.): *IUTAM Symposium on Anisotropy, Inhomogeneity and Nonlinearity in Solid Mechanics.* Proceedings of the IUTAM Symposium held in Nottingham, U.K. 1995 ISBN 0-7923-3594-5
40. J.-P. Merlet and B. Ravani (eds.): *Computational Kinematics '95.* 1995 ISBN 0-7923-3673-9
41. L.P. Lebedev, I.I. Vorovich and G.M.L. Gladwell: *Functional Analysis.* Applications in Mechanics and Inverse Problems. 1996 ISBN 0-7923-3849-9
42. J. Menčik: *Mechanics of Components with Treated or Coated Surfaces.* 1996 ISBN 0-7923-3700-X
43. D. Bestle and W. Schiehlen (eds.): *IUTAM Symposium on Optimization of Mechanical Systems.* Proceedings of the IUTAM Symposium held in Stuttgart, Germany. 1996 ISBN 0-7923-3830-8

Kluwer Academic Publishers – Dordrecht / Boston / London

Mechanics

SOLID MECHANICS AND ITS APPLICATIONS

Series Editor: G.M.L. Gladwell

44. D.A. Hills, P.A. Kelly, D.N. Dai and A.M. Korsunsky: *Solution of Crack Problems*. The Distributed Dislocation Technique. 1996 ISBN 0-7923-3848-0
45. V.A. Squire, R.J. Hosking, A.D. Kerr and P.J. Langhorne: *Moving Loads on Ice Plates*. 1996 ISBN 0-7923-3953-3
46. A. Pineau and A. Zaoui (eds.): *IUTAM Symposium on Micromechanics of Plasticity and Damage of Multiphase Materials*. Proceedings of the IUTAM Symposium held in Sèvres, Paris, France. 1996 ISBN 0-7923-4188-0
47. A. Naess and S. Krenk (eds.): *IUTAM Symposium on Advances in Nonlinear Stochastic Mechanics*. Proceedings of the IUTAM Symposium held in Trondheim, Norway. 1996 ISBN 0-7923-4193-7
48. D. Ieşan and A. Scalia: *Thermoelastic Deformations*. 1996 ISBN 0-7923-4230-5
49. J. R. Willis (ed.): *IUTAM Symposium on Nonlinear Analysis of Fracture*. Proceedings of the IUTAM Symposium held in Cambridge, U.K. 1997 ISBN 0-7923-4378-6
50. A. Preumont: *Vibration Control of Active Structures*. An Introduction. 1997 ISBN 0-7923-4392-1
51. G.P. Cherepanov: *Methods of Fracture Mechanics: Solid Matter Physics*. 1997 ISBN 0-7923-4408-1
52. D.H. van Campen (ed.): *IUTAM Symposium on Interaction between Dynamics and Control in Advanced Mechanical Systems*. Proceedings of the IUTAM Symposium held in Eindhoven, The Netherlands. 1997 ISBN 0-7923-4429-4
53. N.A. Fleck and A.C.F. Cocks (eds.): *IUTAM Symposium on Mechanics of Granular and Porous Materials*. Proceedings of the IUTAM Symposium held in Cambridge, U.K. 1997 ISBN 0-7923-4553-3
54. J. Roorda and N.K. Srivastava (eds.): *Trends in Structural Mechanics*. Theory, Practice, Education. 1997 ISBN 0-7923-4603-3
55. Yu. A. Mitropolskii and N. Van Dao: *Applied Asymptotic Methods in Nonlinear Oscillations*. 1997 ISBN 0-7923-4605-X
56. C. Guedes Soares (ed.): *Probabilistic Methods for Structural Design*. 1997 ISBN 0-7923-4670-X
57. D. François, A. Pineau and A. Zaoui: *Mechanical Behaviour of Materials*. Volume I: Elasticity and Plasticity. 1998 ISBN 0-7923-4894-X
58. D. François, A. Pineau and A. Zaoui: *Mechanical Behaviour of Materials*. Volume II: Viscoplasticity, Damage, Fracture and Contact Mechanics. 1998 ISBN 0-7923-4895-8
59. L. T. Tenek and J. Argyris: *Finite Element Analysis for Composite Structures*. 1998 ISBN 0-7923-4899-0
60. Y.A. Bahei-El-Din and G.J. Dvorak (eds.): *IUTAM Symposium on Transformation Problems in Composite and Active Materials*. Proceedings of the IUTAM Symposium held in Cairo, Egypt. 1998 ISBN 0-7923-5122-3

Kluwer Academic Publishers – Dordrecht / Boston / London